Communications
in Computer and Information Science 240

Elena García-Barriocanal Zeynel Cebeci
Mehmet C. Okur Aydın Öztürk (Eds.)

# Metadata and Semantic Research

5th International Conference, MTSR 2011
Izmir, Turkey, October 12-14, 2011
Proceedings

Volume Editors

Elena García-Barriocanal
University of Alcalá
28871 Alcalá de Henares (Madrid), Spain
E-mail: elena.garciab@uah.es

Zeynel Cebeci
University of Cukurova, Balcalı, Adana, Turkey
E-mail: zcebeci@cukurova.edu.tr

Mehmet C. Okur
Yasar University, Bornova, Izmir, Turkey
E-mail: mehmet.okur@yasar.edu.tr

Aydın Öztürk
Yasar University, Bornova, Izmir, Turkey
E-mail: aydin.ozturk@yasar.edu.tr

ISSN 1865-0929 e-ISSN 1865-0937
ISBN 978-3-642-24730-9 e-ISBN 978-3-642-24731-6
DOI 10.1007/978-3-642-24731-6
Springer Heidelberg Dordrecht London New York

Library of Congress Control Number: 2011938025

CR Subject Classification (1998): H.4, H.3, I.2, H.2.8, H.5, D.2.1, C.2



*Typesetting:* Camera-ready by author, data conversion by Scientific Publishing Services, Chennai, India

Printed on acid-free paper

Springer is part of Springer Science+Business Media (www.springer.com)

# Preface

Metadata and semantics are undoubtedly one of the cornerstones of the Web and its future applications. The wide range of ways to enhance the current Web using metadata and ontologies moves many researchers and enterprises worldwide to the creation and use of conceptual, theoretical and methodological frameworks to offer innovative computational solutions in that direction. Further, metadata in its various forms can be considered the fabric of a new concept of digital library, which spans across systems, and the efforts of the European Union on the creation of the Europeana digital library are a prominent example of this emergent idea. Now the concept of "linked data" has become a key driver of innovation that promises a reorganization of metadata to allow for direct linking between collections, avoiding information silos and enabling a new range of applications that are able to benefit from direct interlinking between resources.

In Accordance with the principles of the original Metadata and Semantics Research Conference (MTSR) event in 2005 and following the steps of its successors MTSR 2007 (Corfu, Greece), MTSR 2009 (Milan, Italy) and MTSR 2010 (Alcalá de Henares, Spain), MTSR 2011 was conceived as a meeting point for researchers in the field and to discuss specialized results as general contributions to the state of the art. The conference attempts to gather a truly multi-disciplinary audience, with research papers coming from different IT-related fields, such as information science and computing (in all its areas, including information systems), but also from application-specific areas.

The conference, as in previous years, was organized in a general/main track and several others: a track on Metadata and Semantics for Open Access Repositories and Infrastructures, a second on Metadata and Semantics for Learning Infrastructures, a third on Metadata and Semantics for Cultural Collections and Applications and finally one on Metadata and Semantics for Agriculture, Food and Environment. All the papers underwent a thorough and rigorous peer review process. The review and selection this year was highly competitive to guarantee that only papers containing substantial novel research results of relevance for the international R&D audience were accepted for publication. Only 36 of 118 submissions were accepted in the category of full research paper, representing around a 30% of the total. Other papers covering interesting and insightful results were accepted in the category of short research paper or project/experience report, totalling 52 contributions accepted.

This proceedings book collects the papers presented in the MTSR 2011 event, which was held at Yaşar University, in the city of İzmir, Turkey. We would like to give our sincere thanks to the administrative staff of the university and the local organizers for their professional and hard work, which made the event a reality. Also, we would like to mention the important support of several sponsors, including Netsis in Turkey, Bilgitürk Teknoloji in Izmir, Agro-Know Technologies in Greece and TSRI (Technology and Sustainability Research Institute) in Spain. Further, the VOA3R project (http://voa3r.eu/ funded by the European ICT PSP program) provided support for the organization and co-located a project meeting at İzmir just a few days before the conference, allowing for the creation of synergies between conference attendants and project partners' representatives.

The conference this year invited five extraordinary keynote speakers: Rob Davies, Yannis Ioannidis, Bram Luyten, Marcia Zeng and Charalampos Karagiannidis. All of them increased the quality of the conference and contributed to discussions on the latest advances in the field. The conference organization is grateful for their willingness to participate.

In addition, attendants to the conference were able to participate in two highly relevant tutorials conducted by Nikos Korfiatis and Hannes Ebner respectively. Korfiatis discussed the use of experimental methodology as a way to learn more about the users, and also the statistical methodologies and techniques that are needed to conduct an online experiment. Ebner discussed Linked Data and some areas of application such as educational metadata and learning repositories, as well as Linked Data in the context of environmental information.

We would like to finish by thanking the people that made the conference possible: the members of the Program Committees (track committees included), the Steering Committee and the Organization Committees (both general and local). Special thanks to Miguel-Angel Sicilia (once again!) and Fernando Álvarez for managing the online submission system and supporting the preparation of the camera-ready version of the proceedings.

August 2011

Elena García-Barriocanal
Zeynel Cebeci
Mehmet C. Okur
Aydın Öztürk

# Organization

## Program Chairs

| | |
|---|---|
| García-Barriocanal, Elena (Chair) | University of Alcalá, Spain |
| Cebeci, Zeynel (Co-chair) | Çukurova University, Turkey |
| Okur, Mehmet C. (Co-chair) | Yaşar University, Turkey |

## Organization Chairs

| | |
|---|---|
| Álvarez, Fernando-Luis | University of Alcalá, Spain |
| Palavitsinis, Nikos | Greek Research and Technology Network, Greece |

## Conference Steering Committee

| | |
|---|---|
| Sicilia, Miguel-Angel | University of Alcalá, Spain |
| Manouselis, Nikos | Agro-Know Technologies, Greece |
| Sartori, Fabio | Università degli Studi di Milano-Bicocca, Italy |

## Local Organization Committee

| | |
|---|---|
| Öztürk, Aydın (Chair) | Yaşar University, Turkey |
| Ercan, Tuncay | Yaşar University, Turkey |
| Karabulut, Korhan | Yaşar University, Turkey |
| Komesli, Murat | Yaşar University, Turkey |
| Koltuksuz, Ahmet H. | Yaşar University, Turkey |

## Program Committee Members

| | |
|---|---|
| Akerkar, Rajendra | Technomathematics Research Foundation, India |
| Altun, Arif | Hacettepe University, Turkey |
| Askar, Petek | Izmir University of Economics, Turkey |
| Athanasiadis, Ioannis N. | Democritus University of Thrace, Greece |
| Bartol, Tomaz | University of Ljubljana, Slovenia |
| Bouquet, Paolo | University of Trento, Italy |
| Budin, Gerhard | University of Vienna, Austria |
| Cagiltay, Kürsat | Middle East Technical University, Turkey |
| Caracciolo, Caterina | Food and Agriculture Organization of the United Nations, Italy |

| | |
|---|---|
| Chebotko, Artem | University of Texas-Pan American, USA |
| Christodoulakis, Stavros | Technical University of Crete, Greece |
| Colomo-Palacios, Ricardo | Universidad Carlos III, Spain |
| Contreras Cino, Jesús | Intelligent Software Components, Spain |
| Costopoulou, Constantina | Agricultural University of Athens, Greece |
| Cunningham, Sally Jo | Waikato University, New Zealand |
| Dietze, Stefan | Open University, UK |
| Dodero, Juan-Manuel | University of Cádiz, Spain |
| Dogdu, Erdogan | TOBB Teknoloji ve Ekonomi Univ, Turkey |
| Drachsler, Hendrik | Open University of the Netherlands, The Netherlands |
| Garoufallou, Emmanouel | Technological Educational Institution of Thessaloniki, Greece |
| Gomez, José Manuel | Intelligent Software Components, Spain |
| Iglesias, Carlos A. | Universidad Politécnica de Madrid, Spain |
| Jaiswal, Pankaj | Oregon State University, USA |
| Kanellopoulos, Dimitris | University of Patras, Greece |
| Keizer, Johannes | Food and Agriculture Organization of the United Nations, Italy |
| Kop, Christian | University of Klagenfurt, Austria |
| Manouselis, Nikos | Agro-Know Technologies, Greece |
| Moen, William | University of North Texas, USA |
| Morshed, Ahsan | Food and Agriculture Organization of the United Nations, Italy |
| Ochoa, Xavier | Centro de Tecnologías de Información Guayaquil, Ecuador |
| Palmonari, Matteo | University of Milan-Bicocca, Italy |
| Papaleo, Laura | University of Genova, Italy |
| Poulos, Marios | Ionian University, Greece |
| Prabhakar, T. V. | Indian Institute of Technology Kanpur, India |
| San Gil, Inigo | Long Term Ecological Research Network, USA |
| Sanchez, Salvador | University of Alcalá, Spain |
| Senkul, Pinar | Middle East Technical University, Turkey |
| Sgouropoulou, Cleo | Technological Educational Institute of Athens, Greece |
| Slavic, Aida | Universal Decimal Classification Consortium, The Netherlands |
| Sugimoto, Shigeo | University of Tsukuba, Japan |
| Suleman, Hussein | University of Cape Town, South Africa |
| Tummarello, Giovanni | National University of Ireland, Ireland |
| Tonkin, Emma | University of Bath, UK |
| Tüzün, Hakan | Hacettepe University, Turkey |
| Ünalır, Murat Osman | Ege University, Turkey |
| Villazón-Terrazas, Boris | Universidad Politecnica de Madrid, Spain |
| Zarraonandia, Telmo | Universidad Carlos III de Madrid, Spain |

## Track on Metadata and Semantics for Open Access Repositories and Infrastructures

### Special Track Chairs

| | |
|---|---|
| Subirats, Imma | Food and Agriculture Organization of the United Nations, Italy |
| Houssos, Nikos | National Documentation Centre, Greece |

### Special Track Program Committee (TBU)

| | |
|---|---|
| Castelli, Donatella | Italian National Research Council, Italy |
| Carr, Les | University of Southampton, UK |
| Dunshire, Gordon | University of Strathclyde, UK |
| Giaretta, David | Science and Technology Facilities Council, UK |
| Ioannidis, Yannis | University of Athens, Greece |
| Jack, Kris | Mendeley, UK |
| Jeffery, Keith | Science and Technology Facilities Council, UK |
| Jörg, Brigitte | German Research Centre for Artificial Intelligence, Germany |
| Krichel, Thomas | Long Island University, USA |
| Luzi, Daniela | Italian National Research Council, Italy |
| Manghi, Paolo | Italian National Research Council, Italy |
| Manola, Natalia | University of Athens, Greece |
| Manouselis, Nikos | Agro-Know Technologies, Greece |
| Matthews, Brian | Science and Technology Facilities Council, UK |
| Plexousakis, Dimitris | University of Crete, Greece |
| De Robbio, Antonella | University of Padova, Italy |
| Rodrigues, Eloy | University of Minho, Portugal |
| Schöpfel, Joachim | University of Lille, France |
| Stathopoulos, Panagiotis | National Documentation Centre, Greece |
| Tzitzikas, Yannis | University of Crete and ICS-FORTH, Greece |
| Wittenburg, Peter | Max Planck Institute for Psycholinguistics, The Netherlands |
| Zeng, Marcia | Kent State University, USA |

## Track on Metadata and Semantics for Learning Infrastructures

### Special Track Chairs

| | |
|---|---|
| Palavitsinis, Nikos | Greek Research and Technology Network, Greece |
| Stoitsis, Giannis | Agro-Know Technologies, Greece |

### Program Committee

| | |
|---|---|
| Arapi, Polyxeni | Technical University of Crete, Greece |
| Broisin, Julien | Université Paul Sabatier, France |

| | |
|---|---|
| Ebner, Hannes | KTH, Royal Institute of Technology, Sweden |
| Hadzilacos, Thanasis | Open University of Cyprus, Cyprus |
| Karampiperis, Pythagoras | Demokritos National Center for Scientific Research, Greece |
| Klerkx, Joris | Katholieke Universiteit Leuven, Belgium |
| Koutoumanos, Anastasios | Technological Educational Institute of Athens, Greece |
| Najjar, Jad | Information Technology Department, Al-Quds University, Israel |
| Stracke, Christian | University of Duisburg-Essen, Germany |
| Vignare, Karen | Michigan State University, USA |
| Wolpers, Martin | Fraunhofer Institute of Technology, Germany |

## Track on Metadata and Semantics for Cultural Collections and Applications

### Special Track Chairs

| | |
|---|---|
| Papatheodorou, Christos | Ionian University, Greece |
| Gergatsoulis, Manolis | Ionian University, Greece |

### Program Committee

| | |
|---|---|
| Christodoulakis, Stavros | Technical University of Crete, Greece |
| Dallas, Costis | Panteion University, Greece |
| Davies, Rob | MDR Partners, UK |
| Dekkers, Makx | Independent Consultant, Spain |
| Ferro, Nicola | University of Padua, Italy |
| Gomez-Pantoja, Joaquín | Universidad de Alcala, Spain |
| Isaac, Antoine | Vrije Universiteit Amsterdam, The Netherlands |
| Lourdi, Irene | University of Athens, Greece |
| Manouselis, Nikos | Agro-Know Technologies, Greece |
| Ore, Christian-Emile | University of Oslo, Norway |
| Roberts, Dave | Natural History Museum, UK |
| Sfakakis, Michalis | National Documentation Centre, Greece |
| Tudhope, Douglas | University of Glamorgan, UK |

## Track on Metadata and Semantics for Agriculture, Food and Environment

### Program Chairs

| | |
|---|---|
| Athanasiadis, Ioannis N. | Democritus University of Thrace, Greece |
| Manouselis, Nikos | Agro-Know Technologies, Greece |
| Keizer, Johannes | Food and Agriculture Organization of the United Nations, Italy |

## Program Committee

| | |
|---|---|
| Antoniou, Grigoris | University of Crete, Greece |
| Beck, Howard | University of Florida, USA |
| Carlson, Jack | Colorado State University, USA |
| Donatelli, Marcello | European Joint Research Centre, Italy |
| Donnelly, Kathryn | Norwegian Institute of Food, Fisheries and Aquaculture Research, Norway |
| Houssos, Nikos | National Documentation Center, Greece |
| Keet, Maria | University of KwaZulu-Natal, South Africa |
| Martini, Daniel | KTBL, Germany |
| Roberts, Dave | The Natural History Museumm, UK |
| Rizzoli, Andrea Emilio | Istituto Dalle Molle di Studi sull'Intelligenza Artificiale, Switzerland |
| Sanchez, Salvador | University of Alcala, Spain |
| Schima, Gerald | Austrian Institute of Technology, Austria |
| San Gil, Inigo | Long Term Ecological Research Network, USA |
| Subirats, Imma | Food and Agriculture Organization of the United Nations, Italy |

# Table of Contents

## Track on Metadata and Semantics for Open Access Repositories and Infrastructures



## Track on Metadata and Semantics for Learning Infrastructures

## Track on Metadata and Semantics for Cultural Collections and Applications

## Track on Metadata and Semantics for Agriculture, Food and Environment

# An Ontology for a Naval Wargame Conceptual Model

Serhan Kars[1] and Halit Oğuztüzün[2]

[1] METEKSAN Defence Industry Inc. Beytepe Koyu Yolu No: 3,
06800 Bilkent, Ankara, Turkey
skars@meteksansavunma.com.tr

[2] Department of Computer Engineering, Middle East Technical University,
Dumlupınar Bulvarı No.1, 06800 Ankara, Turkey
oguztuzn@ceng.metu.edu.tr

**Abstract.** Naval wargames play an important role for both training of naval cadets and trying out tactics for naval warfare. Constructing an ontology can be regarded as a way to formalize the mission space component of a simulation conceptual model. An ontology has been developed to capture the entities involved in a naval wargame and the relationships among them. The ontology is defined in OWL (Web Ontology Language), using the Protégé ontology editor and knowledge-base framework. This paper presents the ontology development process, discusses the decisions made throughout the creation of the ontology, and identifies the lessons learned.

**Keywords:** Ontology, Naval Warfare, Wargame Simulation, Conceptual Model.

## 1 Introduction

Simulation development process starts with identifying user needs and proceeds with describing the problem space accordingly. The accuracy of expected results is largely determined in this phase. In order to achieve interoperability with other systems, boundaries and interfaces between the wargame system and the surrounding execution environment are decided. Conceptual models, based on user requirements, are defined in the next phase [1]. These two steps are not strictly separated in practice because each stimulate and derive from the other.

Simulation context and simulation concept comprise the conceptual model [2]. Simulation context is a collection of information about the constraints and boundaries on the domain of simulation, tactical naval wargaming in our case. Simulation concept is the composition of simulation space and the mission space. Simulation space is a description of the control capabilities for the simulation system, including hardware and software (e.g. number of personnel, data collection and display capabilities, equipment to operate the model). Mission space includes simulation elements (and composites of them) which represent the real world objects and the relationships among them. Simulation elements identify the parameters, attributes, possible states according to the required fidelity and scale of the simulation. Conceptual modeling helps all the parties involved with the project to understand the problem space and user needs clearly. A validated conceptual model helps the

E. García-Barriocanal et al. (Eds.): MTSR 2011, CCIS 240, pp. 1–11, 2011.

development team to determine the system and software requirements without inconsistencies, ambiguities and gaps. A conceptual model also forms an extensive communication environment among all stakeholders, including the development team and users. A conceptual model reflects the common understanding of the stakeholders about the concepts, relationships, assumptions and restrictions within the scope of the simulation. In a typical training scenario, the users of a wargame are trainers and trainees. Having a complete and unambiguous conceptual model is a precondition to ensure "fair fight" between the trainees playing at opposite sides.

The different strata of the ontology spectrum, their usage and some examples of efforts in modeling and simulation are listed in [3]. It is observed that an ontological meaning specified as a part of a model which developers would refer to when building applications has proven useful. Hence we adopt the view that constructing an ontology can be a way to formally define the simulation concept, in particular the mission space.. This paper introduces an ontology approach to describe naval warfare domain for defining a mission space. A summary of naval warfare is given in next section as a part of problem and domain analysis.

## 2 Naval Warfare Domain

Naval warfare involves opposing navies that execute various types of operations (e.g. submarine warfare, electronic warfare, and amphibious warfare) and various types of military units, platforms, and sensor and weapon systems that interact with one another in the battlespace. Battlespace is the environment that has artificial (e.g. ports, buoys) and natural components (e.g. water, sea bottom, bay, lifeforms). It is a dimension in naval warfare since environmental conditions affect the performance and behaviors of naval platforms and systems. A naval combat group consists of components that have specific structures and play different roles in a battle including surface ships, submarines, aircrafts, troops and land platforms. These elements of naval warfare are specialized according to their attributes, determining their abilities and restrictions in battlespace. While weapon, sensor, communication and electronic warfare systems on the entities contribute to these specializations, relations with other types also distinguish some entities from others.

Surface ships can be classified as aircraft carriers, battleships and battle cruisers, cruisers and frigates, corvettes, fast attack craft and amphibious warfare ships [4]. Aircraft carriers are surface ships where aircrafts can take off from and land on. Destroyers which are warships that provide multi-mission offensive and defensive capabilities and can operate independently. Frigates are smaller than destroyers and designed to protect other ships as antisubmarine warfare combatants. Fast attack ships are attack boats of smaller displacement and greater maximum speed comparing with frigates. Naval air power comprises airframe types which are conventional takeoff and landing (CTOL), short takeoff and vertical landing (STOVL) and helicopters. These aircrafts have five main roles of antisubmarine warfare (ASW), strike/attack, air defense, electronic warfare and airborne early warning.

Warships' and aircrafts' combat capabilities are dependent on the weapon and sensor systems they carry. Enemy targets must be detected by sensor systems and engaged by weapon systems in order to defeat them. Sensor systems include radars, sonars, electronic support devices, visual sensors and magnetic anomaly detectors (MAD). Weapons are classified as missiles, guns, mines, depth charges, torpedoes, air bombs. Missiles and torpedoes can be considered in more detail with respect to their guidance systems, types of the platforms they are launched from or types of targets (e.g. anti submarine missiles, surface to air missiles, RF-guided missiles). Guns are weapons that have shorter range and lighter engagement capability compared to missiles, but they are immune to counter measures. The positionings of mines in the sea are one of their classification criteria (e.g. ground mine, moored mine). Depth charges can be used against submarines and explode in preselected depths in the sea. Communication links within members are formed using communication systems and additionally counter-measure systems are used in protection from threats.

The purpose of the wargame is to evaluate the players' decisions and their handling of special events and situations occurring in the simulated warfare [5]. It is useful in order to understand the behavior of the entities (which we represent in our ontology) or evaluate the tactics for the types of naval warfare operations (e.g. submarine warfare, electronic warfare, and amphibious warfare).

## 3 Ontology Development

An ontology is defined as a part of the conceptual model of a naval tactical wargame.

### 3.1 Ontology Development Process

Our ontology development process started by enumerating the important terms in the domain. The terms were listed initially regardless of the relationships between each other, whether they are classes or properties, whether they have common individuals etc. Additionally, items must have counterparts in real world and they are introduced to the model only when it is essential for computational considerations. For example some of the important terms are: platform, its maximum speed, sensor, its frequency, weapon, missile, its maximum effective range. When choosing items, rationale represented in [2] served as a checklist to ensure that conceptual model would be complete and coherent. According to this checklist, elements of simulation must be represented by software requirements and items which are interested to be assessed in simulation.

After defining important terms in the ontology, the next step is to define the classes and class hierarchy. Among the three approaches for class hierarchy development [6], the combination development process was chosen, rather than the top-down or bottom-up processes. One could choose a combination process when the enumerated terms in the ontology cannot be easily deployed into a hierarchy due to the difficulty of deciding levels of concepts between each other and choosing the concepts from the domain. This kind of process can start with some salient top-level concepts and then after continuing with adding new concepts, relating them to appropriate ones in the ontology.

When classes were being introduced to ontology if it was considered that there was a relation between two sets of classes, any object property that had those classes as its domain and range were added to the ontology. After the domains and ranges of the added relation are set, the constraints of classes about being disjoint should be checked and updated. For example, after the *Entity* and *Subentity* classes were added to the ontology, *hasSubentity* object property was defined which has *Entity* class as domain and *Subentity* class as range. If these two classes were not defined as disjoint, *hasSubentity* property could be applied to an *Entity* class and it would be used to infer that the *Entity* must be a subclass of *Subentity*.

In the initial phases of the ontology development, datatype properties were added to the ontology without a hierarchy between properties with the domain and range axioms. All the data properties were only the subproperties of *topDataProperty*. As the ontology development progressed, the number of classes and their data properties increased and it was realized that it was more convenient to organize the properties in a class hierarchy. This new version of data properties resulted in a more clear representation of properties.

## 3.2 Entities

**Class Hierarchy.** Top level classes of the ontology are *SimulationElement* and *Value*. *SimulationElement* name means that the subclasses of that class are the representation of naval warfare's concepts in the simulation and how they are modeled in simulation. When one examines the bottom level classes below the *SimulationElement*, they all have a mathematical model in the simulation software to simulate the real life responses to events and conditions. During the wargame, the user controls entities and apply his tactical decisions by giving orders to them. What user can do with an entity (e.g. attack, sense and move) depends on the parameters of the entity and depends on other entities what the main entity carries. Those entities carried and owned by a main entity, will not be simulated alone, due to design decisions of simulation software. For instance, a radar itself is not a standalone unit in the simulation, it must be owned by a main entity, such as a frigate, in order to be involved in the simulation. Classes named *Entity* and *Subentity* are added to the ontology for this reason.

Sensors, weapons, communication devices and counter measures are the things on a platform that determine the capability of that platform in the naval warfare. Entity instances are related with the *Subentity* instances via *hasSubentity* property. This relation is a universal restriction for entities; it enforces that *Entity* instances have only *Subentity* instances as subentity.

Platforms are the standalone units in the simulation. Platforms are specialized as, *AirPlatform*, *Buoy*, *LandPlatform*, *Submarine*, *SurfacePlatform* and *Troop*. Since troops were planned to be stand-alone entities in the wargame which would be controlled by the users like other platforms and since they have a significant number of common attributes with platforms, *Troop* concept is modeled as a subclass of *Platform* class. *Troop* class should also be associated with subentities (because they carry weapons and sensors) similarly to ordinary platforms.

*SurfacePlatform* is the representation of surface ships. Some of the subclasses of *SurfacePlatform* are defined classes. For instance, being a surface platform and being a base for *CTOL* and *STOVL* airplanes to land on are necessary and sufficient conditions for being an *AircraftCarrier*. The anonymous class is described by an existential restriction with range *Airplane*, not *Helicopter* ("*isABaseFor* some *Airplane*") because *Frigates* can be a base for an *Helicopter* to land on. *Airplatform* has subclasses named *Helicopter* and *Airplane*. These siblings are made disjoint so they can not have common individuals.

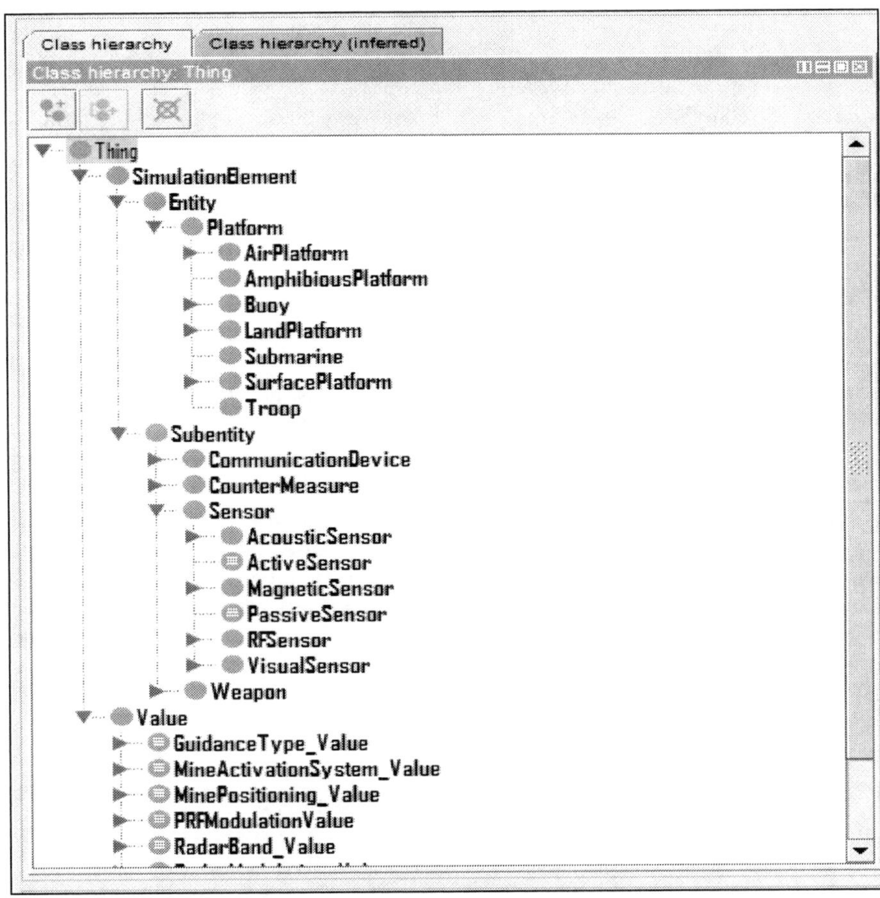


**Fig. 1.** The top level class hierarchy

*Subentity* class has four disjoint subclasses named *Communication Device*, *CounterMeasure*, *Sensor* and *Weapon*. These classes are also detailed by defining subclasses based on them. For example *Weapon* class has disjoint subclasses that refer the weapons of naval warfare. Weapons in the naval warfare can be represented in a taxonomy which contains air bombs, depth charges, guns, mines and the torpedoes. If a missile has a guidance mechanism, it is a guided missile and this necessary and sufficient condition is stated in the ontology. Further, there is a subclass of *GuidedMissile* named *IlluminatedGuidedMissile* with a universal restriction of having only RF or laser illumination as a guidance value. *Mine* has a more specific type that has a *Torpedo* inside for using it against a target, detected by a mine laid. The relation between the *TorpedoMine* class and *Torpedo* is different from the *hasSubentity* property, so a *hasPart* and its inverse *isPartOf* object properties are created.

**Value Entities.** In order to describe some properties of classes, the possible options for these properties are included in the ontology. The *Value* concept of the ontology and its subclasses are for representing the possible values for the regarding object properties. "Value partitions" design pattern [7] is applied for the subclasses of the *Value* class. Value partitions restrict the range of possible values to an exhaustive list, for example radars' frequency band values are restricted to radar bands which are represented by capital letters from "A" to "M". Similarly, types of naval mines according to their positioning in the sea are *Ground* or *Moored* mine.

The covering axiom for value classes is specified in the equivalent classes part rather than the super classes part of Protégé, because being a value option for an individual is sufficient to be a value, while a value class is necessarily a value option.(e.g.: *MinePositioning_value* is necessarly *Ground* or *Moored* and being *Ground* or *Moored* is sufficient in order to be a *MinePositioning_value*.)

**Defined Classes.** A defined class is a class that has not its conditions not only necessary but also sufficient. One of the examples of defined classes in the ontology is the *ActiveSensor*. An active sensor is a sensor which makes transmissions that can be sensed by a passive sensor (which does not make transmissions). As the usage of active and passive sensors is important in the warfare in order to hide from enemy entities, these concepts should be included in the ontology. An active sensor is defined as a *Sensor* which has a relation with a double-precision floating point number value, via one of the data properties named *hasTransmitterPower*, *hasSourceLevel*, *hasInterrogatorTransmitterPower* and *hasTransponderTransmitterPower*. These properties are about transmissions of sensors and their values are used as attributes to calculate the sensing ability of sensors in the simulation. *PassiveSensor* class is also a defined class. They are defined as sensors which are not an *ActiveSensor*.

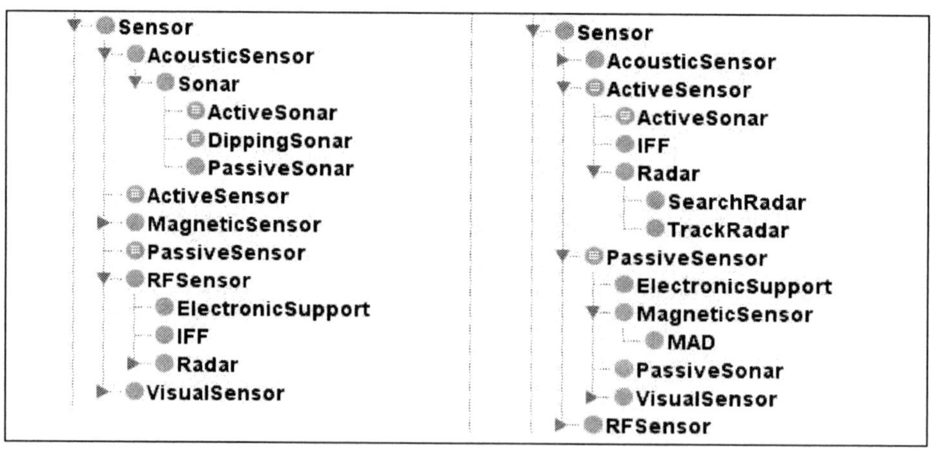


**Fig. 2.** Asserted and inferred class hierarchy

The FaCT++ ,reasoner for Protégé is used to for computing and maintaining the subclass-superclass hierarchy of classes in the ontology. It produces an inferred hierarchy. In the inferred hierarchy, some classes that inherit multiple classes are placed under defined classes which the reasoner determines. It helps to keep ontologies maintainable and well-defined. This feature of ontology modeling gives to the user the ability to define a hierarchy of classes according to their relations with other classes. In our example, sensors are placed under either *ActiveSensor* or *PassiveSensor* according to their definition. While those sensors that are related with transmission properties to some values are placed under *ActiveSensor* class, those that are not active are placed under the *PassiveSensor* class.

### 3.3 Properties

**Object Properties.** Object properties are relationships between two individuals. In our ontology, the recommended naming convention for the property names that property names start with a lower case letter, have no spaces and have the remaining words capitalized is followed. While properties which have subclasses of *SimulationElement* as their domain (e.g *landsOn*, *launches*), have their inverse properties (e.g. *isABaseFor*, *isLaunchedFrom*) in our ontology, other properties do not because they are only used to represent the possible values of properties.

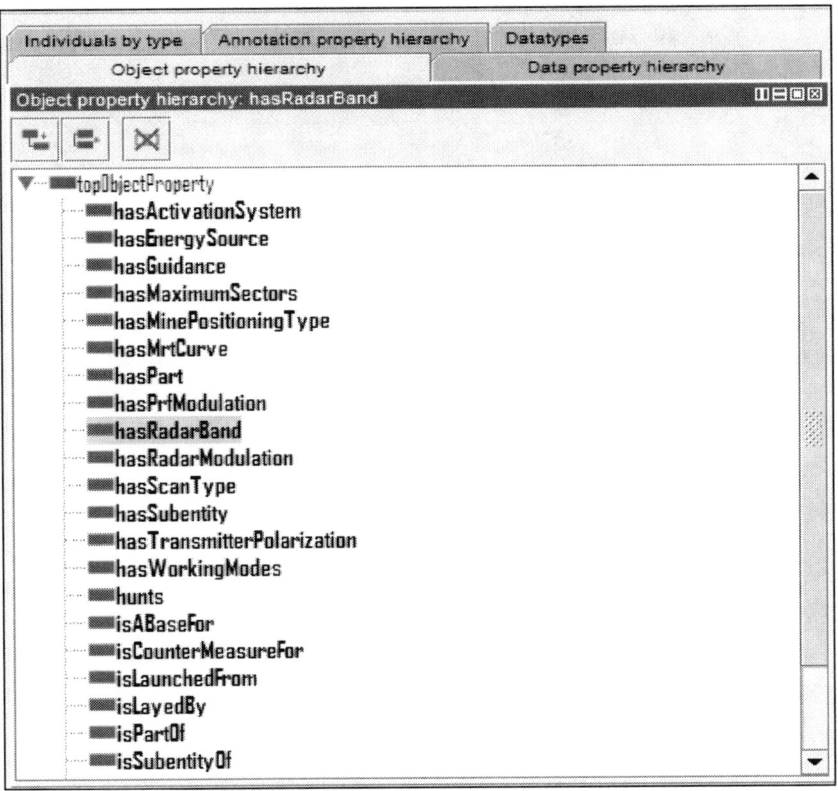


**Fig. 3.** Object Properties. They are relations between individuals of classes.

**Datatype Properties.** Datatype properties link an individual to an XML Schema Datatype value or an RDF [8] literal. They describe relationships between an individual and data values. There are datatype properties more than a number of 200 in our ontology. The same naming convention for the object properties are used for datatype properties as well. Data type properties are kept in an hierarchy according to their domain classes. Properties with only one domain class (e.g. *hasMTIMode*, *processingLoss*, *scanTime*) are subproperties of a data property about that class (e.g. *radarProperty*). Data properties with multiple domains are subproperties of the property named *commonProperty*.

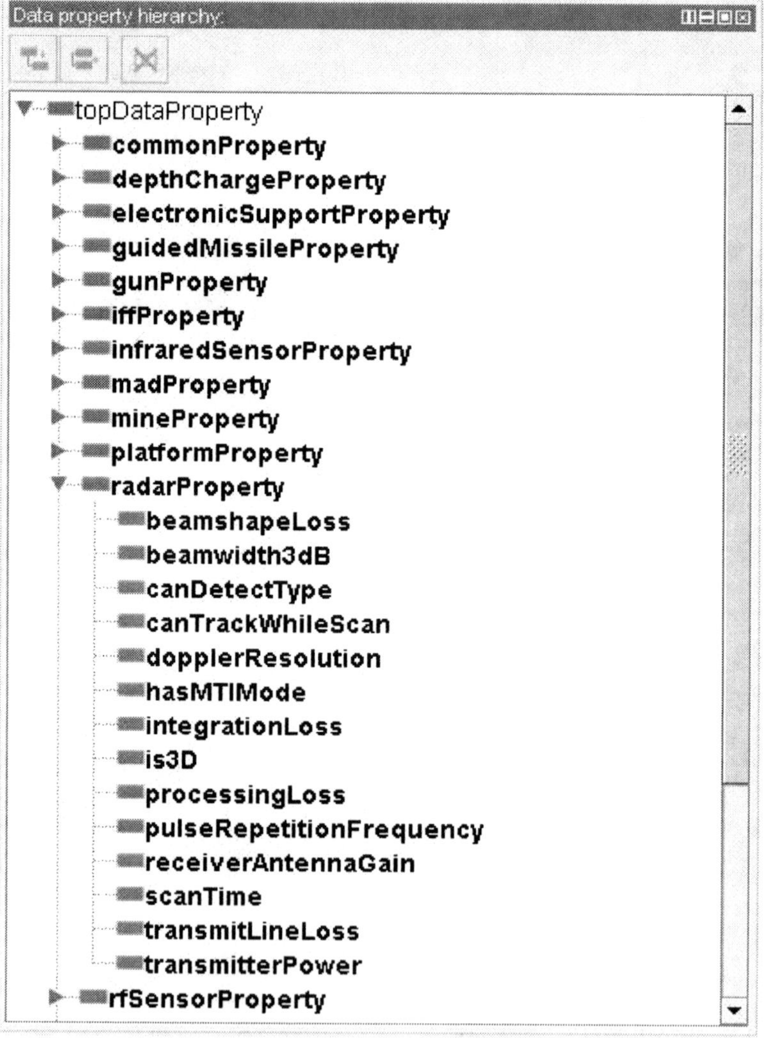


**Fig. 4.** Datatype properties. Object properties were kept in hierarchy according to their base classes.

### 3.4 Other Lessons Learned

When the conceptual model was completed, it served as a basis for initial design decisions. Classes with a significant number of data and object properties have a mathematical behavior model for each in the simulation system (e.g. submarine model, radar model, gun model). These models have parameters, represented by properties which have model classes as their domains in ontology. Ranges of properties determine the parameters' data types in simulation models. While *scanTime* parameter of radar model is one of datatype properties which have *Radar* class as domain, radarBand parameter is also represented by *hasRadarBand* object property in ontology. Those subclasses that are not domain of any properties (or significant number of properties) are represented by their base classes' models by setting different values for parameters according to that concept. Subclasses of *SurfacePlatform* class which include *Frigate*, *FastAttackCraft* and *WaterTanker* are all designed as *SurfacePlatform* model in simulation which have different *maximumSpeed*, *tonnage* and *waterCapacity* parameter values.

Object properties can be separated into two types when the design of the simulation software is considered. The first type of the properties represents the relations of models with other models in simulation. These relations can be implemented as messages between models or order mechanisms in the wargame which can be used by players to control models, and so on. For example, *launches* property which has *Weapon* class as range is the representation of *Launch Weapon* order in wargame. Object property, named *isACounterMeasureFor*, showed the need for countermeasure messages that are sent to the *Weapon* models to calculate the reduction in their effectiveness. Other types of the object properties are the parameters of models in simulation with assigned enumerated values. These enumeration members were subclasses of *Value* classes that are modeled in the ontology using "Value partitions" design pattern (e.g. *radarModulation* parameter has possible values such as 0= pulse, 1= pulse Doppler, 2= continuous wave). Manifest raceability between ontological modeling decisions and design decisions suggest that conceptual modeling prepares the development team well for the design phase.

Ontology building process, as a part of conceptual modeling, encourages the modelers to make tacit domain knowledge explicit and to bring out hidden assumptions. For example, at one point the development team assumed that sonobuoys conveyed the detection information only to the platforms that had laid them. But when discussing the relations of *Sonobuoy* with other classes, it became clear that *Sonobuoy* was related to *Sonar* and *RFCommunicationDevice* via *hasPart* object property. *RFCommunicationDevice* of *Sonobuoy* is related to other *RFCommunicationDevices* via *conveysInfoViaRFSignalTo* property. This relation represented the ability of sonoubuoys to convey information to all individuals of *Entity* that are related with an *RFCommunicationDevice*.

## 4 Evaluation and Validation of Ontology

The main purpose of the ontology is the specification of naval warfare's conceptualization throughout wargame development lifecycle. Measuring the ontology in functional dimension, which is about the intended uses of the ontology, needs intellectual judgment, supported by automatic and semi-automatic techniques [9].

The resulting ontology was reviewed in a series of meetings with domain experts, end-users and conceptual model developers to make sure that it captured the right concepts and properties in the domain of naval wargame with sufficient fidelity and scope with respect to intended uses.

Verification and validation of conceptual models for simulation are crucial, yet challenging tasks [9]. Fortunately, guidance from the literature is available. The evaluation approaches according to levels of ontologies are reviewed in [10]. Sargent[11] presents two paradigms for simulation model verification and validation. He also discusses validation techniques, conceptual model validation and how to document verification and validation results.

## 5 Technical Details

Protégé version 4.1 alpha is used as the ontology editor. The user interface of Protégé can be configured by adding/removing tabs by and adding/removing views in tabs in order to have a more brief and clean screen configuration for developing ontology. For this purpose, before the development phase two plugins were installed: Annotation Template View plugin for concise viewing/editing of annotations of interest, and Matrix Views plugins for having spreadsheet-style views of the ontology when entering values for individuals. Both plugins can be downloaded from [12] and [13].

## 6 Related Works

There are a few published representations of the warfare domain in ontological spectrum. One of the examples is Joint Warfare Simulation Object Library (JWSOL) taxonomy [14]. It was designed in a perspective of Commander in Chief in order to build a repository of warfare simulation classes. It includes the hierarchy of objects (warfighting and environmental objects), events (military, civilian and environmental) and agents (human or organization) which can be parts of any kind of warfare. In addition to this general representation of warfare, there are specialized ontologies in the naval domain as well, e.g. [15].

Silver et al. suggest a method for Ontology Driven Simulation (ODS) supported by an ODS design tool that can map between domain ontologies, modeling ontologies and executable simulation models [16]. The ontology presented in this paper can be regarded as a domain ontology. In [17] the authors discuss several issues regarding military training simulations, including verification and validation, simulation software elements, and critiques and comments of users for a simulator system named Action Speed Tactical Trainer (ASTT).

## 7 Conclusion

In this study, a naval warfare domain ontology for a tactical wargame simulation is introduced. It is noted that the mission space of conceptual model can be defined by ontological approach. Ontology development process was explained which started by enumerating important terms and continued with choosing classes, relating them with object properties and adding data properties according to their domains with a clear representation. Classes in the ontology represent the concepts that are represented by simulation models and the allowable values of these models' attributes. It is indicated that conceptual modeling serves as preparatory work for design by showing the connections between ontology modeling decisions and design decisions for the simulation system. Evaluation and validation of the resulting ontology via meetings was offered which answers the need for the intellectual judgment effort.

## References

1. Smith, R.D.: Military Simulation & Serious Games: Where we came from and where we are going. Modelbenders Press (2009)
2. Pace, D.K.: Ideas About Simulation Conceptual Model Development. John Hopkins APL Technical Digest 21(3), 327–336 (2000)
3. Turnitsa, C., Tolk, A.: Ontology Applied - Techniques employing Ontological Representation for M&S. In: 2006 Fall Simulation Interoperability Workshop (2006)
4. Miller, D., Miller, C.: Modern Naval Combat. Crescent Books (1986)
5. Dunnidan, J.F.: Wargames Handbook, Third Edition How to Play and Design Commercial and Professional Wargames. Writers Club Press (2000)
6. Uschold, M., Gruninger, M.: Ontologies: Principles, Methods and Applications. Knowledge Engineering Review 11(2), 93–136 (1996)
7. Horridge, M.: A Practical Guide To Building OWL Ontologies Using Protégé 4and CO-ODE Tools v1. University Of Manchester (2011)
8. RDF Documentation, `http://www.w3.org/TR/rdf-concepts/`
9. Pace, D.K.: Modeling and Simulation Verification and Validation Challenges. John Hopkins APL Technical Digest 25(2), 163–172 (2004)
10. Brank, J., Grobelnik, M., Mladenic, D.: A Survey Of Ontology Evaluation Techniques. In: Conference on Data Mining and Data Warehouses (2005)
11. Sargent, R.G.: Verification and Validation of Simulation Models (2008)
12. The web page of annotation view plugin, `http://code.google.com/p/co-ode-owl-plugins/wiki/AnnotationTemplate`
13. The web page of matrix view plugin, `http://code.google.com/p/co-ode-owl-plugins/wiki/MatrixViews`
14. Conwell, C.L.: Joint Warfare Simulation Object Library: Joint Warfare Taxonomy (1995)
15. US Navy Ships Ontology, `http://www.daml.org/ontologies/361`
16. Silver, A.G., Hassan, O., Miller, J.A.: From Domain Ontologies to Modeling Ontologies to Executable Simulation Models. In: Henderson, S.G., Biller, B., Hsieh, M.H., Shortle, J., Tew, J.D., Barton, R.R. (eds.) 2007 Winter Simulation Conference, pp. 1108–1117 (2007)
17. Van Veen, M., Van Fenema, P., Grant, T.: User Perspectives on Design Logic in Military Training Simulators. In: 13th International Command and Control Research and Technology Symposium (2008)

# Exploring the Development of Endorsed Learning Resources Profiles in the Connexions Repository

Cristian Cechinel[1], Salvador Sánchez-Alonso[2], Miguel-Ángel Sicilia[2], and Priscyla Waleska Targino de Azevedo Simões[3]

[1] Computer Engineering Course, Federal University of Pampa, Caixa Postal 07, 96400-970, Bagé (RS), Brazil
contato@cristiancechinel.pro.br
[2] Information Engineering Research Unit, Computer Science Dept., University of Alcalá, Ctra. Barcelona km. 33.6 – 28871 Alcalá de Henares (Madrid), Spain
{salvador.sanchez,msicilia}@uah.es
[3] Applied Computational Intelligence Research Group, University of the Extreme South of Santa Catarina, Caixa Postal 3167, 88806-000, Criciúma (SC), Brazil
pri@unesc.net

**Abstract.** Existing learning object repositories are adopting strategies for quality assessment and recommendation of materials that rely on information provided by their community of users, such as ratings, comments, and tags. In this direction, Connexions has implemented an innovative approach for quality assurance where resources are socially endorsed by distinct members and organizations through the use of the so-called lenses. This kind of evaluative information constitutes a referential body of knowledge that can be used to create profiles of endorsed learning resources that, in their turn, can be further used in the process of automated quality assessment. The present paper explores the development of endorsed learning resources profiles based on intrinsic features of the resources, and initially evaluates the use of these profiles on the creation of automated models for quality evaluation.

**Keywords:** Learning objects, Connexions, automated assessment, endorsement mechanisms, repository.

## 1 Introduction

Learning object repositories (LORs) are potential aggregators of communities of practitioners (Brosnan, 2005; Monge, Ovelar, & Azpeitia, 2008), i.e. people who share interests and concerns about something they do and learn through their interactions. As such, they tend to harness the features of such social environments through the adoption of strategies for the establishment of quality that rely on the impressions of usage and evaluations given by regular users and experts that are members of the repository community. Examples of usage of the community

E. García-Barriocanal et al. (Eds.): MTSR 2011, CCIS 240, pp. 12–21, 2011.

impressions inside repositories can be found, for instance, in Connexions[1], MERLOT[2], eLera[3], and Organic.Edunet[4].

Even though such strategies tend to assure quality inside repositories, recent studies (Cechinel & Sánchez-Alonso, 2011; Sanz-Rodríguez, Dodero, & Sánchez-Alonso, 2010) have highlighted that the task of manually reviewing all materials is unfeasible given that the number of resources inside LORs is not just enormous but also rapidly increasing every day. The proposed alternatives for this problem are to pursuit the development of measures of quality that can be automatically obtained from the repository. On that direction, Ochoa & Duval (2008) proposed the development of a set of metrics for ranking the results of learning objects search inside repositories by using information obtained from the learning objects metadata, from the user queries, and from other external sources such as the records of historical usage of the resources.

However, automated assessment will also be applied for resources that still have not been shared to the public and thus will not have any usage information associated to them in this initial moment. Another problem is that metadata may be incomplete in some cases (Sicilia, Garcia, Pages, Martinez, & Gutierrez, 2005) or even contain inaccurate descriptions (Cechinel, Sánchez-Alonso, & Sicilia, 2009). Regarding that, García-Barriocanal & Sicilia (2009) proposed a complementary approach that relies only on the data that can be directly extracted from the learning resources themselves. The idea here is to use the social evaluative metadata stored on the repository as baseline information that allows the classification of resources according to their quality, and to compare these classes of resources in order to identify intrinsic metrics associated to the classes (thus associated to quality). These metrics could then be further used to generate models for automatically classify resources inside repositories according to their quality. This approach was successfully tested by Cechinel, Sánchez-Alonso, & García-Barriocanal (2011) which have developed profiles of highly-rated learning objects in the MERLOT repository. The authors have identified 13 intrinsic features associated to learning objects quality and have generated a Linear Discriminant Analysis (LDA) model able to classify resources between *good* and *poor* with 91.49% of accuracy. Even though the model has achieved very good results, the authors highlighted that the work was context focused, and that the possibility of extrapolating such findings for other repositories other than MERLOT was still a subject of further investigation.

The present work tries to help in answering this question by generating and testing models for automated quality assessment of learning objects stored on Connexions repository. The rest of this work is structured as follows: Section 2 describes the Connexions repository and the most important evaluative metadata available on it. Section 3 explains the methodology followed on the study, and section 4 shows the results of the models generated for the resources of the repository. Section 5 discusses the results found here and points out the main limitations of this work. Finally, section 6 presents the final remarks and possible directions for future work.

---

[1] `http://cnx.org/`

[2] `http://www.merlot.org`

[3] `http://www.elera.net/`

[4] `http://portal.organic-edunet.eu`

## 2 Connexions Repository

Connexions is a repository that allows users to collaborative create and share learning materials and that has presented an exponential growth of contributors in the last years. According to Ochoa (2010), such success can be attributed to the fact that, differently from the traditional LORs, Connexions functions through the "*social interaction for the creation of materials*", where all materials are created by its own community of members. This community can develop materials in two formats: modules (small pieces of knowledge) and collections (groups of modules structured into course notes). In Connexions every material available is free for using, reusing and sharing with others under a Creative Commons[5] license.

Quality in Connexions is approached by a system called *lenses* (see figure 1) that arranges resources according to evaluations provided by individuals and organizations (Kelty, Burrus, & Baraniuk, 2008). In this context, resources are explicitly endorsed by third parties, and gain higher quality assurance as they start to accumulate more endorsements (lenses) from others. Moreover, Connexions also provides mechanisms to sort materials considering their number of accesses over the time and considering the ratings given by users. Recently, Connexions has also integrated in the repository plugins of two popular and well succeeded tools for social interaction (Facebook and Twitter) thus allowing the community of users to recommend and disseminate materials across these other social platforms.

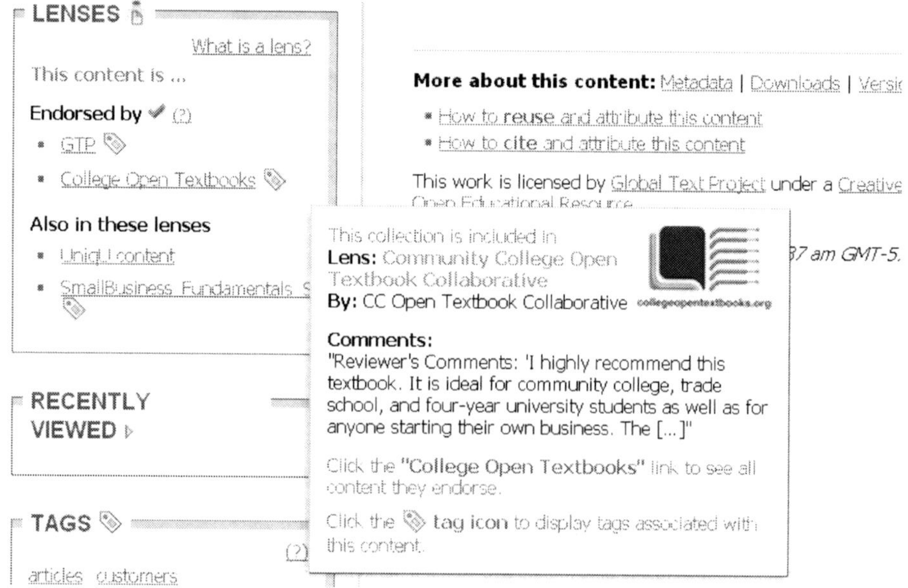


**Fig. 1.** Connexions repository – lenses display

[5] `http://creativecommons.org/`

# 3 Method

The method followed for this study was the development of profiles of endorsed resources in Connexions based on their intrinsic features. This method is based on the methodology applied before by Ivory & Hearst (2002) for the creation of profiles of highly-rated websites, and on the methodology described by García-Barriocanal & Sicilia (2009) and Cechinel, et al. (2011) for the creation of highly-rated learning objects profiles in the context of LORs. The profiles developed here were then used to generate models focused on automatically classify resources according to the endorsement perspective.

## 3.1 Data Collection

Information from a total of 8,855 modules were collected (January 2011) through the use of a crawler that systematically traversed the pages of the Connexions repository. Connexions is basically formed by contents in the form of websites, and due to that, we collected 12 intrinsic metrics that are supposed to appear in such type of material (i.e., link measures, text measures, graphic measures and site architecture measures).

### 3.1.1 The Metrics

The metrics collected for this study (see Table 1) are a subset of the metrics used by Cechinel, et al. (2011).

**Table 1.** Metrics collected for the study

| ***Class of Measure*** | ***Metric*** |
|---|---|
| Link Measures | Number of Links, Number of Unique[a] Links, Number of Internal Links[b], Number of Unique Internal Links, Number of External Links, Number of Unique External Links |
| Text Measures | Number of Words |
| Graphic, Interactive and Multimedia Measures | Number of Images, Total Size of the Images (in bytes), Number of Scripts |
| Site Architecture Measures | Size of the Page (in bytes), Number of Files for downloading |

a The term Unique stands for "non-repeated"

b The term internal refers to those links which are located at some directory below the root site

Some of these metrics have also been mentioned in other works which tackled the problem of assessing quality of resources (sometimes in adjacent fields of e-learning) through the use of measurable aspects. For instance, the Number of Links and the Number of Words have been included in the study of Custard & Sumner (2005) who created a support vector machine model to assess the quality of resources inside an Educational Digital Library. The Number of Words also appears in the research of Blumenstock (2008), who identified the length of an article in Wikipedia as a possible predictor of quality of the articles. The Number of Internal and External Links, the Number of Words, and the Number of Images are used by Stvilia, Twidale, Smith, & Gasser (2005) to automatically discriminate high quality articles inside a

Community-Based Encyclopedia. At last, all metrics shown in Table 1 were used in the study conducted by Ivory & Hearst (2002) to automatically evaluate the quality of websites.

### 3.1.2 Data Sample

The amount of endorsed (one or more lenses) and not endorsed (zero lenses) resources in Connexions is very close (see figure 2). Precisely, 53.55% (4,742) of the resources of the repository are endorsed at least one time, 1.84% (163) of the resources are endorsed from 2 to 4 times, and 44.46% (3,950) of the resources are not endorsed.

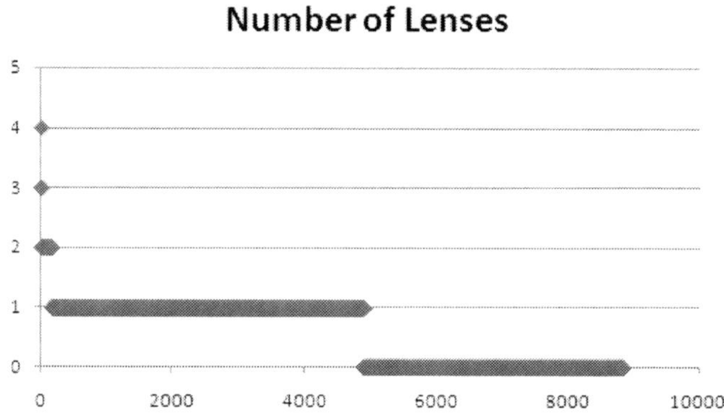


**Fig. 2.** Distribution of the number of lenses per resource

The percentage of explicit preferences given by the community of the repository can be considered extremely high if compared with other learning object repositories. For instance, Cechinel & Sánchez-Alonso (2011) conducted a study about quality assessments on MERLOT repository where only approximately 12% of the resources have received comments and ratings from the community of users. This reinforces the success of social models in LORs (Ochoa, 2010; Sánchez-Alonso, Sicilia, García-Barriocanal, Pagés-Arévalo, & Lezcano, 2011) as one of the best approaches for creating, storing and sharing resources in a sustainable way.

## 3.2 Developing Profiles of Endorsed Resources

The 3,950 resources without endorsements were excluded from this study, and the 4,905 left (the endorsed ones) were divided in two distinct groups: 1) those with just one endorsement, and 2) those with more than one endorsement. These groups and their size are shown in table 2.

**Table 2.** Groups of resources

| *Group* | *Amount* | *Percentage of the Sample* |
|---|---|---|
| One endorsement | 4,742 | 96.67 |
| Two or more endorsements | 163 | 03.33 |
| Total | 4,905 | 100.00 |

Intrinsic features (metrics) of the resources belonging to these two groups were then contrasted in order to evaluate whether they presented significant differences among them. As the samples did not follow a normal distribution, a Mann-Whitney (Wilcoxon) test was performed to evaluate whether the classes presented differences between their medians, and a Kolmogorov-Smirnov test was applied to evaluate if the classes presented distinct distributions. When both distributions and medians presented significant differences, the metric was considered as a potential indicator of the amount of endorsements. The tendency of each metric (whether they influence negatively or positively the amount of endorsements) was observed by comparing the median values of the samples. Table 3 presents the results of this analysis for the evaluated metrics.

**Table 3.** Tendencies of the metrics towards the number of endorsements

| *Class of the Measure* | *Metric* | *Tendency* |
|---|---|---|
| Link Measures | Number of Links | Y↑ |
| | Number of Internal Links | Y↑ |
| | Number of External Links | Y↑ |
| Text Measures | Number of Words | Y↓ |
| Graphic, Interactive and Multimedia Measures | Number of Images | Y↓ |
| | Total Size of the Images | Y↓ |
| | Number of Scripts | Y↑ |
| Site Architecture Measures | Size of the Page | Y↓ |
| | Number of Files for Downloading | N* |

In table 3, Y stands for a significant difference in both median and distribution of the given metric between the two groups of resources at a 95% level and N* stands for a significant difference just in the distribution. Moreover, ↑ means that the metric presents a positive tendency, i.e., as the value/amount of the metric increases, the number of endorsements of the resource also tends to increase; and ↓ stands for a negative tendency of that metric.

As resources in Connexions do not contain any repeated links, the Number of Unique Links, the Number of Unique Internal Links, and the Number of Unique External Links were not included in the analysis. As it can be in table 3, from the nine metrics we have tested, just one (Number of Files for Downloading) did not present significant difference in both median and distribution between the two groups. The analysis shows that resources with more than one endorsement tend to have a higher number of links (internal and external) and a higher number of scripts. Moreover, these resources tend to have a lower number of words and images, as well as they tend to be smaller in size.

## 4 Automated Classification Based on the Profiles

With the exception of the Number of Links[6], all metrics which presented differences between the two groups were used to generate models for automatically classify the

[6] As both tendencies of the Number of Internal Links and the Number of External Links are the same and their sum is equal to the Number of Links, this metric was omitted from the models.

resources according to their number of endorsements, i.e., whether a resource would have just one, or more than one endorsement. The models were created through distinct data mining classification algorithms and using the software WEKA (Hall, et al., 2009). Precisely, we have created and evaluated the following classification algorithms: J48 (decision tree), SimpleCart (decision tree), PART (rules), Multilayer perceptron and Bayesian network. Table 4 presents the results of these tests.

**Table 4.** Results for the models created through data mining algorithms

| *Algorithm* | *Overall Accuracy - Correctly Classified Instances* | *Mean Absolute Error* | *Correctly Classified Instances for Class = One Endorsement* | *Correctly Classified Instances for Class = More than one Endorsement* |
|---|---|---|---|---|
| J48 | 98.3282 | 0.0297 | 99.7047 | 58.2822 |
| Simple Cart | 97.8186 | 0.04 | 99.2619 | 44.7852 |
| PART | 97.9409 | 0.0327 | 99.2619 | 59.5092 |
| Multilayer Perceptron | 96.6769 | 0.0526 | 100.000 | 00.0000 |
| Bayesian Network | 96.371 | 0.056 | 99.4938 | 05.5214 |

As it is shown in table 4, all five algorithms presented very high results if we consider the overall accuracy and the mean absolute errors. However, an in-depth look in the classification accuracy for each group of resources shows that the high performance of the models is concentrated on the classification of resources with just one endorsement. For this class of resources, all models presented accuracy higher than 99%. On the other hand, the classification of resources that have more than one endorsement can be considered very poor. In this work, the best models for classify this group of resources were generated by the algorithms PART (with only 59.5092% of the resources correctly classified) and J48 (with 58.2822% of the resources correctly classified).

## 5 Discussion

This poor performance of the models for classifying resources that have more than one endorsement is probably a consequence of the small size of this group in the sample (only 3.33% of the total sample). Previous works in this same field have adopted a different approach for splitting the sample. Precisely, Cechinel, et al. (2011) and García-Barriocanal & Sicilia (2009) have used the terciles of the existing ratings of the resources to generate their respective classes of quality (poor, average and good), thus avoiding the concentration of resources in a given group (all three groups – poor, medium and good resources – had the same percentage of resources). Unfortunately, it is still not possible to adopt such approach in Connexions given that there is just a small part of the resources that is endorsed more than once.

Moreover, Cechinel, et al. (2011) have stated that in order to generate models for automatically assessing quality in MERLOT, the process of searching for indicators of quality should consider the different aspects of the resources, i.e., the target discipline (Arts, Business, Education, Mathematics and Statistics, Science and Technology, Social Sciences) and the type of the material (Simulation, Animation, Drill, Lecture, etc). This is not the case of the present work. Here we have evaluated the metrics considering the overall sample as a whole and without taking into consideration the possible classifications of the resources. As resources in Connexions are created inside the repository, they present a more homogenous structure than the resources referenced by MERLOT, thus leading us to think that this information is not as important here as it was in the previous mentioned work[7]. However, the target discipline (the subject of the resource) can be crucial information to the creation of the profiles, and due to limitations of our Crawler was not included here. To better illustrate this problem, Table 5 shows a comparison between the tendencies of the metrics found in this study, and the tendencies of the same metrics in some disciplines of MERLOT.

**Table 5.** Tendencies of the metrics in Connexions and in the different disciplines of MERLOT (considering the perspective of the peer-reviewers ratings) (taken and adapted from Cechinel, et al. (2011)

| *Metric* | *Business* | *Education* | *Math & Statistics* | *Science& Technology* | *Social Sciences* | *Connexions* |
|---|---|---|---|---|---|---|
| Number of Links | | ↑ | ↑ | | | ↑ |
| Number of Internal Links | | ↑ | | | | ↑ |
| Number of External Links | | | ↑ | | ↓ | ↑ |
| Number of Images | | ↑ | ↑ | ↑ | ↓ | ↓ |
| Size of the Page | | ↑ | ↑ | | | ↓ |
| Total Size of the Images | | ↑ | ↑ | | | ↓ |
| Number of Scripts | | | | ↑ | | ↑ |
| Number of Words | ↓ | | | | | ↓ |

As it can be seen in Table 5, in MERLOT, the metrics presented different tendencies according to the disciplines (some of the metrics did not even present any tendency for some disciplines). It is interesting to highlight that the tendency of some metrics in Connexions concurs in some disciplines of MERLOT, but not always. For instance, the Number of Links presented positive tendency in Connexions, as well as in MERLOT for the disciplines of Education and Mathematics & Statistics; the Number of Internal Links presented positive tendency in Connexions and positive tendency in the discipline of Education; the Number of Scripts presented positive tendency in both Connexions and the discipline of Science & Technology; and the Number of Words presented negative tendency in Connexions and in the Business discipline. On the other hand, the Number of External Links presented positive tendency in Connexions, but negative tendency in the discipline of Social Sciences; and the negative tendency of the Number of Images in Connexions concurs only with the discipline of Social Sciences, but not with the disciplines of Education, Mathematics & Statistics and Science & Technology. At last, the Total Size of the Images presented negative tendency in Connexions, but positive tendencies in the disciplines of Education and Mathematics & Statistics.

[7] In fact, all resources in Connexions are not distinguished by the type of the material because they all have the same modular structure, i.e., they all belong to the same type called "module".

In resume, it is possible that the associations and tendencies found in the present work could be representative of just some specific categories of disciplines in the repository, and thus are not reflecting their exactly strength in the models we have created.

## 6 Final Remarks

The present work has created profiles of endorsed resources in the Connexions repository. Precisely, we have compared nine intrinsic metrics of resources belonging to two distinct groups divided according to their number of endorsements (those with just one endorsement, and those with more than one endorsement). From the tested metrics, just one did not present any significant difference between the two groups (the Number of Files for Downloading).

The profiles of endorsed resources indicated that resources with more than one lens (endorsements) tend to have more links, a higher number of scripts, less words and images and a smaller size. We have used these profiles to generate and evaluate models for automatically classify resources regarding their number of endorsements. The five models created have achieved very high overall accuracy levels (higher than 96%), but these high performances were limited to classify only one group of resources (those with just one lens). According to our point of view, this was probably caused by the small amount of resources in the sample that had more than one lens (just 3.33%). It is still needed to wait the growth of endorsements in the repository in order to better evaluate the feasibility of creating models for automatically classify resources according to their amount of endorsements.

Moreover, and as mentioned before, previous works have stated that the process of creating learning resources profiles should consider the different types of materials and categories of disciplines. This was not the case of the present study. The results found here could be influenced by this fact, since that some of the metrics that were found as indicators of quality could be related to specific categories of disciplines, and not to the entire population of resources stored on the repository.

Besides working with a larger sample of endorsements, future work will also create profiles taking into consideration the different categories of disciplines, the different kinds of lenses existing in Connexions (members, affiliation, featured), the number of collections in which resources are included, and the new endorsements provided through the use of the recent social plugins implemented in the repository.

**Acknowledgments.** The results presented in this paper have been partially funded by Carolina Foundation through its Mobility Program for Public Brazilian Professors, by the University of Alcalá and the CAM (Comunidad de Madrid), as part of project MARIA (code CCG08-UAH/TIC-4178) and by the Spanish Ministry of Science and Innovation through projects MAPSEL (code TIN2009-14164-C04-01) and MAVSEL (code TIN2010-21715-C02-01).

## References

Blumenstock, J.E.: Size matters: word count as a measure of quality on wikipedia. Paper Presented at the Proceedings of the 17th International Conference on World Wide Web, Beijing, China (2008)

Brosnan, K.: Developing and sustaining a national learning-object sharing network: A social capital theory perspective. In: Williams, J.B., Goldberg, M.A. (eds.) Proceedings of the ASCILITE 2005 Conference, pp. 105–114. ASCILITE, Brisbane (2005)

Cechinel, C., Sánchez-Alonso, S.: Analyzing Associations between the Different Ratings Dimensions of the MERLOT Repository. Interdisciplinary Journal of E-Learning and Learning Objects 7, 1–9 (2011)

Cechinel, C., Sánchez-Alonso, S., García-Barriocanal, E.: Statistical profiles of highly-rated learning objects. Computers & Education 57(1), 1255–1269 (2011)

Cechinel, C., Sánchez-Alonso, S., Sicilia, M.Á.: Empirical Analysis of Errors on Human-Generated Learning Objects Metadata. In: Sartori, F., Sicilia, M.Á., Manouselis, N. (eds.) Metadata and Semantic Research, vol. 46, pp. 60–70. Springer, Heidelberg (2009)

Custard, M., Sumner, T.: Using Machine Learning to Support Quality Judgments. D-Lib Magazine 11(10) (2005)

García-Barriocanal, E., Sicilia, M.Á.: Preliminary Explorations on the Statistical Profiles of Highly-Rated Learning Objects. In: Sartori, F., Sicilia, M.Á., Manouselis, N. (eds.) Metadata and Semantic Research, vol. 46, pp. 108–117. Springer, Heidelberg (2009)

Hall, M., Frank, E., Holmes, G., Pfahringer, B., Reutemann, P., Witten, I.H.: The WEKA data mining software: an update. SIGKDD Explor. Newsl. 11(1), 10–18 (2009)

Ivory, M.Y., Hearst, M.A.: Statistical profiles of highly-rated web sites. In: Proceedings of the SIGCHI Conference on Human Factors in Computing Systems: Changing our World, Changing Ourselves, pp. 367–374. ACM, Minneapolis (2002)

Kelty, C.M., Burrus, C.S., Baraniuk, R.G.: Peer Review Anew: Three Principles and a Case Study in Postpublication Quality Assurance. Proceedings of the IEEE 96(6), 1000–1011 (2008)

Monge, S., Ovelar, R., Azpeitia, I.: Repository 2.0: Social Dynamics to Support Community Building in Learning Object Repositories. Interdisciplinary Journal of E-Learning and Learning Objects 4, 191–204 (2008)

Ochoa, X.: Connexions: a Social and Successful Anomaly among Learning Object Repositories. Journal of Emerging Technologies in Web Intelligence 2(1) (2010)

Ochoa, X., Duval, E.: Relevance Ranking Metrics for Learning Objects. IEEE Transactions on Learning Technologies 1(1), 34–48 (2008)

Sánchez-Alonso, S., Sicilia, M.-A., García-Barriocanal, E., Pagés-Arévalo, C., Lezcano, L.: Social models in open learning object repositories: A simulation approach for sustainable collections. Simulation Modelling Practice and Theory 19(1), 110–120 (2011)

Sanz-Rodríguez, D.J.M., Sánchez-Alonso, S.: Ranking Learning Objects through Integration of Different Quality Indicators. IEEE Transactions on Learning Technologies 3(4), 358–363 (2010)

Sicilia, M.-A., Garcia, E., Pages, C., Martinez, J.J., Gutierrez, J.M.: Complete metadata records in learning object repositories: some evidence and requirements. International Journal of Learning Technology (IJLT) 1(4), 411–424 (2005)

Stvilia, B., Twidale, M.B., Smith, L.C., Gasser, L.: Assessing information quality of a community-based encyclopedia. Paper Presented at the Proceedings of the International Conference on Information Quality - ICIQ 2005 (2005)

# Automatically Specifying Parallel Composition of Matchers in Ontology Matching Process

Marko Gulić[1], Ivan Magdalenić[2], and Boris Vrdoljak[3]

[1] Faculty of Maritime Studies, University of Rijeka,
Studentska ulica 2, HR-51000 Rijeka, Croatia
marko.gulic@pfri.hr

[2] Faculty of Organization and Informatics, University of Zagreb,
Pavlinska 2, HR-42000 Varaždin, Croatia
ivan.magdalenic@foi.hr

[3] Faculty of Electrical Engineering and Computing, University of Zagreb,
Unska 3, HR-10000 Zagreb, Croatia
boris.vrdoljak@fer.hr

**Abstract.** Ontology matching plays an important role in the integration of heterogeneous data sources. In order to find semantic correspondences between entities of different ontologies, various matchers can be used. When multiple matchers are integrated into a matching system, the issue of combining these matchers together and determining the importance of every basic matcher arises. In a parallel composition of basic matchers, several matchers are used independently, and then their results are aggregated. In this paper we propose a method for automatically determining weights of basic matchers in their parallel composition. An evaluation of the method is performed on case study examples concerning the matching of three ontology pairs, and a comparison with other aggregation methods is given.

**Keywords:** Ontology matching, parallel matching systems, weighted aggregation, ontology integration.

## 1 Introduction

An ontology is an explicit specification of a conceptualization [1]. Ontology defines a set of entities and relations between them in a way that both humans and machines understand it. Ontologies are expressed in an ontology language. One of the most popular languages is Web Ontology Language (OWL) [2] that is recommended by W3C organization.

Ontology matching is the process of finding semantic relationships or correspondences between entities of different ontologies [3]. Ontology matching is a key part in the integration of heterogeneous applications. The goal of ontology matching is to reduce heterogeneity between different ontologies.

According to [3] the most obvious types of heterogeneity are:

- Syntactic heterogeneity – when ontologies are not expressed in the same ontology language.

E. García-Barriocanal et al. (Eds.): MTSR 2011, CCIS 240, pp. 22–33, 2011.

- Terminological heterogeneity – it considers variations in names when referring to the same entities in different ontologies.
- Conceptual heterogeneity (also called semantic heterogeneity [4]) – it occurs when differences exist in modelling the same domain of interest.

There are many different matchers that attempt to determine the correspondence between the elements of two different ontologies. Single matcher will not give the best results so a set of different basic matchers are usually integrated into the matching system. The problem is how to put these matchers together and determine the importance of every basic matcher. One approach for integration of multiple matchers into the system is a parallel composition of matchers. The problem of these systems is how to determine the importance of basic matchers.

When ontologies are relatively small, an expert can manually associate entities in an ontology to corresponding entities in another ontology. When ontologies are large (e.g. ontologies contain 1000 different entities), manual mapping is too time consuming. Therefore, there is a need for the automation of the ontology matching process. Of course, the ontology matching process can never be fully automated. However, any increase of automation in the matching process is useful if it facilitates the usage of matching system. The focus of this paper is on automatically determining importance of every basic matcher in a parallel composition, in order to improve the efficiency and quality of the ontology matching process. As each ontology is unique, the efficiency of any basic matcher depends on the implementation of the ontologies that are matching. If the efficiency of a basic matcher is poor for a certain pair of ontologies, the importance of the matcher should be automatically set to a low value.

In this work, we focus only on terminological heterogeneity. Resolving terminological heterogeneity is the basic step in matching two ontologies. For example, when finding correspondences by matchers that compare the structure of ontologies (semantic heterogeneity), if we want to find equal entities based on their parent entities, we need to know whether their parents are equal. The correspondence of parents can be found by using some string matchers that resolve terminological heterogeneity.

Five basic string matchers are chosen to evaluate our automatic method for aggregation. Even the matchers that deal with the same heterogeneity type give different results while matching different pairs of ontologies. Our automatic method automatically defines the importance of various basic matchers in order to improve overall performance of the matching system.

In the method that we propose, the importance of a certain correspondence found within the results of a basic matcher is higher when the same correspondence is found within a smaller number of other basic matchers. The method that finds the same correspondences as all other methods does not provide any new significant information for the matching process. This assumption will be explained in more detail and evaluated on test examples in the rest of the paper.

The paper is organized as follows. In Section 2 we discuss related work. In Section 3 basic terminology of ontology matching and parallel matching systems are introduced. In Section 4 our Autoweight method for automatically determining weights in the weighted aggregation of basic matchers is presented. In Section 5 different aggregation approaches are tested. An evaluation of the Autoweight method is performed on case study examples concerning the matching of three ontology pairs, and a comparison with other aggregation methods is given. Finally, the conclusion is given in section 6.

## 2 Related Work

Ontology matching plays an important role in the integration of heterogeneous data sources [3]. It is usually performed by joint execution of multiple matchers. When multiple matchers are integrated into an automatic matching system, one of the problems is combining these matchers together and determining the importance of every basic matcher. Automatically determining importance of every basic matcher increases the efficiency of the ontology matching process. The implementation of the ontologies that are matching affects the efficiency of any basic matcher. For example, a developer can create a hierarchy of classes using the rdf:subClassOf tag, while another developer can avoid to use it, because of simplicity. Hence, it is not suitable to use the basic matcher based on the rdf:subClassOf tag for comparing a relation between two ontology classes. Therefore, matcher selection, matcher combination and matcher tuning are issues to explore [13]. In this paper we focus on a part of matcher combination, i.e. aggregation of individual matchers.

There are various approaches ([5], [6], [7], [8], [9], [10], [11], and [12]) that use common correspondence which is the result of aggregation of correspondences obtained in basic matchers. However, only in [5], [6] and [12] the authors explicitly define the procedure for defining the importance of basic matchers.

Coma [5] and Coma++ [6] are ontology matching tools based on parallel composition of matchers. Four methods for aggregation of basic matchers are implemented. In the Max value method, selection of the highest correspondence between two elements obtained from basic matchers in many cases could not be accurate, and this is a major limitation. For example, basic matcher Prefix that checks whether a string is the prefix of another string, finds maximal similarity between elements Node and NodePair, although the NodePair element represents a pair of nodes. The Min value aggregation method has similar limitation. Some basic matchers do not find any similarity between two elements although these elements are very similar. The Average aggregation method defines equal weights for every basic matcher in the matching system. All basic matchers have the same importance in this method. However, the Weighted aggregation specified by our Autoweight method has better results than the Average method, as it will be shown in Section 5. The principles of the Weighted aggregation are proposed in [5], but the authors did not use this method, as they did not want to make any assumption about the importance of the basic matchers. In this paper, we propose a method for automatically determining weights of every basic matcher for weighted aggregation.

NOM [7] is also an ontology matching tool based on parallel composition of matchers. It uses weighted aggregation method, but importance for every matcher is defined manually by experts.

FALCON AO [8] combines two element matchers, Edit distance and Statistic analysis. Importance of matcher Edit distance is 0.2 and importance of Statistic analysis is 0.8. The authors did not explain why they determined importance this way, and they stated that the importance is defined based on experience.

RiMOM [9] combines many element matchers like string, instance, structural, and constraint matchers. The weighted aggregation method is also used, but the authors did not mention the way of defining the importance.

H-Match [10] combines two element matchers, linguistic affinity and structural affinity. It was not explained how the importance of matchers is determined.

ASCO [11] combines three string and two structural element matchers. First, the weights for three string matchers are defined and then the aggregation is performed. Then the weights for two structural matchers are defined and the aggregation is performed. At the end the aggregation of previous aggregations is performed. The way of determining the importance of matchers was not explained.

Prior+ [12] is ontology mapping approach that uses parallel matching composition. First, it determines similarity with three basic matchers that calculate element-level and structure-level similarity. These results are aggregated using the Harmony method. The Harmony method automatically determines the importance of every basic matcher based on the results achieved by each matcher. First, the alignments of basic matchers are determined. The harmony measure of every basic matcher is the number of pair of elements that has the highest and the only highest correspondence in its corresponding row and column in the alignment of that matcher.

# 3 Terminology

In this section, definitions of the most important terms related to our research, according to [3], are presented. Matching is the process of finding relationships or correspondences between entities of different ontologies. Alignment ($A$) is a set of correspondences between two ontologies $o$ and $o'$. The alignment is the output of the matching process. Correspondence $\delta\ (e_i,\ e_j')$ is the relation holding between entities of different ontologies. These entities can be as different as classes, relations, properties... Mapping is the oriented, or directed, version of an alignment: it maps the entities of one ontology to at most one entity of another ontology.

## 3.1 Parallel Composition of Basic Matchers

Various architectures for integrating correspondences from diverse basic matchers are used in matching systems. Parallel composition is one of these architectures and is presented in figure 1.

An alignment between two ontologies ($A^n$) is the result of the execution of a basic matcher. Aggregation is the process of merging all the alignments into a final alignment. When all the basic alignments are calculated, the aggregation of these alignments is made and the final common alignment ($A$) is created (figure 2).

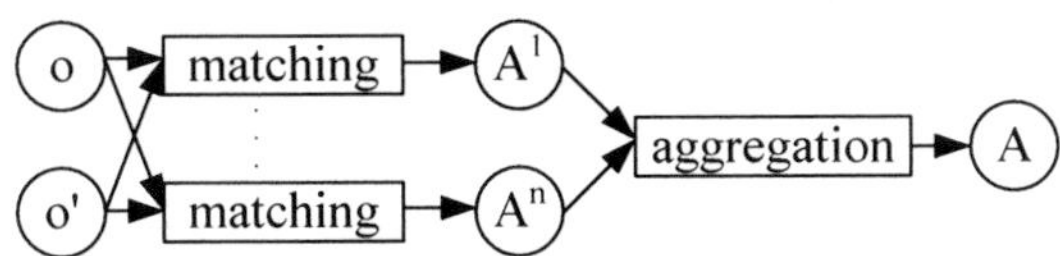


**Fig. 1.** Parallel composition of basic matchers

Each alignment is shown as a correspondence matrix between elements of two ontologies. Correspondence $\delta\ (e_iA^n,\ e_jA^{n'})$ between all individual elements $e_i$ and $e_i'$ of two related ontologies is included in the $A^n$ alignment. Correspondence $\delta\ (e_iA^n, e_jA^{n'})$ is number within 0 and 1 (figure 2). A higher correspondence implicates a higher correlation within two individual elements $e_i$ and $e_j'$.

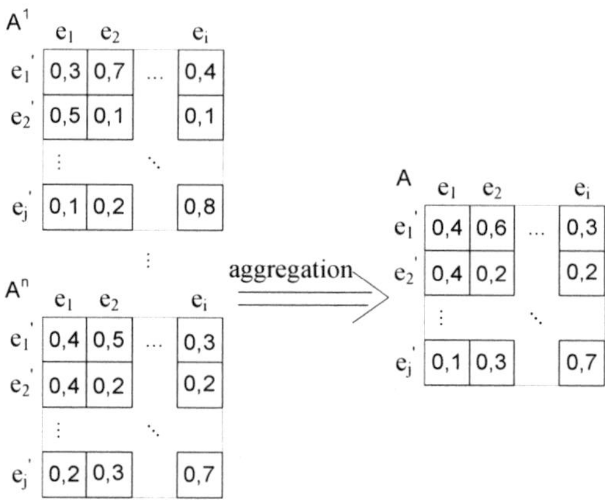


**Fig. 2.** Aggregation of basic alignments

Several methods for aggregation of basic alignments exist [12]:

- Min value – minimal value of all correspondences is taken in the final alignment:

$$\delta\ (e_i^{A}, e_j^{A'}) = \min_n \delta\ (e_i^{An}, e_j^{An'}) \tag{1}$$

- Max value – maximal value of all correspondences is taken in the final alignment:

$$\delta\ (e_i^{A}, e_i^{A'}) = \max_n \delta\ (e_i^{An}, e_j^{An'}) \tag{2}$$

- Average - average of all correspondences is taken in the final alignment:

$$\delta\ (e_i^{A}, e_j^{A'}) = (\textstyle\sum_n \delta\ (e_i^{An}, e_j^{An'}))\ /\ n \tag{3}$$

- Weighted – sum of all weighted correspondences is taken in the final alignment:

$$\delta\ (e_i^{A}, e_i^{A'}) = \textstyle\sum_n w_n * \delta\ (e_i^{An}, e_j^{An'}),\ \sum w_n = 1 \tag{4}$$

- Harmony – sum of all correspondences multiplied by their harmony measure h (weight of every basic matcher) is taken in the final alignment, k is the number of different types of similarity:

$$\delta\ (e_i^{A}, e_i^{A'}) = \textstyle\sum_n h_n * \delta\ (e_i^{An}, e_j^{An'})/k \tag{5}$$

## 4 The Autoweight Method for Automatically Determining Weights of Individual Matchers in Weighted Aggregation

In our approach a part of Harmony method [12] is adapted in process of determining weight of every basic matcher in weighted aggregation. In the Harmony method, the importance of every highest pair of elements in its corresponding row and column is equal. In our Autoweight method, we assume that such a highest correspondence found within several basic alignments has less importance than another highest correspondence that was found within only one basic alignment $A^n$. The assumption rests on the fact that the final mapping only includes those correspondences between the two entities of different ontologies where an entity $e_i$ from ontology $O$ has the greatest correspondence with entity $e_j'$ in ontology $O'$ and vice versa. Therefore, we deal only with the greatest correspondences between two entities in both directions of mapping, as those correspondences are the most relevant. The testing results that evaluate this assumption are provided in Section 5.

In the first step of our Autoweight method, after obtaining the matrices of correspondences $A^n$ (figure 2), a part of the Harmony method is used. The Harmony method determines the number of the pairs of elements $\#s\_max^n$ in every matrix of correspondences $A^n$ that has the highest correspondence in its respective row and column. Figure 3 shows an example of determining $\#s\_max^n$.

For example, the correspondence in the second row and the second column is the only highest correspondence that has the value 1. Hence, one of the $s\_max^n$ in the alignment $A^n$ is found. If any row or column has more than one highest correspondence, the highest correspondence of pair of elements in this row or column does not count in $\#s\_max^n$. It can be seen in the first row and the third column, where the highest correspondence is 0.95, but the first row has two equal values 0.95. Therefore, the found highest correspondence does not count in $\#s\_max^n$. If any row or column has all values 0, then max correspondence in row or column does not exist. The number $\#s\_max^n$ determines the importance of one basic matcher. The higher the number $\#s\_max^n$, the greater the importance of basic matcher.

Our Autoweight method estimates the importance of every highest correspondence based on equal highest correspondence found in other basic alignments. A simple example is shown in figure 4.

The $im_{ij}$ measure represents the importance of the highest correspondence found between elements $e_i$ and $e_j'$. We calculate the importance $im_{ij}$ of every highest correspondence $s\_max\ (e_i, e_j')$ within one basic alignment given the number of equal highest correspondences founded in other basic alignments. If a highest correspondence $s\_max\ (e_i, e_j')$ is found within m basic alignments, then its importance $im_{ij}$ is $1/m$. For example, the highest correspondence $s\_max\ (e_4, e_4')$ is found within two basic alignments, $A^1$ and $A^2$. Therefore, the importance $im_{44}$ of the highest correspondence $s\_max\ (e_4, e_4')$ is $1/2$.

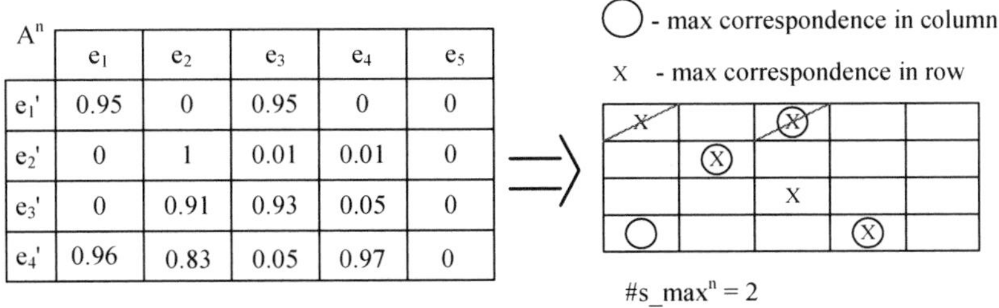


**Fig. 3.** Defining importance of a basic matcher using Harmony method

If a highest correspondence *s_max* $(e_i, e_j')$ is found within all basic alignments, then it should not be included in determining weights of individual matchers, i.e. its importance $im_{ij}$ is *0*. In figure 4, the highest correspondence *s_max* $(e_1, e_1')$ is found within all basic alignments, so the importance $im_{11}$ is discarded. The importance $im_{ij}$ of *s_max* $(e_i, e_j')$ within basic alignment is higher when the number of the same *s_max* $(e_i, e_j')$ found within other basic alignments is smaller. Therefore, only some basic matchers have the ability to find this correspondence. In our example, first basic matcher has the highest common importance $im_1$ (sum of the all $im_{ij}$ in this matcher). When we normalize the importance $im_n$ of every basic matcher, we get the weights that we will use in the weighted aggregation of basic matchers.

The algorithm of Autoweight method consists of the following steps:

1. Find all pairs of elements in every correspondence matrix $A^n$ that has the highest and the only highest correspondence in its respective row and column
2. Calculate the importance $im_{ij}$ of every highest correspondence based on equal highest correspondence found in other basic matchers. If *m* matchers found a specific highest correspondence, the importance of this correspondence is $1/m$
3. Calculate the importance $im_n$ of every basic matcher summing up all the $im_{ij}$ of all found highest correspondences
4. Calculate the weights of every basic matcher for weighted aggregation normalizing the importance $im_n$ of every basic matcher.

## 5 Results of Testing Different Aggregation Methods

To test the alignments considering different aggregation methods we took three pairs of ontologies: networkA.owl and networkB.owl, animalA.owl and animalB.owl and russiaA.rdf and russiaB.rdf. We used the corresponding alignments between these ontologies networkAB, animalAB and russiaAB to compare the results. These

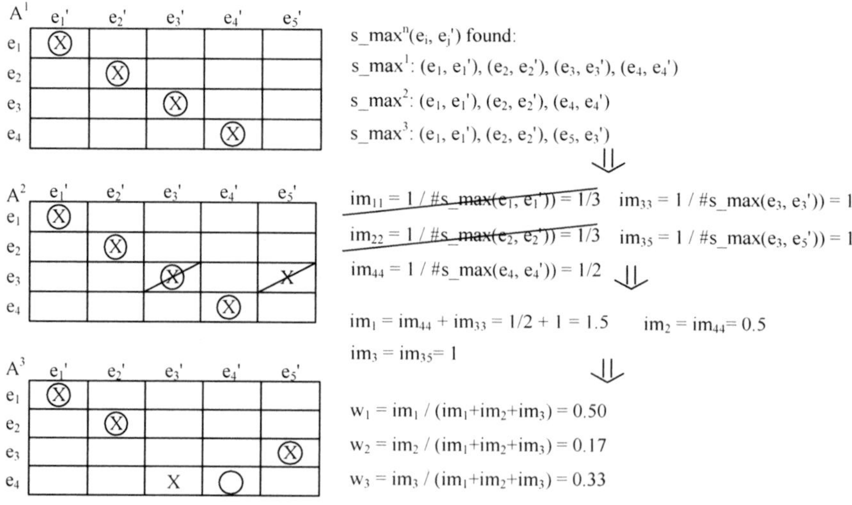


**Fig. 4.** An example of automatically determining weights of individual matchers in weighted aggregation

ontologies and corresponding alignments were used at the I3CON (Information Interpretation and Integration Conference) and can be found in [14]. I3CON was held at the NIST Performance Metrics for Intelligent Systems (PerMIS) Workshop, on August 25, 2004. The goal of I3CON was to provide an open forum and a software framework for a systematic and comprehensive evaluation of ontology/schema interpretation and integration tools.

We used five different string matchers that will be aggregated using various aggregation methods. These string matchers are:

- Prefix – tests whether one string is the prefix of another
- Suffix – tests whether one string is the suffix of another
- Digram – tests how many digrams (substrings of length 2) are the same within two names
- Trigram – tests how many trigrams (substrings of length 3) are the same within two names
- Edit distance – minimal number of operations (insertion, replacement or deletion of character) to be applied to one of the names in order to obtain another one.

Generally, various matchers give different results while matching different ontology pairs. Our automatic method automatically defines dominant matchers for a certain pair of ontologies.

In figure 5 pieces of ontologies NetworkA.owl and NetworkB.owl are displayed. Concerning these ontologies, we assume that the prefix method will be the dominant method as in the NetworkA.owl ontology each name of the child node has the name of the parent node appended at the end, while in the NetworkB.owl this is not the case (e.g. OfficeSoftware is a child of the Software element in networkA.owl, and Office is a child of the Software element in networkB.owl).

We evaluate five different aggregation methods that were discussed before: Min value, Max value, Average, Harmony and finally, the Weighted aggregation that uses our Autoweight method (called AutoWeighted).

Final alignment of the whole matching system should be determined after aggregation of all basic alignments in the aggregated alignment (figure 6). Several procedures of determining final alignment are explained in [5]. First, the direction of selection of candidates for alignment (from one ontology to another, or vice versa) should be decided. If we choose the option Both, all the elements $e_j'$ from ontology $O'$ that are associated with an element $e_i$ from ontology $O$ (and vice versa) will be selected and all this correspondences will be ranked by the value of correspondence.

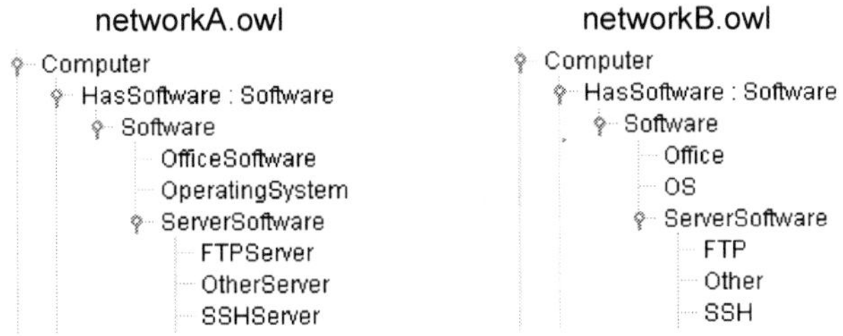


**Fig. 5.** Parts of ontologies networkA.owl and networkB.owl

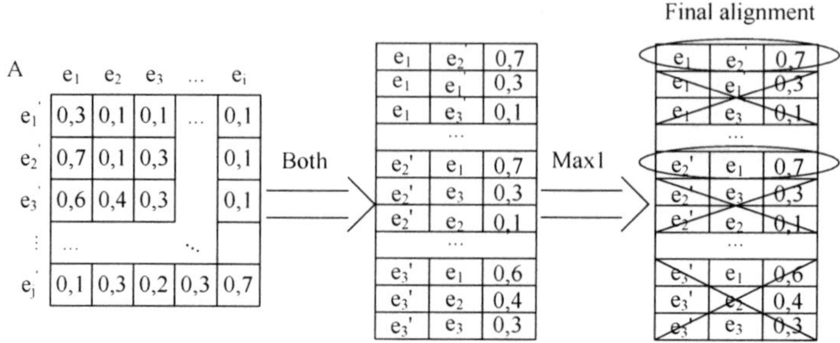


**Fig. 6.** Process of the final alignment

Evaluation measures that we use are:

- Precision - the ratio of correctly found correspondences over the total number of returned correspondences
- Recall - the ratio of correctly found correspondences over the total number of expected correspondences
- F-Measure – harmonic mean of precision and recall, the higher the value $\alpha$, the more importance is given to precision with regard to recall:

$$\text{F-Measure} = (\text{Precision} + \text{Recall})/((1-\alpha)*\text{precision} + \alpha*\text{recall}),\ 0 \leq \alpha \leq 1 \quad (6)$$

In figure 7 the values of the Recall measure are shown.

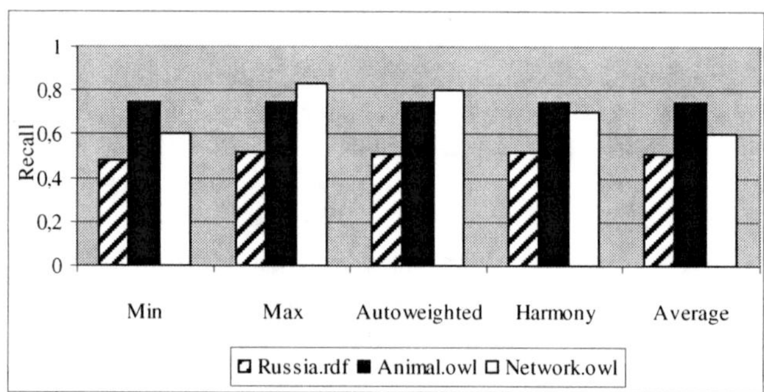


**Fig. 7.** Measure Recall for three pair of ontologies obtained from different aggregation methods

Methods Max value and AutoWeighted outperformed other aggregation methods. Method Max value has high Recall, as this method selects all highest correspondences within all the correspondence matrices for the final mapping. Therefore, the method Max value has more mappings than other aggregation methods. However, in the Max value, a large number of wrong mappings exists, hence it has the lowest Precision.

In figure 8 the values of the Precision measure can be seen.

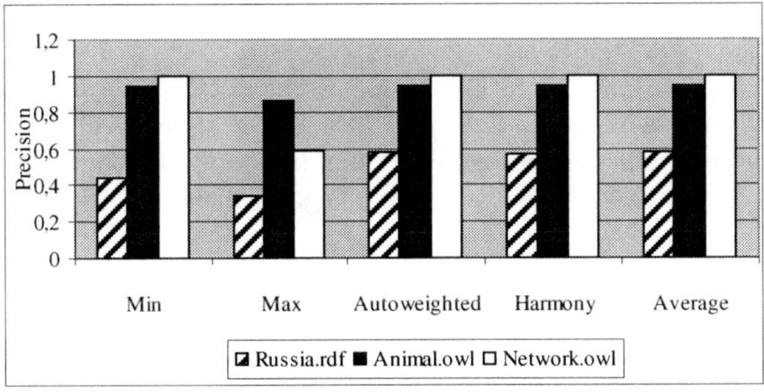


**Fig. 8.** Measure Precision for three pair of ontologies obtained from different aggregation methods

Methods Min value, AutoWeighted, Average and Harmony have similar results. Methods Autoweighted and Average have slightly better results than other methods when mapping the pair of ontology RussiaA.rdf and RussiaB.rdf.

In figure 9 the values of the F-Measure ($\alpha$=0.5) are presented. This measure is a combination of measures Precision and Recall. The F-Measure shows that the Auto-Weighted method outperforms all other aggregation methods.

The Autoweighted method has shown the best results for matching the *network* ontologies. It has the best values for F-Measure. When matching the *animal* ontologies, the Autoweighted method achieves the equal values as the aggregation methods Min value, Harmony, and Average. Of all aggregation methods for matching *russia* ontologies, methods Autoweighted and Average have shown the best results.

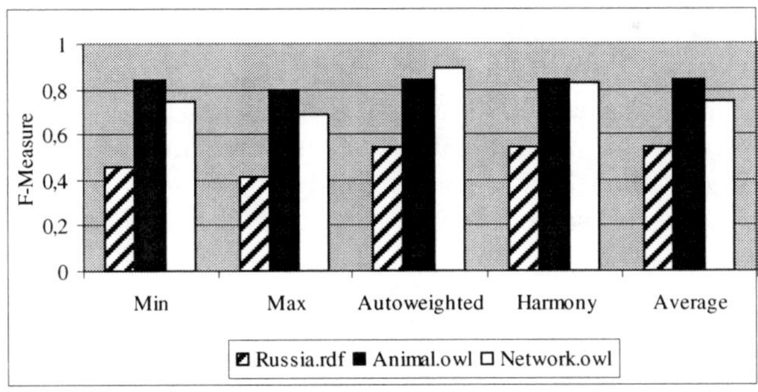


**Fig. 9.** F-Measure for three ontology pairs obtained from different aggregation methods

The F-Measure is the most important evaluation measure, because it combines the values of measures Precision and Recall. Better the Recall value, there is a larger number of correct correspondences. Better the Precision value, there are less false correspondences. Hence, better the F-Measure, which is a harmonic mean of Precision and Recall, there is less additional work for the expert to correct obtained correspondences. The overall results for F-Measure show that our Autoweight method outperforms the other methods.

In the table 1 normalized weights of basic matcher obtained with two aggregation methods, Autoweighted and Harmony, can be seen for the case of matching the *network* ontologies. Our Autoweight method determined the Prefix method as dominant while in the Harmony method the weight of the Prefix method is only slightly greater than other weights. Hence, our previous assumption that the Prefix method is the dominant method of matching *network* ontologies is confirmed.

**Table 1.** Comparison of weights of basic matchers obtained with Autoweighted and Harmony

| | Prefix | Suffix | Digram | Trigram | Edit distance |
|---|---|---|---|---|---|
| Autoweighted | **0,47** | 0,03 | 0,18 | 0,17 | 0,15 |
| Harmony | **0,26** | 0,1 | 0,22 | 0,21 | 0,21 |

# 6 Conclusion

In this paper we proposed a method for automatically specifying a parallel composition of basic matchers in ontology matching process. Aggregation of individual matchers, especially the calculation of the importance of every basic matcher is the main issue in determining a parallel composition of basic matchers. Our method Autoweight automatically determines the weights of basic matchers in a weighted aggregation. To the best of our knowledge, the Autoweight method is the only method that determines the importance of every basic matcher by comparing the alignments of basic matchers.

A software implementation for five aggregation methods has been developed, and these methods were evaluated using five different string element basic matchers. The tests have proven that different weight assignments for every basic matcher lead to better results during the matching process. The weighted aggregation where the weights are determined by using our Autoweight method, showed the best results for aggregation of basic matchers in their parallel composition.

In our future work we will introduce other matchers that deal with other heterogeneity types in the parallel composition. Particular attention will be given to structural matchers that can found correspondences unknown to string matchers.

# References

1. Gruber, T.R.: A translation approach to portable ontology specifications. Knowledge Acquisition 5(2), 199–220 (1993)
2. Antoniou, G., van Harmelen, F.: Web ontology language: Owl. In: A Semantic Web Primer, pp. 110–150. MIT Press, Cambridge (2004)
3. Euzenat, J., Shvaiko, P.: Ontology matching. Springer, Heidelberg (2007)
4. Euzenat, J., Stuckenschmidt, H.: The "family of languages" approach to semantic interoperability. In: Omelayenko, B., Klein, M. (eds.) Knowledge Transformation for the Semantic Web, pp. 49–63. IOS Press, Amsterdam (2003)
5. Do, H., Rahm, E.: COMA – a system for flexible combination of schema matching approaches. In: Proc. 28th International Conference on VLDB, pp. 610–621 (2002)

6. Aumüller, D., Do, H., Masmann, S., Rahm, E.: Schema and ontology matching with COMA++. In: Proc. 24th International Conference on Management of Data (SIGMOD), Software Demonstration, Baltimore (MD US), pp. 906–908 (2005)
7. Ehrig, M., Sure, Y.: Ontology mapping – an integrated approach. In: Bussler, C.J., Davies, J., Fensel, D., Studer, R. (eds.) ESWS 2004. LNCS, vol. 3053, pp. 76–91. Springer, Heidelberg (2004)
8. Jian, N., Hu, W., Cheng, G., Qu, Y.: Falcon-AO: Aligning ontologies with Falcon. In: Proc. K-CAP Workshop on Integrating Ontologies, Banff (CA), pp. 87–93 (2005)
9. Tang, J., Li, J., Liang, B., Huang, X., Li, Y., Wang, K.: Using Bayesian decision for ontology mapping. Journal of Web Semantics 4(1), 243–262 (2006)
10. Castano, S., Ferrara, A., Montanelli, S.: Matching ontologies in open networked systems: Techniques and applications. Journal on Data Semantics, V:25–63 (2006)
11. Bach, T., Dieng-Kuntz, R., Gandon, F.: On ontology matching problems (for building a corporate semantic web in a multi-communities organization). In: Proc. 6th International Conference on Enterprise Information Systems (ICEIS), Porto (PT), pp. 236–243 (2004)
12. Mao, M., Peng, Y., Spring, M.: An adaptive ontology mapping approach with neural network based on constraint satisfaction. Web Semantics, Science, Services and Agents on the World Wide Web 8, 14–25 (2010)
13. Shvaiko, P., Euzenat, J.: Ten challenges for ontology matching. In: Chung, S. (ed.) OTM 2008, Part II. LNCS, vol. 5332, pp. 1164–1182. Springer, Heidelberg (2008)
14. Information Interpretation and Integration Conference (July 22, 2011), http://www.atl.lmco.com/projects/ontology/i3con.html

# Competences and Skills in the Digital Age: Competence Development, Modelling, and Standards for Human Resources Development

Christian M. Stracke

University of Duisburg-Essen, Information Systems for Production and Operations Management,
Institute for Computer Science and Business Information Systems,
45141 Essen, Germany
ISO-Convener ISO/IEC JTC1 SC36/WG5 (www.sc36.org)
Chair CEN TC 353 (www.cen.eu/isss/TC_353)
Christian.Stracke@icb.uni-due.de

**Abstract.** This paper will summarize the potential use cases and impacts of competences and skills in the new area, often called "Digital Age": It will highlight the roles and benefits of standards and metadata for HR development and points out the special support that competence models can provide for the quality development in learning, education, and training. In this regard, the main characteristics of this innovative approach called competence modelling and its relevance in vocational education and training (VET) can only be summarized. A general competence model with a standardized competence structure and taxonomy of levels is introduced and discussed for the general application scenarios and for the specific use cases in vocational education and training. It can be shown that competence modelling and the presented competence model lead to an improvement of the working places, of the organizational and individual development, to an increase of the mobility worldwide as well as to a higher transparency and recognition of competences and skills. Finally leading European initiatives on competence modelling are introduced and the current standardization activities are highlighted.

**Keywords:** Competences, Skills, Knowledge, Human Resources Development, Learning, Education, and Training (LET), Competence Development, Quality Development, Competence Modelling, Competence Model, Vocational Education and Training (VET), WACOM, eCOTOOL, Competence Standards, ISO/IEC JTC1 SC36, ISO/IEC 20006, Generic Reference Model for Competences, CEN TC 353, Metadata for Competence Modelling.

## 1 Introduction

This article deals with the tasks and potentials of the competence development and competence modelling as an innovative and very promising approach and of competence models as appropriate and comprehensive instruments. The key question is: "How to support and improve the quality and outcomes in learning, education, and

E. García-Barriocanal et al. (Eds.): MTSR 2011, CCIS 240, pp. 34–46, 2011.

training?" The answer leads to the innovations and changes which can be realized and provided by learning outcome orientation. Based on the general definition of quality and competence development and of the principles for competence modelling, we will introduce competence models as an appropriate means for quality development. To summarize, this article points out the special support that competence models can provide for the quality development in learning, education, and training and in this regard, we can only highlight the main characteristics of this innovative approach called competence modelling and its special relevance in vocational education, and training (VET).

## 2 Definitions of Competences and Competence Models

In the digital age the importance of competences and competence modelling is increasing dramatically: Not only the European Union by its European Digital Agenda 2020, but also experts and practitioners in human resources (HR) and learning, education, and training (LET) are confirming the importance of skills and competences building [cf. 16]. This is valid for the working places in the globalization times as well as in particular for the vocational education training. The term competence with its general meaning is defined by Stracke [15] as:

> **Competence** is the ability (that cannot be observed directly but only by activities) to adequately and successfully combine and perform necessary activities in any contexts to achieve specific tasks or objectives.

Using this definition the potential (non-observable) competences and (observable) activities performing the competence can be distinguished. That can be expressed by the following simplified representation:

- Competence: = Knowledge + Skills
  (+ individual ability)
- Activities: = Performance of Knowledge + Skills + Competences
  (+ individual ability)

Competences can be built and exist without being demonstrated and performed. It is most important to note that they are non-observable: They are only shown and observable by acting, i.e. by performance and activities. Only activities can be observed and measured.

The Competence Model that will be presented in the following is completely in line and compliant with the unique ISO quality standard for IT-supported LET (ISO/IEC 19796-1) [10] as well as with the international quality management principles of ISO 9001 including the TQM philosophy and the PDCA cycle (Plan, Do, Check, Act): Thus it ensures both, international interoperability as well as flexibility for organizational and individual adaptations (cf. [18]).

For implementation of competence models in human resources and vocational education and training (VET), competence development has to take into account three dimensions as shown in the following figure:

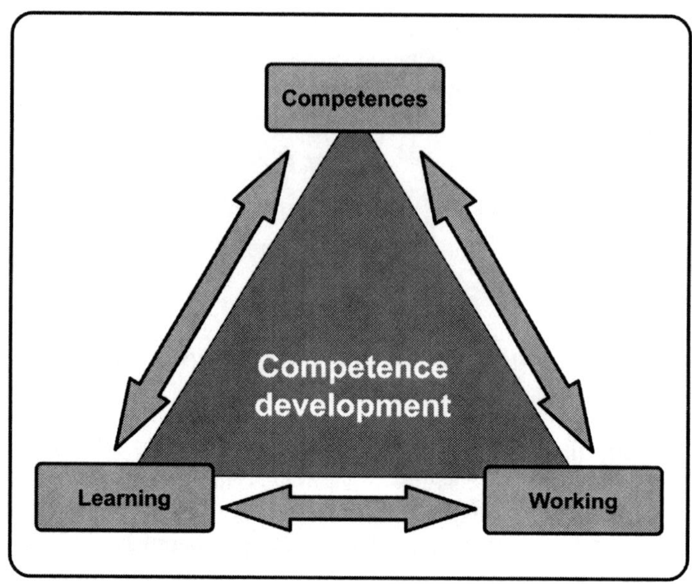


**Fig. 1.** The Triangle of the Competence Development

The following four main target groups using competence models exist in vocational education and training (VET):

1. **Managers:** Managers responsible for hiring new employees or human resource development are interested in enlarging existing and developing needed qualifications. Therefore they are depending on finding out a balance between these two tasks to fix training needs for their employees. Managers have to define requirements for specific working places and job offers to ensure that candidates apply who comply most with the needed requirements.
2. **VET providers:** On the other side there are the vocational education and training providers. They adapt themselves to the needs of enterprises, national institutions and other organisations for a suitable offer.
3. **Learners:** They are the third target group: The same adaptation goal applies for individuals (= the learners) planning their personal development at any age.
4. **Organisations:** Finally there can be organisations like enterprises or public authorities developing their own competence model and competence profiles due to their very specific and extraordinary needs.

A competence model is required for the introduction of competence modelling and harmonization of competence descriptions. The two main components of a competence model are: 1. the competence structure and 2. the competence levels.

With a selection and detailed description of all competences and the definition of the levels it is possible to adapt and implement a Competence Model.

The following table shows the criteria that have to be fulfilled for the standardized description of competences and skills to achieve a consistent and comparable competence structure within Competence Models:

| The Competence Structure | |
|---|---|
| Action Verb | [e.g., "to develop"] |
| Object | [e.g., "project management plan"] |
| Optional elements | [e.g., "for E-Learning projects"] |

**Fig. 2.** The Competence Structure

The competence structure can be used to (1) introduce competence modelling and a competence model for the first time or to (2) integrate it into existing competence models and their pool of competences.

For the levels, the European Commission has issued and supported the European Qualification Framework (EQF): Its eight levels are defined by a set of descriptors indicating the learning outcomes relevant to qualifications at that level in any system. The reduction and adaptation of these eight levels to five levels is proposed for easier application and implementation into practice [15].

In general it is possible to describe all required competences according the following template for competence descriptions:

| Competence "XXX" [e.g., E-Learning PM Planning] | | |
|---|---|---|
| **Structure of the competence "XXX"** | [according the competence structure: e.g., "to develop project management plan for E-Learning projects] | |
| **Definition of the competence "XXX"** | [any written plain text, i.e. free text field] | |
| **Target group of the competence "XXX"** | [e.g., group of employees, single working place] | |
| Knowledge contained in and required for competence "XXX" | | |
| **Name of Knowledge** | **Definition of Knowledge** | **Minimum required level** |
| [e.g., PM Basic Knowledge] | enter your definition here | enter the minimum required level (1 to x) here |
| | enter your definition here | enter the minimum required level (1 to x) here |
| Skills contained in and required for competence "XXX" | | |
| **Name of Skills** | **Definition of Skills** | **Minimum required level** |
| [e.g., Drawing of Plans] | enter your definition here | enter the minimum required level (1 to x) here |
| | enter your definition here | enter the minimum required level (1 to x) here |
| Competences contained in and required for competence "XXX" (optional) | | |
| **Name of competence** | **Definition of competence** | **Minimum required level** |
| [e.g., E-Learning Design] | enter your definition here | enter the minimum required level (1 to x) here |
| | enter your definition here | enter the minimum required level (1 to x) here |

**Fig. 3.** Template for Competence Descriptions

In general it is possible to describe all required competences according this competence structure to set up a "pool of competences": In a specific use case you have only to select these competences that are most important in the case of a job description and that have to be defined.

A competence model includes all competence descriptions of all selected and defined competences for a specific organisation. Thus, the competence model can easily be derived from those existing competence descriptions. In addition a job profile can be developed as the application of the table for competence descriptions for a specific job at a working place within a specific organisation. In this way, job profiles can easily derived from those existing competence descriptions.

## 3 Competence and Quality Development

Competence and quality development are becoming more and more important for success of the organisations and their business: Currently their adaptation and integration into a common approach has been started due to their similar objectives and requirements [cf. 15]. Competence development can benefit from the long-term experiences that have been made in the fields of quality development and that only be summarized here in brief.

Quality development is a crucial task for vocational education and training as well as for human resources, learning, education, and training in general: A long-term debate has focussed the quality development regarding the different quality issues, aspects and approaches (cf. [6]; [7]; [17]; and for an overview [18]). Quality development in its broad sense can be defined as follows (cf. [4]):

> Quality development covers every kind of strategy, analysis, design, realisation, evaluation, and continuous improvement of the quality within given systems.

A long process is needed to establish and integrate quality development throughout a whole organisation. Once started, it has to be a continuous ongoing circle to be successful (cf. [19]; [20]). Quality cannot be described and fixed by a simple definition, because in itself quality is too abstract to have any impact. Thus, quality has to be specified according to the given context and situation considering the perspectives of the involved stakeholders [5]. It is important to identify the relevant aspects and to define the suitable and relevant criteria. To find a consensus amongst the different views and perspectives is necessary to gain a common understanding of quality for the given context and situation due to different and sometimes contradictory needs and definitions of quality by all stakeholders (for detailed explanations on context determinations cf. [5]; [19]; [20]).

Consequently, quality awareness is the basic requirement for the adoption of quality development by all stakeholders from any organisation. On the other side quality awareness will also be raised by the implementation of quality development. For a sustainable integration of quality development within the whole organisation and to ensure the involvement of all stakeholders, it is crucial to build a quality strategy and to integrate the quality objectives into the educational and business processes (cf. [21]). In addition the stakeholders' needs and responsibilities have to be integrated into the overall quality development. The process of the adoption, implementation and adaptation of quality development can roughly be divided into three steps based on three different levels that need to be covered and addressed for a sustainable and long-term quality development (for the three level concept of the introduction of quality development cf. [4]):[1]

- Level of the individual person: to address and convince the employees;
- Level of the organisation: to define and meet the business requirements;
- Integration of quality development involving all stakeholders: to involve all relevant persons including the internal and external suppliers and customers.

The following figure demonstrates the dependence of the three levels of competence development that are building up on the other one:

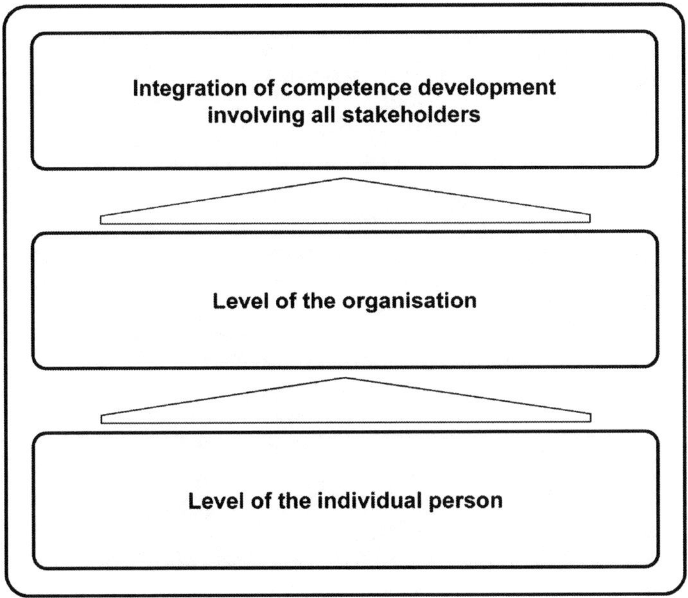


**Fig. 4.** Three levels of Competence Development

## 4 Competence Development and Competence Modelling

The term "Competence development" is used in a broad sense here and covers all processes that are relevant and dealing with the planning, modelling, strengthening, evaluation, and the continuous improvement of the competence of learners and learning organisations. By this definition, the competence development includes as

[1] The presented research findings are partially results of Q.E.D., the flagship initiative for quality in learning, education, and training worldwide, see: <http://www.qed-info.de>.

the general term the competence modelling, the competence building and the competence management.

Competence development is based on the important and influential theory on cognitive development by Jean Piaget. He called his theory "genetic epistemology" to explain the cognitive development of children [22]. The competence development in HR and LET is not yet discussed for a long time and basic theories and approaches are still lacking at the moment. In particular a harmonized and integrated reference model for the adaptation and implementation of quality and competence development is missing that is currently under development [cf. 15]: This article provides general insights and proposals for this ambitious task.

Competence development consists of four processes that are building a continuous improvement cycle following the philosophy of the Total Quality Management plus the analysis and definition of the context conditions and competence strategy [cf. 13]:

0. Competence context and analysis;
1. Competence description;
2. Competence measurement;
3. Competence building;
4. Competence evaluation.

The competence modelling combines the two processes competence description and competence measurement - but not completely: the first definitions of the competence description are related to the general structure and thus they are specified at the beginning and not during the processes of competence modelling.

The relations between the four processes and the context analysis are shown in the following diagram:

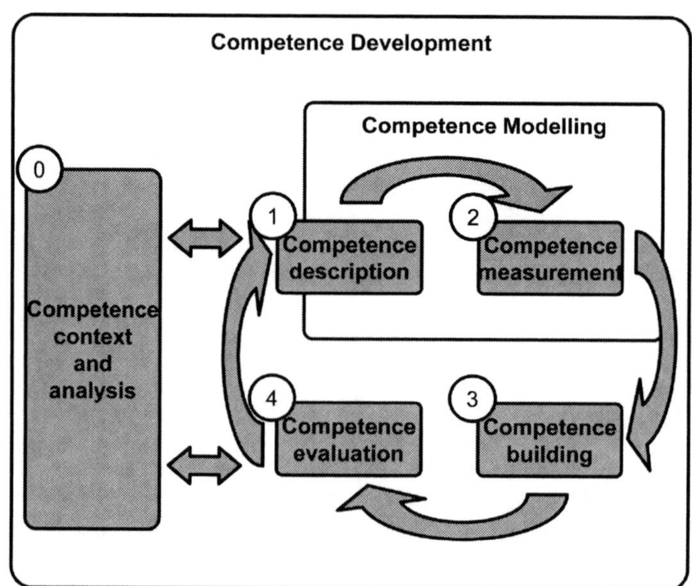


**Fig. 5.** The Phases of Competence Development

In the Phase "Competence Context and Analysis" the general conditions will be identified and a needs analysis with all responsible stakeholders including the decision makers (the top management, the department leaders, etc.) is realized. Thereby the strategic goals and the requirements for the Competence Modelling are investigated and the result is defined and documented in the Competence Strategy.

In the Phase "Competence Description" the (organisation-specific) Competence Model is developed that contains the definitions of the three dimensions of the Reference Framework and the Competence Catalogue next to the Competence Strategy (from the Phase Competence Context and Analysis before). The Competence Catalogue consists of the (organisation-specific) definition of competences and activities that can be developed through top-down processes (e.g. strategy workshops with the management, rating by experts, core competence investigation, prospective orientation, or a combination thereof, etc.) as well as bottom-up processes (e.g. Critical-Incident Technique, rating by experts, structure work-analysis consultation (objective or subjective), employee suggestion scheme or a combination thereof, etc.).

In the Phase "Competence Measurement" the Competence Profile (target and current status) is created. To this end an organizational level (individual, group, or organisation) is chosen and its relevant goals, tasks and situations are determined and described. Thereafter the methods for the observation and measurement of activities are chosen and described. Subsequently, the relevant competences and the activities that constitute them as well as the necessary Competence Levels from the Competence Catalogue are determined for the chosen organizational level. These selections and determinations are documented in the so-called Competence Profile (target status) for the selected organizational level. Afterwards the measurement of the competences (indirectly achieved through the observation and measurement of activities) is carried out, with which their current status are investigated. The analysis of the Competence Measurement is then documented in the so-called Competence Profile (current status) for the chosen organizational level.

In the Phase "Competence Building" the activities for building the competences are created for the chosen organizational level from the basis of the Competence Profile. For this purpose the Competence Development goals are determined first and prioritized on the basis of a target-performance comparison. Finally appropriate activities for Competence Building in the form of opportunities for human resource development and for learning, education, and training are developed and carried out. The desired result is a Competence Change in the chosen organizational level.

In the Phase "Competence Evaluation" the activities for competence measurement and building for the chosen organizational level as well as the Competence Model and Competence Management as a whole are evaluated. The evaluation of activities for Competence Building is based on a second Competence Measurement (indirectly achieved through the observation and measurement of activities). This is particularly aimed at the analysis, assessment and optimization of the opportunities for human resource development and for learning, education, and training for the chosen organizational level. The analysis and evaluation of the competence development and measurements in total, along with the continuous improvement of activities for competence building, particularly serves to create a Competence Balance Sheet on the basis of a target-performance comparison along with the assessment of the development itself. Furthermore, the organisation-wide Competence Management is evaluated on this basis of these results; this serves particularly to analyze, assess and optimize the organisation-wide Competence Model (including the organisation-wide Competence Strategy). The central goal of the Phase Competence Evaluation is therefore the optimization of the entire Competence Development and the Competence Model. (cf. [15])

To summarize, competence models are required and used for describing and measuring competences: Thus, competence models are the core instruments for competence modelling and its implementation. In the following, two leading European projects on competence modelling with different focus and sector approaches will be introduced in brief.

## 5 WACOM - The European Initiative for Water Competences

WACOM is the European research project for harmonizing water competences throughout whole Europe and for integrating competence modelling into the European water sector coordinated by the University of Duisburg-Essen, Germany. The project WACOM (WAter COmpetences Model Transfer) transfers the European Qualification Framework (EQF) and the German Reference Model for the Competence Modelling PAS 1093 into the water sector and its Vocational Education and Training (VET) throughout whole Europe. That enables the identification of the VET needs by employees and learner, of required competences and qualifications at specific working places as well as of the improvement of transparency and comparability of VET opportunities and products.

The development of the WACOM instrument is supported by the transfer of the competence and qualification model into the water sector and by the adaptation for the selected topic management of sewage treatment plants.

WACOM has an impact on vocational training in the whole water management and in additional sectors.

The main WACOM products and impacts are:

- **Competence model for the water sector:** Development of WACOM (the WAter COmpetences Model) as the establishment of a competence model for the water sector based on EQR and PAS 1093: The aim is the foundation of the vocational training on the principle of competence modelling based on the identification of the specific demands and needs of the water management and existing practice concerning competence models in the water sector.
- **WACOM instrument:** Optimization and adaptation of WACOM (the WAter COmpetences Model) to the vocational education and training systems and cultures in Europe. Finally the WACOM instrument is used for the description of the competences and qualifications and for the improvement of their transparency and comparability.

The WACOM Competence Model (WCM) describes the core competences for the employees working in the water sector and can be applied to the Wastewater Treatment Plants (WWTP) and transferred to other branches. In this way WACOM leads to an increase of the mobility throughout Europe as well as to a higher transparency and recognition of qualifications and competences. The competence model is composed of water competences which were developed out of personal interviews with water experts and from an analysing process of the water sector reviewed and refined by the outcome of the WACOM national workshops and the WACOM Online Survey. A detailed description of all collected competences allows a standardised usage of the WACOM competence model.

The Water Competences are the complete list of core competences for the water sector directly related to the requirements and needs from the water sector, the working places and job profiles. The WACOM Competence Model (WCM) includes templates for the application and adaptation by the European water sector.[2]

## 6 eCOTOOL - The European Initiative for Competence Modelling

eCOTOOL is the European research project for harmonizing competence descriptions throughout whole Europe and for integrating competence modelling into European policies, namely the Europass Certificate Supplement (CS) coordinated by the University of Duisburg-Essen, Germany. The two main objectives of eCOTOOL as stated in the work plan are:

1. improving the creation, exchange and maintenance of vocational education and training (VET) certificates and their accessibility and transparency, and
2. increasing the European mobility and transparency of VET systems.

These objectives will be achieved through the creation of a competence model and structure for European policies and in particular for Europass CS. The main goal is the introduction of a pan-European standardized structure to describe competences, skills and knowledge for the harmonization of Europass with other European instruments such as EQF and ECVET and with e-competences. The development of compatible instruments and tools which supports the creation, maintenance and exchange of competence descriptions within VET certificates will ensure its realization. The eCOTOOL project concentrates on the Europass Certificate Supplements (CS) which is the one of the five Europass instruments providing documents and guidelines for the description of qualifications throughout Europe. The Europass CS is a framework for VET providers to describe required qualifications of vocational occupations. The facilitated creation of certificate supplements helps to formulate the requirements for a specific job as well as to discover easily which qualifications job-seekers need for specific employment opportunities. The project restructures the content field three of the Europass CS where the required qualifications for each job have to be listed. In this way, eCOTOOL contributes to the European Lifelong Learning Programme (LLP) by enhancing significantly the transparency throughout the labour market. In current Europass certificate supplements only a couple of phrases describing the required qualifications are listed in the relevant field three. Within eCOTOOL the consortium partners shared their expertise concerning competence modelling, the Europass Certificate Supplement and

[2] For more information see the WACOM website: `http://www.wacom-project.eu`.

the VET-sector and created a structure for describing qualifications consisting of competences, skills and knowledge. This structure was created out of comparing several existing European policies like the EQF, the Key Competences, ECVET and is based on the German PAS 1093 for Competence Modelling [13].[3]

## 7 Vision

Finally we would like to broaden the view on competence and quality development in LET and on competence standardization in the future. What are the main activities today for future development of competence standards?

ISO/IEC JTC1 SC36 is the unique official formal standardization body for IT-supported LET at the international level[4]. The scope of SC36 is defined as: "Standardization in the field of information technologies for learning, education, and training to support individuals, groups, or organisations, and to enable interoperability and reusability of resources and tools" [9].

The first substantial standard that was developed, approved and published by SC36 in 2005 is the quality standard **RFDQ** (Reference Framework for the Description of Quality Approaches), ISO/IEC 19796-1 ("Information Technology — Learning, Education, and Training — Quality Management, Assurance and Metrics — Part 1: General Approach") [10]. It is providing a generic Reference Process Model and the first quality standard of the multi-part ISO/IEC 19796 series. The quality standard has been implemented worldwide[5] and adopted as European Norm EN ISO/IEC 19796-1 by the European standardization Committee CEN TC 353[6]. Currently its adaptation and integration into a common approach combining quality and competence development has been started [cf. 15].

In 2008, SC36 has started its initiative for the development of a **Conceptual Framework for Standards** that will be accompanied by several Technical Frameworks for different use cases and target groups. The presented Generic Reference Model for Standards in the field of IT-supported LET (ITLET) is a helpful and supporting contribution on the long way towards such a generic ITLET framework.

In 2009, SC36 has started another standardization initiative for the development of a **Generic Framework for Competence Standards** acknowledging the increasing

---

3 For more information see the eCOTOOL website: `http://www.ecompetence.eu`

4 The abbreviation stands for: "International Organization for Standardization (ISO)/ International Electrotechnical Commission (IEC) Joint Technical Committee 1 (JTC1) - Information Technology - Subcommittee 36 (SC36) - Information Technology for Learning, Education, and Training (ITLET)". Members of SC36 are National Bodies (NB), i. e. national delegations of appointed experts, and Liaisons Organizations (LO) without voting rights (cf. `http://www.iso.org/jtc1/sc36` and `http://www.sc36.org`).

5 For a detailed description of the quality standard RFDQ and of the Adaptation Model IDEA for the Introduction of Quality Development cf. [11].

6 The abbreviation stands for: "European Committee for Standardization (CEN) Technical Committee (TC) 353: Information and Communication Technologies for learning education and training". Members of CEN TC 353 are National Bodies (NB), i. e. national delegations of appointed experts, and Liaisons Organizations (LO) without voting rights (cf. `http://www.cen.eu/isss/TC_353`).

importance of competence modelling and competence models. It will be developed in three parts by starting with the first part on the general framework: The presented first General Reference Model for Competence Standards is a helpful and supporting contribution on the long way towards such a generic competence framework [cf. 14]. Finally the multi-part ISO/IEC 20006 series on competency will provide a competency model as a General Framework based on Asian, American and European specifications [cf. 14].

Competence and quality development are crucial and indispensable for the long-term success of learning opportunities and in particular of vocational education and training: To reach an economical benefit through competence modelling and building, standards are offering a sustainable support. Their adaptation, implementation and integration can be regarded as one of the main tasks for the future.

## References

1. Luhmann, N.: The Society of the Society [= Die Gesellschaft der Gesellschaft]. Suhrkamp Frankfurt/ Main (1998)
2. Maturana, H.R., Varela, F.J.: The Tree of Knowledge. In: The Biological Roots of Understanding, Shambhala, Boston (1992)
3. Luhmann, N.: Social Systems. Stanford University Press, Stanford (1995)
4. Stracke, C.M.: Interoperability and Quality Development in e-Learning. Overview and Reference Model for e-Learning Standards. In: Proceedings of the Asia-Europe e-Learning Colloquy, e-ASEM, Seoul (2006), http://www.qed-info.de/downloads
5. Donabedian, A.: The Definition of Quality and Approaches to Its Assessment = Explorations in Quality Assessment and Monitoring, vol. 1. Health Administration Press, Ann Arbor (1980)
6. Deming, W.E.: Quality, productivity and competitive position. MIT, Cambridge (1982)
7. Juran, J. (ed.): Quality Control Handbook. McGraw-Hill, New York (1951)
8. Stracke, C.M.: Quality Development and Standards in e-Learning: Benefits and Guidelines for Implementations. In: Proceedings of the ASEM Lifelong Learning Conference: e-Learning and Workplace Learning. ASEM, Bangkok (2009), http://www.qed-info.de/downloads
9. SC36: SC36 Terms of Reference (2002), http://jtc1sc36.org, http://www.iso.org/jtc1/sc36 (retrieved October 28, 2006)
10. ISO/IEC 19796-1:2005: Information Technology - Learning, Education, and Training — Quality Management, Assurance and Metrics — Part 1: General Approach. International Organization for Standardization (ISO), Geneva (2005)
11. Stracke, C.M.: Quality and Standards in Learning, Education, and Training: The Adaptation Model IDEA for the Introduction of Quality Development. In: Proceedings of the International Conference on the Past and Future of e-Learning Standards, Toyo, Tokyo (2010), http://www.qed-info.de/downloads
12. ISO/IEC 19788-1:2010: Information Technology - Learning, Education, and Training — Metadata for Learning Resources — Part 1: Framework. International Organization for Standardization (ISO), Geneva (2010)
13. PAS 1093: Human Resource Development with special consideration of Learning, Education and Training – Competence Modelling in Human Resource Development. DIN, Berlin (2009), http://www.qed-info.de/downloads

14. ISO/IEC 20006-1 (WD): Information Technology for Learning, Education and Training — Information Model for Competency — Part 1: Competency General Framework and Information Model (2010)
15. Stracke, C.M.: Competence Modelling for Innovations and Quality Development in E-Learning: Towards learning outcome orientation by competence models. In: Proceedings of World Conference on Educational Multimedia, Hypermedia and Telecommunication 2011 [= EdMedia 2011], AACE, Chesapeake, VA (in print, 2011)
16. European Commission: European Digital Agenda 2020. EC, Brussels (2010), http://eurlex.europa.eu/LexUriServ/LexUriServ.do?uri=CELEX:52010DC0245:EN:NOT
17. Juran, J.M.: Juran on quality by design. In: The New Steps for Planning Quality into Goods and Services. Free Press, New York (1992)
18. Stracke, C.M.: Process-oriented Quality Management. In: Pawlowski, J., Ehlers, U. (eds.) European Handbook on Quality and Standardisation in E-Learning, pp. 79–96. Springer, Berlin (2006)
19. Crosby, P.B.: Quality is Free. In: The Art of Making Quality Certain. McGraw-Hill, New York (1980)
20. Deming, W.E.: Out of the Crisis. MIT, Cambridge (1986)
21. Stracke, C.M.: Quality development and standards in learning, education, and training: adaptation model and guidelines for implementations. In: Информатизация образования и науки = Информике (Informika), Moscow (Russian Federation), vol. 7(3), pp. 136–146 (2010) ISSN 2073-7572, http://www.qed-info.de/downloads
22. Piaget, J.: The origin of intelligence in the child. Routledge, London (1953)

# OSDBQ: Ontology Supported RDBMS Querying

Cihan Aksoy[1], Erdem Alparslan[1], Selçuk Bozdağ[2], and İhsan Çulhacı[3]

[1] The Scientific and Technological Research Council of Turkey, Gebze/Kocaeli, Turkey
[2] Komtaş Information Management Inc., Ostim/Ankara, Turkey
[3] Turkish Court of Accounts, Söğütözü/Ankara, Turkey
{caksoy,ealparslan}@uekae.tubitak.gov.tr

**Abstract.** Data handling and retrieving has an essential importance where the data size is larger than a certain amount. Storing transactional data in relational form and querying with Structured Query Language (SQL) is very preferable because of its tabular structure. Data may be also stored in an ontological form if it includes numerous semantic relations. This method is more suitable in order to infer information from relations. When transactional data contain many semantic relations inside as in our problem, it is not easy to decide the method of storing and querying. We introduce a new querying mechanism "Ontology Supported RDBMS Querying (OSDBQ)" covering positive sides of both two storing and querying methods. This paper describes OSDBQ and presents a comparison between three querying methods. Results show that most of the time OSDBQ returns fairly better results than the others.

**Keywords:** Ontology, relational database management systems, inference, transactional data querying, semantic matching.

## 1 Introduction

The semantic web introduces "web of data" that enables machines to interpret the meaning of information on the web [1]. This approach was required since one cannot achieve corresponding information belong to the searching criteria from the web, because the web turned into information dump in the last decade. This technology aims at semantically tagging the resources in order to cope with dirty information. One of the most important parts of semantic web is "ontology layer". Ontology is hierarchy of concepts representing the meaning of an information field [2]. It is possible to obtain ontological data from these concepts by instantiating them. Therefore, data includes same relationships between each other like the concepts'. This allows us to infer hidden or deep information that is achieved by using some relationships. On the other hand, data size is increasing between two and three times since the data includes relationships. There are several ontology representations such as OWL [3], RDF [4], N-Triple [5] etc. A widely known query language for RDF is called as SPARQL [8].

Commonly used relational tables are far away from the meaning notion; they only keep the data in a tabular form. Data retrieved from these tables by querying with SQL. Since they only focus on data, their size is smaller than an ontological data store that has same amount of data. That's why querying a relational table returns faster results.

E. García-Barriocanal et al. (Eds.): MTSR 2011, CCIS 240, pp. 47–55, 2011.

Relational tables are designed for tabular data, ontologies well behaved for hierarchical data where semantic relationships exist. "Transactional data" that consist of sales, deliveries, invoices, claims and other monetary and non-monetary interactions, may have many relationships inside. They are usually kept in tabular form. Modeling transactional data is a challenging point when the data size is larger than a certain amount. We should consider the querying performance, data size, extracting hidden or deep information. In this paper, we propose a new querying method that satisfies the lack of relational querying method by benefitting from the powerful side of ontology. Hence, our approach suggests keeping "transactional data" in tabular form and querying with SQL. But while querying, it infers supplementary information from ontologies and adds obtained information in parameterized SQL queries.

The rest of the paper is organized as follows. In Section 2, we give details about the used data models and querying methods. Also we introduce a new querying method and explain its working principle. In Section 3, we apply mentioned methods to the sample data and we show the results. Finally, in section 4, we discuss the limitations of our approach and explain what can be done to solve them as future works.

## 2 Applied Methods

### 2.1 Querying Ontological Data via SPARQL

As a first data model, we made BSBM Tools [10] –a data generator tool provided by Berlin SPARQL Benchmark (BSBM) [11] –generate N-Triple data format to have ontological data. Ontological data includes not only the concepts that belong to a certain domain but also the instances of these concepts. These instances correspond to the tuples of a tabular data model. Moreover, the semantic relations between concepts and between instances are found in this kind of data. That's why ontological data covers larger size on disk compared with relational data.

In order to have a querying interface, we put the generated data in TDB Store [6] by using "tdbloader" component of Jena API [7]. It is observed that the data size on disk is a bit decreasing due to the efficient storing mechanism of TDB Store.

After have stored the data, we sent queries in SPARQL query language from a programmatic interface thanks to Jena API. Also we used JENA's reasoner because SPARQL queries are required to make inferences. That is very suitable for ontological data since they hold relationships inside. Indexing and caching capabilities of TDB Store provides shorter query time after each time for same query as seen in results.

### 2.2 Querying Relational Data via SQL

In order to prepare a relational database as an environment of second method, we again used BSBM Tools and we made it generate an SQL dump in which all necessary commands to form the database tables are found. We executed these commands in MySQL database; we extracted and loaded this data in PostgreSQL.

As we represent the characteristics of the dataset in 3.2, certain points are identical in all types of generated data for the same scale factor. For example, relationships inside the data, amounts of instances, amounts of concepts etc. are same in all types of generated data whereas the value of data may change. After we loaded the data in database, we noticed that there are many relation tables, so that the data represent same relationships which are found in ontological data.

Instead of use the relationships of tables directly; we applied Das et al.'s "Supporting Ontology-based Semantic Matching in RDBMS" approach [9] by taking these relationships into separate tables that represent the ontology. Therefore, as shown in Figure 1, the database can be viewed as it is formed from two main table group; one of them holds the ontologies while the other holds the data.

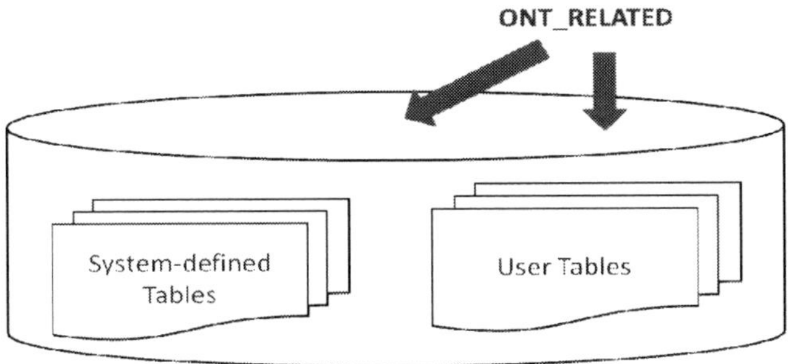


**Fig. 1.** Das et al.'s architecture

In this approach, as mentioned before, there are two groups of tables. In the first group, called system defined tables, the ontological structure and semantic relations are stored. We can infer semantic relations between individuals from system defined data tables. The second group, called user tables, stores the bulk individuals or tuples in tabular form. After inferring semantic rules from system defined tables, prepared rich SQL statement can be executed on user tables. In this way, querying semantic related data by using RDBMS can be achieved without using an inference engine.

After generated the database, we executed the same queries from the same interface except the query language. In this method, we transformed SPARQL queries into SQL queries. Moreover, we inferred relationships from ontology related tables by using stored procedures as explained in Das et al's approach. For example, the operator "ONT_RELATED" returns all parents attached hierarchically to a certain concept or instance. The goal of this method is to represent the ontology in a relational database to be able to benefit from the performance of this kind of database.

### 2.3 Ontology Supported Relational Data Querying

Thirdly, we propose to keep the semantic relations between individuals separate from the tabular transaction data. As we mentioned before, the first approach infers and queries both semantic relations and transactional data from RDF store. On the other hand, the second approach queries both semantic relations and transactional bulk data from RDBMS data store. In this approach we propose to infer semantic relations from RDF store and then querying the transactional bulk data from conventional RDBMS. In other words, we realize the second approach by replacing system-defined tables with RDF store.

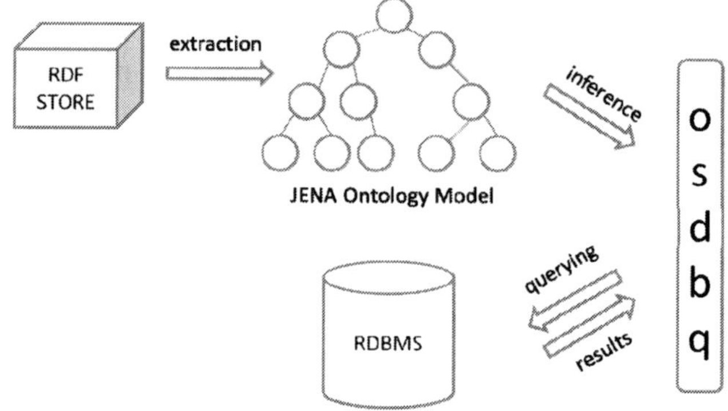


**Fig. 2.** OSDBQ architecture

To realize this approach the same RDF store have been used that we have generated in the first method. Different from first method, this time RDF store is used for querying the domain ontology and inferring semantic relations, not for querying the transactional bulk data. The related ontology file obtained from RDF store is loaded into the memory in order to avoid time losses of file opening while inferring. By using JENA API's reasoner, necessary semantic relations are inferred and then reflected to the second part of this approach. The second part, which queries bulk transactional data, takes the inference results previously presented by JENA as parameters and sends the parameterized SQL query to the RDBMS data store. Therefore a transactional data can be queried in the ontological form and with the RDBMS performance, as seen in Figure 2.

Figure 3 depicts a basic flow of our proposed **O**ntology **S**upported Relational **D**ata**b**ase **Q**uerying Architecture. Complex user query, which may require both semantic inferring and transactional data querying, is given to the system by the user interaction. If the complex query needs inference then the system loads the required ontology RDF files into the memory and realize the semantic inferences by using ontology objects. These inferences basically prepare rich SQL parameters for transactional data querying. In other words, some of the parameters which are used in "WHERE" clauses of transactional SQL queries are prepared by inference engine loaded into the memory. Therefore the system is able to send the SQL queries on the transactional data by using inferred parameters obtained from the inference engine running on memory.

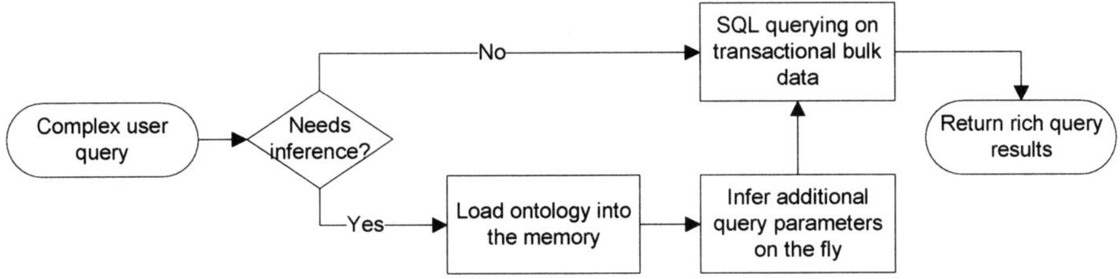


**Fig. 3.** Flowchart of OSDBQ architecture

# 3 Application to Sample Data

It is very important to properly decide for organizations how the data will be stored and retrieved in case of the huge data sizes. Size of the data is not only the considered point, but also the characteristic of the data is taken into account while giving a decision; data may be designed hierarchical, relational, etc. To choose the right architecture, organizations are in need of comparison and benchmarking studies. In this section, firstly we give some information about the environment and utilities that we have used during the tests, secondly we show the dataset on which we have applied proposed methods, thirdly we explain the queries, and finally we represent the comparison results of three proposed methods.

## 3.1 Experimental Setup

We realized the experiments on a HP Workstation (processor: 2 x Intel Pentium 3 Xeon, 2833 MHz; memory: 8GB DDR2 667; hard disk: 320GB 7200Rpm SATA2) with Ubuntu 10.04 LTS 64-bit operating system. Also following utilities were used:

- Jena TDB Version 0.8.9 as RDF storage and query component
- PostgreSQL Version 1.12.2 as database

To measure the performance of these methods, we prepared two different size of dataset for each method so that the results are rendered more consistent. One of these dataset includes 10000 products and related tables whereas the other one has 100000 products. For the same goal, we also repeated the execution of each query 4 times to avoid certain effects which can slightly change the real result, such as caching mechanisms in data stores and in databases. All methods were executed on the same machine in order to avoid network latency. Obtained results are recorded in millisecond.

## 3.2 Dataset

Berlin Sparql Benchmark's dataset [11] are used because it is relevant with transactional data. It is built around an e-commerce use case, where a set of products is offered by different vendors and different consumers have posted reviews about products.

It is possible to generate an arbitrary amount of data where number of product is scale factor. The data generation is deterministic to be able to create different representation of the same dataset.

The dataset is composed of instances of these classes: Product, ProductType, ProductFeature, Producer, Vendor, Offer, Review, Reviewer and ReviewingSite.

All products have between 3-5 textual properties. Each property consists of 5-15 words that randomly selected from a dictionary. Also products contain between 3-5 numeric properties whose values range between 1 and 2000. All products have a product type from the type hierarchy. The depth and width of the product type hierarchy depends on number of products. This hierarchy is set to the dataset even if the data store or database doesn't support RDFS inference. All products have a variable number of product features regarding to its position on the product hierarchy.

All products are offered by vendors. Offers contain the price and the number of days for the delivery, also they are proposed for a certain date interval.

Reviews are published by reviewers. Reviewers have a name, a mailbox checksum, and a country that shows where they live in. Reviews have a title and a review text that consist of between 50-300 words. Also they have four random ratings. Table 1 shows the number of instances of each class in BSBM dataset depending on our choice of product number. We shortly called as small dataset which is generated with 10000 products, and big dataset which is generated with 100000 products. Since ontology hold data as triples, we gave the number of triples below for each amount of data.

**Table 1.** Number of instances in BSBM datasets for different scales

| *Data Type* | ***Small dataset*** | ***Big dataset*** |
|---|---|---|
| Number of Product Feature | 10519 | 47884 |
| Number of Product Type | 329 | 2011 |
| Number of Producer | 206 | 1988 |
| Number of Vendor | 105 | 995 |
| Number of Offer | 200000 | 2000000 |
| Number of Review | 100000 | 1000000 |
| Total Number of Instances | 311159 | 3052878 |
| Total Number of Triples | 3565060 | 35270984 |

### 3.3 Queries

We used 4 queries which include sufficient depths in order to compare the methods. We give attention of depth since we look for an ideal data storing and querying mechanism for "transactional data". As mentioned before, dataset is built around an e-commerce use-case, and queries correspond with the search and navigation pattern of a consumer looking for a product.

In the first query, the costumer searches for products that have a specific type and features. Secondly, the consumer asks for products belong to certain types having several features but not having a specific other feature. Thirdly, the consumer looks for products belong to certain types matching either one set of features or another set. Inference is needed for these three queries since each product type hierarchically belongs to another product type as described in dataset section, thus products that don't match directly with a given type may be returned as result. In the last query, a vendor wants to find out which product categories get the most attention by people from a certain country. A similar inference as the others; product category that attracts most attention of people will increase the popularity of the parental product category.

### 3.4 Results and Interpretation

After have prepared the environment, we started to execute the queries. Firstly we sent the queries on small dataset that consists of 10000 products.

**Table 2.** Results of methods for 10000 products (ms)

| | 1st method | 2nd method | 3rd method | |
|---|---|---|---|---|
| **1st test** | 3009 | 1169 | 1881 | **1st query** |
| **2nd test** | 269 | 35 | 12 | |
| **3rd test** | 179 | 34 | 13 | |
| **4th test** | 117 | 31 | 12 | |
| **1st test** | 1075 | 772 | 1675 | **2nd query** |
| **2nd test** | 146 | 58 | 14 | |
| **3rd test** | 147 | 50 | 13 | |
| **4th test** | 72 | 35 | 12 | |
| **1st test** | 1637 | 104 | 496 | **3rd query** |
| **2nd test** | 150 | 46 | 64 | |
| **3rd test** | 105 | 48 | 62 | |
| **4th test** | 115 | 48 | 61 | |
| **1st test** | 2304 | 3153 | 1912 | **4th query** |
| **2nd test** | 960 | 28 | 28 | |
| **3rd test** | 549 | 30 | 18 | |
| **4th test** | 515 | 29 | 17 | |

In Table 2, each column represents one of data storing and querying method mentioned in the previous section. At the rows, experiments of queries are shown. Each query was performed four times. We took better results for each time we execute the same query. It is observed that results are sufficiently consistent after the first test.

**Table 3.** Results of methods for 100000 products (ms)

| | 1st method | 2nd method | 3rd method | |
|---|---|---|---|---|
| **1st test** | 77034 | 12728 | 10154 | **1st query** |
| **2nd test** | 591 | 286 | 298 | |
| **3rd test** | 484 | 285 | 285 | |
| **4th test** | 468 | 279 | 286 | |
| **1st test** | 514 | 4856 | 12631 | **2nd query** |
| **2nd test** | 448 | 568 | 211 | |
| **3rd test** | 441 | 565 | 212 | |
| **4th test** | 443 | 559 | 211 | |
| **1st test** | 1401 | 1434 | 3233 | **3rd query** |
| **2nd test** | 1367 | 912 | 756 | |
| **3rd test** | 1342 | 912 | 754 | |
| **4th test** | 1323 | 905 | 749 | |

We can easily deduce that the 1st method where ontologies queried gives worse results compared to the others. However, the 3rd method where ontologies were partially used with tabular data, returns better results than 2nd method where only tabular data were queried. This means that supporting relational databases with ontologies and querying them by using semantic reasoners may increase the performance.

In order to be certain from consistency of the results of our small dataset, we performed the same process with 10 times bigger data. In Table 3, where columns and rows represent same points as Table 2, results proved that 3rd method returns usually the best results among three methods. On the other hand, although the 3rd method gives the best result, it may not be always applicable as seen in our experiment. Since the suitable ontology for executing the 4th query was larger than the size of the memory, we couldn't realize last query. This shows the scalability problem of the 3rd method.

## 4 Conclusion

In this paper, a new data storing and querying mechanism, "Ontology Supported RDBMS Querying (OSDBQ)" is introduced by comparing with the most known two data querying mechanisms. SQL is ideal for querying tabular data. SPARQL is preferred for querying and inferring ontological data where relations exist. OSDBQ approach is well behaved for querying tabular data where relations exist like ontological data. Its performance not only relies on the positive sides of the others, but also the necessary inferences are realized with ontologies that are held on memory. However, our approach may be restricted from where it bases on. Large ontologies may not be fit into the memory.

Since the size of the data is increased because of all relationships are included in ontological data, results of $1^{st}$ method are always worse than the others. Most of the time $3^{rd}$ method returned better results than the $2^{nd}$ method. So we can easily say that it should be used as long as the ontologies are fit into the memory.

This paper aims to develop a new querying method handling semantically related transactional dataset. The new OSDBQ method may be applied on huge datasets and large ontologies behind these dataset. Our future work will be coping with handling large ontologies on memory. We will try to predict the necessary parts of ontologies and partially bring them on memory. A huge transactional and semantically relational audit dataset of Turkish Court of Accounts will be adapted to the OSDBQ framework for analytical purposes. Therefore, we will try to overcome the size problem of large ontologies. Ontology merging may be another challenge for improving the performance by facilitating the foresight mentioned above.

## References

1. Berners-Lee, T., Hendler, J., Lassila, O.: The Semantic Web. Scientific American Magazine (2001)
2. Guarino, N.: Formal Ontology and Information Systems. In: 1st International Conference on Formal Ontology in Information Systems (FOIS), Torino, pp. 2–5 (1998)
3. McGuinness, D.L., Harmelen, F.: Owl Web Ontology Language, `http://www.w3.org/TR/owl-features/`
4. RDF, `http://www.w3.org/RDF/`
5. N-Triples, `http://www.w3.org/2001/sw/RDFCore/ntriples/`
6. TDB, `http://openjena.org/wiki/TDB`
7. Jena – A Semantic Web Framework for Java, `http://jena.sourceforge.net/`
8. SPARQL Query Language for RDF, `http://www.w3.org/TR/rdf-sparql-query/`
9. Das, S., Chong, E., Eadon, G., Srinivasan, J.: Supporting Ontology-based Semantic Matching in RDBMS. In: VLDB, Toronto, pp. 1054–1065 (2004)
10. Schultz, A.: BSBM Tools, `http://sourceforge.net/projects/bsbmtools/`
11. Bizer, C., Schultz, A.: The Berlin Sparql Benchmark. Int. J. Semantic Web. Inf. Syst., 1–24 (2009)

# Metadata for Web Ontologies and Rules: Current Practices and Perspectives

Carlos Tejo-Alonso, Diego Berrueta, Luis Polo, and Sergio Fernández

Fundación CTIC
Gijón, Asturias, Spain
{carlos.tejo,diego.berrueta,luis.polo,sergio.fernandez}@fundacionctic.org
http://www.fundacionctic.org

**Abstract.** The Semantic Web contains a number of different knowledge artifacts, including OWL ontologies, RIF rule sets and RDF datasets. Effective exchange and management of these artifacts demand the use of metadata and prompt availability of accurate reference documentation. In this paper, we analyze the current practices in metadata usage for OWL ontologies, and we propose a vocabulary for annotating RIF rules. We also introduce a software tool –Parrot– that exploits these annotations and produces reference documentation for combinations of ontologies and rules.

## 1 Introduction and Motivation

One of the goals of the Semantic Web is to leverage the web infrastructure for exchanging machine-readable knowledge. A full stack of technologies has being developed to this end, including a framework for resource descriptions (RDF [7]), some schema-definition language such as RDF Schema [5] and OWL [14], and the RIF family of rule interchange languages [4]. Some of these W3C standards have gained wide adoption. Particularly, OWL ontologies and RDF Schema vocabularies are being effectively exchanged on the web. A large amount of "linked data" has flourished in the last few years [3], although structured descriptions of the corresponding datasets seem to be one step behind. At the moment of this writing, one year after the RIF specifications reached maturity, rule interchange on the web is still marginal.

We believe that some of the burdens that prevent the take-off of RIF documents interchange are the lack of companion tools and the absence of guidelines for adding metadata to the rules. The purpose of this paper is to make contributions to both fronts: firstly, we propose a metadata scheme for RIF rules; secondly, we introduce Parrot, a software tool that produces human-oriented reference documentation for combinations of OWL and RIF.

The rest of the paper is structured as follows: in the next section, we examine the state of the art regarding vocabularies and tools for web artifacts metadata management. We learn from the study of current practices of metadata usage in Section 3, and we apply our findings to propose a metadata scheme for RIF rules (Section 4). We introduce Parrot in Section 5, and finally, Section 6 closes the paper with conclusions and some insights into future work.

E. García-Barriocanal et al. (Eds.): MTSR 2011, CCIS 240, pp. 56–67, 2011.

## 2 State of the Art

We split our review of the state of the art in two parts. On the one hand, some vocabularies for expressing metadata annotations are listed. On the other one, some software tools that use these vocabularies for generating documentation are evaluated. These two visions are complementary, as tools are supported by vocabularies.

### 2.1 Metadata Vocabularies

Some initiatives have produced schemas to annotate different kinds of resources with metadata, and some of these schemas are available as RDF vocabularies:

*Dublin Core*[1] is the result of an initiative to provide a small and fundamental group of metadata elements for annotating documents. It has two flavors, the older Dublin Core Elements[2] and the newer Dublin Core Terms[3].

*RDF Schema*, or RDFS, is an application of RDF to the description of RDF vocabularies. It includes a basic set of properties for metadata, such as `rdfs:label` and `rdfs:comment`.

*OWL* introduces a few properties for capturing versioning information and compatibility notes. As OWL is built on top of RDF Schema, authors are encouraged to also use RDFS metadata properties.

*SKOS*, the Simple Knowledge Organization System [13], is a common data model for sharing and linking knowledge organization systems on the web. It provides a basic vocabulary for associating lexical labels to any kind of resource. It introduces the distinction to among preferred (`skos:prefLabel`), alternative (`skos:altLabel`) and "hidden" (`skos:hiddenLabel`) lexical labels.

*VANN* introduces terms for annotating descriptions of vocabularies with examples and usage notes [8].

Beyond these vocabularies for general-purpose metadata, there are some others specially designed for describing a concrete domain or artifact, such as datasets. Because of the large amount of data that is becoming available on the web, new issues arise. A common need is to publish meta-descriptions of the data stored on datasets. To this end, some of the most relevant proposals are VoID [1], DCat[4] and voidp[5].

Furthermore, there are some vocabularies, such as FOAF [6] or OpenGraph [15], that are commonly found in metadata annotations even if they were not introduced with this purpose.

We note that there is not any specific vocabulary for rules, at least with a significant adoption. Moreover, the RIF specification[6] suggests 9 properties to be used in annotations. In Section 4 we extend this set to 31 properties in order to expand its coverage to areas such as legal rights or related multimedia objects.

[1] `http://dublincore.org/`
[2] `http://purl.org/dc/elements/1.1/`
[3] `http://purl.org/dc/terms/`
[4] `http://www.w3.org/egov/wiki/Data_Catalog_Vocabulary`
[5] `http://www.enakting.org/provenance/voidp/`
[6] `http://www.w3.org/TR/rif-prd/#Annotation`

### 2.2 Tools

A number of tools can generate reference documentation for RDFS and OWL ontologies:

*OWLDoc*[7] generates JavaDoc-like HTML pages from an OWL ontology. OWLDoc works together with Protégé-OWL.

*SpecGen*[8], is an off-line, ontology specification generator. It combines a template with static text with an index of the vocabulary terms and the detailed views of each one. It has been used to generate the companion documentation of some popular vocabularies, such as FOAF or SIOC.

*VocDoc*[9] is a Ruby script which produces documentation for RDFS/OWL ontologies and vocabularies. It is inspired by SpecGen, and it adds the LaTeX output to make it easier to include the report in larger documents, such as project deliverables or technical reports.

*Neologism*[10] is a web-based RDF Schema vocabulary editor and publishing system [2]. The main goal of Neologism is to dramatically reduce the time required to create, publish and modify vocabularies for the web of data, and to provide companion documentation.

All the aforementioned tools deal exclusively with vocabularies and ontologies. Regarding rules, commercial rule management systems such as IBM WebSphere ILOG JRules or ontoprise OntoStudio, can generate documentation about the rules in their particular proprietary formats. However, to the best of our knowledge, there is not any solution for documenting standard web rules. The RIF specification suggests some properties to be used for metadata, but there is no evidence of any previous tool supporting these properties.

Expanding the horizons of this analysis of the state of the art, we note that generic "linked data" browsers[11] can be used to visualize (and therefore, to document) any kind of resource published on the web. However, due to their general approach and their orientation to instances (as opposed to vocabularies), they provide limited help to grasp an ontology or a rule.

## 3 Analysis of Vocabulary Metadata in the Wild

In this section, we present a survey of the actual usage of metadata in ontologies/vocabularies publicly available on the web. For this study, we examine 23 of the most popular RDFS/OWL vocabularies according to the metrics available from reference web sites. The list is assembled as the union of the top-25 popular vocabulary list from `prefix.cc`[12] and top-18 from `pingthesemanticweb.com`[13].

---

[7] `http://protegewiki.stanford.edu/wiki/OWLDoc`

[8] `http://forge.morfeo-project.org/wiki_en/index.php/SpecGen`

[9] `http://kantenwerk.org/vocdoc`

[10] `http://neologism.deri.ie`

[11] `http://www.w3.org/wiki/TaskForces/CommunityProjects/LinkingOpenData/SemWebClients`

[12] `http://prefix.cc/popular/all`

[13] `http://pingthesemanticweb.com/stats/namespaces.php`

A number of vocabularies are not considered due to their redundancy (some of them belong to a family of vocabularies, such the DBPedia family of namespaces). Additionally, the "time" and "creativecommons" vocabularies have been cherry-picked due to their obvious relevancy to metadata, even if they do not appear in the top positions of the popularity ranking.

We exclude large ontologies from our analysis (e.g., Yago, SUMO or WordNet). At this stage, we focus on small, highly reused vocabularies. We plan to extend our study to these ontologies in the future.

Each one of the RDFS/OWL documents that define these vocabularies has been manually examined, and a comprehensive list of all the metadata properties in use has been collected. The results are captured in Table 1. Metadata properties are described in rows, and are sorted by decreasing usage frequency. Tick marks in this table indicate that the vocabulary of the column uses at least once the metadata property of the row (typically, to annotate one of its classes or properties). For the sake of conciseness, vocabularies in columns are identified by their usual prefix[14].

The results reveal that RDF Schema annotation properties are massively popular. By far, the most frequent metadata associated to vocabulary artifacts are labels and comments. Titles and descriptions are common too, with two namespaces being used for equivalent purposes (Dublin Core Terms and Dublin Core Elements). This duality is also present in other properties, such as the ones used to express attribution (creator and contributor). Legal rights and license information is only present in a minority of the vocabularies.

Versioning information is often limited to simple textual annotations that use `owl:versionInfo` and `dct:hasVersion`, some of them automatically generated by the VCS (version control system) used by the vocabulary authors. It has been observed that some vocabularies convey versioning information in their comments. This practice may be convenient for manual management of the vocabularies, but it is a hindrance to automated management. Some versioning information is sometimes provided by means of time references. Our study reveals that the generic property `dc:date` is commonly used, while more specific properties such as `dct:issued` and `dct:modified` are limited to the vocabularies controlled by Dublin Core.

The absence of some metadata is also interesting. There is a complete lack of multimedia resources associated to the vocabularies, although many vocabularies include very generic pointers (`rdfs:seeAlso`) to other resources. Moreover, the VANN vocabulary, which was designed with the purpose of annotating other vocabularies, is completely absent from the selected sample.

The SKOS vocabulary is sometimes used to introduce definitions, examples and notes. Regarding linguistic information, it is noticeable that SKOS labeling properties are barely used. In fact, even in those cases that indicate preferred labels, there are no alternative labels. The use of SKOS in this context is pointless, as the same semantics could be simply conveyed by `rdfs:label`. Moreover, it has been observed that approximately half of the sampled vocabularies do not

[14] The full namespace URI can be retrieved by means of a query to prefix.cc.

**Table 1.** Metadata properties used in popular web vocabularies, sorted by decreasing frequency of use

| | dbpedia | foaf | dc | rdf | rdfs | owl | geonames | skos | geo | sioc | dct | gr | doap | wot | rss | vcard | cc | mo | time | bibo | geospecies | txn | dctype |
|---|---|---|---|---|---|---|---|---|---|---|---|---|---|---|---|---|---|---|---|---|---|---|---|
| rdfs:label | ✓ | ✓ | ✓ | ✓ | ✓ | ✓ | ✓ | ✓ | ✓ | ✓ | ✓ | ✓ | ✓ | ✓ | ✓ | ✓ | ✓ | ✓ | | ✓ | ✓ | ✓ | ✓ |
| rdfs:comment | | ✓ | ✓ | ✓ | ✓ | ✓ | ✓ | ✓ | ✓ | ✓ | ✓ | ✓ | ✓ | ✓ | ✓ | ✓ | ✓ | ✓ | ✓ | ✓ | ✓ | ✓ | ✓ |
| rdfs:isDefinedBy | | ✓ | ✓ | ✓ | ✓ | ✓ | | ✓ | | ✓ | ✓ | ✓ | ✓ | ✓ | ✓ | | | ✓ | | ✓ | ✓ | ✓ | ✓ |
| rdfs:seeAlso | | | ✓ | ✓ | ✓ | ✓ | | ✓ | | ✓ | ✓ | | | ✓ | | ✓ | | ✓ | ✓ | ✓ | ✓ | ✓ | |
| dc:title | | ✓ | | ✓ | ✓ | ✓ | | | ✓ | | | ✓ | ✓ | ✓ | | ✓ | | ✓ | | | ✓ | | |
| dc:description | | ✓ | | ✓ | | | | | ✓ | | | | ✓ | ✓ | | ✓ | | ✓ | | | ✓ | ✓ | |
| dct:title | | | ✓ | | | | | ✓ | | ✓ | ✓ | ✓ | | | | | ✓ | | | ✓ | | | ✓ |
| owl:versionInfo | ✓ | | | | | ✓ | ✓ | | | ✓ | | ✓ | | | | | | ✓ | | ✓ | | | |
| dc:date | | | | | | | | | ✓ | | | | | ✓ | | ✓ | | ✓ | | | ✓ | ✓ | |
| dc:creator | | | | | | | | | | | | ✓ | ✓ | | | ✓ | | | | | ✓ | | |
| dct:creator | | | | | | | | ✓ | | | | | | | | | | | | ✓ | | ✓ | ✓ |
| dct:description | | | | | | | | ✓ | | ✓ | ✓ | | | | | | | | | ✓ | | | |
| vs:terms_status | | ✓ | | | | | | | | | | | | ✓ | | | | ✓ | | ✓ | | | |
| dct:issued | | | ✓ | | | | | | | | ✓ | | | | | | | | | | | | ✓ |
| dct:hasVersion | | | ✓ | | | | | | | | ✓ | | | | | | | | | | | | ✓ |
| dct:modified | | | ✓ | | | | | | | | ✓ | | | | | | | | | | | | ✓ |
| dc:rights | | | | | | | | | | | | ✓ | ✓ | | | | | | | | | | |
| dct:publisher | | | | | | | | | | | ✓ | | | | | | | | | | | | ✓ |
| foaf:maker | | | | | | | | | | | | | ✓ | | | | | ✓ | | | | | |
| skos:definition | | | | | | | ✓ | ✓ | | | | | | | | | | | | | | | |
| skos:example | | | | | | | | ✓ | | | | | | | | | | | | ✓ | | | |
| skos:note | | | ✓ | | | | | | | | ✓ | | | | | | | | | | | | |
| skos:prefLabel | | | | | | | ✓ | | | | | | | | | | | | | | ✓ | | |
| skos:scopeNote | | | | | | | | ✓ | | | | | | | | | | | | ✓ | | | |
| dc:contributor | | | | | | | | | | | | ✓ | | | | | | | | | | | |
| dc:identifier | | | | | | | | | | | | | | | | | | | | | ✓ | | |
| dc:subject | | | | | | | | | | | | ✓ | | | | | | | | | | | |
| dct:contributor | | | | | | | | ✓ | | | | | | | | | | | | | | | |
| dct:license | | | | | | | | | | | | ✓ | | | | | | | | | | | |
| foaf:homepage | | | | | | | | | | | | ✓ | | | | | | | | | | | |
| skos:changeNote | | | | | | | | | | | | | | | | | | | | ✓ | | | |

explicitly indicate the language of the string literals, which leads to ambiguity. Only a couple of vocabularies contain multilingual metadata.

The results table clearly reflects the fact that some metadata annotations can be captured by different properties, and there is a lack of consensus about which is the preferred one. For instance, the semantics of `rdfs:comment`, `dc:description`, `dct:description` and `skos:definition` are very similar (at least when applied to vocabularies). The choice among them is mainly a matter of the preferences of the author, and it is not exclusive. Some vocabularies use more than one.

Moreover, it has been observed that they are sometimes multivalued (e.g., multiple `rdfs:comments` are attached to the same ontology to separate different aspects of the description).

In the case of the duality between Dublin Core Terms and Dublin Core Elements, it seems that at least for some cases it can be explained by DC Terms being a relatively new specification. It is assumed that newer vocabularies may prefer DC Terms.

## 4 Proposed Vocabulary for Rule Metadata

This section presents our proposal to describe rules and rule sets with medatata, identifying documentation requirements and relevant vocabularies based on the previous work of Section 3.

Rules, like ontologies, are knowledge-based artifacts that capture domain information in a formal way. They declaratively express the dynamic conditions comprising business logic built upon a data model, which describes the entities of the domain. In other words, a ruleset specifies how a system works. In the web, rules and rule sets can be interchanged using RIF, while data models are typically OWL ontologies. The same concerns arising for documenting ontologies apply to rule sets as well. Technical and business people, such as consultants or domain experts, often bear different interests regarding the usage of these artifacts. Moreover, their background may also diverge and logical training cannot be assumed for business-oriented profiles. Metadata provide, on the one hand, a practical mechanism to organize collections of rules without interfering the domain semantics. On the other one, they help lay and nonprofessional users to understand the vision of the world encoded in knowledge-based systems, for instance, by means of natural language expressions.

RIF is an standard for exchanging rules among rule systems, in particular among web-oriented ones. Technically, RIF is a family of languages, called *dialects*, covering different kind of rules: from logic-programming [12] to production rules [10]. The syntax and semantics of each dialect is rigorously and formally specified, trying to reuse as much machinery as possible, such as the mechanism for annotations.

According to the specification, an annotation can be attached to any term and formula within a RIF document (in RIF PRD dialect, this also includes group of rules). Annotations are optional, and only one annotation is allowed per element.

Although XML is the normative syntax for RIF, in this paper we will use the informative, human-readable RIF Presentation Syntax (PS). An annotation is of the form `(* id` $\varphi$ `*)`, where `id` represents the identifier of the annotated syntactic element (an URI), and $\varphi$ is a RIF formula capturing the metadata. In particular, $\varphi$ is a frame (an expression of the form `s[p ->o]`) or a conjunction of frames (i.e., `And(`$s_1$`[`$p_1$` ->`$o_1$`], ... , `$s_n$`[`$p_n$` ->`$o_n$`])`). An example of a RIF annotation is shown in Listing 1.1. Notice that RIF web-oriented design enables the reutilization of existing vocabularies for annotations, such as Dublin Core or even RDFS annotation properties.

**Listing 1.1.** A snippet of an annotated rule in RIF format

```
(* ex:rule ex:rule [
        rdfs:label      -> 'Example Rule'@en
        dc:creator      -> 'Luis Polo'
        dc:date         -> '1981-01-20'
        og:video        -> <http://youtu.be/5h10QHpA5EU>
        dct:publisher   -> <http://ontorule-project.eu>
] *)
```

Nevertheless, the RIF machinery for annotations is very flexible and offers a lot of syntactic freedom, which difficults the correct interpretation of rule and rule sets metadata. For instance, the identifier (`id`) of the rule is an optional element in the annotation expression. Moreover, there could be frames in $\varphi$ not describing the annotated element. Therefore, we propose some additional restrictions on RIF annotations in order to simplify their management, on the one hand, and to guarantee some integrity on rule metadata, on the other:

1. It is mandatory to declare an identifier (`id`) of the rule, providing an identity on the web of data. The identifier of the rule not only enables cross-references between rules and other elements of a RIF document, but also to establish links between rules and any RDF resource (for instance, in the Linked Data cloud).
2. Metadata $\varphi$ must contain at least one frame where the subject is the identifier of the annotation, i.e., the RIF element being described.

Coming back to vocabularies for rules and rule sets annotations, it is worth reminding that it is possible to reuse existent ontologies and vocabularies on the web for this purpose. As both artifacts share requirements with respect to needed metadata, they bring the opportunity to reuse the same resources for both ontologies and rules, without introducing new elements. Table 2 sums up, on the one hand, the kind of metadata required to describe rules and, on the other, our suggestions on which properties can be applied to this end. The reader is encouraged to check the range of the recommended properties in their normative specifications.

One apparent limitation of the normative XML syntax of RIF annotations is that the scope of each metadata expression is constrained to the annotated element, even if the machinery enables references to other rules or terms. This is

**Table 2.** Recommended list of metadata properties for documenting RIF rules

| Metadata | Recommended properties |
|---|---|
| *Labeling* | Rules are usually referenced by a label, such as "rule for identifying defects". These labels can be captured by several properties. An important aspect is to appropriately capture multilingualism.<br>Recommended properties are: `dc:title`, `dct:title`, `rdfs:label`, `skos:prefLabel` and `skos:altLabel`. |
| *Authoring* | Typically several entities are associated to a rule or a ruleset, but playing a different role. For instance, someone in a company is the creator of the rule and people from other department may have contributed to its definition. Finally, the organization itself is the responsible for its publication and distribution.<br>Recommended properties are: `foaf:maker`, `dc:creator`, `dc:contributor`, `dc:publisher`, `dct:creator`, `dct:contributor` and `dct:publisher`. |
| *Description* | Natural language descriptions of rules are useful in order to provide a human-readable expression of its meaning.<br>Recommended properties are: `dc:description`, `dct:description`, `rdfs:comment`, `skos:definition`, `skos:example` and `skos:note`. |
| *Multimedia* | Description of rules may be provided by means of multimedia contents, such as images, videos, graphical tables, etc.<br>Recommended properties are: `foaf:depiction` and `og:video`. |
| *Versioning* | Rules, as other knowledge artifacts, are subject to evolution and timeline modifications, which should be tracked.<br>Recommended properties are: `owl:versionInfo`, `dct:hasVersion` and `skos:changeNote`. |
| *Rights* | Rules are specifications of IT systems, which might be protected by copyright and distributed under a propietary or private license.<br>Recommended properties are: `dc:rights`, `dct:license`. It is also suggested to use RDF descriptions of licenses, such as the ones available from CreativeCommons. |
| *Dates* | Apart from versioning, it is important to capture other temporal stamps relevant for rules, such as rule creation or modification dates.<br>Recommended properties are: `dc:date` (for generic purposes), `dct:issued`, `dct:modified`. |
| *Documentation* | Another aspect about rules is the relationship with the sources from which the knowledge has been extracted, typically business documents. Moreover, a rule can also be linked to other kind of resources that provide additional information about it.<br>Recommended properties are: `dct:source`, `rdfs:seeAlso`, `rdfs:isDefinedBy` |

**Table 3.** Interpretation of RIF annotations as RDF graphs

| Annotation $\varphi$ | $\pi(\varphi)$ |
|---|---|
| `s [p -> o]` | { `s p o` } |
| `s[p`$_1$`->o`$_1$ ... `p`$_n$`->o`$_n$`]` | {`s p`$_1$ `o`$_1$ ; $\cdots$ ; `p`$_n$ `o`$_n$ } |
| `And(F`$_1$`, ... , F`$_n$`)` | $\{\pi(\mathtt{F}_1)\} \cup \cdots \cup \{\pi(\mathtt{F}_n)\}$ . |

a hindrance for reusing descriptions. To overcome this limitation, we propose an RDF interpretation of RIF annotations, so all the annotations of a RIF document are comprehensively collected in a single RDF graph. Moreover, this interpretation makes it possible to execute SPARQL queries over RIF metadata, and it also fosters information reuse based on the principles of "linked data". Table 3 describes the mapping between RIF metadata expressions ($\varphi$) and RDF triples. Notice that $\varphi$ cannot contain variables and that identifiers have a straightforward translation because both sides use URIs for this purpose. It is worth remarking the divergence between annotations translated by $\varphi$ and the RDF syntax for RIF proposed by W3C [11]. In our case, semantically-equivalent $\varphi$ expressions for annotations are provided following [9] (i.e., there exists a direct correspondence between a frame and a triple), while [11] describes an RDF-serialization for its frame-based syntax. Although the latter is more expressive, enabling to capture complete RIF documents, it is notoriously difficult to be queried using SPARQL. The simplicity of common annotations does not justify using this complex RDF syntax for RIF.

## 5 Parrot: Generating Reference Documentation for Ontologies and Rules

One of the applications of ontology and rule metadata is to produce human-oriented reference documentation. We implemented Parrot, a tool that generates documentation for ontologies, rules and combinations of both of them. In this sense, it is a superset of the tools that have been examined in Section 2. To the best of our knowledge, it is the first implementation of a documentation generator for combinations of OWL and RIF.

The input to Parrot are ontology and rule documents compliant with W3C standards, namely OWL and RIF. Typically these documents are available at public web locations and are identified by their URI. Parrot can retrieve them from the web, although it also supports direct file upload.

After parsing the input documents, Parrot builds an in-memory model of the artifacts they describe, mainly ontologies, classes, properties, instances, rules and rule sets. Then, direct and inverse references between the artifacts are established. For instance, classes do not contain references to rules, but rules do use classes in their definitions; therefore bi-directional references are introduced between pairs of classes and the rules. These references manifest as navigable hyperlinks in the final document.

Parrot also builds indexes of the artifacts. These indexes are later transformed into tables of contents, summaries and glossaries in the generated documentation.

The main part of the reference documentation comprises detail views of each artifact. Figure 1 depicts the detailed view of a RIF rule. Note that different aspects of the metadata are visually separated, and can be individually displayed or hidden by the user. Parrot pays special attention to abstract the complexity of the metadata and the underlying OWL and RIF documents, in an effort to make the knowledge accessible to a larger audience. Moreover, Parrot supports user profiles with different skills and interests. For instance, the "business profile" is tailored to users without technical expertise, but operative knowledge of the domain.

**Fig. 1.** Screenshot of the detailed view of a rule in Parrot

Parrot has to deal with the fact that a wide range of properties are used in metadata annotations, as was found in Section 3. Therefore, mechanisms are in place to deal with conflicting and redundant annotations. For instance, in the case of annotations that convey essentially the same semantic information, a priority-driven iterative strategy is implemented. Similarly, Parrot handles multilingual annotations and lets the user choose the language used for the documentation. Currently, Parrot supports most of the properties in Tables 1 and 2 and adds a few more[15].

[15] For a complete listing of all the metadata properties supported by Parrot, please check `http://ontorule-project.eu/parrot/help`

This tool is available as a web service[16] and is distributed as open-source [17]. It has been implemented is Java and reuses a number of components such as Jena[18] and Java-RDFa[19]. One remarkable dependency is RIFle[20], a Java toolkit for managing RIF documents. Among other tasks, RIFle parses the annotations in a RIF document and exposes them as a simple RDF graph, according to the mapping described in Table 3.

## 6 Conclusions and Future Work

We expect this work to have an impact on the quality and quantity of the metadata annotations associated to web ontologies and rules. Firstly, we believe that the guidelines proposed in this paper, as well as the lessons learned from analysing the metadata embedded in publicly available vocabularies, will help the community to be more precise with the metadata they include. This is especially true for rules, because of the current lack of best practices and the vague guidelines provided by the specifications.

Secondly, regarding the quantity of metadata annotations, we hope that the availability of an easy-to-use reference documentation tool will encourage authors to include more metadata. The prompt availability complete reference documentation at no cost should catalyze authors to add metadata. Moreover, it can foster knowledge reuse, by lowering the barrier to gain understanding of ontologies and rules found on the web.

The vocabulary proposed in this paper is not the only one for annotating rules. Some Business Rule Management Systems (BRMS) such as JRules and Drools have their own extensible schemas. Our proposed vocabulary could contribute to exchange metadata between BRMS by suggesting how expressive rule annotations can be captured and interpreted in intermediate RIF documents.

Although the primary target of the vocabulary proposed in Section 4 is RIF, potentially it can be used with any rule language that associates its artifacts to a named RDF resource (URI).

Our perspectives for future work include to extend our analysis to large ontologies. We also plan to assess our proposal for rule metadata by accounting the use of these properties in RIF documents as they gain popularity on the web. To this end, we plan to anonymously monitor the usage trends in the public instance of the Parrot web service. The development roadmap of Parrot also includes extending its coverage to other web resources, such as datasets and queries. Finally we aim to extend rule metadata with new properties to describe other business features, such as their scope and inter-rule relations.

---

16 `http://ontorule-project.eu/parrot/`

17 `http://sourceforge.net/projects/parrot-project/`

18 `http://incubator.apache.org/jena`

19 `https://github.com/shellac/java-rdfa`

20 `http://rifle.sourceforge.net/`

**Acknowledgements.** The work described in this paper has been partially supported by the European Commission under ONTORULE Project (FP7-ICT-2008-3, project reference 231875).

## References


1. Alexander, K., Cyganiak, R., Hausenblas, M., Zhao, J.: Describing Linked Datasets with the VoID Vocabulary (March 2011)
2. Basca, C., Corlosquet, S., Cyganiak, R., Fernández, S., Schandl, T.: Neologism: Easy Vocabulary Publishing. In: Proceedings of the ESWC 2008 Workshop on Scripting for the Semantic Web (SFSW 2008) (June 2008)
3. Berners-Lee, T.: Linked Data design issues (July 2006), `http://www.w3.org/DesignIssues/LinkedData.html`
4. Boley, H., Kifer, M.: RIF Overview. Working Group Note, W3C (June 2010), `http://www.w3.org/TR/rif-overview/`
5. Brickley, D., Guha, R.: RDF Vocabulary Description Language 1.0: RDF Schema. Recommendation, W3C (February 2004), `http://www.w3.org/TR/rdf-schema/`
6. Brickley, D., Miller, L.: FOAF Vocabulary Specification (August 2010), `http://xmlns.com/foaf/spec/`
7. Carroll, J.J., Klyne, G.: Resource Description Framework (RDF): Concepts and Abstract Syntax. Recommendation, W3C (February 2004), `http://www.w3.org/TR/rdf-concepts/`
8. Davis, I.: VANN: A vocabulary for annotating vocabulary descriptions. Technical report (2005), `http://vocab.org/vann/`
9. de Bruijn, J.: RIF RDF and OWL Compatibility. Recommendation, W3C (June 2010), `http://www.w3.org/TR/rif-rdf-owl/`
10. de Sainte Marie, C., Hallmark, G., Paschke, A.: RIF Production Rule Dialect. Recommendation, W3C (June 2010), `http://www.w3.org/TR/rif-prd/`
11. Hawke, S., Polleres, A.: RIF In RDF. Working Group Note, W3C (May 2011)
12. Kifer, M., Boley, H.: RIF Basic Logic Dialect. Recommendation, W3C (June 2010), `http://www.w3.org/TR/rif-bld/`
13. Miles, A., Bechhofer, S.: SKOS Simple Knowledge Organization System Reference. Recommendation, W3C (August 2009), `http://www.w3.org/TR/skos-reference/`
14. Motik, B., Parsia, B., Patel-Schneider, P.F.: OWL 2 Web Ontology Language Structural Specification and Functional-Style Syntax. Recommendation, W3C (October 2009), `http://www.w3.org/TR/owl2-syntax/`
15. Zuckerberg, M., Taylor, B.: The Open Graph Protocol (April 2010), `http://ogp.me/`

# A Reference Ontology for Profile Representation in Communities of Practice

João Luis T. da Silva[1], Alexandre Moreto Ribeiro[1], Elisa Boff[1], Tiago Primo[2], and Rosa M. Viccari[2]

[1] Universidade de Caxias do Sul (UCS), Caxias do Sul, RS, Brazil
{jltsilva,Alexandre.Ribeiro,eboff}@ucs.br
[2] Universidade Federal do Rio Grande do Sul (UFRGS), Porto Alegre, RS, Brazil
tiagoprimo@gmail.com, rosa@inf.ufrgs.br

**Abstract.** Communities of Practice (CoP) integrate people in a group in order to learn in a collaborative way. In these communities it is necessary to establish a formal and standardized representation of the interests and the knowledge that emerges from the CoP. With this scenario in mind, this paper presents a reference ontology for a user profile representation in a CoP framework. The main idea of this framework is to realize that a foundation for the semantic web is feasible to aggregate on virtual communities in the context of collaborative learning. The concern of this work is to establish a reference ontology for the general profile that can be used as guidelines to build a computational CoP framework for realizing such a foundation.

**Keywords:** Communities of Practice (CoP), Ontologies, Collaborative learning, Semantic WEB.

## 1 Introduction

In the knowledge sharing between groups of people who learn collaboratively is necessary to establish a standardized and formal representation of interests and "tacit/explicit" knowledge present in a community. We consider this under the hypothesis of the Semantic Web as the main pattern of collaboration.

This work is part of a proposal of a conceptual framework for the construction and management of CoP. The term CoP is defined as a community of people who share the interest regarding an issue or problem and learn from regular interactions [23,24]. This contact between community members may occur virtually or in person. It should enable the exchange of information and knowledge which, when put into practice by other members may help them in finding solutions and the best practices, thus promoting group learning [22].

The main objective of this framework is to provide a specification and a platform for creating and managing CoP, providing features of knowledge recommendation and management by means of collaborative tools and interaction between its members. For this goal we specified a reference ontology to represent

E. García-Barriocanal et al. (Eds.): MTSR 2011, CCIS 240, pp. 68–79, 2011.

the users' and the communities' profiles. In this framework, a definition of member profile is intended to provide all information identifying the main structural characteristics of the participants in a specific CoP. The main purpose of these requirements is to define an approach to manage virtual communities and to find similarities between profiles of interest (*interest matching*), supporting the recommendation of partners and/or communities in a particular field of interest.

Nowadays there is an increasing number of ontologies available on the web to represent data and the Web itself. Manipulation of this kind of heterogeneous information is quite complex, but on the other hand, the Web is crucial in carrying out information sharing and reusing it through these multiple ontologies. A Reference Ontology can be used to attenuate problems generated by information heterogeneity. According [26], it is possible to use a reference ontology for each knowledge domain and gather a certain number of domain ontologies to integrate our CoP framework.

Despite several definitions concerning reference ontologies, in this work we consider that the closest definition to the proposed framework is based on [3,4]: a reference ontology represents knowledge concerning a particular part of the world in an independent way from specific application purposes; it also represents the theory of a domain in accordance with strict knowledge representation principles belonging to ontologies, and may build extensions, specializations or instantiations to specific domain ontologies.

Usually in any virtual community we have the representation of a member's profile related to personal and professional information, as well as interests that may identify a usage and behavior profile in the context of service-oriented applications, especially in recommendation systems. The traditional formalization of a profile in this context is used to define not only the superficial identification and preferences of the users, but also *(1)* their expertise in a specific area of interest, *(2)* the relevance of their contributions in collaborative interactions and, *(3)* the evolution of their learning, which is promoted by interactions.

Formalizing the learning interactions in a group such as a CoP provides support for knowledge management, in a manner to capture the three mentioned aspects. To achieve this level of knowledge management, the member of a CoP needs to register and share its knowledge and practice, and hence its intellectual evolution in the community domain along with its practices.

Our goal is to use a profile reference ontology in order to establish a standard of sufficient knowledge for modeling a CoP. This approach can detect *expertise* in its environment and automatically (or at least semi-automatically) link participants and other communities related to a problem, as well as the practice that can solve it. It is also a fundamental step that an application or domain ontology could be instantiated to build CoP for specific domains.

This article is organized as follows, the next Section presents some related works for ontology-based user profile representation and details about CoP. In the Section 3 we present our proposal to a profile reference ontology for CoP, the contextualization inside the overall framework proposal and the components built in the profile reference ontology. The final considerations are presented at the Section 4.

## 2 Related Works

In general, an user profile definition reflects only the momentary user's interest about a particular subject in a specific domain. Some authors [16,5] claim that every expression in the user profile represents a characteristic obtained directly from the user and/or eventually inferred during interaction on the Web. The data is stored in a traditional database including user identification, interests and preferences, which can be maintained dynamically [18,16].

According to Donath [7], the definition of *User Profile* in terms of internal identity and social reputation is part of the community formation. While the internal identity is defined by the user, the convention is that the social reputation is an extension of the user profile. Considering that reputation can be defined in terms of what is said or believed about a person or object [12], Donath considers that the social reputation uses the same kind of information stored in the user profile, but under the point of view of other users. In this case, we need to consider the interactions that occur among members of the community.

The authors in [25] present UMMO (*User Modeling Meta-Ontology*), a general ontology for user modeling mainly used to define a uniform interpretation of user models distributed in the semantic Web environments. One of the authors's goal is to make UMMO a central point of reference as a structural basis for capabilities of the user modeling, providing a means of comparison or relationship between various concepts and definitions used in the user modeling. Another UMMO purpose is to serve as an educational tool, helping in to treat the problem of synonymy in the area.

The UMMO general model is presented in several dimensions that define the characteristics of the user, such as *BasicUserDimensions, Emotional States, Characteristics*, and *Personality.* In *BasicUserDimensions*, personal traits as professional, demographical, emotional and psychological, among others, define the basic personality of a user model. Among them is the definition of role (***Role***) as structuring part of the user profile.

GUMO [11] is another solution to identify basic user model dimensions while lets the more general world knowledge open for already existing ontologies, such as [14] and UbisWorld ontology[1], to model intelligent environments. GUMO is defined by an OWL[2] language and its key feature is its modular approach. Some basic user dimensions are Emotional States, Characteristics and Personality.

A typical generic user modeling system only delivers a small portion of the services listed by Kobsa [13], such as: the representation of relevant common characteristics of users pertaining to specific user subgroups of the application system (stereotypes); the recording of users' behavior, particularly their past

[1] `http://www.ubisworld.org`

[2] *Web Ontology Language* (`http://www.w3.org/TR/owl-ref/`).

interaction with the system; the generalization of the interaction histories of many users into stereotypes.

Kobsa's work presents the characteristics of generic user modeling systems: generality, including domain independence; expressiveness and strong inferential capabilities; support for quick adaptation; extensibility; import of external user-related information; management of distributed information; support for open standards; load balancing; failover strategies (fallback mechanisms in case of a breakdown); transactional consistency and privacy support.

Two Semantic Web projects SIOC (*Semantically-Interlinked Online Communities*) [1] and FOAF (*Friend of a Friend*) [2] are also used as standards for communities and user representation. The FOAF vocabulary represents people and their social networks through their relationships and information using the Web. There are currently thirteen classes that compose the standard FOAF. Among this set, there are classes related to personal identification such as *name* and *organization* besides more technical classes such as *Agent* and *Person.* Properties such as *topic_interest* and *knows* are also important to be taken into consideration, because of their important semantic inference and social characteristic. The SIOC project allows the integration of information from the online community providing a Semantic Web ontology for representing data from the Social Web in RDF[3]. The SIOC vocabulary is an open format used for expressing user-generated content in an interoperable way.

Different from SIOC and FOAF approaches, our proposal does not intend to describe only the site or personal identification, but also to aggregate all interaction events and user contents in the CoP profile. It will integrate what we call a "dynamic profile", for instance, adding/commenting a post forum, a file, an image, a video, searching for context/person, adding articles, editing profile infos or posting blog contents, which will trigger a capture for those posting details into collaboration registers. This activity could generate more entries into interests, skills or interaction components in the profile which will be linked to the content in the collaboration registers. Also, our proposal takes into account those services of [13] in our model, as presented in the next section.

According to Wenger [24], CoP have three fundamental characteristics: *Domain, Community* and *Practice.* **Domain** represents the topic or subject about which community members share knowledge and practices and learn from each other, recognizing the value of their collective experiences. The domain should be explicit and present in the CoP as well on the individual's side, which draws a strong link between individuals and communities. The **Community** is the formal relationship between a group of people who participate in discussions and group activities, helping each other and sharing knowledge and experiences of the domain. **Practice** is the set of shared resources and solving problem techniques.

The conceptual framework defined for the management of the CoP relies heavily on these three characteristics, and also in the life cycle that defines the various stages a CoP goes through.

---

[3] *Resource Description Framework* (`http://www.w3.org/TR/2004/REC-rdf-mt-20040210/`).

## 3 Profile Ontology for Communities of Practice

The CoP Framework is built on three layers: the CoP layer, responsible for the relationship among people with common interests, with emphasis on the CoP life cycle; the Virtual Environment Layer, which provides technological collaboration tools needed by the community; and, the Activity layer, which serves as a link with the previous layers through the management activities of the CoP and the relationships between the use of collaborative tools and the CoP life cycle.

In the CoP Layer, the issues and information addressed in the community are contained in several cores of knowledge (see Figure 1): Interest Domain, Profiles, Collaboration Registers. The interest domains should consider the collective construction of the fields, perhaps by a group of editors/mediators, and predefined domain ontologies. Such ontologies can also be used within the portal built by the participants themselves (using mediators) and/or otherwise generated semi-automatically. The interest domains also integrate the repository of ontologies and address broader issues, unifying concepts around the community domains.

The framework proposal uses shared concepts from several existing ontologies such as FOAF [2] and SIOC [1], and uses concepts related to the domain of the CoP. The main idea in this framework is to reuse multiple domain ontologies in order to represent the conceptual classes described in this specification. It is beyond the scope of this paper to describe the overall conceptual CoP framework, thus we have concentrated on the user and the community profiles modeling within this framework. Figure 1 illustrates the *Profile* class related to the *User*, that defines the participants profile in the context of a CoP.

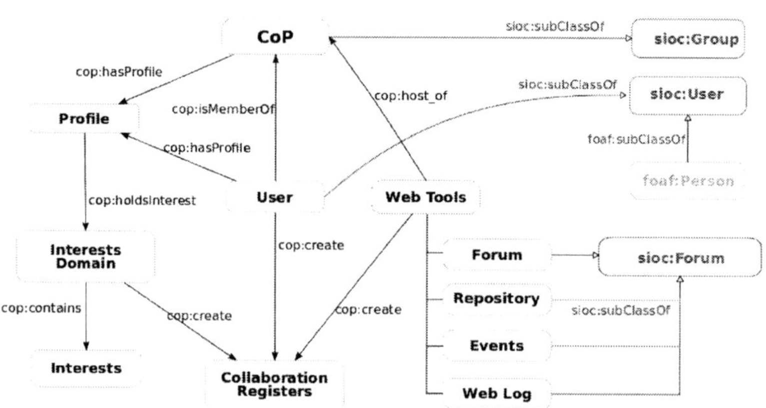


**Fig. 1.** Schema of the Community of Practice Framework

The profile (*Profile class*) of the participant is made up of some basic information *slots* that defines personality. We propose a description in two levels of details: static profile and dynamic profile. Basically this dichotomy is established based on explicit and implicit models by Rich [17] that allows users to provide individual pre-defined data and the system complements it with inferred user data by monitoring its behavior.

The work in [19] also introduces distinction between the explicit and implicit profiles. They aggregate in the explicit profile identification information (username, role, personal settings, etc.), social-economic information (age, gender, hobbies, etc.), reviews (products, standpoint, items of information), information on relationships with others and comments/opinions (text, images, videos and others). They classify the implicit profiles in a transaction profile (transaction logs, purchased products related to product metadata), an interaction profile (user clickstreams, page views, etc.) and external data (information obtained from weather forecasting, local news, events, credit analysis, and more).

In our proposal, the static profile represents information traditionally provided by the user as its personal and professional data, interests, curriculum vitae, etc. In short, the basic information that allows the definition of entries in a "yellow pages" service. The dynamic profile consists of information captured from its interaction with the community at all levels of knowledge, such as files and posts, practices, contributions in problem-solving activities, and through the use of collaborative tools.

The diagram in Figure 2 illustrates the main concept classes defined as reference in the user's profile ontology. A *Profile* has an *Identity, Interactions, Interests, Roles* and *Skills.* In the static context, all these concepts can be informed, and in a dynamic context information is permanently added derived from interactions with the members of communities, as well as from interactions with the tools inside the community.

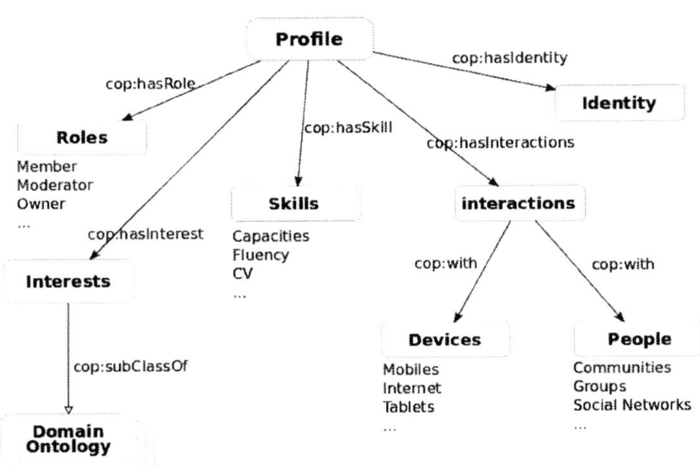


**Fig. 2.** CoP Profile Ontology

### 3.1 CoP:Roles

In the definition of a CoP, the ***Roles*** can be immediately assigned to users. The creator of the community has features and privileges that are formalized in the *Owner* role assigned to the creator immediately upon the creation of the CoP. All guests interested in the community are registered as *Members*. Other roles can be negotiated and awarded for participation, such as the definition of *Moderators* for the community. Typically, each role defines a type of action/interaction in the community, but it can also carry some types of competence, expertise and skills in their field of activity/interest. In this case, the concept of role, for instance, serves as a repository of specific parameters in the instantiation of a domain ontology.

The more general domain provides the basic set of roles *Owner, Member* and *Moderator*. This set can derive and instantiate other behaviors like "Guest", "Beginner", "Regular", "Leader" and "Senior", for example in training areas or working groups. The domain of *e-learning* can instantiate "Teachers", "Tutors", "Specialists", "Students", "Monitors", and others.

In a more general sense, a role has been defined as a collection of necessary features, interests, expectations and behaviors in relation to a particular system [6]. In a narrower sense, the role of a user can be represented by its performance in the context/environment, according to the characteristics and criteria of action/interaction. This means that the characteristics of its performance are influenced by the context. Within the CoP, the definition of roles can standardize the use and the choice of collaborative tools according to the activity developed in the community. In this case, it is important to keep a record of patterns of interaction related to each type of user, or each type of role assumed in a specific collaborative practice.

To illustrate this approach, we can relate the role in CoP to the activity model that represents the practices expressed by community members. The representation of the activities is part of the framework model, whose detailing is not within the scope of this paper. In means for an illustration, Figure 3 shows the representation ontology of practices in the CoP in accordance with the formal components of an activity. In this case, the roles may be linked to specific goals (*Goal*) with a set of tasks (*Task*) specific to each role. It is also possible to link metrics to evaluate practices (*Evaluation*) which can also be specific to each role in the community.

### 3.2 CoP:Interests

The profile presents a relation of interests (***Interests***) representing the user's preferences and knowledge about a domain. Traditionally, interest has been represented by lists of keywords or using folksonomy through tags [21] and incremental [20] or collaborative categorization.

In the present CoP framework, the set of *Interests class* related to the *Profile class* contextualizes a set of domain ontologies related to the community the user takes part in. These ontologies represent the characteristic that defines a domain

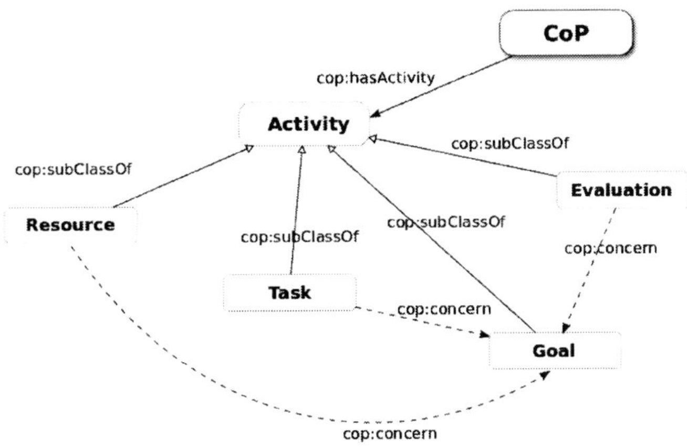


**Fig. 3.** Partial Ontology for the practices of a CoP based on an Activity model

CoP [24], i.e. the explicit knowledge that brings together community members in collaborative learning. Therefore, the set of user interests in the proposed *framework* can be inferred through its links with several domain ontologies (*dynamic model*), besides the predefined set of the user tags (*static model*).

User preferences can be represented as a set of static and dynamic interests related to domain ontologies, making possible its instantiation in the creation of CoP, in accordance with the general interest of the community that gathers its participants. In this model, the *Interests class* can be adapted both to represent community contextual interests as to set metrics in order to evaluate the context proximity. This metrics will be used for representing the several posts, embedded files, comments and any communicative interaction in the CoP. An example of a possible instantiation from the reference ontology Profile is an adaptation of a weighted hierarchy of interests, as represented by the community context, and illustrated in Figure 4.

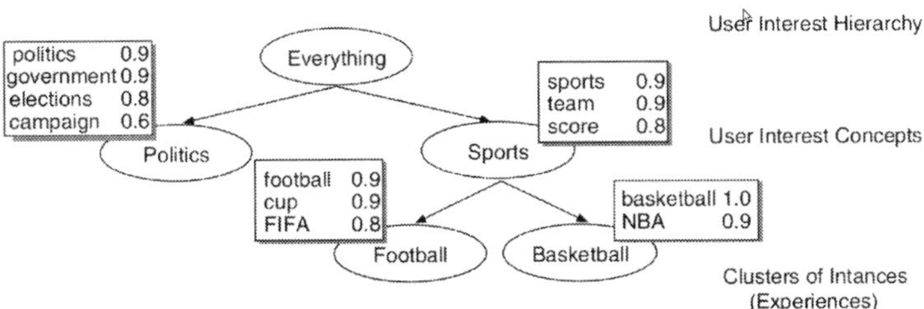


**Fig. 4.** Example of an ontological representation from a community domain context, extract from [10]

The reference ontology models the ***Interests*** class as a subclass of domain ontologies (***Domain Ontology*** class in Figure 1). These, in turn, are defined according to the creation of the community and their description is beyond the scope of this paper. The basic idea is to connect the CoP to a newly created domain ontology, and also promote the construction of shared ontology from interactions among members.

### 3.3 CoP:Skills

The *Cop:Skills* class represents the skills and competencies that the participant has at the CoP. This level of detail provides important information for the recommendation of people, expertise, and discovery of relationships between communities across related skills.

The area of Human Resources Management has a basic and typical issue to find the right person for a job or a position. Normally the area is also responsible for planning the development of these skills in case of gaps in the company. Intellectually, finding an appropriate team for a project or a group of teachers for an appropriate course also derives from that issue: the Skills Management. Human resource systems maintains a database with specific skills or expertise from individuals with some interface for maintenance and search.

There are several definitions for competence where many sources relate competence to a colletion of knowledge (according The Webster Dictionary[4]) and the skills [8] or abilities [15] which people need in its roles and responsibilities, to solve specific problems. Our work is based on Fleury's definition [9] and we relate competence with the identity of the individual and its educational and professional backgrounds, so that the framework can identify relationships in the practices used for a given field and the skills and/or knowledge that the user has mobilized for its implementation.

From the reference ontology *cop:skills*, one can instantiate an application ontology representing a set of skills demonstrated by participants in a community. For example, this can be an OWL scheme representing the user's curriculum vitae along with an ontological structure classifying an area of expertise that can be instantiated by this class.

From this representation, the recommendation of CoP can filter the domain from the skills of its participants or recommend participants according to the dominant skills in a community. By taking advantage of this standardized structure it can still relate to topics of interest (*the CoP domain*) with schemes to solve problems or actions carried out to respond to challenges from the community (*the CoP practice*). In addition, it is possible to recommend experts and to make available some dynamic yellow pages that can be updated from a topic posted by an individual, in order to update its list of skills.

[4] `http://www.webster-dictionary.net/definition/Skill`

### 3.4 CoP:Interactions

The *cop:Interactions* class represents the relationships that may exist in intra- and inter-communities that represents the type of support for multiple platforms that an user can exploit. CoP members need to share domain knowledge and practices in pursuit of their common goals, but they also keep explicit and implicit relationships with other communities and individuals, and interoperate across multiple technological platforms. Thus, *Users* who have *Profiles* in a CoP maintain interactions (***Interactions***) with devices (***Devices***) and individuals (***People***) through a relationship *cop:with.*

Relationships among individuals can be directly described here through the property *foaf:knows* as, for instance, to connect members of a same community or from other communities. In the social domain of the individual, it simply refers to the interactions existing among the various communities in which the user unifies the identity by using an *OpenId*[5] or from a *owl:sameAs* relationship to integrate various sites of the same user.

In the context of recommendation on communities it becomes important to define a more specialized type of interpersonal relationship that may involve more particular kinds of interaction. This applies to the possibility of instantiating an ontology through the FOAF property *foaf:knows*, *rel:acquaintanceOf*, etc. A recommendation algorithm may, for example, infer that two individuals are collaborators or colleagues through this Interaction class. If a community has a document that uses the relation Interaction *cop:with* People, which can be described by the *foaf:knows* property, the recommendation may include the skills and interests from the relationship of users and their communities.

The same happens for ***Devices*** when considering the use of various devices by the same individuals who have *foaf:knows* or *rel:acquaintanceOf* relationships. Inference can establish the use/preference of specific protocols or to identify a profile from mobile behavior in such cases.

### 3.5 CoP:Identity

As seen in [7], the user profile and social reputation are very important to define the user's identity in order to anticipate needs and user behaviors in a virtual community. The understanding and evaluation of interactions in these communities depend on the user's identity, which is also important to motivate people to actively participate in community discussions just like the search for information and the input in discussions to help improve the reputation and establish an recognizable identity.

The user's identity is defined in a traditional way in this class through the unique identification using personal details like name, surname, business and residential information. Many FOAF properties can be invoked at this level as a representation of OpenId (*foaf:openid*), associating a Identity URL. Thus, it is possible to establish the information about the users from any document or resource semantically linked through other sources of knowledge (e.g. RSS, Atom,

[5] `http://openid.net/`

vCard). Private information can also be aggregated to this class, such as date of birth, gender, marital status, and more.

## 4 Conclusions

In this paper, the use of reference ontologies advocates for a user profile representation in a CoP framework. The main idea in the CoP Framework is to realize that a foundation for the semantic Web is feasible to aggregate on virtual communities in the context of collaborative learning. The concern with this work is to establish a general profile reference ontology that can be used as guidelines to build a computational CoP Framework for carrying out such a foundation.

This work is part of a larger proposal to build collaborative learning environments based on CoP. For this approach, we advocate the need for an ontological knowledge representation for the user and the community itself so that software agents can address the intersection of knowledge and recommendations for communities and individuals.

Although significant research issues must be solved before the framework can become a reality, such a reality seems increasingly viable due to a convergence of multiple well established domain ontologies that can be orchestrated by the proposed framework.

**Acknowledgements.** Funded by Pos-Doctoral CAPES/REUNI grant and OTICS Project ICICT/Fio-cruz grant ENSP 060 LIV 09.

## References

1. Bojars, U., Breslin, J.G.: SIOC core ontology specification. Tech. rep., SIOC project (March 2010), `http://rdfs.org/sioc/spec/`
2. Brickley, D., Miller, L.: FOAF vocabulary specification. Tech. rep., FOAF project (August 2010), `http://xmlns.com/foaf/spec/`
3. Brinkley, J., Suciu, D., Detwiler, L., Gennari, J., Rosse, C.: A framework for using reference ontologies as a foundation for the semantic web. In: AMIA Annual Symposium Proceedings, p. 96. American Medical Informatics Association (2006)
4. Burgun, A.: Desiderata for domain reference ontologies in biomedicine. Journal of Biomedical Informatics 39(3), 307–313 (2006)
5. Carreira, R., Crato, J.M., Gonçalves, D., Jorge, J.a.: Evaluating adaptive user profiles for news classification. In: Proceedings of the 9th International Conference on Intelligent User Interface - IUI 2004, p. 206. ACM Press, New York (2004)
6. Constantine, L.: Users, Roles, and Personas. In: Pruitt, J., Adlin, T. (eds.) The Persona Lifecycle, ch. 8, pp. 498–519. Morgan Kaufmann, San Francisco (2006)
7. Donath, J.: Identity and deception in the virtual community. In: Communities in Cyberspace 1996, pp. 29–59 (1999)
8. Dutra, J.S., Hipólito, J.A.M., Silva, C.M.: Gestão de pessoas por competências: o caso de uma empresa do setor de telecomunica cões. Revista de Administra cão Contemporânea 1(4), 161–176 (2000)

9. Fleury, M.T.L., Fleury, A.: Construindo o Conceito de Competência. Revista de Administra cão Contemporânea SPE(5), 183–196 (2001)
10. Godoy, D., Amandi, a.: Modeling user interests by conceptual clustering. Information Systems 31(4-5), 247–265 (2006)
11. Heckmann, D., Schwartz, T., Brandherm, B., Schmitz, M., von Wilamowitz-Moellendorff, M.: GUMO- the general user model ontology
12. Josang, a., Ismail, R., Boyd, C.: A survey of trust and reputation systems for online service provision. Decision Support Systems 43(2), 618–644 (2007)
13. Kobsa, A.: Generic User Modeling Systems. In: Brusilovsky, P., Kobsa, A., Nejdl, W. (eds.) Adaptive Web 2007. LNCS, vol. 4321, pp. 136–154. Springer, Heidelberg (2007)
14. Pease, A., Niles, I., Li, J.: The suggested upper merged ontology: A large ontology for the semantic web and its applications. In: Working Notes of the AAAI 2002 Workshop on Ontologies and the Semantic Web (2002)
15. Perrenoud, P.: Compétences, Langage et Communication. In: Collès, L., Dufays, J.L., Fabry, G., Maeder, C. (eds.) Didactique des Langues Romanes, le développement de compètences chez l'apprenant, pp. 15–104. De Boeck Université, Bruxelles (2001)
16. Poo, D., Chng, B., Goh, J.: A hybrid approach for user profiling. In: Proceedings of the 36th Annual Hawaii International Conference on System Sciences, p. 9. IEEE, Los Alamitos (2003)
17. Rich, E.: Users are individuals: Individualizing user models. International Journal of Man-Machine Studies 18(3), 199–214 (1983)
18. Rousseau, B., Browne, P., Malone, P., Foster, P., Mendis, V.: Personalised Resource Discovery Searching over Multiple Repository Types. In: Seruca, I., Cordeiro, J., Hammoudi, S., Filipe, J. (eds.) Enterprise Information Systems VI, Netherlands, pp. 316–324. Springer, Heidelberg (2006)
19. Schubert, P., Koch, M.: Collaboration platforms for virtual student communities. In: Proceedings of the 00(C) 36th Annual Hawaii International Conference on System Sciences, p. 10 (2003)
20. Sieg, A., Mobasher, B., Burke, R., Prabu, G., Lytinen, S.: Representing user information context with ontologies. In: Proceedings of HCI International Conference, pp. 210–217. Citeseer (2005)
21. Szomszor, M., Alani, H., Cantador, I., O'Hara, K., Shadbolt, N.R.: Semantic modelling of user interests based on cross-folksonomy analysis. In: Sheth, A.P., Staab, S., Dean, M., Paolucci, M., Maynard, D., Finin, T., Thirunarayan, K. (eds.) ISWC 2008. LNCS, vol. 5318, pp. 632–648. Springer, Heidelberg (2008)
22. Terra, J.C.C.: Comunidades de Prática: conceitos, resultados e métodos de gestão. Terra Forum (2005)
23. Wenger, E.: Communities of Practice: learning, meaning and identity. Cambridge University Press, Cambridge (1998)
24. Wenger, E., Mcdermott, R., Snyder, W.: Cultivating communities of practice: A guide to managing knowledge. Harvard Business School Press, Boston (2002)
25. Yudelson, M., Gavrilova, T., Brusilovsky, P.: Towards user modeling meta-ontology. In: Ardissono, L., Brna, P., Mitrović, A. (eds.) UM 2005. LNCS (LNAI), vol. 3538, pp. 448–452. Springer, Heidelberg (2005)
26. Zemmouchi-Ghomari, L., Ghomari, A.R.: Reference ontology. In: International IEEE Conference on Signal-Image Technologies and Internet-Based System, pp. 485–491 (2009)

# PACE: A General-Purpose Tool for Authority Control

Paolo Manghi and Marko Mikulicic

Istituto di Scienza e Tecnologie dell'Informazione "Alessandro Faedo",
Consiglio Nazionale delle Ricerche, Pisa, Italy
{name.surname}@isti.cnr.it

**Abstract.** Curating the records of an authority file is an activity as important as committing for many organizations, which have to rely on experts equipped with so-called authority control tools, capable of automatically supporting complex disambiguation workflows through user-friendly interfaces. This paper presents PACE, an open source authority control tool which offers user interfaces for (*i*) customizing the structure (ontology) of authority files, (*ii*) tune-up probabilistic disambiguation of authority files through a set of similarity functions for detecting record candidates for duplication and overload (*iii*) curate such authority files by applying record merges and splitting actions, and (*iv*) expose authority files to third-party consumers in several ways. PACE's back-end is based on Cassandra's "NOSQL" technology to offer (*i*) read-write performances that scale up linearly with the number of records and (*ii*) parallel and efficient (MapReduce-based) record sorting and matching algorithms.

## 1 Introduction

*Authority control* is the combination of software resources and human actions required to maintain *authority files*, which are lists of *authoritative records* uniquely and unambiguously describing corresponding sets of entities in an information system. In the library world, traditional examples of such lists are names of authors and titles of items. Lately the problem has been extended to other sets of entities, such as research organizations, journals, conferences and, more generally, to so-called *vocabularies*, which are authoritative lists of terms used to classify items into categories (e.g., subjects, languages, countries).

Authority control is essential to provide information systems with certified digital representations for entities and thus leverage exact interlinking, search and browse functionalities. It is a data curation process, whose goal is to addresses two important aspects of information representation: *entity disambiguation* and *entity association*. Disambiguation is the action of making sure that distinct entities, with different identities, are represented by distinct records, with different identifiers. Association is the action of interlinking different records to indicate they are "variations" of the same entities, e.g., two author names are pseudonyms of the same person. The function of authority files is more organizational than informational, that is they contain the minimal information required to disambiguate and associate entities and do not generally include information useful for other purposes, such as end-user reading.

E. García-Barriocanal et al. (Eds.): MTSR 2011, CCIS 240, pp. 80–92, 2011.

Managing authority files is typically a semi-automatic process which requires significant human effort and technology cost, to the point that the trade-off with the benefits is always a valid debate. It is a continuous process, whose quality is affected by the expertise of the data curators, e.g., librarians for catalogues, as well as by the overall evolution of knowledge in the field of application. Its picks of complexity are reached, for example, feeding an authority file with bulk record ingestions, integrating multiple and possibly overlapping authority files, or initiating an authority file from a non-disambiguated set of records. To tackle the dimension of this problem, advanced data mining tools have to be realized, which exploit similarity matching and data processing algorithms and offer administrative interfaces matching the specificities of the authority file at hand.

Motivated by worldwide needs of information sharing through standards and infrastructures, several attempts to construct and maintain in synergy authority files have been made [15]. Among the most known initiatives is VIAF (Virtual International Authority File) [13], LEAF [17], the Digital Author Identifier (DAI) and the projects OCLC-CORC (Connexion), NACO and SACO (Cornell University Library) investigated solutions on how to tackle authority control on global scale.[1] All such initiatives are supported by tools with which authority files about people or other kinds of entities are maintained and offered to access-authorized communities in order for them to uniformly populate their information systems. Unfortunately, the technology underlying such tools is mostly proprietary, bound to given application scenarios, targeting only a portion of the problem, or based on licensed products. It is therefore difficult for organizations willing to undertake similar initiatives to get hold of open source and reusable authority control software. In general, these end up implementing from scratch such software as combinations of existing software packages and applications.

In this paper we present the architecture and the current implementation of PACE (Programmable Authority Control Engine), an open source tool for authority control. PACE offers administrative user interfaces for customizing the structure of authority files, tuning-up the associated disambiguation algorithms, managing their life-cycle, and exposing them to third-party consuming systems through various APIs and formats. PACE's back-end is based on Cassandra's distributed technology [11], capable of offering read-write performance rates that scale up linearly with the number of records and enable parallel MapReduce-based implementations of traditional $O(n^2)$ deduplication algorithms. The purpose of PACE is to support a framework where disambiguation algorithms can be easily plugged-in and authority files can be easily personalized, populated and curated with no development effort and optimal performance. PACE is today used to curate the authority files of persons and bibliographic records of the EC projects European Film Gateway and OpenAIRE, respectively.[2]

*Outline of the paper.* The paper is organized as follows. Section 2 details the main features behind authority control tools, with references to existing solutions, pointing out the lack of reusable and open source tools for authority control. Section 3 presents

[1] `http://www.narcis.nl`, `http://www.oclc.org/connexion`, `http://www.loc.gov/catdir/pcc/naco/naco.html`

[2] `http://www.europeanfilmgateway.eu`, `http://www.openaire.eu`

functional requirements, data model and architecture of PACE, while Section 4 describes the features of its current first release. Finally, section 5 concludes the paper.

## 2 Authority Control Tools

Nowadays, the strong need of sharing information across independent systems, makes the construction and maintenance of authority files more and more important [10]. Not only, such files are not necessarily focused on names or titles as it was in traditional digital library cataloguing settings, but can involve any set of entities of an information system's data model (e.g., journal titles, research organizations, geographic profiles, subjects [6]), as long as such set requires disambiguation to improve search precision and valid references. More generally, an authority file can be described as a set of authority records conforming to a uniform flat structure $(l_1 : T_1, \ldots, l_n : T_n)$, where $l_i$ is a field label and $T_i = (M_i, U_i)$ expresses the "type" of the field in terms of its multiplicity $M_i$ (i.e., $[0 \ldots 1]$, $[1]$, $[0 \ldots m]$, $[1 \ldots m]$) and its usage $U_i$, namely an *identifier*, *heading*, *context*, or *explanation* field:

- *Identifier field*: each record must have one such field, containing a code that uniquely identifies the record within the authority file;
- *Heading fields*: fields describing properties of the entity represented by the record (e.g., for person entities: name and date of birth; for publication entities: language and script);
- *Context fields*: fields capturing information about the context of the entity in a given application scenario (e.g., for person entities: list of own's publication titles; for publication entities: citation references to other publication entities); one record may feature context fields relative to different contexts;
- *Explanation fields*: in some cases, for the sake of cataloguing, it can be necessary to introduce fields whose purpose is to hold information, automatically or manually generated, aimed at justifying the current content of the heading fields. For example, why the name of an author is kept in the form "D. Alighieri", rather than "Dante Alighieri"; this may be due to the original source of the name, an article or book citation.

In some cases, authority control also keeps track of so-called *syndedic associations* between disambiguated records in order to maintain links between *variations* of authority records. A "see" variation links an authority record with another record that is not authoritative, but whose content is a valid alternative to the authoritative record (e.g., records relative to pseudonyms of a given author). A "see also" variation links an authority record with records that have been deprecated in favor of the former because of deduplication (i.e., invalid alternatives). Depending on the application domain's context, other types of relationships may be introduced.

Authority control is an activity operated by a group of experts, whose task is to ensure disambiguation of the set of records by avoiding and resolving *record duplication*, when two records redundantly represent the same entity, and *record overload*, when two (or more) entities are erroneously represented with one record. Duplication and overload issues are addressed by *merging* (also referred as *deduping*) and *splitting* actions, whose consequences are to collapse two records into one or create two records out of one, respectively.

Disambiguation is a complex task which requires adequate machinery. Consider for example the problem of a research institute which stored millions of records relative to authors and co-authors of the papers published by its researchers. Turning such a collection into an authority file, thus into a disambiguated list of records, is not an easy challenge for humans. A similar challenge occurs when merging independent and large authority files into one (Fayad and Uthurusamy [8] stated that possibly above 90% of the work in these scenarios is associated with cleaning up the duplicates). Due to the complexity dimension of the problem, management and disambiguation of an authority file is typically aided by systems referred as *authority control tools* (figure 1). These implement *candidate identification* algorithms capable of suggesting candidate record merges and splittings in an authority file [19] and, based on these, provide "experts" with a set of functionalities for *drafting*, i.e., disambiguating and committing, a newer version $N + 1$ of the authority file. Finally, they offer third-party information systems with APIs for *consuming* a version $M(\leq N)$ of an authority file.

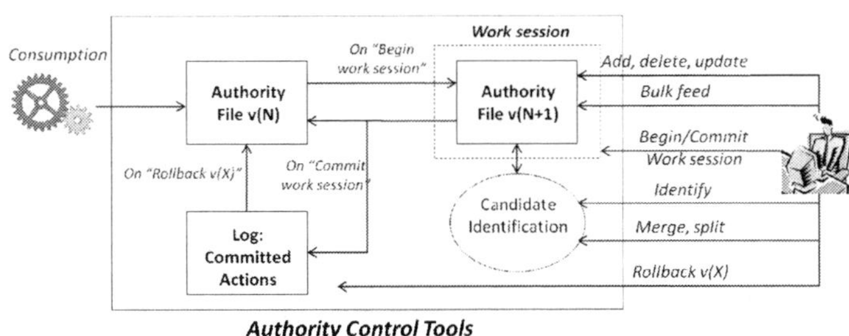


**Fig. 1.** Authority Control Tools functionalities

## 2.1 Candidate Identification

Candidate records are those who might allegedly entail merge or splitting actions. Typically two records are regarded as *candidates for merging* if some "similarity measure" $0 \leq F_S \leq 1$ calculated over the fields of the two goes beyond a given threshold; the action is also called *record linkage* [19], *object identification* [14], and *entity resolution* [1]. One record is instead *candidate for splitting* when an "overload measure" $0 \leq F_O \leq 1$ goes beyond a given threshold.

Designers and implementers of candidate identification functionality must face two main challenges, algorithmic and implementative. The algorithmic challenge is to find the $F_S$ and $F_O$ capable of returning meaningful measures of similarity and overload based solely on the record informative fields. The implementative challenge is to realize tools that scale with the number of records while still performing record matching and analysis in an efficient way.

*Similarity measure.* The similarity measure of two records is a function $F_S$ that calculates the "distance" between the two, i.e., the likelihood for these records of being mistakenly used to represent two distinct entities while they are instead representing the same one. Approaches split in "deterministic" ones, i.e., yes/no rule-based detection of duplicates, or "probabilistic" ones, where a "similarity measure" is assigned to each pair of records, quantifying their likelihood of duplication [7]. Typically, the latter is obtained by combining similarity measures specific to the field domains of the records, be them dates, person names, subjects, etc. In other variants, $F_S$ also depends on information inferred from the information system in which the authority file is immersed, based on some relationship with the records (i.e., context). Examples of $F_S$ are string-based similarity functions focusing on variants of the "typographical error" problem (e.g., Jaro-Winkler [18], Cohen's variants of the TFIDF metrics [5], Edit distance, Biagram distance) or the "person name and address matching" problem [3] [4]. Other methods focus on special cases of the problem, such as identifying short texts duplicates (SimFinder [9]); when richer textual information is available, near-duplicate functions, typically used for documents and web pages duplicates (Simhash [2], [12]) can be used.

In principle, the complexity upper bound for identifying duplication is $O(n^2)$, where $n$ is the size of the file, since each record must at least be $F_S$-matched against all other records. Due to the amount of records involved, possibly scaling up to hundreds of millions, the time for such calculation might be very long and thus hinder a smooth authority file management life-cycle, made of frequent cycles of candidate identification. The technical challenge is therefore dual: ($i$) finding heuristics capable of reducing complexity while identifying candidates with minimal false positive/negatives, and ($ii$) deliver storage implementations capable of scaling up with the number of records. Heuristics are intended to reduce computation associated with comparing pairs from $O(n^2)$ to $O(n)$ or $O(nlogn)$. A well-known example of such heuristics is *blocking*, that is clustering records with similar field values into one or more "passes", in order to reduce the similarity checking window. Each pass corresponds to a different sort ordering of the files and record pairs are only considered in a sliding window of fixed size. Realizing such heuristics requires choosing the adequate storage technology, in order to offer input and output rates in line with the expected usage of the authority control activities at hand. Dedicated indexing structures, distributed databases, sorted storage techniques are examples of solutions in this context.

*Overload measure.* The overload measure of a record is a function $F_O$ that implements domain specific analysis to deliver a given notion of "record overload", i.e., the likelihood for that record of representing two or more entities rather than only one. Its input parameters are the record, thus its context fields if any, plus possibly a selected set of other records, and the value it returns expresses the degree of overload of the record. For example, an authority file of persons which includes a context field for "research subjects of interest", could be equipped with an $F_O$ which exploits such values to identify splitting candidates. Persons whose research subjects diverge "too much" w.r.t. to a distance function, might be amenable to splitting. Performance issues here depend on the complexity of the function. It is therefore not possible to generalize the problem as for the case of merging.

### 2.2 Drafting

In most cases, drafting a new version of an authority file is an operation that is performed in a temporary *work session*. In such protected settings, experts can do/undo a number of actions, which include addition, deletion, update and bulk-feed of records, candidate identification over the record set and, as a consequence of this, merging and splitting. When satisfied, experts may decide to *commit* the new version, that is make the changes authoritative. Some tools keep track of the actions performed since the last commit, in order to support *roll-back* of committed authority files, incremental consumption (as described above), and automatic disambiguation by reapplying merging and splitting actions if the relative candidates are found again in later sessions.

### 2.3 Consumption

Depending on the tool's functionality, third-party systems may consume an authority file in two ways: by fetching and materializing its content locally or by dynamically accessing it when required. In the former case an authority file can be requested in-full, i.e., fetching the full authority file as available at a given commit time, or incrementally, i.e., by fetching the actions executed after a given time.

## 3 PACE (Programmable Authority Control Engine)

The literature and the market offer a number of authority control tools. Their strength often lays in the techniques devised to precisely and efficiently identify groups of records candidate for merging or splitting based on a few informative fields and out of millions of records. Unfortunately, such tools are often proprietary products and, when open source, they they tend to be either application specific or to tackle sub-problems of authority control. This is the case for record linkage tools, which are open source, can be customized to one's authority file, but focus on the specific sub-problem of duplicate identification (and sometime deduplication) across several data sources; e.g., FRIL, Link King, and LinkageWiz.[3]

In this section we present PACE (Programmable Authority Control Engine), which is, to our knowledge, the first open source authority control tool designed to offer out-of-the-box and fully-fledged authority file management functionality. The tool has been designed to offer the functionalities described in Section 2 in a general-purpose and personalizable fashion. As such, it leaves administrators the ability to customize (the structure of) their authority files, configure their preferred candidate identification settings and exploit these to curate their life-cycle.

### 3.1 Drafting

PACE introduces two levels of drafting, based on the high-level data model depicted in Figure 2. On the higher-level, experts (i.e., *owners*) can create one or more authority files and grant to a set of colleagues (i.e., *administrators*) the right to curate such

[3] `http://fril.sourceforge.net`, `http://www.the-link-king.com`, `http://www.linkagewiz.com`

files. In particular, given an authority file, administrators can define one or more *candidate identification configurations*, i.e., similarity or overload functions, to be used to identify possible merges and splittings. On the lower-level, administrators of one authority file can apply the actions illustrated in Figure 1. They can open multiple *work sessions*, each associated to a given candidate identification configuration. In a work session they draft a new version of the authority file by adding, deleting, updating and bulk-feeding records and by accepting merge and splitting actions suggested by PACE. When a session is *committed*: (*i*) all parallel sessions are canceled, (*ii*) all actions executed in the session are applied to the last committed authority file, and (*iii*) the actions are saved into the *authority file log*. Such logs keep track of all actions executed on committed sessions and by whom, thus enabling incremental consumption and automatic re-disambiguation as described in Section 2. Optionally, committing a session may require the approval of a minimal number of the administrators in charge of the authority file ("voting mode").

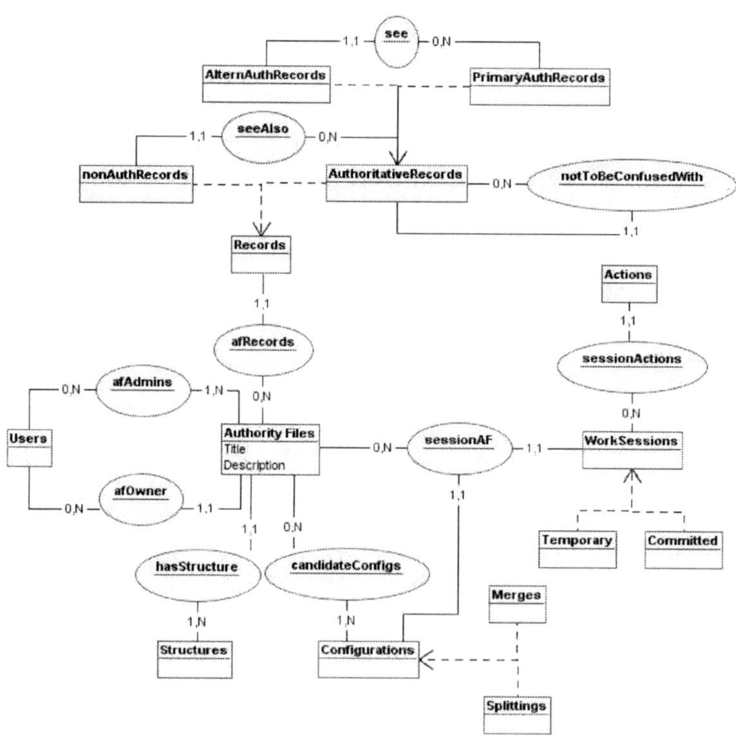


**Fig. 2.** High-level data model

## 3.2 Candidate Identification Configurations

Users can create and thus become owners of authority files, by providing a name, a description and providing the relative flat record structure $(l_1 : T_1, \ldots, l_n : T_n)$, as defined in Section 2. Owners can give the permission to maintain files they own to a number of administrator users, which can manage a set of candidate identification configurations for such files and operate maintenance actions over them. In particular, configurations are of two kinds: similarity for record merging and overload for record splitting.

*Merging.* PACE implements a similarity identification algorithm based on a clustering (also referred as "blocking") and probabilistic approach. The approach exploits a sorted-record $K$-window heuristic, to reduce the overall record matching complexity, and a notion of record similarity based on a weighted mean. In practice, records are firstly sorted by applying a *sorting function* to special record *clustering values*, obtained by applying a *clustering function* to each record. The aim of such values is to represent records in a way that potentially similar records may have similar values, i.e., close w.r.t. the sorting function. For example, in the case of person records, a naive clustering function may be the one that returns the normalized value of the field "name" of the records. When sorted alphabetically, persons with similar names may fall close to each other. The method is rather naive, for example names with typos may cause the records not fall close to each other. A more complex function, as described in section2, could implement near-duplicate fingerprints methodologies (e.g., SimHash, SimFinder) and returns a bit-map that is representative of all values within the records.

Once the records are sorted, the function $F_S$ is applied to a "window" of $K << n$ records in order to drop the complexity to $O(n * K) = O(n)$.

In PACE, given two records $r_1$ and $r_2$ of an authority file with structure $(l_1 : T_1, \ldots, l_n : T_n)$, their similarity function $F_S(r_1, r_2)$ is defined as the weighted mean:

$$F_S(r_1, r_2) = \sum_{i=1}^{n} (f_i(r_1.l_i, r_2.l_i) * w_i) / \sum_{i=1}^{n} w_i$$

where $0 \leq f_i \leq 1$ and $w_i$ are respectively the similarity functions and the weights w.r.t. the overall similarity measure associated to the field $l_i$. $F_S(r_1, r_2)$ is the result of combining a set of similarity measures, calculated over one or more record fields and possibly with different weights. For example person names are more important than birth dates when trying to identify candidates in a person authority file, but similarity measures for both fields may be considered to establish the overall similarity of two person records.

In summary, similarity configurations consist of: the constant $K$, the clustering function, the sorting function, the record fields $l_i$'s to be used by the clustering function, the $f_i$ to be applied to each $l_i$, together with the weight $w_i$. PACE offers administrators a range of pairs clustering-sorting functions and, once a combination is selected, they need to specify the record fields to used as input and the constant $K$. Similarly, administrators select the $f_i$'s out of a list depending on the value domains of the $l_i$'s (which may range from dates, to given vocabulary terms, ontologies, person names, or generic

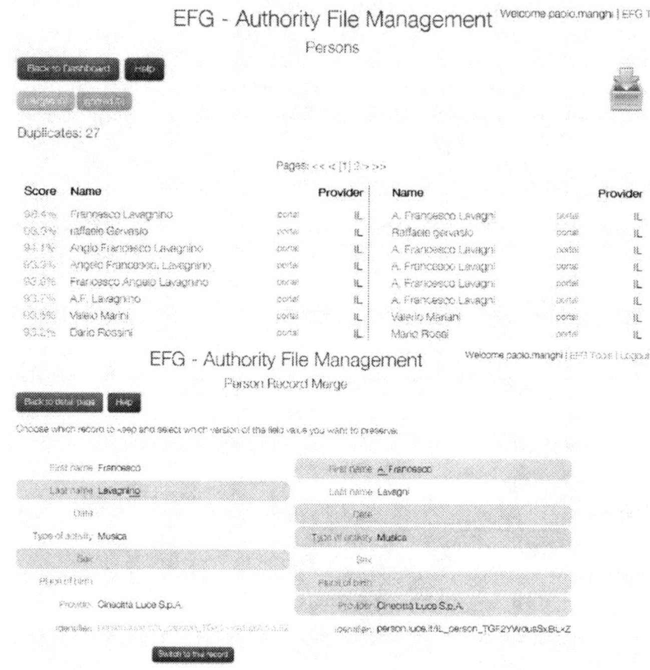


**Fig. 3.** PACE screen shots: candidate records list and merge preparation

strings) and specify the relative weight $w_i$, in order to configure the $F_S$ which captures at best the notion of similarity for the authority file at hand. PACE is designed to allow developers to introduce new clustering-sorting function pairs and new similarity functions $f_i$'s, to be offered for selection to administrators.

*Splitting.* PACE implements an overload identification algorithm based on the assumption that any information that could be used to identify record overload should be encoded as a value of a context field in the authority file structure. Such an assumption allows to generalize, at least to some extent, over the specificities of different application domains. Administrators select the $F_O$ they are willing to use out of a list of possibilities and specify the contextual fields that are to be used for the calculation. PACE is designed to allow developers to introduce new overload functions, to be subsequently offered for selection to administrators.

*Authority records and syndedic associations.* In PACE records can be *authoritative*, in the sense they are currently considered non-candidate records by authority control, or *non-authoritative*, in the sense they have been deprecated by merge actions. In turn, authoritative records may be of two kinds: *primary records*, to be considered as the effective representative of an entity, and *alternative records*, which are valid alternatives of a primary record, e.g., the author pseudonym and the author. Merge and splitting actions cause the authority file to change, that is records to move from one category to another and to be connected by relationships capturing and keeping track of such changes:

- Merge actions cause one record to prevale, hence its identifier to be used as authoritative. When merging, the administrator has to select which field-value pairs of the deprecated record must integrated in the authoritative record and which fields-pair of the authoritative records are to be replaced or removed (see figure 3). The action creates a "seeAlso" relationship between the two records.
- Splitting actions generate a new authoritative record, hence a new identifier to be generated. When splitting, the administrator has to specify which fields have to join which record and which have to be replicated in both. The action creates a "notToBeConfusedWith" relationship between the old and the new record.
- "see" relationships are created by administrators to specify equivalences between two records. The action also determines which of the two records has to become primary w.r.t. the relationship: if the primary record was alternative, the relationship is created between its primary record and the new alternative record; if the alternative record was primary, then all its "seeAlso" relationships are inherited by its new primary record. As a result, "seeAlso" relationships define clusters of alternative records equivalent to one primary record.

### 3.3 Consumption

PACE exposes APIs for (authorized) third-party consumers to access authority files of interest. Authority files (referred by a unique identifier) can be accessed in two modalities: *full access*, i.e., the list of records at a given commit time $t$, and *incremental access*, i.e., the list of actions committed after commit time $t$, obtained by accessing the relative logs. The list of records can be exported in XML proprietary format or RDF (Linked Data). The list of changes are exported as an XML file or as a text file.

## 4 PACE v1.0

PACE v1.0 implements the following functionality: creation of authority files; creation of merging configurations (with a selection of standard matching algorithms); opening of work sessions together with a merging configuration of reference; merge actions in work sessions; bulk-feeds of records; work session commits; logs of committed actions and incremental consumption.

Efficient data management and storage technologies are important and often underestimated issues in authority control. PACE v1.0 experiments on a Cassandra [11] "NOSQL" storage back-end (e.g., GFS, HBase). The framework manages data replicas automatically and offers read-write performance that scale linearly on the number of records, allowing fast storage, caching and access of records. Cassandra is in use at Digg, Facebook, Twitter, and many other companies that have large, active data sets.

[4] In combination with the MapReduce's package Hadoop [5] it supports the parallelization of algorithms over millions of records. For example, record sorting, required for clustering/blocking a data set and for ordering the list of candidates, can be executed in $O(nlogn)/H$ where $H$ is the number of machines in the Cassandra cluster.

In the future, besides completing the implementation of PACE with the missing features, we shall experiment with Hadoop's solution in an attempt to avoid clustering function heuristics and achieve true $n$ to $n$ record matching [16]. Of interest are also experimentations with the construction of authority files out of a independent sets of records who do not carry unique identifiers or cannot guarantee their persistence. The absence of persistent identifiers complicates the procedures of logging administrator's merging and splitting actions in order not to request them again in the future when the very same duplication or overload issues occur (e.g., when the same authority file is bulk-fed again in the authority file). In this case, identifiers are to be created from input records with hash functions and the challenge is to identify forms of contextual information (possibly relationships with other records or other entities in the original authority file) to be stored with the merge or splitting actions in order to automatically detect and repeat them in the future.

PACE is today used in the European Film Gateway EC project, whose aim is to aggregate content from European movie archives in Europe. The tool was adopted to reorganize the authority files of persons in the local archives and then to build a centralized authority file out of them. Due to the privacy of the data, access to the tool can be temporarily granted by contacting the authors of this paper. PACE is currently under deployment in the OpenAIRE EC project whose aim is to offer statistics over the ratio open access/licensed scientific publications funded by EC projects. The tool will be used to disambiguate the set of bibliographic records of papers partly collected from a set of European institutional repositories and partly manually "claimed", i.e., ingested, by authors.

## 5 Conclusions

This paper presented PACE, an open source and general-purpose tool for managing authority files. PACE offers a framework where administrators can create (the structure of) their authority files, configure their preferred candidate identification settings and exploit them to curate the authority file life-cycle. Its implementation, thanks to the adoption of Cassandra's "NOSQL" technology, features high store-write performance over millions of records. In combination with its framework, PACE is an ideal environment for merging independent authority files or perform "fast" disambiguation experimentations with different similarity matching and overload functions. Future issues include the completion of the implementation according to the specification in Section 2 and the experimentation with the Hadoop framework.

---

[4] A. Anand, `http://developer.yahoo.com/blogs/hadoop/posts/2008/07/apache_hadoop_wins_terabyte_sort_benchmark`

[5] `http://hadoop.apache.org/`

**Acknowledgements.** This work was partially funded by the EC projects European Film Gateway (Best Practice Networks project, grant agreement: ECP-517006-EFG, call: FP7 EU eContentplus 2007) and OpenAIRE (grant agreement: 246686, call: FP7-INFRASTRUCTURES-2009-1). Our gratitude also goes to Dejan Kolundzija for his initial studies and developments on this topic.

## References

1. Benjelloun, O., Garcia-Molina, H., Su, Q., Widom, J.: Swoosh: A generic approach to entity resolution. Stanford University technical report (March 2005)
2. Charikar, M.: Similarity estimation techniques from rounding algorithms. In: 34th Annual Symposium on Theory and Computing, Montreal, Quebec, Canada (May 2002)
3. Christen, T., Churches, P., Zhu, J.: Probabilistic name and address cleaning and standardization. In: The Australian Data Mining Workshop (November 2002)
4. Churches, T., Christen, P., Lu, J., Zhu, J.X.: Preparation of name and address data for record linkage using hidden markov models. BioMed Central Medical Informatics and Decision Making 2(9) (2002)
5. Cohen, W.W., Ravikumar, P., Fienberg, S.E.: A comparison of string metrics for matching names and addresses. In: International Joint Conference on Artificial Intelligence, Proceedings of the Workshop on Information Integration on the Web (August 2003)
6. Dalrymple, P.W., Young, J.A.: From authority control to informed retrieval: Framing the expanded domain of subject access. College & Research Libraries 52, 139–149 (1991)
7. Elmagarmid, A., Ipeirotis, P., Verykios, V.: Duplicate record detection: A survey. IEEE Transactions on Knowledge and Data Engineering 19(1), 1–16 (2007)
8. Fayad, U., Uthurusamy, R.: Evolving data mining into solutions for insights. Communications of the Association of Computing Machinery 45(8), 28–31 (2002)
9. Gong, C., Huang, Y., Cheng, X., Bai, S.: Detecting near-duplicates in large-scale short text databases. In: Washio, T., Suzuki, E., Ting, K.M., Inokuchi, A. (eds.) PAKDD 2008. LNCS (LNAI), vol. 5012, pp. 877–883. Springer, Heidelberg (2008)
10. Gorman, M.: Authority control in the context of bibliographic control in the electronic environment. In: International Conference Authority Control: Definition and International Experiences, Florence, February 10-12 (2003)
11. Lakshman, A., Malik, P.: Cassandra: a decentralized structured storage system. SIGOPS Oper. Syst. Rev. 44, 35–40 (2010)
12. Manku, G., Jain, A., Sarma, A.D.: Detecting near-duplicates for web crawling. In: 16th International World Wide Conference, Banff, Alberta, Canada (May 2007)
13. Rick, B., Hengel-Dittrich, C., O'Neill, E.T., Tillett, B.: Viaf (virtual international authority file): Linking the deutsche nationalbibliothek and library of congress name authority files. International Cataloging and Bibliographic Control 36(1), 12–19 (2007)
14. Tejada, S., Knoblock, C., Minton, S.: Learning object identification rules for information extraction. Information Systems 26(8), 607–633 (2001)
15. Tillett, B.T.: Authority control: State of the art and new perspectives. In: Authority Control International Conference, Florence, Italy (2003)
16. Wang, C., Wang, J., Lin, X., Wang, W., Wang, H., Li, H., Tian, W., Xu, J., Li, R.: Mapdupreducer: detecting near duplicates over massive datasets. In: Proceedings of the 2010 International Conference on Management of Data, SIGMOD 2010, pp. 1119–1122. ACM, New York (2010)

17. Weber, J.: Leaf. linking and exploring authority files. In: International Conference Authority Control: Definition and International Experiences, Florence, February 10-12 (2003)
18. Winkler, W.E.: String comparator metrics and enhanced decision rules in the fellegi-sunter model of record linkage. In: Proceedings of the Section on Survey Research Methods, American Statistical Association, pp. 354–359 (1990)
19. Winkler, W.E.: Overview of record linkage and current research directions. Technical report, Research Report Series, RRS (2006)

# Ontologies for Emergency Response: Effect-Based Assessment as the Main Ontological Commitment

Leopoldo Santos[1], Miguel-Angel Sicilia[2], and Sergio Padrino[2]

[1] Emergency Military Unit, CG J6
28850 Torrejón de Ardoz Air Base, Spain
lsantos@et.mde.es
[2] University of Alcalá, Computer Science Department,
Polytechnic Building, Ctra. Barcelona km. 33.6
28871 Alcalá de Henares (Madrid), Spain
{msicilia,sergio.padrino}@uah.es

**Abstract.** Emergency management involves the coordination of agencies in a complex evolving situation, as they usually affect several domains and infrastructures. This requires complex models representing detailed knowledge about the types of adverse events, their potential impact and the means and resources that are best suited for response. Existing models for emergency management do not address a detailed typology of incidents that cover their relationships and how they should be managed. In ontologies explicitly engineered to such kind of effect-based assessment, the possible consequences of an adverse event become the main representational commitment. This paper describes the main representational issues associated with ontologies for emergency response that are engineered to represent knowledge for effect-based assessment, and provides concrete examples. The integration of such ontologies in existing emergency response systems is also sketched.

**Keywords:** Emergency events, critical infrastructures, ontologies, OWL, reasoning, situation awareness.

## 1 Introduction

Emergency management (EM) is the continuous process by which individuals, groups, and communities manage hazards in an effort to avoid or ameliorate the impact of disasters resulting from them. Emergency response is triggered by the occurrence of geospatially localized adverse events or incidents of a very diverse kind and that typically chain or combine with others as the situation evolves (Fan and Zlatanova, 2010). The kind of these events determines its potential impact, being thus critical to achieve an awareness of the situation as complete as possible in order to assess the possible courses of action of the unfolding emergency situation. Othman and Beydoun (2010) found references in major models for the concept of adverse event or incident, identifying them as the triggers of coordinated actions. Also, the nature of adverse events is the key determinant of the kind of resources and actions required for an organized response. There exist several adverse event taxonomies

E. García-Barriocanal et al. (Eds.): MTSR 2011, CCIS 240, pp. 93–104, 2011.

proposed by some standards and specifications, and the need for taxonomies of disasters has been recognized long ago (Kreps, 1989). While they are a useful resource for achieving a degree of interoperability, they differ significantly and the ontological criteria they use in the breakdown into event categories is not made explicit. This is hampering their extension and reuse. Also, these taxonomies are not associated to the kinds of actions and resources that are known to be effective in handling them, which represents a critical kind of expert knowledge in unfolding situations. Formal ontologies (Gruber, 1993) can be used to provide richer models encoding also these relationships. Indeed, several proposed ontologies for EM have included incidents in their structures. However, the relation of incidents to particular kinds of other elements has not been subject to formalization in most of them (Fan and Zlatanova, 2010; Galton and Worboys, 2011). Sicilia and Santos (2009) described the *Basic Formal Infrastructure Incident Assessment Ontology* (BFiaO), which provides a basic model covering the common high-level aspects of infrastructures (electrical, gas, etc.), causes of incidents (be them natural or caused by humans) and the incidents themselves as situations that in some cases evolve or are qualified as emergencies. Rules expressed in SWRL could be used to combine the structure of events representing the evolving situation with the knowledge of important infrastructure *nodes* and *connections* in BFiaO to infer potential situations of risk associated to infrastructures. However, the BFiaO is a metamodel for handling infrastructure-related situations, and it is not providing an actual catalogue for adverse events and their associated response means. In this paper, extensions to the BFiaO are described that account for these aspects. Our discussion first goes to the key ontological commitments required for developing extensible typologies of adverse events that are driven by the effects rather than by other aspects as causes or facilities affected. Then, a schema for connecting adverse event types to the types of actions and resources required for mitigation are presented. The proposed model allows for the codification of EM knowledge in a reusable way that could later be used for the planning of actions accounting for available resources and other contextual circumstances.

The rest of this paper is structured as follows. Section 2 provides background information on existing models and ontologies for emergency management. Then, a way of structuring a typology of adverse events is presented in Section 3 that follows criteria that are significant for EM. Section 4 describes a concrete case for the relation of types of adverse events, effects and the resources required for response. In Section 5, the practical aspects of the integration of the models with an open framework for emergency management are sketched. Finally, conclusions and outlook are provided in Section 6.

## 2 Background

Events are any kind of spatiotemporal wholes that have differentiated beginnings and endings corresponding to real discontinuities. Incidents are a kind of event that are considered "adverse". The CEN Workshop Agreement (CWA) code CWA 15931-2 "Disaster and emergency management - Shared situation awareness - Part 2: Codes for the message structure" provide an example of a taxonomy of adverse events. The categories specified in the CWA as `/EVENT/TYPE/CATEGORY` represent types, e.g. `/FIR/CLA` category for "class A" fire. When examining these adverse event taxonomies, it becomes apparent that they have significant differences in their organization and level of detail. While they are useful in capturing important event types, the ontological criteria applied for differentiating types of events are not made explicit, and they are not associated to the kind of resources or actions that are known to be effective for handling them.

One of the first organizations that have tried to describe events related to emergencies and disasters is OASIS (Organization for the Advancement of Structured Information Standards[1]). The result of this initiative is CAP[2] (Common Alert Protocol). The CAP 1.1 includes a `category` element in its message structure a codification of kinds of eleven events, e.g. "Geophysical", "Meteorological" or "Fire". Then, element `responseType` can be used to encode the type of action recommended by the sender party. Responses are expressed via a verb, e.g. "evacuate" or "prepare" that refers to the element instruction in which a natural language description for the action plan proposed. However, natural language descriptions are free text targeted to human readers, and the category structure in CAP is of a very high level, oversimplifying the complex nature of adverse events.

If we consider disaster types, there are classifications as the one used in the disaster database of the Centre for Research on the Epidemiology of Disasters (Wirtz, Below and Guha-Sapir, 2009) that divide the categories in natural and technological, and further subdivide the natural category into subtypes (Biological, Geophysical, Meteorological, Hydrological, Climatological and Extra-Terrestrial). These are further subdivided, e.g. *Forest fire* is considered a sub-subtype under the classification path *Natural disaster / Climatological / Wild fire*. These classifications are built for the description of data sets, and are thus based on the specificities of groups of disasters. In Ianella (2006) we can see a case study of a model for Severe Weather Advisories (SWA). In this paper Ianella compares Australian model to USA model, including in the Australian model, a classification for Threats.

Several proposed ontologies for EM include some form of adverse event concept or consider events in their information model, e.g. (Galton and Worboys, 2011). Fan and Zlatanova (2010) also include them in their general terminological study. Yu, Wang and Rong (2008) divide elements into four groups: Event, Resource, Subject and Task. They consider seventeen kinds of concrete events grouped hierarchically in the two broad categories `Disaster` and `Disease`. Li et al (2008) developed another approach based on the four different phases of an emergency: Preparation, Response, Rescue and what they called Aftermath Handling (usually called Recovery by others authors). They have defined this ontology as "Emergency Response Ontology", but they have centred in tasks or actions independently of events that have produced the emergency. Wang et al (2005) stratified their system ($CE^2M$)[3] into three levels: Event, Process and Action and also take into account that the event is divided into pre-state and final-state. It emphasizes the spatio-temporal relationship included in the ontology for events and actions. Sotoodeh and Kruchten (2008) describe in two

[1] `http://www.oasis-open.org`

[2] Common Alerting Protocol, v. 1.1 OASIS Standard CAP-V1.1, October 2005.

[3] $CE^2M$: Crime Emergency Event Model.

papers a conceptual model of disasters affecting infrastructures but do not further describe the events that can damage these infrastructures, which is important to identify actions and resources to minimize the effects of the events. In spite of the existence of several proposed ontologies covering events, these ontologies do not provide comprehensive taxonomies or catalogues of adverse events as those that can be found in taxonomies as the CWA. Further, they are not covering the specificities of connecting these taxonomies to actions and resources for response.

There are important works related to upper ontologies for situation awareness, like one by Baumgartner and Retschitzegger (2006) where we can find a survey of several upper ontologies like SAWA, Situation Ontology, SOUPA and CONON. Little and Rogova (2005) specifies an upper ontology framework about situation awareness related to disaster management and model a taxonomy of relations-types of interest for higher-level fusion that can be useful to design domain ontologies. These upper ontologies provide interesting formal representations for relating events to events and to actual means, but they are not specifically targeted to emergency response.

## 3 Typology of Adverse Events

In ontologies, `Event`[4]s are ocurrences having some temporal and spatial extent, and that are first level entities, as they are known to have happended. However, assessing what may happen in EM requires also statements about the future, which are by nature not necessary. The `PossibleEvent` concept introduces a category of future events that serve as the basis for those short-term future statements. They should be interpreted in terms of epistemic possibility, i.e. "given the available information" such element is possible.

Ceusters et al. (2009) studied the concept of "adverse event" as applied to health in medical care. Even though the domain dealt with here is very different, their distinction of different levels of reality of the knowledge to be represented still applies. At a first level, we have actual events like a *flood,* which are first-order reality and are the triggers of the response. However, when an operator identifies a risk of the flood affecting some roadway, this is a cognitive representation over first order entities. Finally, a third level is concerning universals, e.g. when we are encoding the resources that are known to be effective for particular events (or the expected interaction between events). An example is the knowledge that a particular type of aircraft is effective in mitigating fire events. We are concerned here basically with the first and third levels, the latter representing a kind of "state of the art" of general knowledge on the domain of EM. In consequence, the emphasis on the effects of the adverse events lead to a criteria for the structuring of the typology of events that can be summarized in the following: "distinct adverse event types must differ in some way of effectively responding to them or in some characteristics that are important for preparedness". This naturally brings up the need to represent further knowledge related to the event type, which is not present in current taxonomies. For example, CWA 15931-2:2009 (E) `/WIN/CYCL` event type is defined as "The atmospheric pressure distribution in which there is a low central pressure relative to

[4] In this document, `Courier` font is used to differentiate ontology elements or elements in classifications.

the surroundings. Cyclonic circulation is anticlockwise round the centre in the northern hemisphere and clockwise in the southern hemisphere [...]." while `/WIN/TRST` (tropical storm) is defined as "A tropical cyclone having winds ranging from approximately 48 to 121 kilometres or 30 to 75 miles per hour.". The problem with the differentiation of both types relies in that they are not formally related via subsumption, and the ontological difference from the definition is regarding some particular property (wind speed). The differences regarding response and/or preparedness are not made explicit if any. Also, some other cyclone types (e.g. polar cyclones or mesocyclones) are not represented, so the criteria for including only one subtype become even more difficult to grasp. Further, there are different ways of representing the same, e.g. `/FLD/TID` (included in `EVENT/ETYPE/CATEGORY`) is referring to a tsunami but also we can define a tsunami with `/DIS/TSNAMI` that it is described in `EVENT/ETYPE/ENV`.

Following a similar reasoning, there are also other categorizations that are not useful for response management using formal semantics. For example, upper categorizations as "meteorological" incidents are generic labels indicating the originating cause, but have little value in deciding on responses as they cover different phenomena, i.e. response to a cyclone has little in common to response to snow shower or haze. These categorizations are still useful for browsing taxonomies but do not provide relevant computational semantics. There are also cases in which several terms refer to the same concept as may be the case of "tropical cyclone" and "hurricane". These are useful for searching but refer to the same concept and should be represented as a single ontology element with different terms that refer to the same.

In consequence, our proposed ontology distinguishes between *event types*, which have some unique response difference from other types, and *event categories*, that are labels with no differentiated response characteristics. It should be noted that emergency management is a discipline that evolves with time, so the differentiation of event types should also be considered open ended and determined by our current state of methods and techniques. For example, it may be that a new mitigation device for fire is engineered that is only useful for some particular kind of fire. In this case, the advance in the materials available determines a change in the ontology. Adverse event types following the rationale presented can be represented in some cases as concept expressions, if they are not required as a new type because they are not known to be better handled in a different way from a more general class of events. For example, the description logics expression `FireEvent • ∃affected.Forest` is subsumed by `/FIR/CLA` in CWA, representing "class A" fire events that require water for mitigation. Other classes are mitigated with other kinds of resources. Geospatial entities as `Forest` can be reused from existing ontologies, e.g. that concept can be found in NASA SWEET ontologies[5].

In some cases, adverse event subtypes are important for preparing response. This is the case for example for events that are specific to some geographical areas. For example, it is known that most cyclones form a band of activity known as the Intertropical Convergence Zone. In Mexico, storms track near and along the western Mexican coastline primarily between July and September. This information is essential to preparedness, so it is the basis for a new adverse event type, justified in itself by its occurrence characteristics.

---

[5] `http://sweet.jpl.nasa.gov/`

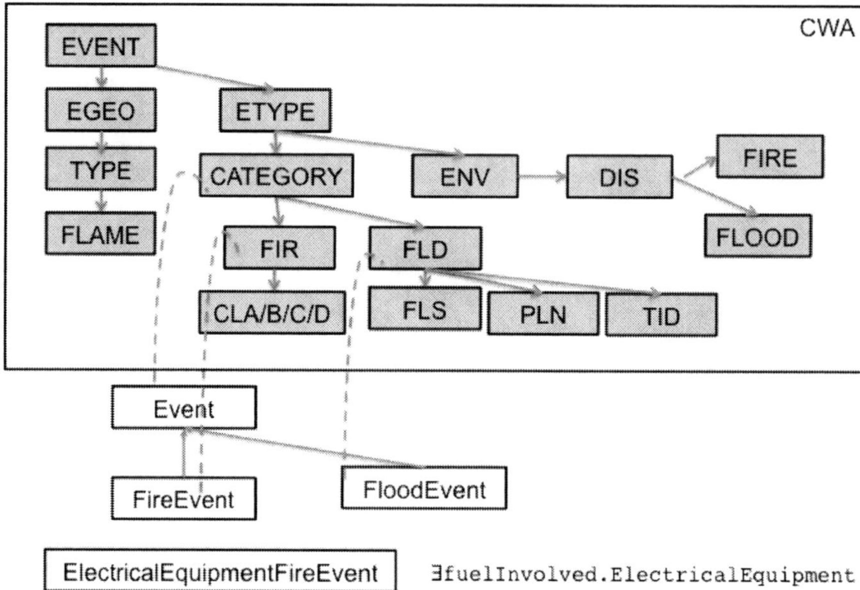


**Fig. 1.** Refactoring of the flood and fire categorization from CWA

In what follows, we describe a concrete case on the refinement of a part of the adverse event typology taking as a basis the CWA. We take the case of floods and fires.

The `EVENT/EGEO/TYPE` taxonomy is used to give a type to the geographical location of TSO messages. Subcategory `/FLAME` is defined as "area in combustion". However, the actual typology of fire events is in `ETYPE/CATEGORY` in which a subcategory `/FIR` appears with subcategories for `/CLA`, `/CLB`, `/CLC`, `/CLD`. These categories refer to existing classification systems that designate the fuel involved in the fire, and thus the most appropriate extinguishing agent. In Figure 1, the primitive typology of event types only covers `FireEvent` and `FloodEvent`. The rest are defined concepts that encode the differences. For example, when the fuel involved in the fire includes electrical equipment (class C in CWA) CO2 or dry powder can be used for extinguishing the fire. In the case of floods, CWA is using the cause as the differentiating criterion (extreme weather conditions, water rising over an area or impact of a wave of water as a tsunami). In this case, the differences are not evident so the introduction of subtypes is not justified at this point (but may be when introducing additional knowledge in the definitions). Response to the flood itself is in principle the same for water rising or a wave of water. Differences are in the responses for the previous, originating adverse event.

The `/ENV/DIS/FLOOD` category (and also the similar subcategory for fire) is not relevant for our purposes, it responds in TSO as a different information item representing the general context of the event.

In another direction and relevant to our discussion below, the CWA has also some mix of events and missions. For example, in `EVENT/ETYPE/CATEGORY` we can find missions instead of events. There are examples like "Rescue from a height" or "Rescue Ice", and some like for "Alarm" that are actions deriving from an adverse event rather than the event itself.

## 4 Representing Knowledge on Events, Resources and Responses

The second step in effect-based representation of adverse events is the dealing with the knowledge on recommended actions and resources that are most appropriate for each kind of adverse type. Here we start with an examination of the CESAR (Coordination of Emergencies and Tracking[6] of Actions and Resources) model, that focuses in the effects caused by the event and the connections with the missions that are recommended or known to be effective to minimize them. Figure 2 shows an overview of the CESAR model main entities. The `EVENT` concept can be assimilated to the `Event` concepts discussed in the previous section. In CESAR, the main modelling construct is the concept of `MISSION`, which captures different forms of organized response that are related to particular kinds of events. Some of these missions are generic as they can be associated to different kinds of events like `EvacuationMission`, `ZoneCleaningMission` or `ReconnaissanceMission`, while others like `FirebreakMission` or `DropWaterMission` are strongly coupled with the event. It should be noted that in a complex situation, several events will be the component of the current event, and in consequence, missions will also mix in the recommendations for response. Missions are also linked to particular kinds of `RESOURCE` instances. These detail the kind of materials and equipment that are required for deploying the missions.

From a representational perspective, it should be noted that both `Events` and `Missions` and `Resources` (the latter adapted directly from CESAR) are representing actual concrete events, i.e. missions and resources and their types are represented using the subsumption hierarchy. Typically, when the events are unfolding, an EM support system will give recommendations on mission types and resource types. These recommendations will eventually taken in consideration, and the actual missions and resources will be planned and deployed. This gap requires the introduction of a separate concept representation for mission types and resource types. With this separation, it is possible to use rules for the recommendation process. For example, the following SWRL rule is trivially associating fire events with an air tanker, reusing the `Forest` class from the SWEET ontology.

```
  FireEvent(?f) • affected(?f,  ?x) • sweet:Forest(?x) ->
recommendedResourceType(?f, airtanker)
```

These rules can be stated at several levels of the hierarchy. For example, the above applies to any kind of wildfire (not only forests), and the rule is not taking into account the type of fire as it is somehow implicit in the context.

[6] Tracking in Spanish is "seguimiento", which gives sense to the acronym.

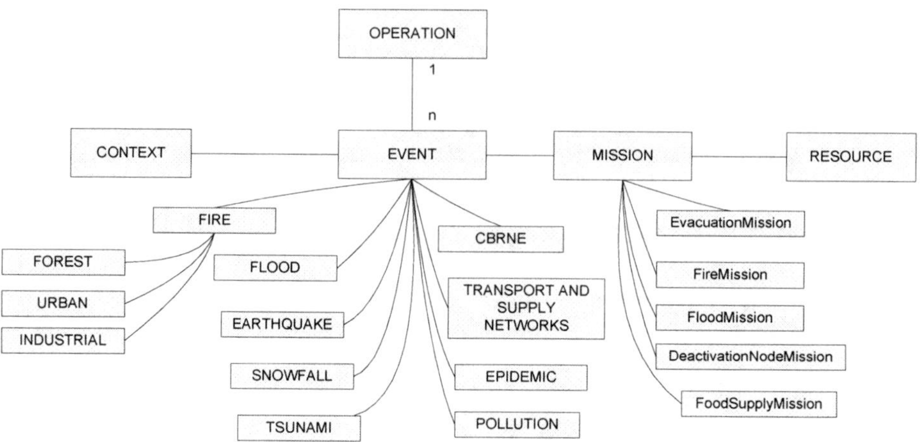


**Fig. 2.** An overview of CESAR

This use of types instead of events can be extended to missions and resources. In what follows, a concrete scenario us described in terms of events.

**Event 1:** A forest fire starts near several towns, some of them with closed industrial areas. Trivially, a firebreak mission type is recommended, which has some resource types associated. Then, this forest fire is getting worse, due to changes in wind direction and it is needed to call more resources to mitigate the effects of this fire. From the type of event, a knowledge model based on CESAR can be used to recommend operators mission types and actions. Fig. 3 provides examples of these connections.

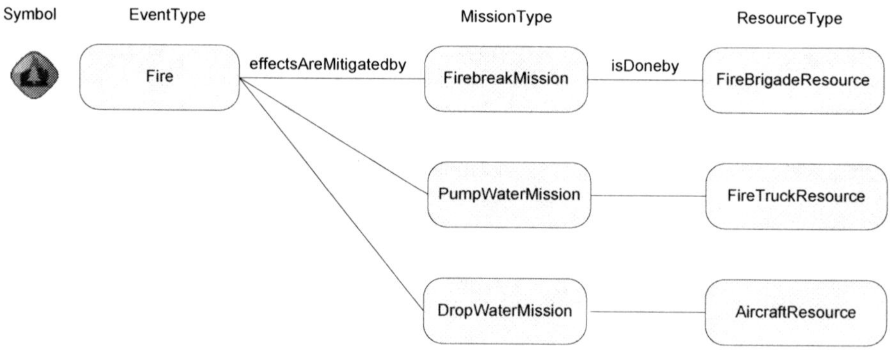


**Fig. 3.** Example association of even types to mission and resource types

**Event 2:** Due to the intensity of the fire, it is reached an industrial area where there is an industrial warehouse with chemical compounds.

The second event will be detected by a rule like the following one, which is reusing part of the infrastructure definitions in the BFiaO (Sicilia and Santos, 2009).

```
FireEvent(?f) ∧ geolocation(?f, ?g) ∧
ChemicalFacilityNode(?n) ∧ physicalLocation(?n, ?g2) ∧
overlaps(?g, ?g2) ->
IndustrialFireEvent(?f2) ∧ physicalLocation(?f2, ?g2) ∧
hasPart(?f, ?f2)
```

The rule is spawning a new sub-event of a different kind, in this case with the difference in the kind of fire depending on the fuel. This technique allows for a separate handling of adverse events as an aggregate, which is better than multi-classifying the same event. As the event type is different, the event type association with mission types is also different. Figure 4 depicts some example associations that would be used for recommending actions and resources.

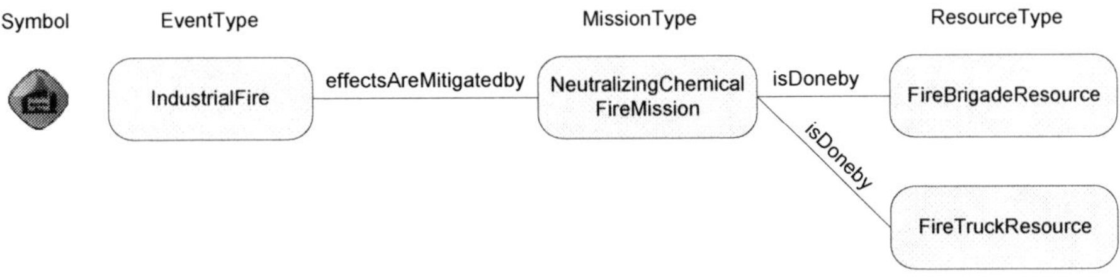


**Fig. 4.** Example association of even types to mission and resource types (ii)

The schema and combination of rules with explicit associations provides a flexible way of connecting event types with response means. This partially automates the tracking of the evolving situation and enables reusing knowledge on response that is useful for enhancing the awareness of the situation.

## 5 Integrating Ontology-Based Assessments into Emergency Management Systems

Emergency management systems are software pieces that provide tools to handle exceptional situations provoked by terrorism, natural disasters, accidents, etc. Sahana[7], the Free and Open Source Disaster Management System, is an example of this (Careem et al., 2006): a web-based application compound of several modules useful to this purpose like a missing person registry, an organization registry ("who is doing what and where") or a situation awareness map.

Ontology-based assessment could be used in conjunction with those systems to automatize some deductions that have to be made before taking decisions when managing an emergency situation. Thus, the emergency management system provides a user interface which is easily understandable by the operator, with several specialized tools, while ontology-based assessment helps him to make predictions about the situation and receiving suggestions about what to do to mitigate the potential incidents.

[7] Sahana Foundation `http://sahanafoundation.org/`

GANESHA[8] is an Open Source ontology-based assessment framework for emergency situations that is oriented to reasoning and inferring effects on critical infrastructures during unfolding incidents. An integration module was created to give Sahana the ability of communicating with GANESHA, as shown in Figure 5: an operator using Sahana sends to GANESHA some incidents taking place in an emergency situation. GANESHA will take those incidents and, by means of its ontology-based reasoning core, will tell Sahana what could happen next and how can those incidents be mitigated. After that, the operator can see and manage in Sahana the information retrieved by GANESHA.

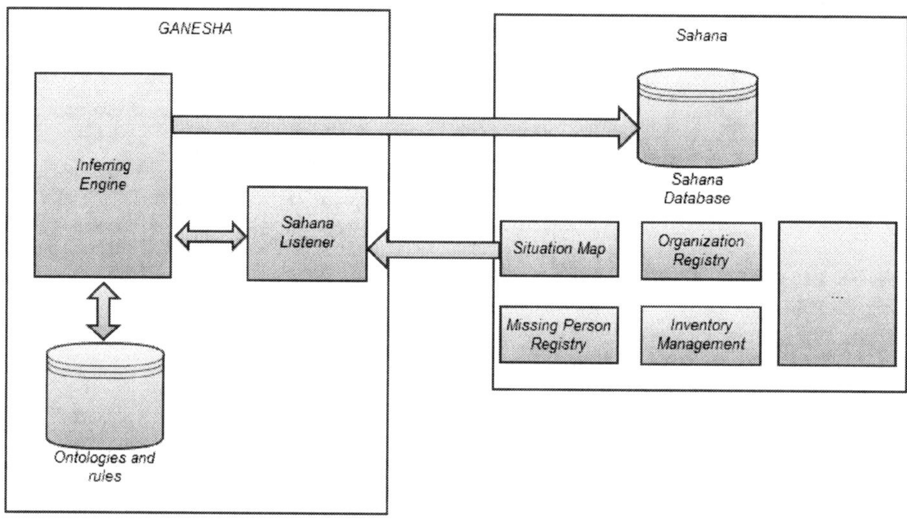


**Fig. 5.** Diagram of communication between GANESHA and Sahana

This kind of integration can turn an emergency management system into a more complete and automatized tool with the ability of saving time when taking decisions, which is crucial in disaster situations.

## 6 Conclusions and Outlook

The domain of emergency or disaster management requires complex models integrating adverse events with representations of the knowledge about resources and types of actions that are better suited to handling the unfolding situation. While there have been some proposals of ontologies for disaster management so far, they stay at the level of general schemas, but do not account for some well-defined criteria to characterize different classes of adverse events, that can then be related formally to knowledge regarding how to deal with them while accounting for the contextual situation. This paper has described an effect-based criterion for creating adverse event

[8] GANESHA `http://www.ieru.org/projects/ganesha/`

typologies, and how these can be related to missions and resources. The link between the events and these latter categories is a kind of expert knowledge that is extremely valuable for situation awareness in response phase. The details on how these models can be used with a concrete disaster management software framework have also been briefly described.

Future work should provide a complete refactoring of existing adverse event taxonomies accounting for how their effects have to be handled. This would eventually provide a bridge from taxonomies and proposed ontologies that are currently unrelated to an integrated, comprehensive schema that can be shared and refined. Also, there is a need to represent the associated expert and technical knowledge and make it reusable using the same formalism.

## References

1. Baumgartner, N., Retschitzegger, W.: A survey of upper ontologies for situation awareness. In: Proc. of the 4th Intl. Conf. on Knowledge Sharing and Collaborative Engineering, pp. 1–9. ACTA Press (2006)
2. Careem, M., Silva, C.D., Silva, R.D., Raschid, L., Weerawarana, S.: Sahana: Overview of a disaster management system. In: IEEE International Conference on Information and Automation (2006)
3. Ceusters, W., Capolupo, M., De Moor, G., Devlies, J.: Introducing realist ontology for the representation of adverse events. In: Eschenbach, C., Gruninger, M. (eds.) Formal Ontology in Information Systems, pp. 237–250. IOS Press, Amsterdam (2008)
4. Fan, Z., Zlatanova, S.: Exploring ontology potential in emergency management. In: Proceedings of the Gi4DM Conference - Geomatics for Disaster Management, Torino, Italy (February 2010)
5. Galton, A.P., Worboys, M.: An Ontology of Information for Emergency Management. In: 8th International Conference on Information Systems for Crisis Response and Management (ISCRAM 2011), Lisbon (2011)
6. Gruber, T.R.: A translation approach to portable ontology specifications. Knowledge Acquisition 5(2), 199–220 (1993)
7. Iannella, R.: Modeling and Integration of Severe Weather Advisories for Situational Awareness. In: Proceedings of the 9th International Conference on Information Fusion, Florence, pp. 1–8 (July 2006)
8. Kreps, G.A.: Future directions in disaster research: The role of taxonomy. International Journal of Mass Emergencies & Disasters 7(3), 215–241 (1989)
9. Krutchen, P., Woo, C.C., Monu, K., Sootedeh, M.: A conceptual model of disasters encompassing multiple stakeholder domains. International Journal of Emergency Management 5, 25–56 (2008)
10. Li, X., Liu, G., Ling, A., Zhan, J., An, N., Li, L., Sha, Y.: Building a Practical Ontology for Emergency Response Systems. In: Proceeding of International Conference on Computer Science and Software Engineering, Wuhan, Hubei (2008)
11. Little, E.G., Rogova, G.L.: Ontology meta-model for building a situational picture of catastrophic events. In: 8th International Conference on Information Fusion, July 25-28, vol. 1, p. 8 (2005)
12. Othman, S.H., Beydoun, G.: Metamodelling Approach To Support Disaster Management Knowledge Sharing. In: ACIS 2010 Proceedings, Paper 97 (2010)

13. Sicilia, M.A., Santos, L.: Main Elements of a Basic Ontology of Infrastructure Interdependency for the Assessment of Incidents. In: Lytras, M.D., Damiani, E., Carroll, J.M., Tennyson, R.D., Avison, D., Naeve, A., Dale, A., Lefrere, P., Tan, F., Sipior, J., Vossen, G. (eds.) WSKS 2009. LNCS, vol. 5736, pp. 533–542. Springer, Heidelberg (2009)
14. Sotoodeh, M., Kruchten, P.: An Ontological Approach to Conceptual Modeling of Disaster Management. In: 2nd Annual IEEE Systems Conference, Montreal (2008)
15. Wang, W., Guo, W., Luo, Y., Wang, X., Xu, Z.: The Study and Application of Crime Emergency Ontology Event Model. In: Khosla, R., Howlett, R.J., Jain, L.C. (eds.) KES 2005. LNCS (LNAI), vol. 3684, pp. 806–812. Springer, Heidelberg (2005)
16. Wirtz, A., Below, R., Guha-Sapir, D.: Disaster Category Classification and Peril Terminology for Operational Purposes. In: 2009 CRED: Brussels, MunichRe: Munich (2009)
17. Yu, K., Wang, Q., Rong, L.: Emergency Ontology construction in emergency decision support system. In: Proc. of the IEEE International Conference on Service Operations and Logistics, and Informatics, IEEE/SOLI 2008, pp. 801–805 (2008)

# Open Linked Data Model Revelation and Access for Analytical Web Science

Juan Manuel Dodero, Iván Ruiz-Rube,
Manuel Palomo-Duarte, and Juan Vázquez-Murga

Department of Computer Languages and Systems,
University of Cádiz
{juanma.dodero,ivan.ruiz,manuel.palomo}@uca.es,
juan.vazquezmurga@alum.uca.es

**Abstract.** The extension of regular web applications with linked data and the provision of specific web services to exploit their datasets is still a challenge. In this paper we describe a method to generate linked data and reveal them in a controlled manner for open source web applications. Revelation is carried out either at the controller or the model components of web applications that are based in the model-view-controller architecture. The approach facilitates the definition of access control policies and other non-functional requirements on the application objects and services, as well as the mapping of the application model to standard RDF schemata and vocabularies.

**Keywords:** Linked data, web science, web application frameworks.

## 1 Introduction

The large amount of information in the Web paves new ways for doing data analysis [16]. Web Science [6] emerges as a new research field that views the Web as a source of data to be analysed with relevance in diverse fields. Information of the Web is often endowed with semantic conceptualizations [26] and enriched with metadata to build up *linked data* [9]. Semantic and linked representations of information opens the way to enhanced analysis of the Web structure and behaviour [22].

Web-based applications and information systems that hold valuable information and data to be analysed are more and more built on the basis of linked data [4,17]. In this vein, emergent open data challenges[1] have met web developers to foster the debate about the need for public institutions to open their data to citizens' analysis in an accessible and reusable way. Such open linked data applications often provide an usable, mashed-up interface to exploit some public dataset for a specific purpose (e.g. awareness of war materials trade by territory[2], or analysis of energy production and demand depending on the source of energy

[1] examples can be seen on opendatachallenge.org or www.abredatos.es

[2] www.eldisparate.de

E. García-Barriocanal et al. (Eds.): MTSR 2011, CCIS 240, pp. 105–116, 2011.

and its producer or consumer region[3]). The development of these applications is grounded on open formats, such as XML, RDF(S) and OWL, which are used to expose web application data. A deeper analysis of linked data sources and applications involves the following steps:

1. defining a model or schema of the relevant data to be analysed;
2. to fill the data model in with open data instances that are compiled from external data sources; and
3. programming an application-specific user interface view to aggregate and exploit the data.

Participants in such challenges invest a great effort in developing the user interface, assuming that the data model and data instances are readily available. Nevertheless, this assumption is not always accurate. There is still a lack of open data sources in the Web, partly due to the difficulty of extending and wrapping regular web sites with linked data [18] to enable its exploitation for some sort of application, and partly due to the necessary programming and publication of specific web service interfaces to exploit datasets and manage their lifecycle [23]. The provision of application architectures and methodologies that take into account linked data for such software development issues is a major research challenge [17,20].

In this paper we advocate for the use of Open Source Software (OSS) and application development frameworks to re-engineer traditional web applications. Applications are thus refactored to extend them with open linked data capabilities. Linked data extensions are based on the controlled revelation of the application data model by means of a generative procedure that exposes all or part of the relevant data model and instances to external applications. The approach enables inner analyses of web applications' data. We provide an example of how the controlled revelation method is applied to a common OSS web application.

## 2 Developing Linked Data Applications

Web Science is a merge of the synthetic and analytic paradigms of research, in the sense that human interactions need to be designed (i.e. engineered), but also be analysed and understood. The synthetic face of the semantic web has to do with the engineering of formalisms, such as RDF and OWL, that represent the Web structure and support for a rich behaviour of web applications. There are several methods to process data from existing web applications to synthesize linked data, depending on the availability of the application source code.

### 2.1 Non-OSS Applications

When you have not the control over the source code of a web content publisher application, the usual way of creating linked data is to scrape web contents, perform a latent content analysis and annotate them. This approach carried

[3] e.nergy.es

out on unstructured web contents has examples in OpenCalais, Zemanta and other semantic content management systems [13]. Specialized services, such as Swirrl[4], look after the publication of open linked data, but do not go into the technical details of how they manage it. In addition, Virtuoso Sponger[5] is a middleware component that generates Linked Data from a wide variety of data sources that publish data in non-RDF formats. Other sources [7,10] approach is to harvest semi-structured data sources, such as the Wikipedia, to convert its contents into structured linked data [3]. For instance, Freebase deploys a graph-shaped linked database that is collaboratively built by the users; and DBpedia is built up from automatically harvesting sources such as Wikipedia or Freebase. DBpedia datasets can be either imported into third party applications or accessed online using a variety of DBpedia interfaces. These datasets can be extracted, analysed, and the results be published for later utilization. These linked data sources provide an extension API to develop specialized applications that can be used to analyse the linked information they hold. For instance, Freebase supports the development of specialized applications to analytically exploit the linked database (e.g. Thinkbase [19]).

## 2.2 OSS Applications

In contrast to the former view, OSS-licensed web applications significantly eases code integration [15]. The well-known Model-View-Controller (MVC) architecture of web applications can turn to developers' advantage. This way linked data can be generated and managed at the level of either the model, view, or controller components. In the following we examine such options separately.

First, linked data can be provided at the application view level. This is made by the addition of rich snippets —i.e. microformats, RDFa or HTML5 microdata— to the HTML content that is to be rendered by the view components [2]. This annotation approach can be applied to web contents and services. For instance, SA-REST and MicroWSMO annotation mechanisms can support the description of web service properties [27]. Based on SA-REST, hRESTS microformats also provide machine-readable annotations to HTML [21]. Second, linked data can be also generated from the application data model, either by translating its relational database schema (e.g. D2R [8]), or supported by the Object-Relational Mapping (ORM) framework, usually based on the ActiveRecord pattern [14], that allows mapping the application objects to the relational database. This pattern enables to align the application data model with a given RDF schema. It has been used by ActiveRDF [24] to exploit RDF-based data models that replace relational data models. Last and surprisingly, the controller components of MVC-based applications have not been explored as the key components where managing linked data provision and access.

In the approaches described above, we find two major issues to open linked data access. First, there must be a procedure to harvest, scrape or annotate

[4] www.swirrl.com

[5] www.uriburner.com

non-linked data and web services before exploiting them for analysis. Data that is compiled from the application must be mapped to the appropriate RDF metadata schemata, which are usually defined on the basis of standard ontologies and RDF vocabularies, such as FOAF, SIOC, SKOS, DOAP and so forth [9]. Secondly, extending the non-functional requirements of the web application such as security —e.g. based on Role-Based Access Control (RBAC) [1] mechanisms— is more challenging in linked applications. Access permissions are given to user roles only as a function of defined authorization over the data items, but not over the controller operations. For instance, you can provide free access at the model level for user IDs or other data (e.g. cities, addresses and phone numbers). But if a new controller operation is provided that links a user ID with its personal data, some authorization requirements should be provided at the controller level that bear on required legal security restrictions.

In the following we describe a new method to generate linked data and reveal them in a controlled manner. Revelation is carried out either at the controller or the model components of a framework-based OSS web application. This approach can facilitate the definition of security policies and other non-functional requirements on the application objects and services, as well as the mapping of the application model to standard RDF schemata and vocabularies.

## 3 Controlled Revelation of Linked Data

In this section we describe a new approach to providing linked data and metadata of an existing web application. The approach is based on the automated generation and controlled revelation of RDF metadata, as well as enrichment of current web contents with RDFa datasets, as generated from the application data model. This procedure can be developed as long as: (1) the application source code is available; and (2) the application is designed using the Model-View-Controller architecture or equivalent pattern [11]. The procedure consists of the following steps:

1. *Revealing the underlying application data model*: The RDF model that is equivalent to the application Object-Relational Mapping (ORM) [14] data schema is generated and published as metadata. In addition, when the application receives a request, RDFa annotations to elements on the delivered HTML are generated from the application data model. This way, the linked data model instances that describe the datasets can be easily generated.
2. *Linking data instances*: The datasets retrieved from the application database are explored to link the items of the data model. Yet automated item linkage is not straightforward over external resources, they can be readily provided for the internal items generated from the application dataset. A browse can be done around these linked data for any analytical purpose that involves harvesting or scraping semantics from the contents.
3. *Publishing the service model API*: A service-based API to exploit the linked data model is generated, following a Create/Read/Update/Delete (CRUD)

pattern, and published using SA-REST or hRESTS. Similarly to dataset annotation, extended service descriptions are generated containing SA-REST annotations that describe the semantics of each service call. Service annotations are created from the inspection of the application data model. The generated service model is not restricted to the original use cases intended for the application. Instead, any query can be done about each element of the application model for further analysis.

4. *Control security authorizations*: A set of authorization permissions can be defined for data items, data types and/or service calls that require to be secured. Permissions are defined as cross-cutting issues, following an RBAC approach, to control access to data, metadata and services.
5. *Enabling the analytic API*: This step provides a hook to define extended analytic operations that are relevant to the application. These can be defined as extended services above the basic CRUD API, so extending the analytical possibilities to query each element of the data model, including those that were not planned in the original use cases of the application. SA-REST annotations included in the service descriptions can then be browsed —in a analogous manner to harvesting data semantics— to build extended service operations for the aims of a particular analysis.

As an example of the approach, let us consider an issue previously studied, consisting in making explicit as RDF the links between software artifacts found in software development [20]. Some examples of web tools commonly used to host and manage software projects include Redmine, Google Code, JIRA, Trac, Assembla, etc. [12]. We focus on *Redmine* because is an OSS web application, used for project management, issue and time tracking, as well as other general-purpose functions (e.g. file repository, news, wiki, calendar, etc.)

Redmine provides an API[6] that enables to query a repository about a limited number of elements of the application model, namely issues, projects, users and time-entries; news and wiki pages are planned but yet not included in the current API. Fig. 1 depicts the Redmine application's model of elements that are actually accessible through the Redmine API. Public operations are included in the `RedmineService` interface. In the figure, the `getIssues(s:Status, p:Project, u:User)` method is used to get all the issues within project $p$ that are in $s$ status and are assigned to user $u$. If some parameter values are omitted, all the items of that type are filtered out.

Through the Redmine API you can find all the issues in a project and all the issues that have been assigned to a certain user, e.g. by providing different parameter values to `getIssues(s,p,u)`. You can find out also the project to which an issue belongs (e.g. calling `getProject(i)` on issue $i$), or which user reported an issue (e.g. calling `getUserReporter(i)` on issue $i$). With this API you cannot find out, however, all the projects reported by a user, since there is not a specific operation in the model and the public `RedmineService` façade defined for that aim.

[6] www.redmine.org/projects/redmine/wiki/Rest_api

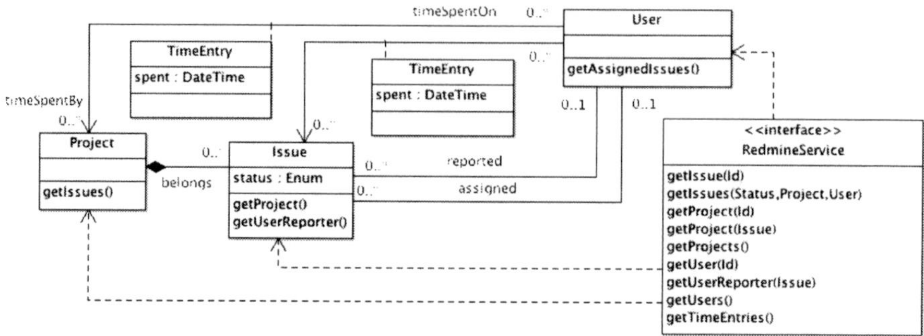


**Fig. 1.** Redmine application service data model expressed in UML, including all model elements used to implement the available Redmine service API

Listing 1.1 is an RDF description of the classes and object properties managed by the Redmine API service. They represent an OWL service description equivalent to the UML model of Fig. 1.

**Listing 1.1.** Class model of the ontology that represents the data objects that can be actually known through the Redmine service API

```
@prefix owl: <http://www.w3.org/2002/07/owl#> .
@prefix rdf: <http://www.w3.org/1999/02/22-rdf-syntax-ns#> .
@prefix rdfs: <http://www.w3.org/2000/01/rdf-schema#> .
@prefix : <http://my.example.org/ontologies/2011/4/OntologyRedmine.owl#> .
<http://my.example.org/ontologies/2011/4/OntologyRedmine.owl>
   rdf:type owl:Ontology .
:Issue rdf:type owl:Class .
:Project rdf:type owl:Class .
:TimeEntry rdf:type owl:Class .
:User rdf:type owl:Class .
:issuesWithin rdf:type owl:ObjectProperty ;
   rdfs:domain :Project ;
   rdfs:range :Issue ;
   owl:inverseOf :belongsToProject .
:issuesAssigned rdf:type owl:ObjectProperty ;
   rdfs:domain :User ;
   rdfs:range :Issue ;
   owl:inverseOf :assignedToUser .
:projectIssuesAssigned rdf:type owl:ObjectProperty ;
   rdfs:subPropertyOf :issuesAssigned , :projectIssues .
:assignedToUser rdf:type owl:FunctionalProperty , owl:ObjectProperty ;
   rdfs:domain :Issue ;
   rdfs:range :User .
:belongsToProject rdf:type owl:InverseFunctionalProperty ,
 owl:ObjectProperty ;
   rdfs:domain :Issue ;
   rdfs:range :Project .
:issuesAssigned rdf:type owl:ObjectProperty ;
   rdfs:domain :User ;
   rdfs:range :Issue ;
   owl:inverseOf :assignedToUser .
:issuesReported rdf:type owl:ObjectProperty ;
   rdfs:domain :User ;
   rdfs:range :Issue .
:projectIssues rdf:type owl:ObjectProperty ;
   rdfs:domain :Project ;
   rdfs:range :Issue ;
```

```
    owl:inverseOf :belongsToProject .
:reportedByUser rdf:type owl:FunctionalProperty , owl:ObjectProperty ;
    rdfs:domain :Issue ;
    rdfs:range :User ;
    owl:inverseOf :issuesReported .
:timeSpentByUser rdf:type owl:ObjectProperty ;
    rdfs:domain [ rdf:type owl:Class ; owl:unionOf ( :Issue :Project ) ] .
    rdfs:range :TimeEntry ;
    owl:inverseOf :timeSpentOnIssue , :timeSpentOnProject ;
:timeSpentOn rdf:type owl:InverseFunctionalProperty , owl:ObjectProperty ;
    rdfs:domain :TimeEntry ;
    rdfs:range [ rdf:type owl:Class ; owl:unionOf ( :Issue :Project ) ] .
:timeSpentOnIssue rdf:type owl:ObjectProperty ;
    rdfs:domain :User ;
    rdfs:range :TimeEntry .
:timeSpentOnProject rdf:type owl:ObjectProperty ;
    rdfs:domain :User ;
    rdfs:range :TimeEntry .
```

Object properties `projectIssues`, `issuesAssigned` and `projectIssuesAssigned` enable implementing the `getIssues (Status,Project,User)` method to find out, respectively, all the issues in a project, all issues assigned to a user, or both. You can ask the API for the time spent by a user working on an issue (`timeSpentOnIssue`) or a project (`timeSpentOnProject`), as well as about the user that reports (`reportedByUser`) or is assigned to an issue (`assignedToUser`).

Our approach can be used to extend the service API with new operations that provide further information about the application, therefore facilitating the analysis of the foregoing application behaviour. Listing 1.2 describes the object property used to augment the service API with the new `getProjectsReported()` operation. Although it might be not initially considered by the designer of the service, this operation can be afterwards created from its object property and the original source code of the application.

**Listing 1.2.** Object property that can be generated to provide the Redmine API with a new operation

```
:projectsReported rdf:type owl:ObjectProperty ;
    rdfs:domain :User ;
    rdfs:range :Project .
```

From such OWL specifications, new filtering `findXByZ()` method implementations can be generated and added to the application controllers to find only those objects of type $X$ that are associated to a given object of type $Z$.

The limitations of the Redmine API that are explained above are also found in other project management web tools. For instance, both Redmine and Google Code do not provide an API to query the set of projects of the repository that fulfill a condition (e.g. be associated to a given user); with Google Code you cannot even inquiry for all the projects in a repository (i.e. the `getProjects()` method). By applying the controlled revelation method, we can re-engineer the application to generate a `findProjects()` service implementation that extends the MVC controller's logic for enabling to query for all projects of the repository. If we are only interested in the projects in which a user works, we can readily generate a `findProjectsByUser()` operation along with the required controller method that enables to call this service from an external analysis application.

The ReST-based architecture design would enable to issue requests to simple URIs such as `http://demo.redmine.org/projects` or `http://demo.redmine.org/user/` *:userId* `/projects`, where *:userId* is the identification of the user in whose projects we are interested.

After publishing the application data model as RDF, a basic service exploitation API has to be published. The most basic service implementation conforms to the CRUD pattern. Because of the analytical purpose of applications, only *read* operations need to be published in the service interface. These can be implemented as HTTP GET methods following the ReST paradigm. Besides the ReST-based service implementation, a service description based in SA-REST [21] annotations to the service operations is generated, as depicted in listing 1.3. For the Ruby on Rails implementation, SA-REST annotations to the arguments of the service operations are generated after inspecting the data model with the `eval` function to know the service argument types.

**Listing 1.3.** SA-REST specification of new **getTimeSpentOn** service operations

```
<html xmlns:sarest="http://lsdis.cs.uga.edu/SAREST#">
...
<p about="http://my.example.org/redmine/getTimeSpentOnProject/">
The logical input of this service is an
<span property= sarest:input >
   http://my.example.org/ontologies/2011/4/OntologyRedmine.owl#User
</span> object
and an
<span property="sarest:input">
   http://my.example.org/ontologies/2011/4/OntologyRedmine.owl#Project
</span> object.
The logical output of this service is a list of
<span property="sarest:output">
   http://my.example.org/ontologies/2011/4/OntologyRedmine.owl#TimeEntry
</span> objects.
This service should be invoked using an
<span property="sarest:action">
   HTTP GET
</span>
<meta property="sarest:lifting"
 content="http://my.example.org/api/lifting.xsl"/>
<meta property="sarest:lowering"
 content="http://my.example.org/api/lowering.xsl"/>
<meta property="sarest:operation"
 content="http://my.example.org/ontologies/2011/4/
 OntologyRedmine.owl#timeSpentOnProject"/>
</p>

<p about="http://my.example.org/redmine/getTimeSpentOnIssue/">
The logical input of this service is an
<span property="sarest:input">
   http://my.example.org/ontologies/2011/4/OntologyRedmine.owl#User
</span> object
and an
<span property="sarest:input">
   http://my.example.org/ontologies/2011/4/OntologyRedmine.owl#Issue
</span> object.
The logical output of this service is a list of
<span property="sarest:output">
   http://my.example.org/ontologies/2011/4/OntologyRedmine.owl#TimeEntry
</span> objects.
This service should be invoked using an
<span property="sarest:action">
   HTTP GET
```

```
</span>
<meta property="sarest:lifting"
 content= h t t p ://my.example.org/api/lifting.xsl"/>
<meta property="sarest:lowering"
 content= h t t p ://my.example.org/api/lowering.xsl"/>
<meta property="sarest:operation"
 content= h t t p ://my.example.org/ontologies/2011/4/
  OntologyRedmine.owl#timeSpentOnIssue"/>
</p>
```

## 4 Discussion

The controlled revelation of linked data operates either at the model or the control layer of the web application architecture. A complete RDF model of the application schema can be generated, along with linked data instances. If a standard RDF schema (e.g. FOAF or DOAP) has to be mapped or aligned against the application data model, the ORM mechanism can manage it. For instance, using the Ruby on Rails framework, the Redmine application model can be re-factored as shown in listing 1.4.

**Listing 1.4.** Ruby specification of the Redmine application data model using the Ruby on Rails' ORM

```
Namespace.register(:doap, "http://usefulinc.com/ns/doap#")
Namespace.register(:foaf, "http://xmlns.com/foaf/spec")
Namespace.register(:owltime, "http://www.w3.org/TR/owl-time")
class Project < ActiveRecord::Base
   has_many :issues
   set_rdf_model_name "doap:Project"
end
class Issue < ActiveRecord::Base
  @status = IssueStatus::OPEN
  belongs_to :project
end
class TimeEntry < ActiveRecord::Base
  @spent  = 0
   set_rdf_model_name "owltime:DurationDescription"
end
class ProjectTimeEntry < TimeEntry; end
class IssueTimeEntry < TimeEntry; end
class User < ActiveRecord::Base
  has_many :projects, :through => :projectTimeEntries
  has_many :issues, :through => :issueTimeEntries
  set_rdf_model_name "foaf:Person"
end
```

The example above maps Redmine `Project` objects to DOAP projects, `User` objects to FOAF persons and `TimeEntry` to OWL-Time durations. Issues are not mapped to any standard vocabulary or ontology.

The controlled revelation approach enables to add further security to items defined at the level of the model or the controller. Authorizations can be defined for data items or data types at the model as well as for service calls of the controller. Security framework extensions can be used to define access permissions on each ReST-based CRUD method of the controller. Data in the model can be also annotated with additional security permissions to inhibit, for example, reading password-type data or other sensitive data items or types.

**Listing 1.5.** Ruby security specification of access control to the Redmine application model and controllers using the declarative_authorization plugin

```
# model
class Project < ActiveRecord::Base
   has_many :issues
end
class Issue < ActiveRecord::Base
   using_access_control
   @status = IssueStatus::OPEN
   belongs_to :project
end
class TimeEntry < ActiveRecord::Base
   using_access_control
   @spent  = 0
end
class ProjectTimeEntry < TimeEntry; end
class IssueTimeEntry < TimeEntry; end
class User < ActiveRecord::Base
  using_access_control
  has_many :projects, :through => :projectTimeEntries
  has_many :issues, :through => :issueTimeEntries
end
# controllers
class ProjectController < ApplicationController
   filter_access_to :all
   filter_access_to :getIssues, :require => :read
   def getIssues ... ;  end
end
class UserController < ApplicationController
   filter_access_to :all
   filter_access_to :getAssignedIssues, :require => :read
   def getAssignedIssues ... ;  end
end
# authorization_rules.rb
authorization do
   role :admin do
      has_permission_on :projects, :to => [:create, :read, :update, :delete]
      has_permission_on :issues, :to => [:create, :read, :update, :delete]
      ...
   end
   role :analyst do
      has_permission_on :projects, :to => [:read, getIssues]
      has_permission_on :issues, to => [:read, :getAssignedIssues]
   end
end
```

## 5 Conclusion and Future Work

For the analytical aims of Web Science, it is necessary to implement methods of providing open linked data in web contents and applications. In this work we present a procedure of linked data provision for web applications that are designed following an MVC or equivalent pattern, as long as the application source code is available.

We have described the benefits of our approach using a common OSS project management web application that provides a default service-based API to deal with projects and issues. Using our method we explain how to re-engineer an open linked data version of the application, which additionally extends the API. The application data and the extended API are self-described using the RDFa and SA-REST annotations that are generated. We also discuss how the application model can be mapped to existing ontologies thanks to our method, as well as how to secure the access to linked data resources. The linked data API extension described in this work is purposed for web application data analysis. However, the approach is not only useful for data analysis on the web, but is applicable for other aims requiring web application integration based on linked data.

As a future work, we aim at extending our framework to provide data provenance and make web applications writable [5]. Data provenance can be provided as a non-functional aspect that enriches the queries that are issued by the application's controller implementation [25]. Besides, going beyond the original purpose of analytical applications, we can generate the rest of HTTP methods (i.e. POST, PUT and DELETE) to provide writable endpoints to linked data. Also, we will publish the Ruby on Rails gem component that implements our framework.

**Acknowledgments.** This work has been sponsored by a grant from the ASC-ETA project (P09-TIC-5230) of the Andalusian Government, Spain.

## References

1. Role-based access control. ANSI INCITS 359-2004, American National Standards Institute (2004)
2. Adrian, B., Hees, J., Herman, I., Sintek, M., Dengel, A.: Epiphany: Adaptable RDFa Generation Linking the Web of Documents to the Web of Data. In: Cimiano, P., Pinto, H.S. (eds.) EKAW 2010. LNCS, vol. 6317, pp. 178–192. Springer, Heidelberg (2010)
3. Auer, S., Lehmann, J.: What Have Innsbruck and Leipzig in Common? Extracting Semantics from Wiki Content. In: Franconi, E., Kifer, M., May, W. (eds.) ESWC 2007. LNCS, vol. 4519, pp. 503–517. Springer, Heidelberg (2007)
4. Berners-Lee, T., Chen, Y., Chilton, L., Connolly, D., Dhanaraj, R., Hollenbach, J., Lerer, A., Sheets, D.: Tabulator: Exploring and Analyzing linked data on the Semantic Web. In: Proceedings of the 3rd International Semantic Web User Interaction Workshop, Athens, Georgia, USA (2006)
5. Berners-Lee, T., Hollenbach, J., Lu, K., Presbrey, J.: Pru d'ommeaux, E., Schraefel, M.C.: Tabulator Redux: Writing Into the Semantic Web. Tech. rep., Electronics and Computer Science, University of Southampton (2007)
6. Berners-Lee, T., Weitzner, D.J., Hall, W., O'Hara, K., Shadbolt, N., Hendler, J.a.: A Framework forWeb Science. Foundations and Trends inWeb Science 1(1), 1–130 (2006)
7. Bizer, C., Lehmann, J., Kobilarov, G., Auer, S., Becker, C., Cyganiak, R., Hellmann, S.: DBpedia-A crystallization point for the Web of Data. Web Semantics: Science, Services and Agents on the World Wide Web (2009)
8. Bizer, C., Cyganiak, R.: D2R Server Publishing Relational Databases on the Semantic Web. In: Proceedings of the 5th ISWC, Athens, Georgia, USA (2006)
9. Bizer, C., Heath, T., Berners-Lee, T.: Linked Data - The Story So Far. International Journal on Semantic Web and Information Systems 5(3), 1–22 (2009)
10. Bollacker, K., Evans, C., Paritosh, P., Sturge, T., Taylor, J.: Freebase: a collaboratively created graph database for structuring human knowledge. In: Proceedings of the 2008 ACM SIGMOD International Conference on Management of Data, pp. 1247–1250. ACM, New York (2008)

11. Buschmann, F., Meunier, R., Rohnert, H., Sommerlad, P., Stal, M.: Pattern-Oriented Software Architecture, 1st edn. A System of Patterns, vol. 1. Wiley, Chichester (1996)
12. Cabot, J., Wilson, G.: Tools for Teams: A Survey of Web-Based Software Project Portals. Dr. Dobb's, 1–14 (October 2009)
13. Dotsika, F.: Semantic APIs: Scaling up towards the Semantic Web. International Journal of Information Management 30(4), 335–342 (2010)
14. Fowler, M.: Patterns of Enterprise Application Architecture. Addison-Wesley Longman, Boston (2002)
15. Gabriel, R.P.: The commons as new economy & what this means for research. In: International Workshop on Emerging Trends in FLOSS Research and Development, p. 10 (2007)
16. Hall, W., De Roure, D., Shadbolt, N.: The evolution of the Web and implications for eResearch. Philosophical Transactions. Series A, Mathematical, Physical, and Engineering Sciences 367(1890), 991–1001 (1890)
17. Hausenblas, M.: Exploiting Linked Data to BuildWeb Applications. IEEE Internet Computing, 68–73 (July/August 2009)
18. Hausenblas, M., Halb, W., Raimond, Y., Heath, T.: What is the Size of the Semantic Web? In: International Conference on Semantic Systems (2008)
19. Hirsch, C., Grundy, J., Hosking, J.: Thinkbase: A visual semantic wiki. In: Sheth, A.P., Staab, S., Dean, M., Paolucci, M., Maynard, D., Finin, T., Thirunarayan, K. (eds.) ISWC 2008. LNCS, vol. 5318. Springer, Heidelberg (2008)
20. Iqbal, A., Ureche, O., Hausenblas, M., Tummarello, G.: LD2SD: Linked Data Driven Software Development. In: 21st International Conference on Software Engineering and Knowledge Engineering, Boston, USA, pp. 240–245 (2009)
21. Kopecky, J., Gomadam, K., Vitvar, T.: hRESTS: An HTML Microformat for Describing RESTful Web Services. In: IEEE/WIC/ACM International Conference on Web Intelligence and Intelligent Agent Technology. IEEE, Los Alamitos (2008)
22. O'Hara, K., Berners-Lee, T., Hall, W., Shadbolt, N.: Use of the Semantic Web in e-Research. The MIT Press, Cambridge (2010)
23. Oren, E., Delbru, R., Catasta, M., Cyganiak, R., Stenzhorn, H., Tummarello, G.: Sindice.com: a document-oriented lookup index for open linked data. International Journal of Metadata, Semantics and Ontologies 3(1), 37–52 (2008)
24. Oren, E., Gerke, S.: ActiveRDF: Object-Oriented Semantic Web Programming. In: Proceedings of the WWW Conference, Banff, Alberta, Canada, pp. 817–823 (2007)
25. Ruiz-Rube, I., Dodero, J.M., Stoitsis, J.: Non-functional aspects of information integration and research for the web science. Procedia Computer Science (4), 1631–1639 (2011)
26. Shadbolt, N., Hall, W., Berners-Lee, T.: The Semantic Web Revisited. IEEE Intelligent Systems, 96–101 (May/June 2006)
27. Sheth, A.P., Gomadam, K., Lathem, J.: SA-REST: Semantically Interoperable and Easier-to-Use Services and Mashups. IEEE Internet Computing 11(6), 91–94 (2007)

# Connectivity and Semantic Patterns in Automatically Generated SNOMED-CT Subsets

Leonardo Lezcano and Miguel-Angel Sicilia

Information Engineering Research Unit,
Computer Science Dept., University of Alcalá,
Ctra. Barcelona km. 33.6 – 28871 Alcalá de Henares (Madrid), Spain
{leonardo.lezcano,msicilia}@uah.es

**Abstract.** The importance of enabling a consistent development and use of modern clinical terminologies is nowadays unquestionable by the health informatics community. SNOMED-CT is one of such terminologies whose adoption is fostered by the IHTSDO with a worldwide scope. However, the large scale of SNOMED-CT is a major barrier to its progress while there is evidence that only a small fraction of its content is being used. This paper proposes a series of graph-based methods to analyze and evaluate the fitness and expressiveness of automatically generated SNOMED-CT subsets vis-à-vis the clinical context they were generated for. Such evaluation methods are then applied to the subsets obtained from previous experiments that used clinical guidelines' glossaries as seeds for the automatic generation of subsets. Current research provides new means to judge if clinical applications of SNOMED-CT can go beyond the traditional uses of a controlled vocabulary or not.

**Keywords:** SNOMED-CT, subsets, graphs measures, connectivity.

## 1 Introduction

According to the SemanticHEALTH report published by the European Commission [1], the achievement of level 3 or full Semantic Interoperability (SIOp) between heterogeneous healthcare systems requires improvements in the precision of meaning, consistency, understandability and reproducibility during the exchange of information. In the particular case of Electronic Health Record (EHR) systems, it is essential to enable the consistent use of clinical terminologies, for example, by defining terminology bindings for each element of the data structures included in the Information Models supporting the (EHR).

SNOMED-CT[1] is one of such terminologies whose adoption is fostered by the IHTSDO with a worldwide scope. It provides the core general terminology for the EHR and contains more than 311,000 active concepts with unique meanings and formal logic-based definitions organized into hierarchies [2]. The large scale of SNOMED-CT is a major barrier to its progress while there is evidence that only a small fraction of its content is being used [1].

[1] http://www.ihtsdo.org/

E. García-Barriocanal et al. (Eds.): MTSR 2011, CCIS 240, pp. 117–125, 2011.

OpenGALEN[2] is a not-for-profit organization providing another medical terminology. Currently available open source resources include a sophisticated ontology development environment and a large open source description logic-based ontology for the medical domain. The Unified Medical Language System (UMLS)[3] is a compendium of three knowledge sources in the biomedical sciences. One of them is the UMLS Metathesaurus, a large, multi-purpose and multilingual vocabulary database that currently comprises more than 1.5 millions biomedical terms from over 100 sources. Other extensive terminologies are the International Classification of Diseases (ICD)[4] and the Logical Observation Identifier Names and Codes database (LOINC)[5].

As a result new approaches are being conceived for the purpose of selecting and extracting subsets or ontology parts focused on a specific domain, facilitating the bindings or mappings definition between terminologies and the EHR elements as well as providing a semantically sound context. These subsets can be manually defined by expert-driven processes which are ad-hoc and usually effort demanding or they can be automatically generated in a more general way.

Given that SNOMED-CT connects its concepts in a graph like structure by means of four kinds of relationships with different semantics (see section 2), the purpose of the present research is to provide some methods to analyze and evaluate the SNOMED-CT subsets according to traditional measures of the graph theory. Such methods are here used to analyze the SNOMED-CT subsets obtained from five clinical guidelines of different clinical domains. The extraction mechanism, described in [3], has been used to generate 13 subsets from each clinical guideline, as explained in the next section.

Section 2 introduces the background of SNOMED-CT required for the analysis and briefly explains the subset generation mechanism. Then section 3 includes the description of the graph methods and measures, and the results from the studied subsets, as well as an explanation of the patterns found in each case. Finally, section 4 provides the conclusions of the current research.

## 2 Background and Subset Generation

This section provides definitions for the elements involved in the subsets analysis, as well as a brief description of the automatic process that extracted the subsets from the whole terminology.

According to [4], a SNOMED-CT subset is a group of Concepts, Descriptions or Relationships that are appropriate to a particular realm, language, dialect, country, specialty, organization, user, context or domain. In the present research we work with a narrower definition inherited from the research described in [3], where a subset is simply a set of SNOMED-CT concepts that belong to a clinical domain defined by a given clinical guideline.

[2] http://www.opengalen.org/

[3] http://www.nlm.nih.gov/research/umls/

[4] http://www.who.int/classifications/icd/en/

[5] http://loinc.org/

A SNOMED-CT concept is a clinical meaning identified by a unique numeric identifier or *ConceptID* that never changes. They are represented by a unique human-readable *Fully Specified Name* (FSN). For example, the *Flu* concept has *ConceptID = 6142004* and *FSN = influenza (disorder)*. Then, SNOMED-CT concepts are linked through four types of relationships: *defining*, *qualifying*, *historical* and *additional* [4]. We will be particularly interested in defining relationships like *IS-A*, *FINDING SITE*, *CAUSATIVE AGENT*, etc., as they are used to model concepts and create their logical definitions. For example, the following assertions can be found in SNOMED-CT: *Flu IS-A Viral respiratory infection* and *Flu FINDING SITE Structure of respiratory system*.

SNOMED-CT concepts are organized into hierarchies where subtypes (or "children") concepts are the descendant concepts of supertypes (or "parents") concepts. Hierarchies are defined by means of the *IS-A* relationship that, according to [5], provides the basis for the "inheritance of properties": when a concept is more specific than some other concept, it inherits the properties of the more general one. SNOMED-CT includes 19 top-level concepts that subsume the rest of concepts. For example, the top-level *Clinical Finding* has *Bleeding* as a child because it inherits properties like *severity* and *episodicity* which are required to describe every clinical concept classified as *Finding*.

The subsets analyzed in this paper were derived from clinical terms collected from clinical guidelines' glossaries. The extraction process was developed by Rodríguez-Solano, Cáceres and Sicilia, and is described [3]. The process contains the following main iterative steps:

i. Retrieve the SNOMED-CT concepts that match the glossary terms in a given clinical guideline.
ii. Consider the result as a new SNOMED-CT subset.
iii. Connect each concept in the subset to other SNOMED-CT concepts using a given SNOMED-CT relationship.
iv. Expand the subset to contain discovered SNOMED-CT concepts. Go to step ii.

The above procedure depends on two key parameters: the set of relationships used in step iii, and the number of iterations performed. It should be noted that the direction of asymmetric relationships must be also specified in step iii. For example, looking for children through the *IS-A* relation increases the level of granularity after each iteration, while looking for parents adds more general and broad concepts to the subset.

In the study case carried out by Rodríguez-Solano et al. in [3], final subsets were the result of several iterations of *IS-A* look up, followed by an iteration of every other relationship (i.e. a *not-IS-A* look up). The subsets in that research were derived from glossaries in two clinical guidelines from NICE[6]. Then current research has extended the application of such extraction method to three more clinical guidelines, to work on a varied set of data that allows detecting patterns in the subsets:

[6] NICE - National Institute for Health and Clinical Excellence. `http://www.nice.org.uk/`

- The recognition and initial management of **Ovarian Cancer**[7].
- **Parkinson**'s disease: diagnosis and management in primary and secondary care[8].
- **Hypertension**: management of hypertension in adults in primary care[9].
- **Glaucoma**: diagnosis and management of chronic open angle glaucoma and ocular hypertension[10].
- Clinical diagnosis and management of **Tuberculosis**, and measures for its prevention and control[11].

The names of the studied subset have the following syntax:

*<ClinicalGuideline> - <expansion direction> - <number of iterations>*

For example, using the **Tuberculosis** guideline glossary and performing three iterations of *IS-A* traversing towards more specific concepts will generate a subset named:

*Tuberculosis-Children-H3*

In order to study such patterns, subsets are here represented as directed graphs each of one can be defined as $G = \{V, A\}$, where $V$ is the set of SNOMED-CT concepts and $A$ is a set of ordered pairs or arcs representing every relationship between the concepts in $V$. At this point it is important to emphasize that, for example, a subset generated by traversing the *IS-A* relationship from parents to children will become into a directed graph where $V$ will contain the concepts retrieved through the *IS-A* relationship and $A$ will include not only the *IS-A* arcs but also every other SNOMED-CT relationship connecting two concepts in $V$. In other words, new vertices or concepts can only be added to the subset by traversing the IS-A relationship, but existing concepts can be interconnected with any SNOMED-CT relationship. The reason for not to include the concepts found through any kind of relationship in the subsets is described in section 3.1.

## 3 Subset Patterns

This section illustrates the results of the graph measures and metrics taken from the subsets. There are some traditional metrics first and then the following subsections introduce several methods that have been adapted to suite the requirement of the SNOMED-CT context.

The traditional metrics in Table 1 show that the number of concepts and relations vary between different subsets despite the fact that they have all been generated with four *IS-A* iterations. Another straightforward conclusion is that upward expansion (parents) adds less new concepts to subsets than downward expansion (children). On the other hand, average degree and diameter reveals a more constant behavior.

[7] http://guidance.nice.org.uk/CG122
[8] http://guidance.nice.org.uk/CG35
[9] http://guidance.nice.org.uk/CG34
[10] http://guidance.nice.org.uk/CG85
[11] http://guidance.nice.org.uk/CG117

**Table 1.** Graph metrics in the subsets obtained after four *IS-A* iterations

| | | OvarianCancer | Tuberculosis | Parkinson | Glaucoma | Hypertension |
|---|---|---|---|---|---|---|
| Children-H4 | Vertices (Concepts) | 21124 | 1089 | 161 | 189 | 7620 |
| | Arcs (Relationships) | 37070 | 1614 | 151 | 219 | 11647 |
| | Average Degree | 3,5 | 3,0 | 1,9 | 2,3 | 3,1 |
| | Diameter | 9 | 6 | 4 | 4 | 7 |
| Parents-H4 | Vertices (Concepts) | 329 | 149 | 105 | 183 | 163 |
| | Arcs (Relationships) | 529 | 185 | 126 | 273 | 211 |
| | Average Degree | 3,2 | 2,5 | 2,4 | 3,0 | 2,6 |
| | Diameter | 9 | 10 | 8 | 9 | 8 |

### 3.1 Density of Qualifying Relationships

When expanding the subset by means of every type of relationships (IS-A and not-IS-A expansion) we run the risk of introducing general concepts or classes of concepts having so many relations with the rest of the graph that the understanding of connectivity and semantic patterns is dramatically blurred. That is the case of *qualifying* relationships, as they introduced concepts like *Severities*, *Courses* and *Episodicities*, among others, which are related to every concept in the *Disorder* SNOMED-CT top-level concept.

As *Disorder* specializations are very common in clinical guidelines, the *Severities*, *Courses* and *Episodicities* concepts and their relations with the rest of the graph provide no other significant information than classification. At the same time, they are responsible for a high percentage of the arcs as shown in Fig. 2 for a **Parkinson-Children-H3** enriched with a not-IS-A expansion while Fig. 1 contains the original subset. If included, these three concepts were responsible for 30% to 50% of the arcs in every studied subset at H3 level. In addition, considering *qualifying* relationships distorts the analysis of the subsets' diameter as, for example, the *Severities* concept acts like a hub that connects every pair of *Disorder* concepts with a two jump path. It influences the average degree[12] of the vertices as well.

Therefore, and given that *IS-A* relationship is more appropriate to fulfill the purpose of classification, ***the graphs in the subsets analysis of the present research only include non IS-A relations when connected concepts can be also reached through IS-A***.

[12] The degree of a vertex of a graph is the number of edges incident to the vertex, with loops counted twice [6]. In a directed graph, the number of head endpoints adjacent to a vertex is called the indegree of the vertex, and the number of tail endpoints is its outdegree.

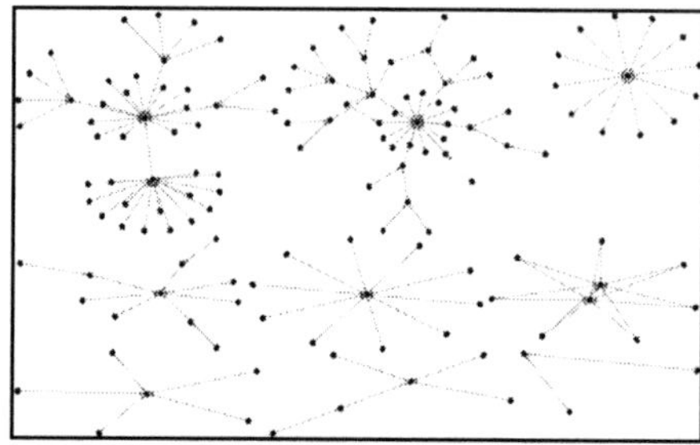

**Fig. 1.** Parkinson-Children-H3 without not-IS-A expansion

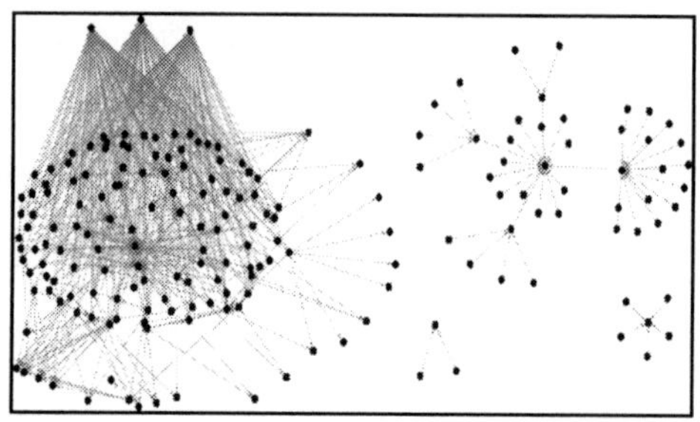

**Fig. 2.** Parkinson-Children-H3 with not-IS-A expansion. The three concepts above the biggest component are *Severities, Courses* and *Episodicities.*

### 3.2 Analyzing the Number of Components

Given a directed graph *G* = *{V, A}*, a *component* is a maximal connected subgraph of *G*. A *subgraph* is said to be connected if every pair of vertices in the subgraph is connected. Two vertices *u* and *v* are *connected* if *G* contains a path from *u* to *v*. As it is a directed graph, we can accept a path only when it follows the arcs direction (i.e. *strong component*) or we can consider the arcs as edges without direction in which case the resulting components are called *weak components* [7]. Current research uses the latter definition as it is appropriated for a tree resulting from *IS-A* traversing to be recognized as a single component.

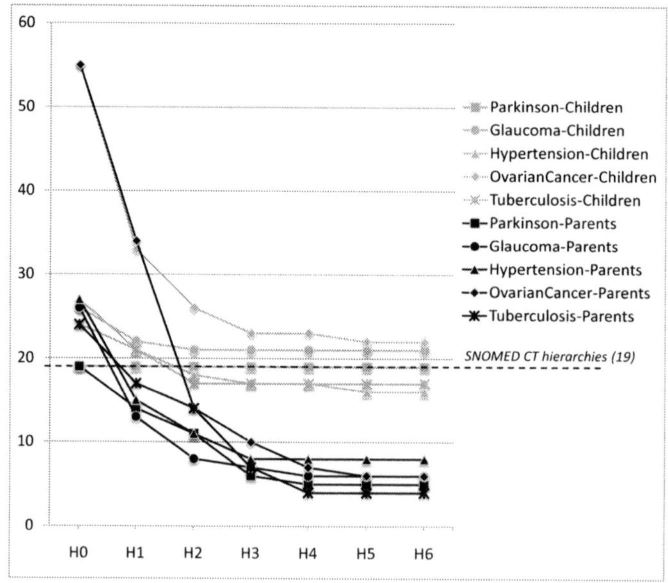


**Fig. 3.** Number of components in the subsets as we expand them up (black lines) or down (gray lines)

In addition, it should be noted that the root concept named *SNOMED-CT Concept*, that is related to all the top-level concepts through the *IS-A* relation (see section 2), has been ignored during *IS-A* traversing as it is not a clinical concept by itself and does not generalizes any clinical property. Its purpose is to classify all SNOMED-CT concepts as terminology elements, which is a different use of the *IS-A* relationship when compared to the rest of the terminology.

**Table 2.** Percentage of glossary terms which are descendants in Children-H4 subsets

| | |
|---|---|
| Parkinson | 5% |
| Tuberculosis | 27% |
| OvarianCancer | 35% |
| Glaucoma | 36% |
| Hypertension | 47% |

So, it would be expected that expanding the subsets up to more general concepts, to what is called "parents subsets", should take the number of components toward the total number of SNOMED-CT hierarchies (19). However, the chart in Fig. 3 reveals that this is not the case. The group of black lines represents the evolution of the number of components in the subsets as we expand them up toward general concepts. The results show that, after four levels of expansion (H4 in the x-axis), all parent subsets stabilize their components between four and eight, allowing us to conclude that: **a)** less than the half of SNOMED-CT hierarchies is usually enough to cover all the terms in clinical guidelines' glossaries and **b)** performing four iterations using the *IS-A* relationship in the parents direction is usually enough to link every glossary concept to the top level concept it belongs to.

The group of gray lines in Fig. 3 represents the evolution of the number of components in the subsets as we expand them down toward narrower concepts (i.e. "children subsets"). As well as the parents subsets, it can be appreciated that these subsets decrease their components as we perform the expanding iterations. Such behavior can only be due to two reasons, according to the clinical guidelines glossaries: either extracted terms include both ancestors and descendants, or they are initially unrelated concepts that can be connected through a common descendant in the SNOMED-CT terminology. It should be noted that SNOMED-CT allows for multiple inheritance. The stats in Table 2 show that both reasons are present as the amount of glossary terms which are descendants of another term in the same glossary goes from 5% to 47%. Thus, the lower this value, the more glossary terms will there be acting as local root concepts in the children subsets.

Another interesting result is that the number of components of the children subsets stabilizes after three *IS-A* expansions (H3 in Fig. 3), allowing us to conclude that three or four levels of expansion is enough for both parents and children expansions of the SNOMED-CT subsets. ***Therefore, the rest of measures in this paper are taken from H3 or H4 subsets***.

### 3.3 Analysing the Size of the Subsets Components

Typical measures to evaluate the size of components include counting the vertices (i.e. SNOMED-CT concepts) and arcs (i.e. SNOMED-CT relationships) within each component. However, the significance of such results would be hard to asses given the high variability of vertices degrees between different SNOMED-CT subareas. For example, disorders like *Poisoning* and *Cerebrovascular Disease* have 73 and 84 direct children respectively while *Obesity* and *Vitiligo* have only 9 children each.

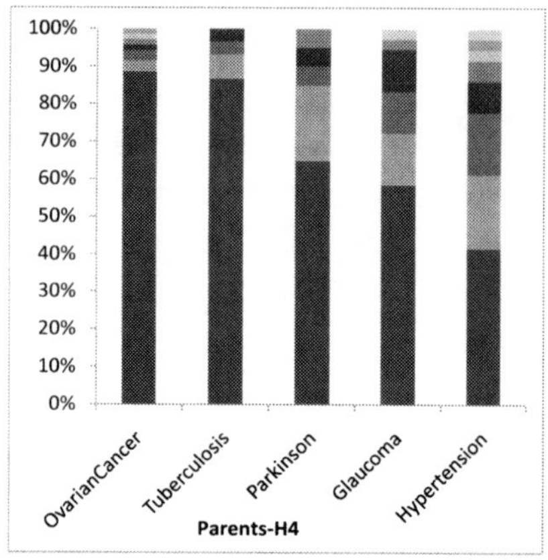


**Fig. 4.** Percentage of H0 subsets found in each parents-H4 component

Therefore, instead of using absolute values, we propose a relative measure for the size, based on the percentage of the initial subset (i.e. H0 or guideline glossary) present in each component after four up expansions (parents-H4) or down (children-H4). The chart in Fig. 4 shows distribution of the initial concepts in the components of parents-H4. The bigger the bar section with the same gray tone, the larger the number of initial concepts present in that component.

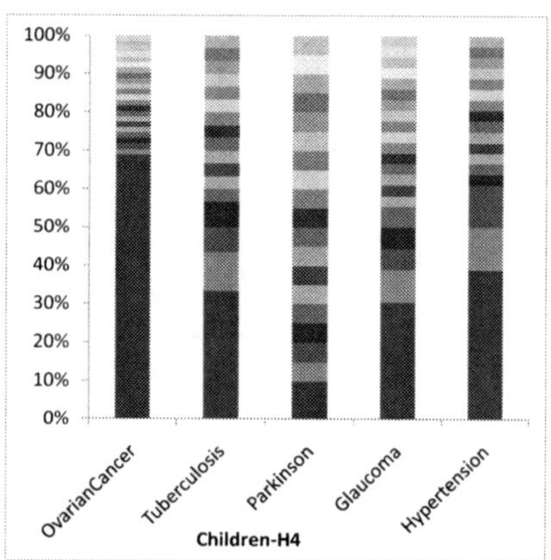


**Fig. 5.** Percentage of H0 subsets found in each childre-H4 components

The results reveal that most of the initial concepts belong to the same component which, in addition, is the component that includes the core concepts of the clinical guideline. The rest of the initial concepts is distributed in small "secondary domain" components. Although no with the same intensity, a similar effect takes place in down expansion, as illustrated by the chart in Fig. 5.

## 4 Conclusions

This paper proposed a series of graph-based results to analyze and evaluate the fitness of automatically generated SNOMED-CT subsets vis-à-vis the clinical context they were generated for. The evaluation methods were applied to 65 subsets obtained from previous experiments that used clinical guidelines' glossaries as seeds for the automatic generation of subsets.

The results show that different subsets share similar values in a particular set of characteristics and measures, therefore enabling the existence of patterns. On the other hand, there are measures whose values are quite different between subsets. Future work will include the study of such patterns in many more SNOMED-CT subsets, in order to establish whether they are due to the SNOMED-CT intrinsic structure, the features of each clinical domain and/or the subset extraction mechanism.

## References

[1] Stroetmann, V.N., et al.: Semantic Interoperability for Better Health and Safer Healthcare. Luxembourg: Office for Official Publications of the European Communities: European Commission (2009)

[2] Schulz, S., Suntisrivaraporn, B., Baader, F., Boeker, M.: SNOMED reaching its adolescence: Ontologists and logicians health check. International Journal of Medical Informatics 78, S86–S94 (2009)

[3] Rodríguez-Solano, C., Cáceres, J., Sicilia, M.A.: Generating SNOMED CT subsets from clinical glossaries: an experiment with clinical guidelines. In: International Workshop on Health and Social Care Information Systems and Technologies (HCist), CENTERIS 2011, Algarve, Portugal (2011)

[4] SNOMED Clinical Terms® User Guide, International Health Terminology Standards Development Organisation, CVR #: 30363434 (January 2010), http://www.ihtsdo.org/fileadmin/user_upload/Docs_01/Publications/doc_UserGuide_Current-en-US_INT_20100131.pdf

[5] Baader, F., Calvanese, D., McGuinness, D., Nardi, D., Patel-Schneider, P.: The description logic handbook: theory, implementation, and applications. Cambridge University Press, New York (2003)

[6] Diestel, R.: Graph Theory, 4th edn. Springer, Heidelberg (2010)

[7] Nooy, W., Mrvar, A., Batagelj, V.: Exploratory Social Network Analysis with Pajek, Illustrated edn. Cambridge University Press, Cambridge (2005)

# OWL Ontology Development for Destination Marketing

Murat Komesli, Tuncay Ercan, and Gökçe Özdemir

Yaşar University, Üniversite Cad. 35,
35100 Bornova, İzmir, Turkey
{Murat.Komesli,Tuncay.Ercan,Gokce.Ozdemir}@yasar.edu.tr

**Abstract.** Internet, as a marketing medium in the tourism industry, is a very valuable destination marketing tool. Different tourism authorities and organizations serve as marketing and service providers. They benefit from the use of Internet within the concept of marketing efforts. A new technology, Semantic Web, brought incremental changes by bringing machine-readable descriptions to current web interfaces of these organizations. This paper provides necessary framework for the applicability of Semantic Web technology on Destination Marketing. It creates a methodology on how web pages are created and analyzed with the value-added tourism marketing purposes. Semantic Web technology achieves such interoperability by referring to ontology models. The ontology presented here determines the value of this integration in destination marketing organizations.

**Keywords:** Ontology, RDF, OWL, Semantic Web, Destination Marketing.

## 1 Introduction

Internet, as a popular distribution channel for information, changed the way of communication in the tourism industry and improved the process from service providers to consumers. As the Internet mutually offers beneficial advantages from the point of both producer and customer sides; tourists use the Internet to acquire the necessary detailed information to make a good decision in a timely manner and producers use the Internet to reach more tourists in a cost-effective way. Therefore, destination marketing organizations attempt to market a destination through Internet by their appealing and informative websites. The increasing number of destination marketing organizations invests in the improvement and revision of the web sites' content in terms of information and technology. In fact, a user-friendly destination web site can make a destination be considered by potential tourists as an alternative among other destinations. Hence, creating an effective and attractive web site is of paramount to market the destination in order to stand out from its competitors.

Web pages on the Internet are designed for human readability and understanding while the computers are not able to do the same thing. Consequently, the information on the Web cannot be reused and integrated among different computer systems, applications and their databases. One way to enable machine-to-machine exchange and automated processing is to provide the information that computers can understand. The Semantic Web allows the representation and exchange of information

E. García-Barriocanal et al. (Eds.): MTSR 2011, CCIS 240, pp. 126–136, 2011.

in a meaningful way, facilitating automated processing of descriptions on the Web Semantic Web is defined as "An extension of the current web in which information is given in a well-defined meaning to better enabling computers and people to work in cooperation" by the founders of current web [1]. Web pages of the tourism authorities and organizations are queried by the users in order to retrieve required documents if they use the same domain of interest which is currently handled by ontologies.

The Semantic Web and its technologies offer a new approach to manage information processes and the fundamental principles for the creation and use of semantic metadata. Semantic metadata can exist at two levels: They may describe a document such as a web page and a part of any written document or entities within the document such as a person or company. The difference between the metadata and today's web encoded in Hyper Text Markup Language (HTML) is that HTML describes the presentation format and specifies that a given string can be displayed as bold and red font, but cannot specify that the string denotes a product price, or an author's name, and so on. However, metadata can handle intelligent software agents to make the best use of the resources available on the web [2]. It is possible to organize the metadata and find the necessary meaningful information on words and phrases rather than the textual meaning.

Using semantics also enables to merge information from all relevant documents, removing redundancy, and summarizing where appropriate. Relationships between key entities in the documents can be visually represented in semantic graphs. When existing web services are provided with semantic metadata describing their function and context, new web services can be automatically composed by the combination of these services. This is the ability to create new knowledge by drawing inferences from the existing information sources. Semantic web is based on the Resource Description Framework (RDF) language, its extension RDF Schema (RDF-S) and Web Ontology Language (OWL), which integrates a variety of applications using XML (eXtended Markup Language) for syntax and Uniform Resource Identifiers (URI) for naming. "Concepts" in an ontology may be thought as a resource identified by URI and may either exist on the web (e.g. a document) or be presented on the web (e.g. a person) [3].

Destination Marketing has an important value on Internet for the vast and heterogeneous population of users and an increasingly available supply of travel and tourism resources [4]. Jun, Vogt and MacKay argued that the Internet enhanced the information search process and encouraged consumers to decide before going on vacation [5]. Heung points out the two main considerations for Internet users that are the time saving and cost effective convenience [6]. According to Mitsche, accommodation and destinations are the categories that users mostly search through the Internet [7]. Buhalis define destinations as places towards which people travel and where they choose to stay for a while in order to experience certain features or characteristics-a perceived attraction of some sort [8]. In a similar vein, destinations can be defined as amalgams of tourism products offering an integrated experience to consumers [9]. According to Echtner and Ritchie, one major implication of pressure on being more competitive is the need for destinations and their stakeholders to take a more strategic perspective in planning, development and marketing [10].

As globalization triggered the increasing competition of destinations, the need for marketing of destinations particularly has become essential. Therefore, marketers have lately been working on new strategies to promote destinations more efficiently and professionally. As Hanlan and Kelly imply, entrepreneurs and destination marketing organizations need to consider business innovations that create a great travel experience that is engaging, robust, compelling, memorable and newsworthy which stands out from the mundane, it needs to be a great experience [11]. However, marketing a tourist destination is often difficult to organize, as there are often many stakeholders involved, all with their own aims, goals and motivations, which have to co-exist [12]. According to Kozak and Rimmington, new destinations become established, some existing ones make further progress, and others decline [13]. According to MacKay and Fesenmaier the mandate of a destination marketing program is to control image [14]. Thus, creating and managing an appropriate destination image are critical to effective positioning and marketing strategy [15]. Destination marketing is not only about imaging, branding and positioning of destinations but also segmenting the target market. Segmentation is an important component of destination marketing since marketing strategies can be developed accordingly. Leisen also emphasize that marketers need to target their promotional campaigns to the tourist markets that are the most likely to choose the given destination [16]. On the other hand, considering the intangibility of destination experience, image acts as the most significant aspect for marketing destinations. Therefore, destination marketers try to generate or adjust a favorable destination image and hold branding practices consequently in order to influence the decision-making process of the tourists. As Mohsin stresses in his study, a lack of information, a lack of marketing efforts and in consequence a failure to produce a strong brand and a strong image will inevitably lead to a lack of interest and an inability to attract tourists [17]. As a result, destinations consisting of many resources and attractions which assure the viability and thus, enhance the destination's competitiveness in the international marketplace should be marketed in a professional manner. Therefore, these unique and distinguished features of a destination should be communicated through various channels in order to attract the tourists to the destination.

This paper provides significant information for the applicability of Semantic Web technology on Destination Marketing. It contributes the previous literature with two new objectives. 1) a semantic architecture for destination marketing terminology, and 2) a representation of this architecture in an OWL ontology. This will create a common understanding for the destination marketing web sites. It also differs this work from the other related ontologies specified in Section 3. The paper is structured as follows: Section 2 outlines an intensive literature review on Destination Marketing and its uses on Internet. Section 3 outlines related ontologies and their basic specifications. Section 4 proposes a semantic architecture and its ontology for destination marketing purposes. Section 5 concludes and reflects our ideas what can be done in the future.

## 2 Literature Review

Internet is a powerful tool in marketing the destinations internationally and reaching the customers from all around the world. So, the web holds the promise of advancing tourism practices such as dissemination of data and information, transactions processing, promotion of products and services globally [18]. People use Internet for various purposes such as searching for destinations and their tourism products and experiences within. Thus, a destination web site is the most appropriate tool in promoting the destination to both potential and current tourists. So and Morrison reports that the web site users more likely visit and be more willing to take future trips to destinations [19]. Buhalis describe the distribution or marketing channels as sets of independent organizations involved in the process of making a product or service available for use or consumption [8]. Additionally, Mortorell stresses the introduction of the Internet as a commercial tool and the importance it provides as an alternative channel for communication, marketing and distribution [20].

As mentioned by Buhalis, the availability of information on the Internet and the emergence of electronic intermediaries revolutionized the global distribution [8]. Accordingly, the advent of the Internet and e-commerce presents new challenges for Destination Marketing Organizations (DMOs) as it changes how consumers are searching for travel information and destination choice [21]. Predominantly, Park and Gretzel state that DMOs invest a considerable amount of money in the development of web sites as part of their overall promotion efforts [22]. Herewith, perceived usefulness of technology reflects the degree to which a DMO considers using Internet technology as a factor that enhances or improves the organization's overall performance [23]. On the other hand, tourists are often overwhelmed by the huge amount of available information online, and cannot locate what they intended to find [24]. Fesenmaier points out that the reason for the insufficient usability of the Internet is the mismatch between the mental models of information providers and information users, which are based on their understanding of information structure and information content on the Internet [25].

Tourism has changed in fundamental ways, and new developments in technology and society are expected to continuously redefine what the future of tourism will be [26]. Fesenmaier also argues that successful destination marketing strategies require a substantial integration and coordination of Internet marketing efforts as well as the development of a favorable organizational environment that supports innovation [27]. While the Internet invalidates the assumption that small organizations cannot market on a global cost effective business basis, it has been adopted slowly and hesitantly by many Regional Tourism Authorities (RTAs) and DMOs [28]. In response to the increasing use of Internet, DMOs created web sites of their own in order to benefit from the advantages it offers. As the number of website user groups increases, software applications have been introduced to many websites to help capture bookings and keep track of inquiries [29]. Some DMOs are more advanced and sophisticated with their web technologies to make business activities more effective and efficient by providing e-commerce related capabilities such as making online reservations, selling attraction and event tickets [23]. Wang, Hwang and Fesenmaier highlights the need for DMOs to reconfigure their marketing activities to develop marketing strategies that effectively capture data about existing and potential customers, enabling them to generate actionable marketing activities in their study[30]. However, DMOs should seek a better understanding of what information tourists need, want and expect to find on the websites related to a destination and hence, how marketers can benefit the Internet technology.

Fodness and Murray suggest that tourists typically use more than one information source [31]. External information search has two dimensions: search from personal sources and destination-specific search [32]. Professional accompaniment along the entire decision process up to departure generally leads to a decline of importance for additional information [33], but people go on to look for details through Internet which provides easy and quick access to information. To the extent that the Internet enables e-travelers to easily arrange and purchase their own services/products, the future of travel agencies – the traditional intermediary – becomes uncertain [34]. Some travelers that are "planners" may have visited Web sites, including general travel Web sites for a destination, or searched the Web for specific attractions and activities, transport, and accommodation options [35]. A research indicates that the travelers search for information on the Internet for many reasons and refer to it in different decision-making stages [24]. On the other hand, inexperienced travelers need simple, understandable general information while experienced travelers need detailed information about the destination and attributes to make their vacation decisions [32]. Therefore, information quality which ontologies provide has emerged as a major research topic and meaningful information (semantic) search is perceived essential for the success of tourism organizations [36].

## 3 Related Ontologies

Many of the existing ontologies help to manage the heterogeneity of the data in tourism and destination marketing domains. Main properties of each are identified to better compare with the ontology developed at this work and explained in Section 4. Briefly, the hierarchy and relation among the tourism organizations such as RTA, DMO, CVB, etc. are showed within the scope of the work explained in this paper.

**e-Tourism Ontology** is an OWL ontology developed by Digital Enterprise Research Institute (DERI). This ontology describes the domain of Tourism. In its current version it focuses on accommodation and activities [37].

**HarmoNET.** The Harmonization Network for the Exchange of Travel and Tourism Information (HarmoNET) is an international effort aiming to exchange information and data in the domains of travel and tourism by utilizing a centralized ontology. The HarmoNET ontology includes especially on events (conferences, artistic performances, sport, etc.), accommodation (e.g. hotels and guesthouses) but neglecting self-catered accommodation (e.g. camping, holiday apartments), attractions (museum, a waterfall or a cultural district within a city, etc.), and gastro (food and drink entities). Its end users are the tourism organizations, National Tourism Organizations (NTO) and commercial players in the tourism domain [38].

**OpenTravel.** The OpenTravel Alliance is a non-profit organization aiming to plan and design specifications to ensure traveler and supplier information flow smoothly in the domains of travel, tourism and hospitality. OpenTravel creates, manages open specifications, by utilizing XML, for the electronic exchange of business information among all sectors of the travel industry [39].

**Tourism.owl** is an OWL ontology for creating Semantic Web sites related to tourism. It includes general terms in tourism domain [40].

**Travel.owl** is a tutorial OWL ontology for a Semantic Web of tourism. It includes rating information [41].

Some characteristic information about other related ontologies such as Tourism Ontology developed by the University of Karlhouse, the EON Travelling Ontology and the TAGA Travel Ontology can be accessed through [42].

# 4 Methodology for Destination Marketing

## 4.1 Semantic Architecture

The proposed framework helps the applicability of Semantic Web technology on Destination Marketing. It explains how to build the ontology which will facilitate the destination marketing information on the various active touristic web sites. Users who are travelers and potential tourists can easily reach the touristic information on the web according to their choices by the help of this ontology. We defined the basic terms and the relationships among these terms on Destination Marketing specified in OWL by utilizing "Protégé" [1]ontology development tool software [43]. These terms are destination, country, city, village, resort, region, Destination Marketing Organizations (DMO), Convention & Visitors Bureau (CVB), Regional Tourism Authority (RTA), and National Tourism Organization (NTO). The relationship among these terms and the proposed destination marketing ontology is shown at Figure 1.

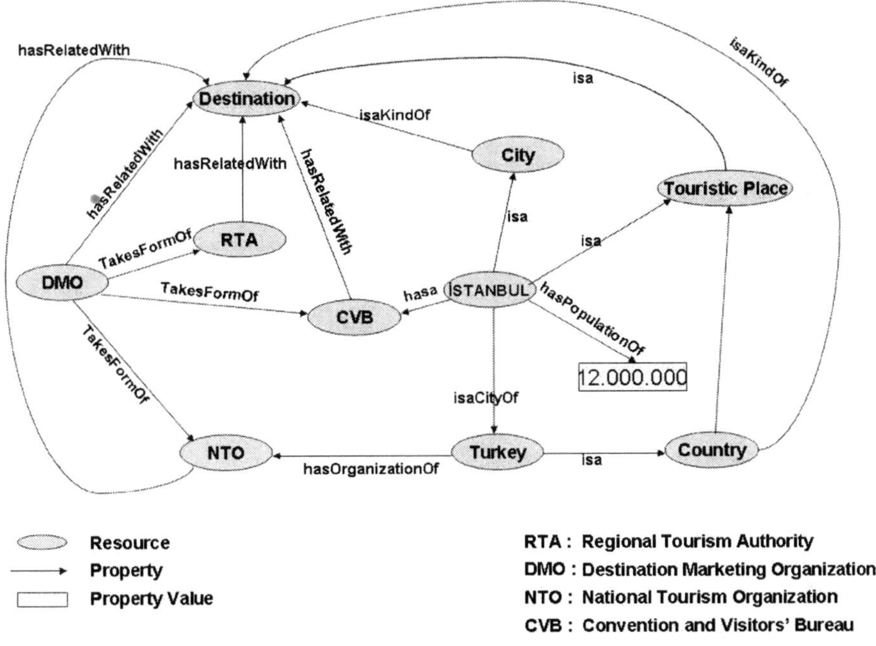


**Fig. 1.** Semantic Graph of the Destination Ontology

[1] This work was conducted using the Protégé resource, which is supported by grant LM007885 from the United States National Library of Medicine.

Concepts are specified to support computer interpretation in the ontology. The ellipses represent concepts or resources (country, city, touristic place, destination etc.). The arcs/arrows between the concepts show relationships that can be defined as attributes or properties (IsKnownAs, hasOrganizationOf, isA, etc.) of which values are other resources, and the rectangles show the attributes or property values (a string, an integer value etc.)

## 4.2 Developed Ontology

The main purpose for developing Destination Ontology is to establish a common understanding among various Destination Marketing web sites on the Internet. The web sites (actually the agent software) will be able to speak with the others by using the same language specified in the ontology. In this way, every web site on Destination Marketing or Tourism will be able to process the information such as CVB, RTA, DMO, city, country, destination, etc.

One of the main features of ontologies is that they can be referred to other ontologies. An ontology can base new concepts and relationships defined in another ontology. This means that all the universal and shared concepts can be reused. In ontology terminology, new knowledge can be added to what has already existed as long as it is consistent. In accordance with the semantic graph in Figure-1, the ontology in Figure-2 has been developed in OWL utilizing Protégé. Figure-2 shows the ontology as a snapshot of the related Protégé screen. Class hierarchy tree is on the left panel. On its right, the relations between the classes are shown in the OntoGraf tab.

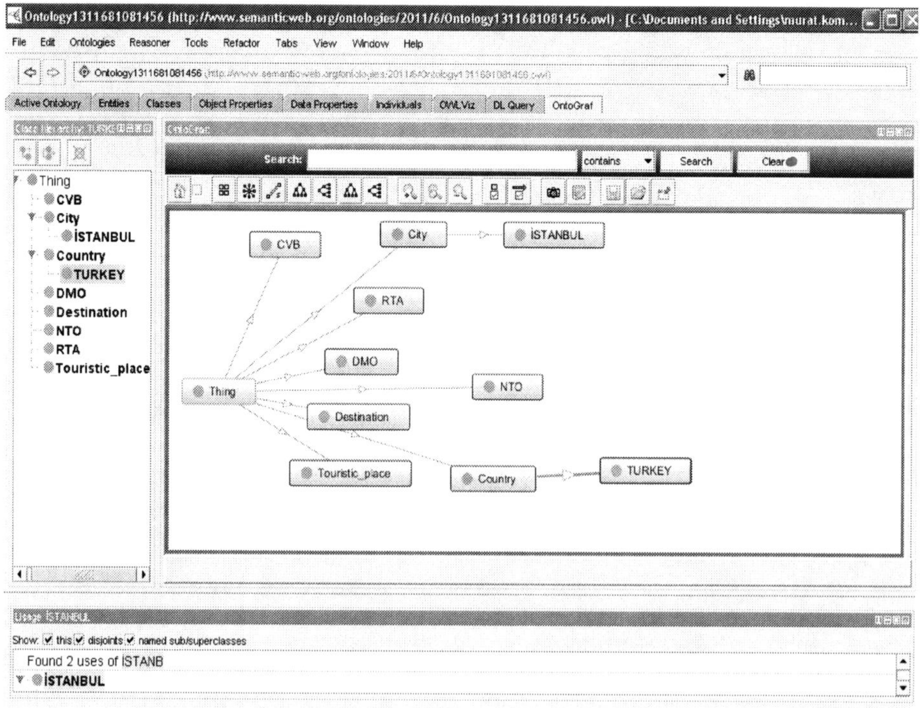


**Fig. 2.** Destination Ontology developed with Protégé

The OWL ontology developed within this project consists of Individuals, Properties, and Classes. Individuals represent objects in the domain. Individuals can be referred to as being *"instances of classes"*. Properties are binary relations on individuals, i.e. properties link two individuals together. As depicted in Figure-1, the property "TakesFormOf" links the DMO to the individual CVB. Properties are also known as resources/roles in description logic, relations in UML (Unified Modeling Language) and other object oriented notions. In some other formalism, they are called as attributes. OWL classes are interpreted as sets that contain individuals. Classes may be organized into a superclass-subclass hierarchy, which is also known as taxonomy. A reasoner can compute these hierarchical relationships automatically.

## 5 Conclusions and Future Works

After emerging semantic web technologies, destination marketing web sites will be able to exchange information automatically by means of agent software. This will simply be achieved by exchanging RDF files based on the prepared ontology. It will also save resources while preparing queries. By utilizing ontology based semantic web, tourists may not only reach touristic information easily and fast, but also purchase their travel tickets easily. For example, the user of the web based destination marketing application software can type a query on the user interface such as "*I would like to go to Istanbul from Izmir.*" The software will understand that Istanbul is a touristic city in Turkey, and check the user's calendar to look for a suitable time interval for the journey. After finding the time interval, it will also check and negotiate with travel agencies to purchase destination ticket. Last, the agent program will send user a pop-up message saying that "*Your travel to Istanbul is planned between ... (dates), and x amount of TL is withdrawn from your credit card account. Have a pleasant journey.*" This is a simple example showing the future of web based destination marketing in the light of the emerging technology "Semantic Web".

Domain specific destination marketing ontologies allow the use of well-known tourism information for destination queries incorporated with semantic web capabilities. This work is a contribution in reaching more meaningful information in the tourism context than the existing approaches aforementioned in Section-3.

It is hard to implement semantic web. The concept is complex and not easy to deploy. Tools are not supportive and stable, although more and more commercial vendors are adapting or extending their products in order to support semantic web (e.g. Oracle has developed plug-ins in order to use Oracle Database as an OWL repository). By gradually applying to practical cases, it will be understood what can be done and what else is needed in order to make it work. The semantic technology is not yet mature, although it shows immense promise for the application point of view. For instance, some initiatives are going on within the World Wide Web Consortium (W3C) [44] regarding rules. It has been identified that OWL is not expressive enough to reason and for chained relationships in some cases, and it might be necessary to define inference rules [45]. Although the new version of OWL as OWL2 has been proposed, this is one of the fields to investigate mostly in the coming years, along with how W3C initiatives can be used and intertwined with RDF and OWL. It is necessary to be in tune with standardization bodies. Ontologies need to change due to knowledge and its usage change. Thus, the evaluation of ontologies is of key importance, namely, there are two different approaches; one reflecting knowledge change and other for usage change. The emphasis of ontologies is increasingly evolving.

Internet is considered as both a distribution channel and a successful marketing tool for the destinations. Internet allows reaching both necessary and unnecessary information of the tourism products that make the consumers lose time during the information search part of the buyer's decision-making process. Thus, by creating an ontology, it is important to eliminate the unnecessary information in terms of decreasing the complexity and shortening the time spent on searching about the tourism products through Internet.

The creation of ontologies requires human intervention with a specific methodology. Previous methodologies for introducing knowledge technologies into the organization assumed a centralized approach that is inconsistent with flexibility in which modern organizations operate. The need for today is a distributed evolution of ontologies. Typically, individual users may create their own variations on a core ontology, which then needs to reflect the best of the changes introduced by users. This means that the descriptive and human-understandable content should always be provided and the data types should be explicitly stated. This study may be repeated periodically to gauge how DMOs are developing in their Internet marketing efforts. The study is universally applicable to DMOs and tourism organizations worldwide. The ontology presented in Section-4 provides a simple framework for further analysis of a range of destination sites. Equally, it provides a practical model that can be used by tourism organizations to develop or modify their destination marketing efforts. A challenge for future research is to set up an evaluation framework to assess the contribution of these techniques for specific tasks and phases of the ontology construction process.

## References

1. Berners-Lee, T., Hendler, J., Lassila, O.: The Semantic Web. Scientific American (2001)
2. Davies, J., Fensel, D., Van Harmelen, F.: Towards the Semantic Web: Ontology-Driven Knowledge Management. John Wiley & Sons, Ltd., Chichester (2003) ISBN: 0470848677
3. W3C, RDF Recommendation, 'RDF Primer', http://www.w3.org/TR/2004/REC-rdf-primer-20040210
4. Walle, A.H.: Tourism and the Internet: Opportunities for Direct Marketing. Journal of Travel Research 35(1), 72–77 (1996)
5. Jun, S.H., Vogt, C.A., MacKay, K.J.: Relationships Between Travel Information Search and Travel Product Purchase in Pretrip Contexts. Journal of Travel Research 45(3), 266–274 (2007)
6. Heung, V.C.S.: Internet Usage by International Travelers: Reasons and Barriers. International Journal of Contemporary Hospitality Management 15(7), 370–378 (2003)
7. Mitsche, N.: Understanding the Information Search Process within a Tourism Domain-specific Search Engine. In: Information and Communication Technologies in Tourism, Vienna, pp. 183–193. Springer, Heidelberg (2005)
8. Buhalis, D.: Marketing the Competitive Destination of the Future. Tourism Management 21, 97–116 (2000)

9. Davidson, R., Maitland, R.: Tourism Destinations. Hodder & Stoughton, London (1997)
10. Ritchie, R.J.B., Ritchie, J.R.B.: A framework for an industry supported destination marketing information system. Tourism Management 23, 439–454 (2002)
11. Hanlan, J., Kelly, S.: Image formation, information sources and an iconic Australian tourist destination. Journal of Vacation Marketing 11(2), 163–177 (2005)
12. Grangsjo, Y.: Destination networking: Co-opetition in peripheral surroundings. International Journal of Physical Distribution & Logistics Management 33(5), 427–448 (2003)
13. Kozak, M., Rimmington, M.: Tourist satisfaction with Mallorca, Spain, as an off-season holiday destination. Journal of Travel Research 38(3), 260–269 (2000)
14. MacKay, K.J., Fesenmaier, D.R.: Pictorial element of destination in image formation. Annals of Tourism Research 21(3), 537–565 (1997)
15. Echtner, C.M., Ritchie, J.R.B.: The measurement of destination image: An empirical assessment. Journal of Travel Research 31, 3–13 (1993)
16. Leisen, B.: Image segmentation: the case of a tourism destination. Journal of Services Marketing 15(1), 49–66 (2001)
17. Mohsin, A.: Tourist attitudes and destination marketing—the case of Australia's Northern Territory and Malaysia. Tourism Management 26, 723–732 (2005)
18. Rachman, Z.M., Richins, H.: The Status of New Zealand Tour Operator Web Sites. The Journal of Tourism Study 8(2), 62–77 (1997)
19. So, S.I., Marrison, A.M.: Destination Marketing Organizations' WEB Site Users and Nanusers: A Comparision of Actula Visits and Revisit Intentions. Information Technology & Tourism 6, 129–139 (2003)
20. Martorell, G.A.: The Internet as a Marketing Tool for Tourism in The Balearic Islands. Information Technology & Tourism 5, 91–104 (2002)
21. Cai, L.A., Feng, R., Breiter, D.: Tourist Purchase Decision Involvement and Information Preferences. Journal of Vacation Marketing 10(2), 138–148 (2004)
22. Park, Y.A., Gretzel, U.: Success Factors for Destination Marketing Web Sites: A Qualitative Meta-Analysis. Journal of Travel Research 46, 46–63 (2007)
23. Yuan, Y.L., Gretzel, D., Fesenmaier, D.R.: Internet Technology Use by American Convention and Visitors Bureaus. Journal of Travel Research 41(3), 240–255 (2003)
24. Pan, B., Fesenmaier, D.R.: Online Information Search Vacation Planning Process. Annals of Tourism Research 33(3), 809–832 (2006)
25. Pan, B., Fesenmaier, D.R.: A Typology of Tourism Related Web Sites: Its Theoretical Backgound and Implications. Annals of Tourism Research 3, 155–176 (2001)
26. Gretzel, U., Fesenmaier, D.R., Formica, S., O'Leary, J.: Searching for the future: Challenges faced by destination marketing organizations. Journal of Travel Research 45(2), 116–126 (2006)
27. Fesenmaier, D.R.: Introduction: Challenging Destination Promotion. Journal of Travel Research 46, 3–4 (2007)
28. Benckndorff, P.J., Black, N.L.: Destination Marketing on the Internet: A Case Study of Australian Regional Tourism Authorities. The Journal of Tourism Study 11(1), 11–21 (2000)
29. Hudson, S., Lang, N.: A Destination Case Study of Marketing Tourism Online: Banff, Canada. Journal of Vacation Marketing 8(2), 155–165 (2001)
30. Wang, Y., Hwang, Y.H., Fesenmaier, D.R.: Futuring Internet Marketing Activities Using Change Propensity Analysis. Journal of Travel Research 45(2), 158–166 (2006)
31. Fodness, D., Murray, B.: A Typology of Tourist Information Search Strategies. Journal of Travel Research 37(2), 108–119 (1998)

32. Gursoy, D., McCleary, K.W.: Travelers' Knowledge and Its Impact on Their Information Search Behaviors. Journal of Hospitality & Tourism Research 8(1), 66–93 (2004)
33. Beiger, T., Laesser, C.: Information Sources for Travel Decisions: Toward a Source Process Model. Journal of Travel Research 42(4), 357–371 (2004)
34. Law, R., Leung, K., Wong, J.: The Impact of the Internet on Travel Agencies. International Journal of Contemporary Hospitality Management 16(2), 100–107 (2004)
35. Hyde, K.F., Lawson, R.: The Nature of Independent Travel. Journal of Travel Research 1(42), 13–23 (2003)
36. Kim, D.Y., Lehto, X.Y., Morrison, A.M.: Gender Differences in Online Travel Information Search: Implications for Marketing Communications on the Internet. Tourism Management 28(2), 423–433 (2007)
37. E-Tourism Working Group, `http://e-tourism.deri.at/`
38. The Harmonisation Network for the Exchange of Travel and Tourism Information, `http://www.harmonet.org/`
39. OpenTravel Alliance, `http://www.opentravel.org/`
40. Tourism.owl contributed by Tonya Kalinka, `http://www.bltk.ru/OWL/tourism.owl`
41. Travel.owl contributed by Holger Knublauch from Stanford University, `http://protege.cim3.net/file/pub/ontologies/travel/travel.owl`
42. Prantner, K., Ding, Y., Luger, M., Yan, Z., Herzog, C.: Tourism Ontology and Semantic Management System: State-Of-The-Arts Analysis. In: IADIS International Conference WWW/Internet, Universidade de Trás-os-Montes e Alto Douro, Vila Real, Portugal (2007)
43. Protégé Web Site, `http://protege.stanford.edu`
44. W3C, Semantic Web Introduction, `http://www.w3.org/2001/sw`
45. W3C, OW2 Recommendation, 'OWL Overview', `http://www.w3.org/TR/owl2-overview/`

# New Process Ontology-Based Character Recognition

Aicha Eutamene, Hacene Belhadef, and Mohamed Khireddine Kholladi

Mentouri University of Constantine, Department of Computer Sciences,
Road of Ain El Bey, Constantine 25017, Algeria
{aicha.eutamene,hacene_belhadef,kholladi}@umc.edu.dz

**Abstract.** We present in this paper, the general description of a classical process of character recognition, and subsequently we describe our contribution for a new process that is characterized by a step of creating of domain ontology, representing the Latin alphabet in the form of concepts, where each concept represents a grapheme (features). In this work we consider that there is a robust extraction module of primitives that will serve us to instantiate our ontology, by the different graphemes and their spatial relationships that exist in the processed document. Our ontology is generic and can support other languages by enriching it by new specific spatial relationships. Our process represents a new approach in the domain of character recognition, we have not yet developed a final product but our idea is promising in this domain.

**Keywords:** Annotation, Characters recognition, Grapheme, Ontology, Pattern recognition, Semantic, Typographical primitives.

## 1 Introduction

The general goal of the recognition of documents; whether printed or manuscript, is to transform them into understandable and usable representation by machine. The process of recognition is not always easy as long as the content of documents can have multiple representations. In the case of printed documents, size, style (Bold, Italic... etc.), the cast of characters and other factors play a crucial role in this process. As for handwritten documents, the conditions of safeguarding are not often adequate. In our days, a large number of books and old manuscripts are preserved in museums and archives are in danger of disappearing due to several factors such as moisture, acidity. This requires digitization of these documents in order to preserve the heritage and exploit it more effectively. The digitization of documents is the most effective and speedy remedy, it consist to convert a paper document in the form of a digital image. Transcription is another solution but it is less used and limited to manuscripts documents not-long.

The result image of such operation of digitization is used as raw material in the recognition process, to decorticate the content and extract the necessary primitives for the identification and characters recognition, also the entire contents of the document, to use it in a lot of area such as the restoration of national heritage or world, classification, indexing and archiving.

E. García-Barriocanal et al. (Eds.): MTSR 2011, CCIS 240, pp. 137–144, 2011.

No matter the rich content of the documents, but this wealth is insufficient to help the process of character recognition. All OCR systems interested in the content of documents is usually based on the step of classification forms but they are not interested in what's behind this, as meta-information or information semantics. In this paper we realized that a step annotation of the document is necessary to add additional information to help this process to accomplish its task. The annotation of an image through the construction of ontologies is the main tool for associating semantics to an image and allows the use of more powerful search methods and able to answer complex queries. The association between data and ontologies then allows software agents to take advantage of the knowledge represented in ontologies to better use of images.

The present paper is organized as follows. In the first part, we present an overview on the two concepts closely linked to our process, as the ontology and grapheme. In the following paragraphs, we present a description of the steps of a classical recognition process, dice digitization until the last stage which is the post-treatment, and in the third section we illustrate our contribution in such process and we end by an illustrative example to show how our approach works by using ontologies.

## 2 Ontology

All The term "ontology" comes from the field of philosophy that is concerned with the study of being or existence. In philosophy, one can talk about ontology as a theory of the nature of existence. In the context of computer and information sciences, ontology defines a set of representational primitives with which to model a domain of knowledge or discourse. The representational primitives are typically classes (or sets), attributes (or properties), and relationships (or relations among class members). [2]

The preceding definition leads to a set of definitions that can be used as a basis for algebraic formulation of the term Ontology and its components [1]:

**Definition 1. A term** is a triple $\tau = [\eta, \delta, A]$ , $\tau \in T$, where $\eta$ is a string of characters containing the name of the term, $\delta$ is a string of characters containing its definition and A is a set of *attribute domains* $A_1$, $A_2$, ..., $A_n$, each associated to a value set $V_i$.

**Definition 2. A relation $\phi$**: $T \rightarrow T$, $\phi \in \Phi$, : is a function from T to T such that for every term $\tau_1 \in T$ , there is a term $\tau_1 = \phi(\tau_1), \tau_2 \in T$.

**Definition 3. A semantic relation $\sigma$** between two terms is a relation that belongs to the set of semantic relations $\Sigma =$ {*Hypernymy, Hyponymy (is-a), Mereonomy (part-of), Synonymy* }, $\Sigma \subset \Phi$ .

**Definition 4. A spatial relation $\rho$** between two terms is a relation that belongs to the set of spatial relations P = {*adjacency, spatial containment, proximity, connectedness*}, $P \subset \Phi$.

**Definition 5. An ontology** is a pair $\Theta = [T, \Phi]$ , where $T = \{\tau_1, \tau_1, ...., \tau_n\}$ is a set of terms, and $\Phi = \{\Phi_1, \Phi_2, ......, \Phi_n\}$, and $\exists \phi_i \in (\Sigma \cup K)$ .

## 3 Grapheme

The grapheme is the fundamental unit of a writing data; it is the smallest unit of meaning graph whose variation changes the value of the sign in writing[1].

For ideographic scripts, it can represent a concept. In phonographic writing, it represents an element of achieving sound (syllable, consonant, and letter). So in alphabetic writing, the grapheme is commonly referred letter[2]. In our work, we show the interest of using the graphemes as features for describing the individual properties of Handwriting (Part of character).

Each grapheme is produced by the segmentation module [3] of a recognition system. At the end of a segmentation step, the document can be viewed as the concatenation of some consecutive graphemes (**see example section**). The handwritten document D is thus described by the set of graphemes $X_i$

$$D=\{ X_i , i:1 \text{ to } n\}$$

A subset of successive graphemes, may construct a word $W_j$, or a single character $C_k$:

$$Wj=\{X_i, i:1 \text{ to } m, m<n\}$$

$$C_k=\{X_i, i:1 \text{ to } p, p<m<n\}$$

## 4 Classical Character Recognition Process

The main steps of a recognition process are [5] (see figure 1):

The acquisition enables the conversion of paper document in the form of a digital image (bitmap). This step is important because it is concerned with preparing documents to be seized, the choice and parameterization of hardware input (scanner), and the format for storing images.

Pre-processing whose role is to prepare the document image processing. Pre-processing operations are related to the recovery of the image, remove noise and redundant information, and finally the selection of appropriate treatment areas.

Recognition of the content, that often leads to the text recognition and extraction of logical structure. These treatments usually accompanied by preparatory operations of block segmentation and classification of media (graphics, tables, pictures, etc...).

Post-processing or correction of recognition results to validate the digitization process. This can be done either automatically by the use of dictionaries and linguistic methods of correction, or manually through dedicated interfaces.

## 5 Our Contribution

The main goal of image analysis ontology (IAO) development is formal description of knowledge on the processing, analysis, and recognition of images accumulated and used by experts [7].

Our approach is based on the architecture of a new system of character recognition manuscript, which was described in [6] [8].

---

[1] Wikipedia.

[2] `http://alis.isoc.org/glossaire/grapheme.fr.htm`

Our contribution (see figure 1) is located between the step of features extraction and recognition step, it is expressed by instantiating the domain ontology already created by a domain expert (see figure 2). In this figure we show just an excerpt of the global ontology representing all the Latin alphabet. In this ontology, we implemented all possible relationships, which can be found between the different graphemes that build the characters, these relationships are of spatial type (Just-below, Next-left, etc. ...). For example, the grapheme Tal-trunk, can have several relationships with graphemes: Bar, Arc, etc.... to form respectively the characters **L** and **P** (see figure 3, for more examples).

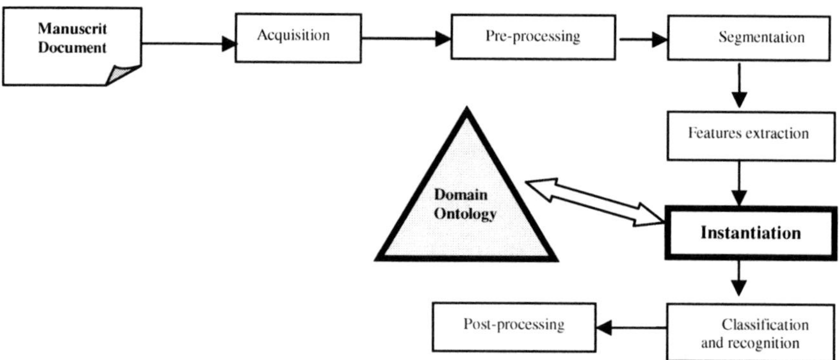


**Fig. 1.** Architecture of new system of character recognition

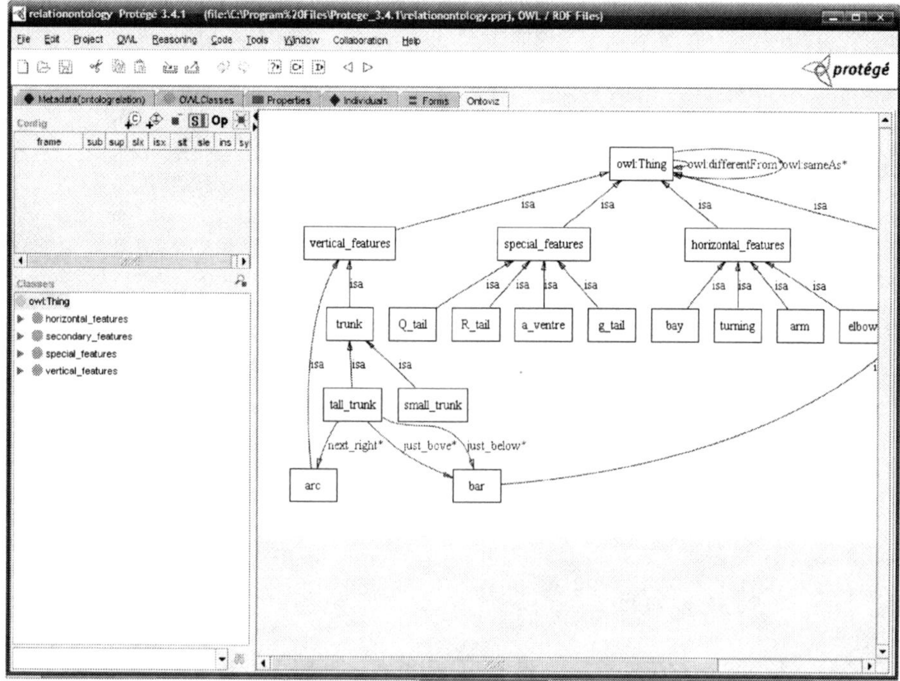


**Fig. 2.** Ontology of the Latin alphabet

Figure 2, presents our ontology that created under protégé editor[3] . "Protégé» is the best known editor and most widely used, it is open source, developed by Stanford University, it has evolved since its first versions (Protégé-2000) to integrate from the 2003 web standards including OWL semantics. It offers many optional components: reasoners and graphical interfaces.

In the figure 4, we show an example of the different graphemes that form the characters D, L, P and T, for example the character "L" is composed by two graphemes such as **Tall-trunk** and **Bar**, respectively numbered by number 1 and 2.

The relationships between all graphemes, that compose these characters, are clearly depicted in figure 5, and the corresponding code of these relationships is shown in figure 4. This description is based on RDF[4] formalism.

In this example, we can notice that a given grapheme may be figured in many character, same for relationships. The only thing that can eliminate this ambiguity, is the numbering of each grapheme to locate its position in the document, which is why we created an attribute "hasNumber" (see Figure 3 (a)) for each concept, to distinguish it during treatment.

```
   <rdf:Description rdf:about="#hasnumber">
       <rdfs:range rdf:resource="http://www.w3.org/2001/XMLSchema#int"/>
       <rdfs:domain rdf:nodeID="A0"/>
       <rdf:type
rdf:resource="http://www.w3.org/2002/07/owl#DatatypeProperty"/>
     </rdf:Description>
```

(a)

```
     <rdf:Description rdf:about="#tall_trunk_1">
       <hasnumber
rdf:datatype="http://www.w3.org/2001/XMLSchema#int">1</hasnumber>
       <hasname
rdf:datatype="http://www.w3.org/2001/XMLSchema#string">taltrunk1</hasname>
       <just_next rdf:resource="#long_arc_1"/>
       <rdf:type rdf:resource="#tall_trunk"/>
     </rdf:Description>
```

(b)

**Fig. 3.** Description of the attribute "hasnumber"

The example in Figure 5, shows that the characters D, P, L and T, share the same grapheme is "Tall_trunk" by an instantiation operation to each occurrence of the grapheme in the document, by distinguishing it by a number, representing its position (see Figure 3 (b)). For example the concept "Tall_trunk" can be instantiated several times "Tall_trunk_1", "Tall_trunk_2", "Tall_trunk_3" and "Tall_trunk_4" where each instance represents an occurence in the document.

[3] `http://protege.stanford.edu/`

[4] RDF: **R**esource **D**escription **F**ramework.

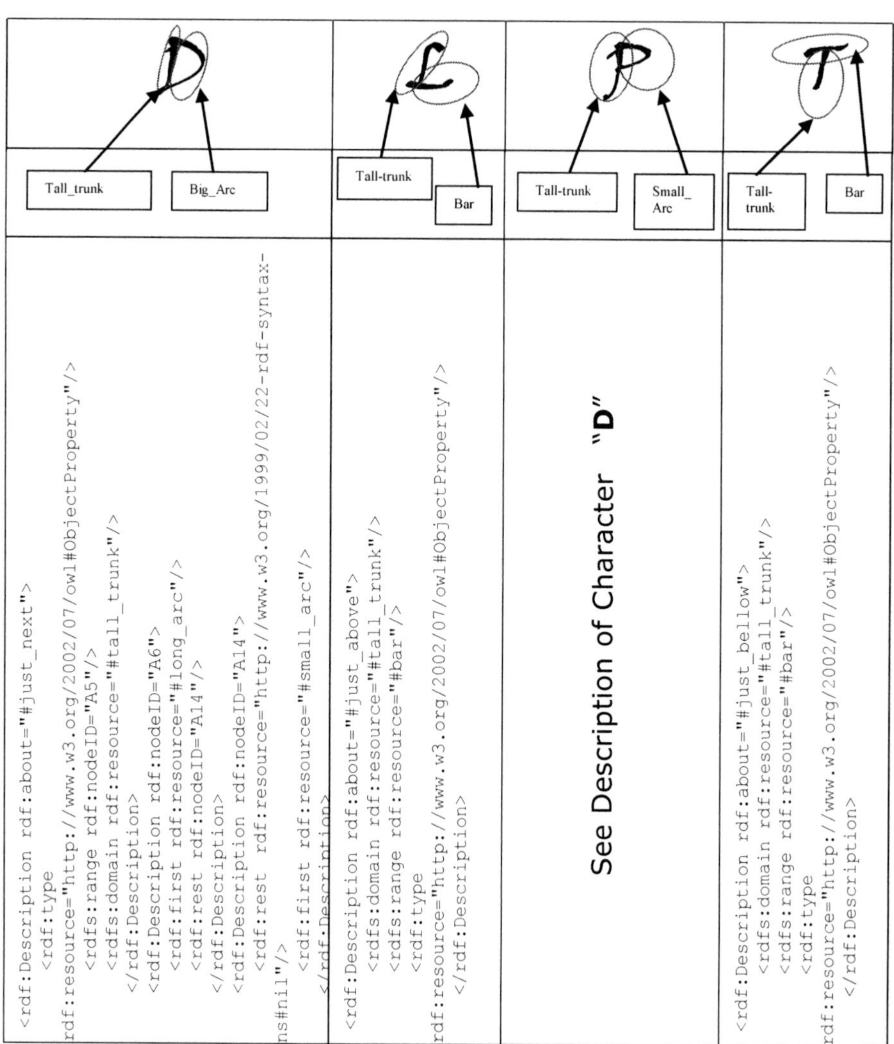


**Fig. 4.** Segmentation of some characters in graphemes features

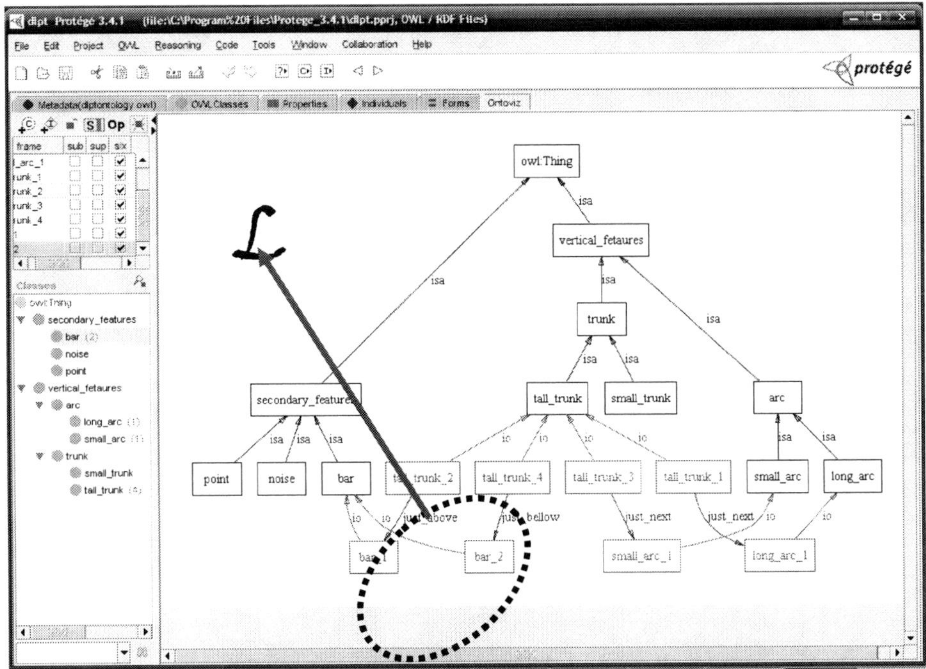


**Fig. 5.** Relationships betwen different intances for Characters of figure 3

## 6 Conclusion

The work presented in this paper presents a new and original vision, to solve the problem of character recognition, our idea is expressed through the creation of an ontology that represents the contents of a document written manually or automatically (by seizure). This ontology transforms the image-version of document to a representation of concepts and spatial relationships. The idea here is not limited to the representation of the content but rather the exploitation of side semantic of this content, taking advantage of all the benefits of ontologies, including the formulation of local queries which are necessary for making intelligent decisions, or creating the web services that can work above. The development of ontologies in this area can be used to provide image analysis automation support and efficient use of modern methods and techniques for image analysis and pattern recognition.

In this paper we have presented a new approach in the domain of character recognition, we have not yet developed a final product but our idea is promising in this domain. The authors of this paper are trying to validate their approach on printed materials. The results are not yet ready but this idea opens a large door in front of researchers for using the ontologies in this field and to solve the problem of recognition but of a semantic point of view. In a future work, we plan to use the matching operations with external resources such as WordNet[5], using dedicated similarity measures.

[5] WordNet : A lexical database of English: http://wordnet.princeton.edu/

## References


[1] Frederico, F., Clodoveu, D., Gilberto, C.: Bridging Ontologies and Conceptual Schemas Geographic Information Integration. Geoinformatica 7(4), 355–378 (2003)

[2] Thomas, R.G.A.: Translation Approach to Portable Ontology Specifications. Knowledge Acquisition 5(2), 199–220 (1993)

[3] Bensefia, A., Paquet, T., Heutte, L.: Information retrieval based writer identification. In: 7th International Conference on Document Analysis and Recognition (ICDAR 2003), pp. 946–950 (2003)

[4] Document pour l'école Jeunes Chercheurs CNRS : INTERACTION HOMME-MACHINE, Luminy (1997)

[5] Belaid, A.: Reconnaissance automatique de l'écriture. Pour la science (2001)

[6] Eutamene, A., Belhadef, H., Kholladi, M.K.: Thinking about a new process of handwritten characters recognition based on ontology. In: SNIB 2010, 7th National Seminar in Computer Science. Biskra-Algeria (2010)

[7] Gurevicha, I.B., Salvettib, O., Trusovaa Yu, O.: Fundamental Concepts and Elements of Image Analysis Ontology. Pattern Recognition and Image Analysis 19(4), 603–611 (2009)

[8] Belhadef, H., Kholladi, M.K., Eutamene, A.: A General Description of a New Character Recognition Process Based on Ontology. In: Proceedings of the International Conference, ICDIPC, Part II. CCIS, vol. 189. Springer, Heidelberg (2011) ISBN 978-3-642-22409-6

# Smart Semantic Content for the Future Internet

Muriel Foulonneau

Tudor Research Centre, Luxembourg
muriel.foulonneau@tudor.lu

**Abstract.** The reuse of metadata in open environments where content can be shared is a challenge. The Internet is undergoing major changes as the vision of objects publishing content (Internet of Things) gives birth to a new paradigm where the amount of data, metadata, and paradata about digital objects will exceed whatever our current systems are ready to deal with. We propose to bridge two semantic models currently proposed for the Future Internet: Content Objects and Content Centric Networks. We show how the corresponding metadata can be used in Future Internet networks and we present strategies to extend the model with existing semantic models.

**Keywords:** Metadata, ontology, content object, Future Internet, semantic rules, Content Centric Networks.

## 1 Introduction

While most metadata models are focused on serving particular types of functions, such as the IFLA FRBR (Functional Requirements for Bibliographic Records) functions (Find, Identify, Obtain, Select) (IFLA, 2009), the current changes on the Internet will make the current models very incomplete. Already, the reuse of metadata in open environments where content can be shared, exchanged, even misused is a challenge. The Internet is undergoing major changes as the vision of objects publishing content (Internet of Things) (Kortuem et al., 2010) gives birth to a new paradigm where the amount of data, metadata, and paradata about digital objects will exceed whatever our current systems are ready to deal with. This has led to conceive the Future Internet, as the major infrastructure transformation of the current Internet at the network level, as well as at the middleware and the content level.

In order to cope with the current evolution, Content Centric Networks have gained interest. The network level deals with content instead of packets. Most importantly, a minimal level of semantic processing is transferred to the network level. In addition, the future of media resources as presented by the Networked and Electronic Media platform (Janse et al., 2009) requires rethinking the way in which content is modeled in order to be efficiently created, managed and distributed on future networks. Zahariadis et al. (2010) have therefore proposed the design of Content Objects, which embed multiple levels of semantics. Overall, semantic layers will cover different levels of the Future Internet architecture.

This will allow the creation of new user experiences, with personalized content delivery and context aware applications. In the scope of our current work on the

E. García-Barriocanal et al. (Eds.): MTSR 2011, CCIS 240, pp. 145–154, 2011.

optimization of data collection for context aware applications, we propose to bridge both semantic models: Content Objects and Content Centric Networks. We show how the corresponding metadata can be used in Future Internet networks to shape a semantic layer across different levels of architecture and we present strategies to extend the model and use metadata not only as passive data but also as dynamic rules.

## 2 Models for Content on the Future Internet

Both Content Centric Networks and Content Objects have been developed in different ways. However, their interactions are still under-specified. In this section, we present the models and their semantic layer.

### 2.1 Content in Content Centric Networks

The concept of Content Centric Networks (CCNs) has gained interest in the recent years (Jacobson et al., 2009). Content Centric Networks aim to focus on the delivery of content to users, seamlessly, whichever the content and user location. The network itself is in charge of finding the content and delivering it from any number of locations where the content may be copied/cached. This model is different from the classic conversational model, where content stored at the content provider location is delivered to user devices.

CCNs have gained interest in particular in the context of the Future media topic of the European Future Internet Initiative (EFII)[1]. This implies that content is mainly perceived as multimedia content. Multiple copies of a content item are stored in the network nodes, at router level and delivered according to the needs, so that very popular contents can benefit from a large bandwidth, whereas contents with a limited audience do not need the same resources. Popular content is likely to be copied many times on the network, as opposed to scarcely used content only available from few locations. Van Jacobson stated that "requests for content from a terminus do not have to go all the way to the host, only just as far as the first stored copy." (Tselentis et al., 2010).

In this context, content is stored at the level of network nodes in the form of data packets. "A small change in router algorithms could remember the most requested packets rather than flushing all the traffic" (Tselentis et al., 2010). Zahariadis et al. (2010) add "the Search Engines crawl the Internet or inspect the routed packets to find, classify and index content or services."

Interest Packets represent the request for content, through the content name, selector information, and a random nonce value used to de-duplicate Interest packets. The matching of Data Packets with Interest Packets is based on the Content Name property.

### 2.2 Content Objects as Content Mashups

Zahariadis et al. (2010) have focused on content in Content Centric Internet, specifically on how media objects would live in Content centric Internet. Objects should not only be adapted but recomposed on the fly. Content is perceived as mashups.

---

[1] http://initiative.future-internet.eu/

They define Content Objects as "the smallest addressable unit" in the Autonomic Layer-Less Objects Architecture *(*ALLOA*)*. "Content is any type and volume of raw information that can be combined, mixed or aggregated to generate new content and media. Content may be prerecorded, cached or live, static or dynamic, monolithic or modular." (Zahariadis et al., 2010).

If a person draws a 2D scene, then it can be populated with people, objects etc, retrieved from a CCN. The content viewed by end users may be a combination of content distributed across the network and recomposed dynamically based on personalization of context awareness mechanisms.

Content is fetched into a Content Object, which contains the *media* as well as metadata. The *media* in this context is typically a multimedia file. Metadata is descriptive metadata (characteristics), but also behavioral information, relations to other resources and rules. Content objects therefore contain sufficient information to be composed dynamically, in particular through Relations, Behavior, and Rules.

Both mechanisms, although complementary and built on a semantic layer are not conceived in an integrated manner. Indeed, CCNs do not specify the content of data packets and Content Objects do not specify the mechanisms for their delivery. However, we can foresee that the integration of both raise new opportunities to reinforce he semantic layer of the Future Internet.

## 3 Model for Content Objects on Content Centric Networks

We propose a model to integrate both frameworks: the Content Centric Networks proposed by Van Jacobson (Jacobson, 2009) and the Content Object Model proposed by Zahariadis et al. (2010). It aims to support the delivery of different types of content in Content Centric Networks and develop a model which crosses boundaries between architectural layers (Foulonneau et al., 2010). We implemented the model using semantic technologies (RDFS) for the content transferred across the networks.

The model reuses in particular the concepts of Data and Interest Packets from the Content Centric Networks, and the concepts of Content Objects from the model proposed by Zahariadis et al. (2010). It also harmonizes the basic concepts of Content, Data, Information, and Metadata.

### 3.1 Content, Data, Information, and Metadata

The first difficulty is to adopt of coherent model for data, content and information, since these concepts are used in a different way in the network environment and in the multimedia environment. Indeed, data is a semiotic unit but perceived in a different manner according to the context of use (Tuomi, 1999). We therefore had to define a model for the concepts of Content, Data, Information, and Metadata.

In our context, Content is defined as the representation of anything that has to be transferred over the network. This includes multimedia content but also an alert, sensor data or user feedback on a recommendation for instance. In order to transfer content over the network, it is downgraded to data packets. In order to store and process data, it is downgraded to data. The organization of data allows preserving relations between the data (i.e., its context). Information can be retrieved by locating data and its context (i.e., the set of relations to other data).

Domain specific content (such as a car crash alert) can be defined as subclasses of media (which we renamed with the more generic term content) in the model defined by Zahariadis et al. (2010). The model represents the content and metadata. We create the concept of DataRelation, which represents the data structure (Figure 3). Together with the data, it allows creating information. It is therefore the elements added in the knowledge pyramid (Tuomi et al., 1999) between Data and Information.

Information (I) is contextualized data, i.e. data (D) and the relations (R(D)) between data and other data:

$$I = \{D, R(D)\}$$

We introduce the concept of Metadata as data about data, i.e. data that should be related to other data to allow making sense of them (Foulonneau et al., 2008). However, in RDF, data can be a part of a RDF statement (e.g., the Object of the statement) or it can be the full RDF statement. For instance, in Figure 1, Data is represented by statements (D1, D2, and D3). In order to understand the meaning of D1, it is necessary to know D2 and D3. The related Data D2 and D3 are considered Metadata of D1. Nevertheless, 37 can also be considered a data and be a metadata of Sensor1. A resource is a metadata of another resource if they can be related by a property, this includes statements.

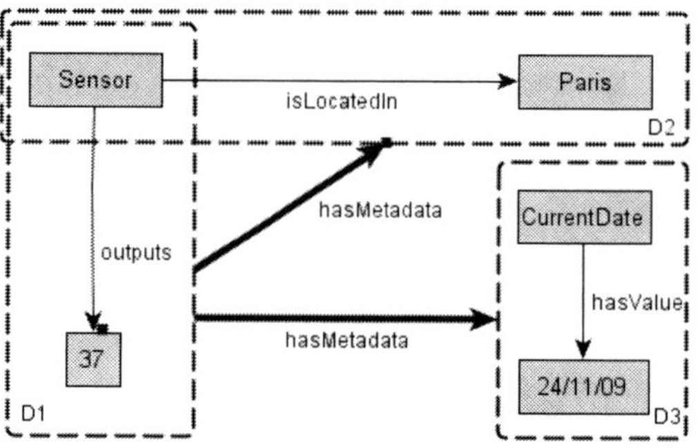


**Fig. 1.** Relations between Data

At this point, the definition of Metadata (MD) is not considered strictly equal to R(D), but rather a subset of R(D).

$$MD \in R(D)$$

Indeed, if we relate D1 with a data D4 consisting of *Sensor2 outputs 37*, then, this can be considered a contextualization of D1, to provide the information that multiple sensors have output the same measure. However, it is not a metadata, since it does not provide any information to make D1 understandable. A data can be metadata of any other data. In the RDF environment, it becomes a metadata with the creation of a predicate between two data.

### 3.2 The Content Object Model

Content Objects are containers. They gather metadata in addition to the content itself, as illustrated on Figure 2. They include both ContentMetadata and Content.

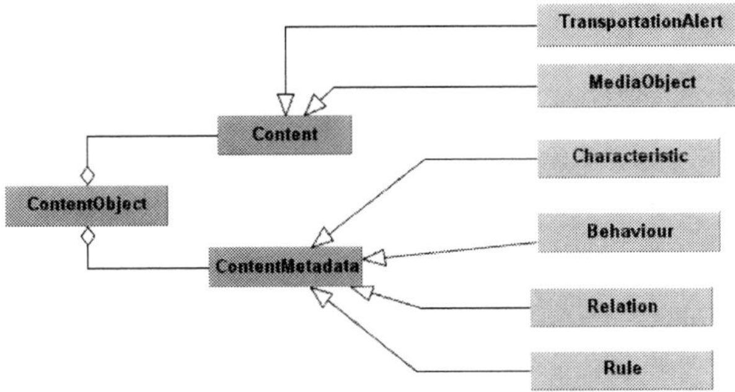


**Fig. 2.** Modified Content Object Model

The Content Object Model presented in Figure 2 is adapted from Zahariadis et al. (2010). In the original model defined by Zahariadis et al. (2010), the Content Objects contain Media and Metadata (as characteristics, behaviors, relations, and rules).

### 3.3 Linking Content Objects and Data Packets

In CCNs, Content is stored as data packets, i.e., it is downgraded to data and managed using packets. Data packets must hold sufficient information (properties) to be retrieved dynamically. Content Objects must hold sufficient information (properties) to create adaptive content mashups dynamically.

In traditional models, content can be decomposed to create processable data, then embedded in packets for the purpose of transfer. However, in CCNs, DataPackets are defined for Content or Content Objects, as presented in our model on Figure 3. They include Data (to represent Content Objects) and metadata associated to the Content Objects. In addition, InterestPackets contain the requests sent to the network. The ContentObjectMetadata is added. Together, they form the Data Packet. The InterestPacket which contains information about the requested content uses InterestMetadata. Collections are dynamic aggregations of DataPackets matching a given InterestPacket (Jacobson et al., 2009).

Anything that is transferred across the network is decomposed into processable data, i.e. data which can be handled by computer systems (machine semiotic unit). Therefore, in **Figure 3**, Data Packets are composed of Content Objects expressed as Data.

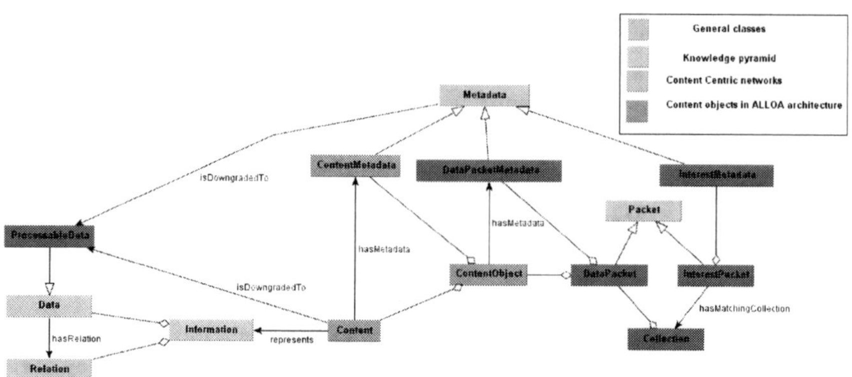


**Fig. 3.** Integration of Data - Information - Content for content centric networks

### 3.4 Packet Properties

The CCN level is characterized by the semantic elements associated to packets and the implementation of a minimal level of retrieval based on that semantic layer. That semantic layer is directly related to the Content Object layer.

DataPackets and InterestPackets both have a ContentName. The look up process for retrieving content is based on this property. Content naming holds semantics, for instance on versioning. The look up process scans COContentName to find out whether it contains InterestContentName (e.g., a match will found if InterestContentName is /food/chocolate, whereas COContentName is /food/chocolate/whiteChocolate). COContentName may or may not contain InterestContentName. If so, then it belongs to the InterestContentName Collection.

This model shows the interaction between semantic layers of both CCNs and Content Objects. We then illustrate the additional semantics included in Content Objects in order to allow its efficient handling by a middleware layer.

## 4 Content Object Properties

Figure 2 represents the categories of metadata and the structure of the Content Object packages. These are divided in Characteristics, Behavior, Relations, and Rules.

ContentMetadata should fit in the Content Object container. A Content Object container must therefore be designed in accordance with the Content Object Model presented on Figure 2. Figure 4 describes the type of metadata that can be added in the Content Object. In most cases, it is not necessary to create metadata elements from scratch. It is sufficient to include concepts already developed for specific purposes (from the ITEA2 WellCom[2] and ITEA2 CAM4Home[3] projects for instance, as illustrated below).

---

[2] http://en.wikipedia.org/wiki/Eureka_ITEA2_WellCom_Project

[3] http://www.cam4home-itea.org/

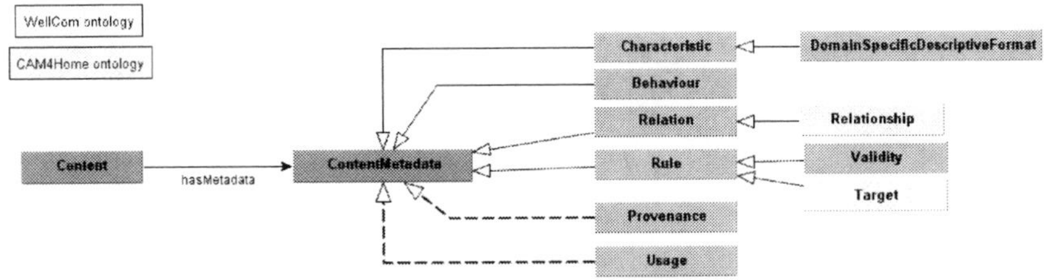


**Fig. 4.** Non exhaustive set of Content Metadata for Clairvoyant

In addition, two categories of metadata were added. Provenance is intended to ensure content traceability across the network. UsageData aims to record information on the actual usage and impact of resources (Foulonneau, 2010). This information is the basis of recommendation mechanisms in collaborative filtering for instance.

## 4.1 Characteristics

Characteristics are domain specific metadata. In the case of media files, a model must be adopted. The representation of media content for instance can be done through the Ontology for Media Resource[4] proposed by the W3C group or through the TV-Anytime inspired model developed in the scope of the Movies project (Naudet et al., 2010). Many metadata models have been developed to describe digital resources in a variety of contexts, including LOM, MARC, Dublin Core for instance. The recent initiative by Yahoo!, Microsoft and Google to create Schema.org as well as other metadata registries[5] and ontology repositories can help provide metadata models for particular types of content.

## 4.2 Behaviour and Rules

Behaviour is defined as "the way the object affects other objects or the environment" (Zahariadis, 2010). Rules on the opposite describe the way in which an object is impacted by its environment.

It is necessary to represent the way in which content should be consumed by specific services for instance (e.g., can be resized). Display constraints should be represented (e.g. should always be displayed on the left hand corner of a screen).

It is possible to attach a Target rule, in order to specify to whom or in which context content should be usefully delivered (Naudet et al., 2010). For instance, a particular representation of information may be targeted to visually impaired users.

A validity stamp can be attached to the content for a specific usage: e.g., a message providing sensor data may have to be consumed in the next 3 hours, otherwise its information is not valid any longer.

We implemented these rules as semantic statements. They can be implemented as semantic rules using SWRL for instance. However, this requires extending the semantic models to support actions on the content and external known or unknown applications and services.

---

[4] http://www.w3.org/TR/2010/WD-mediaont-10-20100608/

[5] http://metadataregistry.org/

### 4.3 Relations

Relations must be defined according to the needs for adaptation and re-composition of content. The CAM4Home project has provided an example of such relations for multimedia content and services.

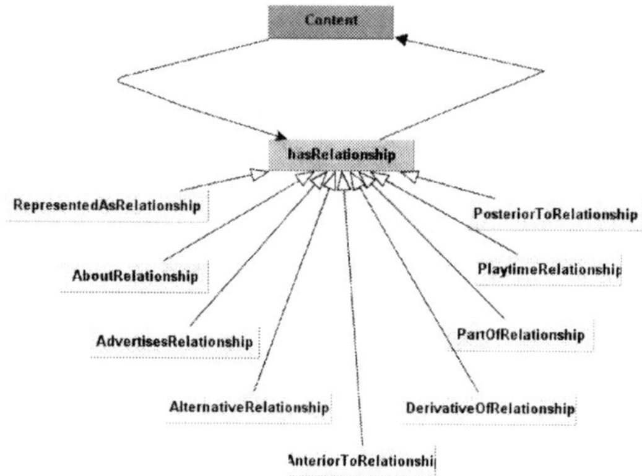


**Fig. 5.** Content Relations

Figure 5 shows the integration of the relation defined in the scope of the CAM4Home project (Bilasco et al., 2010). Other relations can be defined according to different types of content and the context of use.

### 4.4 A Provenance Node

It is possible to add a provenance node to the Content Object Metadata, in order to keep track of the events which created or affected the content. This would ensure the traceability of the processes and sensors which led to provide a specific content.

Different frameworks have been defined to handle provenance. Provenance is used in long term preservation environments (e.g. PREMIS), as well as scientific environments (e.g. the Open Provenance Model OPM). The W3C has opened an incubator group on provenance information[6].

The most elaborate model is the Open Provenance Model, which is the result of a series of workshops and challenges (Moreau et al., 2010).

Provenance information in the context of context-aware applications is necessary if the system needs to trace the chain of events which ended in the provision of a specific content, in order to assess the trust in that content in the following cases: for conflict resolution, or for under-weighting specific statements in the recommendation process, or for processing a feedback on the recommendation.

Accordingly, a provenance node was added to the existing model (Figure 4).

[6] http://www.w3.org/2005/Incubator/prov/wiki/W3C_Provenance_Incubator_Group_Wiki#Current_Participants

### 4.5 Usage Data

The usage data may include Contextualized Attention Metadata (Wolpers et al., 2007) or data on actual consumption of content (e.g. people who have bought a book or the context in which people have viewed a picture or what people have done with a specific resource). Usage data can be used to derive rules for the distribution of content to specific Target populations for instance.

Mechanisms allow storing this type of information, e.g. IMS-QTI (Question and Test Interoperability) for e-assessment resources in the educational environment, the Contextualized Attention Metadata schema[7] or the W3C EmotionML markup language[8] for instance. The challenge is to create aggregates of the usage data, so that a metadata set does not have to include all the logs and traces left by users when they used the content.

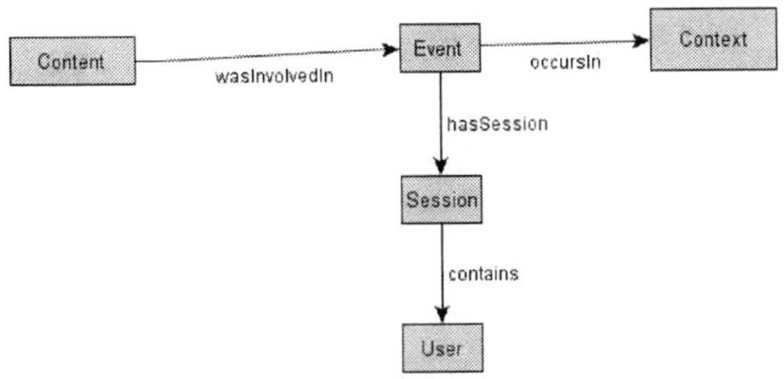


**Fig. 6.** Usage data model

The Usage data model presented on Figure 6 reuses classes from the Contextualized Attention Metadata Model[9].

## 5 Conclusion and Future Work

Overall, the model presented in this paper aims to bridge different semantic layers defined for the Future Internet. This is important in order to optimize the way in which content is handled and to support the creation, management, and delivery of dynamic content, which can be recomposed on the fly according to the user profile and context. In the context of our work, we need to aggregate data from sensors, for instance on the weather, the user location or car speed. The semantic layer built at different level of the infrastructure supports the data retrieval, aggregation and reasoning processes to adapt applications and content to users and their context. Indeed, digital content is more and more adaptive. The translation of this adaptivity on networks requires extensive semantics, the development and use of metadata and semantic models not only for the description of resources, but also to describe their interactions with their environment.

---

[7] http://www.ariadne-eu.org/index.php?option=com_content&task=view&id=39&Itemid=55

[8] http://www.w3.org/TR/emotionml/

[9] http://www.fit.fraunhofer.de/services/mobile/caple/cam/v15_CAM-schema-diagram.pdf

Our future work will be dedicated to the exploration of metadata as rules rather than passive pieces of information attached to a resource. Indeed, the management of the expected amount of information provided by sensors and objects on the Internet will require enhancing the level of intelligence embedded in networks. This can be accomplished through the development of a coherent semantic layer that facilitates the interactions with network nodes able to act upon content (e.g., to create links with other resources or distribute it according to their creator's guidance).

**Acknowledgments.** The work presented here has been realized in the framework of the CLAIRVOYANT project (FNR C09/IS/12), supported by the National Research Fund, Luxembourg.

## References

1. Bilasco, I.M., Amir, S., Blandin, P., Djeraba, C., Laitakari, J., Martinet, J., Gracia, E.M., Pakkala, D., Rautainen, M., Ylianttila, M., Zhou, J.: Semantics for intelligent delivery of multimedia content. In: SAC ACM 2010, Sierre, Switzerland, March 22-26 (2010)
2. Foulonneau, M., Arnould, G., Devooght, K., Naudet, Y.: Linking Data on the Future Internet - Semantic networks and intelligent objects. In: Proceedings of the Workshop on Linked Data in the Future Internet at the Future Internet Assembly, Ghent, Belgium, December 16-17 (2010)
3. International Federation of Library Associations and Institutions Study Group on the Functional Requirements for Bibliographic Records. Functional Requirements for Bibliographic Records. IFLA (2009)
4. Jacobson, V., Smetters, D.K., Thornton, J.D., Plass, M.F., Briggs, N.H., Braynard, R.L.: Networking named content. In: Proceedings of the 5th International Conference on Emerging Networking Experiments and Technologies, pp. 1–12 (2009)
5. Janse, M.D., Sesena, J.: Vision 2020. "Networked and Electronic Media" European Technology Platform (2009)
6. Kortuem, G., Kawsar, F., Fitton, D., Sundramoorthy, V.: Smart objects as building blocks for the Internet of things. IEEE Internet Computing 14(1), 44–51 (2010), doi:10.1109/MIC.2009.143
7. Moreau, L., Clifford, B., Freire, J., Futrelle, J., Gil, Y., Groth, P., Kwasnikowska, N., Miles, S., Missier, P., Myers, J., Plale, B., Simmhan, Y., Stephan, E., Van den Bussche, J.: The open provenance model core specification (v1.1). Future Generation Computer Systems (July 2010)
8. Naudet, Y., Schwartz, L., Mignon, S., Foulonneau, M.: Applications of user and context aware recommendations using ontologies. In: Human Computer Interaction, IHM Conference 2010 (2010)
9. Tuomi, I.: Data is More Than Knowledge - Implications of the Reversed Knowledge Hierarchy for Knowledge Management and Organizational Memory. Journal of Management Information Systems 16(3), 107–121 (1999)
10. Tselentis, G., Galis, A., Gavras, A., Krco, S., Lotz, V., Simperl, E., Stiller, B., Zahariadis, T.: Towards the Future Internet - Emerging Trends from European Research. IOS Press, Amsterdam (2010)
11. Wolpers, M., Najjar, J., Verbert, K., Duval, E.: Tracking Actual Usage: the Attention Metadata Approach. International Journal Educational Technology and Society (2007) ISSN: 1436-4522
12. Zahariadis, T., Daras, P., Bouwen, J., Niebert, N., Griffin, D., Alvarez, F., Camarillo, G.: Towards a Content-Centric Internet. In: Tselentis, G., et al. (eds.) Towards the Future Internet. IOS Press, Amsterdam (2010)

# A Model for Ripple Effects of Ontology Evolution Based on Assertions and Ontology Reverse Engineering

Ousmane Sall[1,2], Mouhamadou Thiam[1,2], Moussa Lo[2], and Henri Basson[3]

[1] UFR Sciences Et Technologies, Université de THIES, BP A967 Thies, Senegal
{osall,mthiam}@univ-thies.sn

[2] Laboratoire d'Analyse Numérique et d'Informatique – UFR SAT, Université Gaston Berger, Saint-Louis, Senegal
{moussa.lo}@ugb.edu.sn

[3] Laboratoire d'Informatique, Signal, et Image de la Côte d'Opale-EA 4491, Université du Littoral Côte d'Opale, Calais, France
basson@lil.univ-littoral.fr

**Abstract.** Ontology undergoes successive phases of evolution in the context of engineering. Taking into account successive modification impacts is of a particular importance, the least change causing an imbalance of the ontology badly taken into account can involve a dysfunction of the whole artifacts which depend on it. In this paper we focus on an approach allowing to priori discover ontologies modification impacts and changes control by the detection of impact propagation paths based on a set of assertions and reverse engineering according to a lexical ontology model for controlling ripple effects.

**Keywords:** ontology, evolution, maintenance, change propagation, ripple effects, reverse engineering.

## 1 Introduction

In [6], ontology evolution is defined as the adaptation - in time - to the needs for change and the coherent propagation of the changes to the dependent artifacts. In the context of ontology engineering, the evolutionary maintenance of ontologies is interested in the various operations carried out on the elements of ontology.

In the application context of agricultural data, we proposed in [8] a flexible approach to use a set of documents produced by experts in a field and the existence of a standard ontology describing the area to generate an ontology conceptualizing the vocabulary of the data provider. Due to the use of a semantic mediation focused on ontologies and the specific ontology construction approach, we must develop a methodology for the management of evolution cycles. Whenever a concept is added or removed, this can lead to change impacts throughout the integration system. It is therefore necessary to establish a system to manage the evolution of ontologies in terms of change impacts propagation with at first step, in the context of ontological engineering management, side effects of a given operation on an ontology component over others. We thus associate with each evolution phase of ontology an additional

E. García-Barriocanal et al. (Eds.): MTSR 2011, CCIS 240, pp. 155–162, 2011.

layer modeling the upgrading capabilities. That makes it possible to more easily check if an ontology is coherent compared to its model, which in this case, if and only if, it satisfies the constraints specified by his model [9]. It is thus enough to check according to the constraints of balances posed as a preliminary in the model if the new state of the system is coherent compared to the model of reference. That stresses the importance of having such a model outlined in [3] where the author makes the observation that an ontology evolution system should be enhanced by a meta-model independent from the language which permits to model generic guidelines change management and can be used for different evolutionary processes. A state of the art of the approaches management of the evolution of ontologies is presented in the second section with a highlighted of impacts flow of ripple effects.

## 2 Ontology Evolution

With the advent of the semantic Web, there exists a great interest and a revival in research on the evolutionary maintenance of ontologies. Crucial questions in this context are (1) to be able to detect the changes as well as their consequence on the system, (2) to detect the inconsistencies generated because of the ontological changes and (3) to finally solve inconsistencies. There exist several approaches of ontology evolution management [1], [3], [4], [7]–[9] which can be characterized as being of "what-if" type.

In [3] the control of the evolution is based on the definition of owners of management of change making it possible to have a "library" of scenarios of possible change. Thus with the detection of each type of change it is possible to propose a procedure to be followed to control its impacts and to solve possible inconsistencies. In [7], the author proposes a methodology of ontology evolution based on a versions log. It is thus defined, in the language, several operations of changes in terms of conditions having to be filled before and after the change operation requested, but it does not support change propagation. In [4], and also [9] where changes requirements are specified by ontology engineer, one finds also the same type of approach consisting in studying with represented OWL and RDF the whole elements which can undergo impacts of modification resulting from the consequences of the evolution of another element. In [1] one also finds an approach of global ontology evolution management.

One of the problems arising from these types of approaches is that the analysis works only for the whole of the defined scenarios, the system will be unable to control all after effects of a modification not taken into account. Moreover, it is not impossible but it is complex to envisage all the scenarios of possible modification and to carry out a logical reasoning for an ontological graph whose nodes and relations can amount in millions. Moreover it will be appropriate, in the case of species, to resound on tests of satisfiability often based on description logic for coherence checking, which is complex enough for the logical systems. Moreover, as underlined in [5], the problems of satisfiability for certain categories of DLs are in the worst case of at least NP-Difficult and EXPTIME-complete problem for most expressive DLs.

Change propagation is globally treated in the literature around those between ontologies and not as the context of internal component of the ontologies. In that context, there are two approaches of change propagation: push-based and pull-based approaches [9]. The question in our work is focusing around that how to control internal ontologies change that makes it inconsistent. The approach that we propose allows in the ontological context, by modeling ontology like representing an oriented graph, to consider that any modification of a node can make its neighbors inconsistent. The impacts propagation phase consists in this case to mark, gradually, the inconsistent nodes. Thus, each evolution of a node gives the signal to a control to detect inconsistencies on its neighbors in order to mark them if needed. Moreover, we substitute to the problems of satisfiability those of occurrence matching in the lexical model of restructuration.

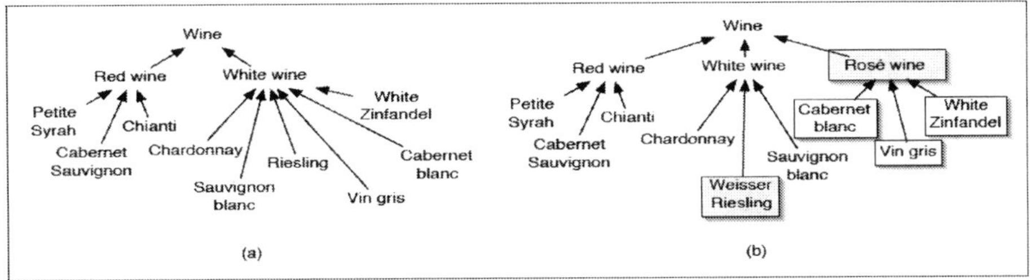


**Fig. 1.** Two versions of an ontology (Extract from [4])

Figure 1 shows an example of evolution of ontology of the wine, where on the level of the second version a new subclass was added with the class "white wine" and three classes changed superclass for "rosy wine" which changes it also superclass for the class "wine". The detection of the changes will be done by each operation. The evolution is justified by the differences between the two structures and it is advisable after each operation to check if the system is always coherent with reference to a whole of rules that predefine the system's state of balance.

## 2.1 Formal Model for Lexical Ontology Reverse Engineering

In the preceding sections, we presented work which argues for the need of having a model of the evolution and ontology. We choose to use the structures of light ontologies for that. A model of lexical ontology is formalized like the structure O=<S, L> where S is the structure and L the lexical level. Thus, the structure of an ontology O with which a lexicon is associated is the tuple:

$$S = \{C, R, A, T, CAR_R, H^C, \sigma_R, \sigma_{CARR}, \sigma_A, \sigma_T\} \text{ where:}$$

- C, A, T, $CAR_R$ are respectively sets containing, the concepts of ontology, the relations of attribute, the types of attribute and characteristics of associative relations which are inter alia synonymy, transitivity, symmetry and the inversion and functional calculus;
- $R \subseteq (C \times C)$ is associative relations set. It makes it possible to define the semantic types of relations connecting the concepts of ontology in (C x C);
- $H^C$ hierarchy (taxonomy) of concepts: $H^C \subseteq (C \times C)$, $H^C(C_i, C_j)$ means that $C_i$ is a sub-concept of $C_j$, for subsumption relations between ontology concepts;

- $\sigma_R$: R → C x C is the signature of an associative relation. We will note R($C_i$, $R_k$, $C_j$) the signature of the associative relation $R_k$ between the concepts $C_i$ and $C_j$;
- $\sigma_A$: A →C x T is the relation of attribute signature, T is composed of the simple types. It is noted as $\sigma_A$($C_i$, $A_k$, $T_j$) specifying the relation of attribute between a concept $C_i$ and a $A_k$ attribute having values of the $T_j$ type;
- $\sigma_T$ : A → T is the signature of the relation associating with an attribute $A_k$, the $T_j$ type in the form $\sigma_T$($A_k$, $T_j$) specifying that the $A_k$ attribute is associated with values of the $T_j$ type;
- $\sigma_{CARR}$: R → $CAR_R$ is the relation specifying the characteristic of an associative relation. We will, thus, note an associative relation $R_k$ transitive by signature $\sigma_{CARR}$($R_k$, Trans).

For an ontology structure S, a lexicon L is defined as the quadruplet

$$L = \{L^C, L^R, F, G\} \text{ where:}$$

- $L^C$ is the set of lexical entry of concepts;
- $L^R$ is the set of lexical entry of relations;
- $F \rightarrow L^C$ the relation of reference for the concepts such as
  - $\forall\, l_c \in L^C$:$F(l_c)=\{c \in C/(l_c,C) \in F\}$ and $\forall\, c \in C : F^{-1}(c) = \{l_c \in L^C /(l_c,c) \in F\}$
- $G \rightarrow L^R$ the relation of reference for the relations such as:
  - $\forall\, l_r \in L^R$: $G(l_r)=\{r \in R/(l_r,R) \in G\}$ and $\forall r \in R : G^{-1}(r)=\{l_r \in L^R / (l_r, r) \in G\}$

We can thus distinguish five components that can be targets of modifications and five relations that can propagate the modifications to the other elements. Let us notice that it is essential to know the direction of flow to have a knowledge a priori direction of propagation of the impacts for each type of relation. The relation of subsumption between two classes $C_i$ and $C_j$ such as $H^C$($C_i$, $C_j$) induced that $C_i$ uses the $C_j$ class and by consequence any modification of $C_i$ does not influence a priori $C_j$ but any modification of $C_j$ has consequences on $C_i$. We will note in this case that any operation Δ of modification $\Delta(C_j) \mapsto \Delta(C_i)$. The study of the various semantic relations suggested as a whole R, enables us to advance the postulate according to which one can suppose that for the greater part of the definite semantic relations such as $\sigma_R(C_i ,R_i, C_j)$ are conducting of impact of modification in the direction of $C_j$ to $C_i$ since in such a case, except for the relations of opposition and equivalence it is the class $C_i$ which uses $C_j$. Thus, we have a defined flow direction such as $\Delta(C_j) \mapsto \Delta(C_i)$. In the phase of ontology reorganization, we will ensure that the relations are made up according to this model.

For the relation of attribute between a concept of unit $C_i$ and an attribute $A_i$ of the unit has such as has $\sigma_A(C_i, A_i, T_k)$, we can distinguish three cases from possible figures. It is obvious that a modification of concept $C_i$ does not have an influence on the $A_j$ attribute nor its $T_k$ type. However a modification of the attribute $A_i$ will involve consequences on the concept $C_i$ which uses it, therefore $\Delta(A_j) \mapsto \Delta(C_i)$. It is as commonplace as an operation on the involving type of ripple effects will have consequences on the attribute and thus also on the concept. Thus in indirect manners an operation of modification on a type $T_k$ will gradually involve impacts propagation $\Delta(T_k) \mapsto \Delta(A_j) \mapsto \Delta(C_i)$.

## 2.2 Change Definition and Model

By generating a ripple effect, one can in a picturesque way say that the system was balanced, the operation unbalanced it and the effects of wave must be contained and controlled for a return to the state of balance. A priori, we can define three essential assertions for the realization of an operation and the control of the range of its ripple effects:(1)what it is really needed for the operation to be carried out, for example to remove a concept, it is necessary that it exists;(2)conditions having to be checked after execution of the operation and which take part in its validation, for example in the case of a suppression of a concept, the concept will not have to exist anymore in the structure of ontology; (3)conditions having to be checked before and after the operation. Example, in the case of a concept suppression, there should not exist and remain no form of relation between the target class and another class or attribute after the operation.

**Table 1.** Basic and general assertions

| Id | Assertion | Signification |
|---|---|---|
| 1 | $+C_i$ | $\exists (C_i \in C)$ |
| 2 | $+R_i$ | $\exists (R_i \in R)$ |
| 3 | $+A_i$ | $\exists (A_i \in A)$ |
| 4 | $+T_i$ | $\exists (T_i \in T)$ |
| 5 | $+CAR_{Ri}$ | $\exists (CAR_{Ri} \in CAR_R)$ |
| 6 | $-C_i$ | $\neg (C_i \in C)$ |
| 7 | $-R_i$ | $\neg (R_i \in R)$ |
| 8 | $-A_i$ | $\neg (A_i \in A)$ |
| 9 | $-T_i$ | $\neg (T_i \in T)$ |
| 10 | $-CAR_{Ri}$ | $\neg (CAR_{Ri} \in CAR_R)$ |
| 11 | $+H^C(C_i, C_j)$ | $\exists (C_i \in C \wedge C_j \in C) / H^C(C_i, C_j)$ |
| 12 | $+\sigma_R(C_i, R_k, C_j)$ | $\exists (C_i \in C, C_j \in C \wedge R_k \in R) / \sigma_R(C_i, R_k, C_j)$ |
| 13 | $+\sigma_{CARR}(C_i, CAR_{Ri})$ | $\exists (C_i \in C \wedge CAR_{Ri} \in CAR_R) / \sigma_{CARR}(C_i, CAR_{Ri})$ |
| 14 | $+\sigma_T(A_i, T_k)$ | $\exists (A_i \in A \wedge T_k \in T) / \sigma_T(A_i, T_k)$ |
| 15 | $+\sigma_A(C_i, A_j, T_k)$ | $\exists (C_i \in C, A_j \in A \wedge T_k \in T) / \sigma_A(C_i, A_j, T_k) \wedge \sigma_A(A_j, T_k)$ |
| 16 | $-H^C(C_i, C_j)$ | $\forall (C_i \in C) \wedge \forall (C_j \in C) : \neg H^C(C_i, C_j)$ |
| 17 | $-\sigma_R(C_i, R_k, C_j)$ | $\forall (C_i \in C) \wedge \forall (C_j \in C) \wedge \forall (R_k \in R) : \neg\sigma_R(C_i, R_k, C_j)$ |
| 18 | $-\sigma_{CARR}(C_i, CAR_{Ri})$ | $\forall (C_i \in C) \wedge \forall (CAR_{Ri} \in CAR_R) : \neg\sigma_{CARR}(C_i, CAR_{Ri})$ |
| 19 | $-\sigma_T(A_i, T_k)$ | $\forall (A_i \in A) \wedge \forall (T_k \in T) : \neg\sigma_T(A_i, T_k)$ |
| 20 | $-\sigma_A(C_i, A_j, T_k)$ | $\forall (C_i \in C) \wedge \forall (A_j \in A) \wedge \forall (T_k \in T) / \sigma_A(A_j, T_k) : \neg\sigma_A(C_i, A_j, T_k)$ |
| 21 | $-H^C(*, C_i)$ | $\forall (C_k \in C) : \neg H^C(C_k, C_i)$ |
| 22 | $-H^C(C_i, *)$ | $\forall (C_k \in C) : \neg H^C(C_i, C_k)$ |
| 23 | $-\sigma_{CARR}(R_i, *)$ | $\forall (CAR_{Ri} \in CAR_R) : \neg\sigma_{CARR}(R_i, CAR_{Ri})$ |
| 24 | $-\sigma_T(A_i, *)$ | $\forall (T_k \in T) : \neg\sigma_{CA}(A_i, T_k)$ |
| 25 | $-\sigma_A(*, A_i, T_j)$ | $\forall (C_k \in C) \wedge \forall (A_i \in A) \wedge \forall (T_j \in T) / \sigma_T(A_i, T_j) : \neg\sigma_A(C_k, A_i, T_j)$ |
| 26 | $-\sigma_A(*, A_i, *)$ | $\forall (C_k \in C) \wedge \forall (T_j \in T) : \neg\sigma_A(C_k, A_i, T_j)$ |
| 27 | $-\sigma_A(C_k, *, *)$ | $\forall (A_i \in A) \wedge \forall (T_j \in T) : \neg\sigma_A(C_k, A_i, T_j)$ |
| 28 | $-\sigma_A(*, *, T_j)$ | $\forall (C_k \in C) \wedge \forall (A_i \in A) : \neg\sigma_A(C_k, A_i, T_j)$ |
| 29 | $-\sigma_R(C_i, *, *)$ | $\forall (C_k \in C) \wedge \forall (R_j \in R) : \neg\sigma_R(C_i, R_j, C_k)$ |
| 30 | $-\sigma_R(*, R_k, *)$ | $\forall (C_i \in C) \wedge \forall (C_j \in C) : \neg\sigma_R(C_i, R_k, C_j)$ |

A change is modeled like a triplet Δ=<Op, Args, Assert> representing the operation, its arguments and Assert =< pre, inv, post> for pre-conditions, invariant conditions, post-conditions as in table 2.

By adapting the definitions given in [3], the basic changes correspond to simple and indivisible changes, which modify only one characteristic of knowledge of ontology. It is the case of a change induced by subsumption relation suppression. The basic operations of change presented by table 2 are expressed by the basic assertions of table 1. The sign "+" in front of an identifier of an assertion and which specifies that the fact posed (a criterion or state of the truth of the proposal posed) must be checked or present as a whole of membership a priori. The sign "-" in front of an identifier of a negative assertion and which specifies that the fact posed should not be checked or must miss as a whole of membership a priori.

**Table 2.** Examples of Basic Operations For A Phase Φ

| Id | Basic Operations | Pre-condition | Invariant | Post-condition |
|---|---|---|---|---|
| 1 | CreateConcept($C_i$) | $-C_i$ | $-H^C(*,C_i)$<br>$-H^C(C_i, *)$<br>$-\sigma_A(C_i, *, *)$<br>$-\sigma_R(C_i, *, *)$ | $+C_i$ |
| 2 | DeleteConcept($C_i$) | $+C_i$ | $-H^C(*, C_i)$<br>$-H^C(C_i, *)$<br>$-\sigma_A(C_i, *, *)$<br>$-\sigma_R(C_i, *, *)$ | $-C_i$ |
| 3 | CreateAssociativeRelation($R_i$) | $-R_i$<br>$-\sigma_{CARR}(R_i, *)$ | $-\sigma_R(*, R_i, *)$ | $+R_i$<br>$+\sigma_{CARR}(R_i, CAR_{Ri})$ |
| 4 | DeleteAssociativeRelation($R_i$) | $+R_i$ | $-\sigma_{CARR}(R_i, *)$<br>$-\sigma_R(*, R_i, *)$ | $-R_i$ |
| 5 | CreateProperty($A_i$, $T_i$) | $-A_i$ | $-\sigma_A(*, A_i, T_i)$<br>$-\sigma_T(A_i, *)$ and $+T_i$ | $+A_i$<br>$+\sigma_T(A_i, T_j)$ |
| 6 | DeleteProperty($A_i$) | $+A_i$ | $-\sigma_A(*, A_i, T_i)$<br>$-\sigma_T(A_k,*)$ and $+T_i$ | $-A_i$ |

## 2.3 Ripple Effects Propagation

We specified it a little higher; ontology is no more coherent since one of the constraints associated with the model with evolution is not satisfied any more. Each operation being associated with a whole of assertions, the appropriateness propagation is declined in three possible cases. Thus, while basing itself on "Change-and-Fix" approach in context of evolutionary maintenance of the software proposed in [2]:

- If the pre-condition and the invariant are checked, then the operation can be carried out without propagation of impacts;
- If the pre-condition is not checked, then the operation is not checked and there is no impact on the ontological components;
- If the pre-condition is checked and that the invariant is not then checked the operation is carried out and there is a process of propagation of impact which will consist in marking the facts which are linked with the invariant. I.e. the components which use the component responsible for the modification. It is for example the case of suppression of a concept having sub-concepts. The assertion

to mark consists in identifying the components and the relations which will be affected by the modification. To mark (*Comp*$_i$) consists in marking a component *Comp*$_i$ affected $a_i$ by an operation of modification, then to mark all the relations $Rel_i \in \{R, H^C, \sigma_R, \sigma_{CARR}, \sigma_A\}$ that have as source of flow impact the latter. To mark ($Rel_i$) consists in propagating the modification with the node source of the relation of $Rel_i$. Two cases of figure can occur: first, the source node of impact of the relation $Rel_i$ is the modified node, involving to stop the impacts propagation; if not, if the source node of the impact of the relation $Rel_i$ is different from the modified node, that makes it possible to propagate the effect of the modification to the modified node neighbors.

**Table 3.** Basic General Assertions for Ripple Effects Propagation

| Id | Assertions | Signification |
|---|---|---|
| 31 | markConcept($C_i$) | $\forall (C_j \in C)$ If $H^C(C_j, C_i)$ then markRelation($H^C(C_j, C_i)$)<br>$\forall (C_j \in C)$ and $\forall (R_k \in R)$ If $\sigma_R(C_j, R_k, C_i)$<br>then markRelation($\sigma_R(C_j, R_k, C_i)$) |
| 32 | markRelation($R_k$) | $\forall (C_i \in C)$ and $\forall (C_j \in C)$ If $\sigma_R(C_j, R_k, C_i)$<br>then markRelation($\sigma_R(C_j, R_k, C_i)$) |
| 33 | markProperty($A_k$) | $\forall (C_i \in C)$ If $\sigma_A(C_i\ A_k, T_j)$ then markRelation($\sigma_A(C_i, A_k, T_j)$) |
| 34 | markType($T_k$) | $\forall (A_i \in A)$ If $\sigma_A(A_i, T_k)$ then markRelation($\sigma_T(A_i, T_k)$) |
| 35 | markType($CAR_{Rk}$) | $\forall (R_i \in R)$ If $\sigma_{CARR}(R_i, CAR_{Rk})$<br>then markRelation($\sigma_{CARR}(R_i, CAR_{Rk})$) |
| 36 | markRelation($H^C(C_j, C_i)$) | If ($C_i$ not marked) then markConcept($C_i$) |
| 37 | markRelation($\sigma_R(C_j, R_k, C_i)$) | If ($C_i$ not marked) then markConcept($C_i$) |
| 38 | markRelation($\sigma_A(C_i, A_k, T_j)$) | If ($C_i$ not marked) then markConcept($C_i$) |
| 39 | markRelation($\sigma_T(A_i, T_k)$) | If ($A_i$ not marked) then markConcept($A_i$) |
| 40 | markRelation($\sigma_{CARR}(R_i, CAR_{Rk})$) | If ($R_i$ not marked) then markConcept($R_i$) |
| 41 | markRelation($H^C(*, C_i)$) | $\forall (C_k \in C)$ If($H^C(C_k, C_i)$) and ($C_k$ not marked)<br>then markConcept($C_k$) |
| 42 | markRelation($H^C(C_i, *)$) | $\forall (C_k \in C)$ If($H^C(C_i, C_k)$) and ($C_k$ not marked)<br>then markConcept($C_k$) |
| 43 | markRelation($\sigma_{CARR}(*, CARR_{Ri})$) | $\forall (R_k \in R)$ If($\sigma_{CARR}(*, CAR_{Ri})$)and($R_k$ not marked)<br>then markRelation($R_k$) |
| 44 | markRelation($\sigma_T(A_i, *)$) | $\forall T_k \in T$, If($\sigma_T(A_i, T_k)$) and($T_k$ not marked)<br>then markType($T_k$) |
| 45 | markRelation($\sigma_A(*, A_i, T_j)$) | $\forall C_k \in C$, If($\sigma_A(C_k, A_i, T_j)$) and($C_k$ not marked)<br>then markConcept($C_k$) |
| 46 | markRelation($\sigma_A(*, A_i, *)$) | $\forall C_k \in C, \forall T_j \in T$ : If ($\sigma_A(C_k, A_i, T_j)$) then If ($C_k$ not marked)<br>then markConcept($C_k$)<br>If ($T_j$ not marked) then markType($T_j$) |
| 47 | markRelation($\sigma_A(C_k, *, *)$) | $\forall A_i \in A, \forall T_j \in T$ : If($\sigma_A(C_k, A_i, T_j)$) then<br>If ($A_i$ not marked) then markConcept($A_i$) |
| 48 | markRelation($\sigma_A(*, *, T_j)$) | $\forall C_k \in C, \forall A_i \in A$: If($\sigma_A(C_k, A_i, T_j)$) then<br>If ($C_k$ not marked) then markRelation($\sigma_A(C_k, A_i, T_j)$)<br>If ($A_i$ not marked) then markRelation($\sigma_T(A_i, T_j)$) |
| 49 | markRelation($\sigma_R(C_i, *, *)$) | $\forall R_j \in C, \forall C_k \in C$ : If($\sigma_R(C_i, R_j, C_k)$) then<br>If ($C_k$ not marked) then markConcept($C_k$) |
| 50 | markRelation($\sigma_R(*, R_k, *)$) | $\forall C_i \in C, \forall C_j \in C$ : If($\sigma_R(C_i, R_k, C_j)$) then<br>If ($C_j$ not marked) then markConcept($C_j$) |

Table 3 presents the assertions which make it possible to propagate the change of impact in case of inconsistency due to an operation. It is what thus enables us to propose an analysis of priori effects of each type of simple or complex modification but also the preaching or the posteriori analysis of modification effects.

## 3 Conclusion

Ontology engineering is one of the semantic web's most important activities. In that context, different changes are likely to occur on ontology and their consequences on its structure and its subordinate artifacts. We proposed in this paper a model of impacts propagation of ontology evolution ripple effects. It consists in ontology restructuring according to a lexical model and set of assertions allowing to know the conditions of implementation of an operation as well as the conditions having to be respected before and after. The impact of modification of a component will be thus propagated by the relations with the other components. This model of ripple effects propagation allows us currently working on a more global transfer flows model between ontologies evolution of an integration system with testing the use of Gallois lattice and infomorphism modeling approach pertinence.

## References

1. Bloehdorn, S., Haase, P., Sure, Y., Voelker, J.: Ontology Evolution. In: Davies, J., Studer, R., Warren, P. (eds.) Semantic Web Technologies — Trends and Research in Ontology-Based Systems, pp. 51–70. John Wiley & Sons, Chichester (2006)
2. Deruelle, L.: Analyse d'impact de l'évolution des applications distribuées multi-langages et à bases de données hétérogènes. Thèse de doctorat. Université du Littoral Côte d'Opale (2001)
3. Jedidi, R.: Approche d'évolution d'ontologie guidée par des patrons de gestion de changement. Thèse de doctorat de l'Université Paris-Sud XI Orsay (Novembre 2009)
4. Klein, M., Noy, N.F.: A component-based framework for ontology evolution. In: Proceedings of the (IJCAI 2003) Workshop on Ontologies and Distributed Systems, CEUR-WS, vol. 71 (2003)
5. Le Pham, A.: De l'optimisation à la décomposition de l'ontologique dans la logique de description. Thèse de doctorat, Université de Nice Sophia-Antipolis, January 21 (2008)
6. Maedche, A., Motik, B., Stojanovic, L.: Managing multiple and distributed ontologies in the Semantic Web. VLDB Journal 12(4), 286–300 (2003), doi:10.1007/s00778-003-0102-4
7. Plessers, P., De Troyer, O.: Ontology change detection using a versioning log. In: Gil, Y., Motta, E., Benjamins, V.R., Musen, M.A. (eds.) ISWC 2005. LNCS, vol. 3729, pp. 578–592. Springer, Heidelberg (2005)
8. Sall, O., Lo, M., Gandon, F., Niang, C., Diop, I.: Using XML data integration and ontology reuse to share agricultural data. Int. J. Metadata, Semantics and Ontologies 4(1/2), 93–105 (2009)
9. Stojanovic, S.L., Maedche, A., Motik, B., Stojanovic, N.: User Driven ontology evolution management. In: Gómez-Pérez, A., Benjamins, V.R. (eds.) EKAW 2002. LNCS (LNAI), vol. 2473, pp. 285–300. Springer, Heidelberg (2002)

# The Need of a Semantic Layer between UMLS and Biomedical Information Systems

Birsen G. Özdemir[1] and Nazife Baykal[2]

[1] Doğuş University,
İstanbul, Turkey
bgozdemir@dogus.edu.tr
[2] Middle East Technical University,
Ankara, Turkey
baykal@ii.metu.edu.tr

**Abstract.** Since biomedical information is scattered among a number of semantically or syntactically incompatible independent systems, a contemporary pragmatic approach is proposed in this study to make use of a semantic middle layer and common standards for information exchange between these systems. Biological and medical terminologies and ontologies take vital part in the background of life sciences information systems and the Unified Medical Language System (UMLS) is an inclusive source for biomedical vocabulary. Taking this fact into account, the need of new tools, which transform whole or part of UMLS content to a semantically sharable and reusable format, is also revealed to support knowledge-intensive biomedical actions.

**Keywords:** Interoperability of biomedical systems, Web Ontology Language, Unified Medical Language System.

## 1 Introduction

Biomedical information is scattered among a number of semantically or syntactically incompatible heterogeneous and independent systems. Those systems typically define information using biomedical vocabularies in dissimilar representation format. In addition to usage of different formats, the same meaning can also be presented with diverse terminologies. This reality usually causes communication gaps and faulty transmission of information between biomedical systems.

In this study, a contemporary approach of the use of a semantic middle layer utilizing common standards is proposed to exchange information between biomedical systems. Utilization of a semantic middle layer will not require changing the current applications and technologies; rather, the new layer will make use of the already existing infrastructure. The candidate common standard for this semantic middle layer is suggested to be Web Ontology Language (OWL).

A comprehensive resource of biomedical terminology is started by National Library of Medicine and called as the Unified Medical Language System (UMLS), which tries to unify the representation format of biomedical domain knowledge

E. García-Barriocanal et al. (Eds.): MTSR 2011, CCIS 240, pp. 163–169, 2011.

spread to different knowledge sources (e.g. SNOMED, ICD10, etc.). The UMLS [1] consists of Metathesaurus, which includes biomedical concepts from various source terminologies and Semantic Network (SN), which is an upper-level abstraction of the biomedical domain. Even the UMLS includes valuable information for biomedical systems, it is too comprehensive [1]. The use of local coding-scheme of knowledge sources do not cumulate in ways useful to research [16] and makes harder to semantically share and reuse of information. Therefore, it is seen that there is a need for tools that transform part of or whole UMLS to semantically reusable and sharable format. These tools transforming the UMLS content into suitable format can give countenance to the proposed semantic middle layer for supporting knowledge-intensive biomedical actions. Information retrieval, text mining, natural language processing, knowledge management, software agents and semantic web applications in biomedicine can take advantage of semantic representation of UMLS data in OWL for validating relations among entities, searching/aligning UMLS concepts or semantic types, and other similar tasks.

The following two sections provide brief information on the Unified Medical Language System and Web Ontology Language, respectively. Section 4 describes the proposed semantic middle layer for interoperability of biomedical systems. In section 5, the new tools role in migration to contemporary use of UMLS for biomedical developer(s) and application(s) is presented. Section 6 contains the major conclusions of this work.

## 2 Unified Medical Language System

Until now various biomedical knowledge sources (e.g. SNOMED, ICD10, etc.), which generally have different representation format and contains biomedical vocabularies playing a critical role in the context of biology and medicine information systems, have been independently developed from one another. Developers need to spend time to learn each formal representation of interested knowledge sources and if necessary need to translate into a suitable format while using them in biomedical applications. This complicates the integration of biomedical knowledge sources into real-world biomedical information systems. The Unified Medical Language System (UMLS), which is a comprehensive resource of biomedical terminology, tries to unify the representation format of valuable biomedical domain knowledge spread to different knowledge sources. In general, a set of terms representing the same meaning are referred to as a concept. In the current release (2011AA) of UMLS there are more than 2.4 million concepts and 8.8 million unique concept names from over 160 source vocabularies (e.g. SNOMED, MeSH, ICD10, etc.) [7].

The UMLS [1] consists of Semantic Network (SN) [2, 5], Metathesaurus [4], SPECIALIST Lexicon and Lexical Tools. The biomedical knowledge in the SN identifies semantic relations between the semantic types and offers a semantic framework for the UMLS vocabularies [3]. The Metathesaurus editors assign each

Metathesaurus concept to one or more semantic types in the UMLS Semantic Network. The categorization of Metathesaurus concepts allows user conceptualizing and sub-setting of the domain [6]. The lexical information needed for the SPECIALIST Natural Language Processing System (NLP) is supplied by The SPECIALIST Lexicon, comprising lots of biomedical terminologies. The Lexical Tools deal with abstraction away from the sort of variability in natural language vocabularies [8].

## 3 Web Ontology Language

In order to give information precise meaning and make machines automatically process and integrate it, the Semantic Web [9] layers structure have been proposed by Tim Berners-Lee. In this new structure, which is a vision of the future Web, Ontology Web Language (OWL) [13] is used as meta-data for achievement of modeling and enrichment of data represented by HMTL/XML. Being the de facto language for the Semantic Web, OWL is a standard for expressing ontologies, which includes inference rules and taxonomy to formally describe the relations among concepts to capture domain knowledge [10]. OWL offers strong syntax, machine interpretability and a large vocabulary. Ontology Web Language is being used to enable interoperability in distributed computer systems by encoding knowledge processing. For achieving interoperability and representation of biomedical knowledge, OWL has recently been used in the National Center for Biomedical Ontology (NCBO) [12] and Semantic Health European project [11]. Leonardo Lezcano et.al. [14] worked on the Archetype Definition Language translation into OWL formal representation to contribute full healthcare semantic interoperability. The transformation into OWL is also done in OBO ontologies [16].

## 4 Semantic Middle Layer for Interoperability of Biomedical Systems

The usage of biomedical information usually demands for comparing and integrating biomedical data that are naturally deployed in varied independent systems having syntactic and semantic incompatibility. It is predictable that an ideal world expectation of biomedical information representation by common standards in all related systems is overly idealistic as each corresponding domain or system might have diverse information representation requirements necessitating different formalism. Therefore, the use of a semantic middle layer utilizing common standards is proposed for a contemporary approach to accurately exchange information among technologically independent biomedical systems. The primary advantage of utilization of a semantic middle layer is that this new layer will not require changing the current applications and technologies while integrating them; rather, the new layer will make use of the already existing infrastructure. (Fig. 1) In Fig. 1, dissimilar shapes of arrows denote the different formalism notations of biomedical data.

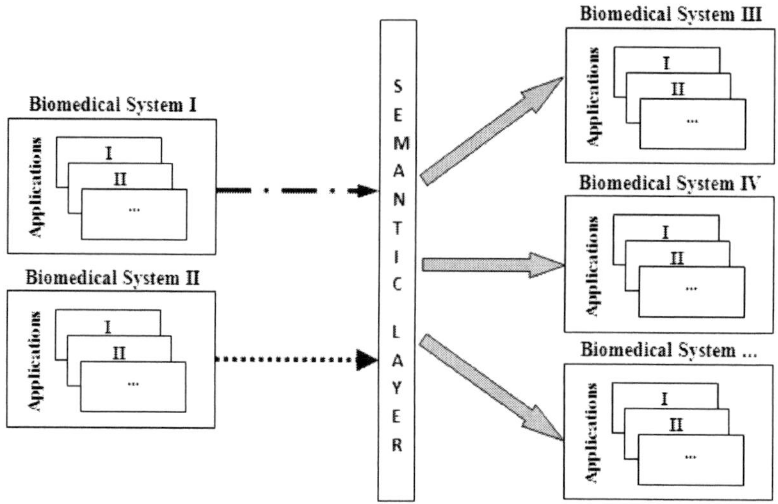


**Fig. 1.** Proposed architecture for interoperability of biomedical systems

## 5 Migration to Contemporary Use of Unified Medical Language System

Having a critical role in the context of biomedical information systems, the UMLS semantics are distributed among different files [8]. The utilization of those files for any semantic activity still needs human interpretation. Each application developer(s), in the ongoing use of UMLS, must interpret the semantics of UMLS and then use different subsets in an interested software formalism (represented as different shapes in Fig. 2).

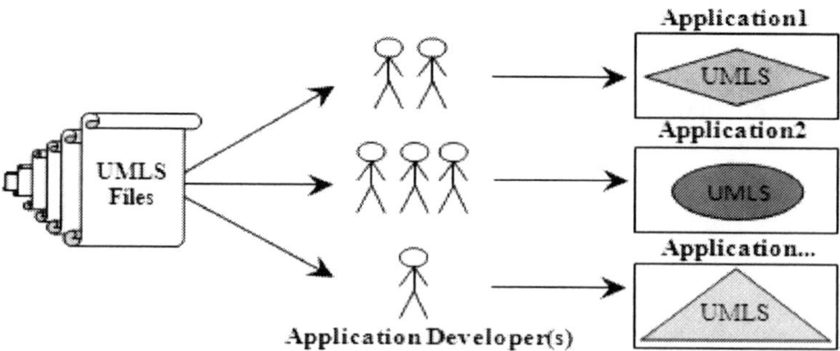


**Fig. 2.** Ongoing uses of UMLS

For contemporary use of UMLS, there is a need for new tools that permit a standardized semantic representation (showed as same shapes in Fig. 3) instead of different representation of semantic formalism on a per-application basis as it is now.

Application developers can use the tools to transform whole or part of UMLS into OWL to be compatible with Semantic Web standards while achieving semantic machine interoperability between biomedical information systems. By the help of these tools, which will offer an agreed-upon vocabulary for expressing semantics in a machine-processable way, applications and developers will have less chance of misinterpretation. (Fig. 3)

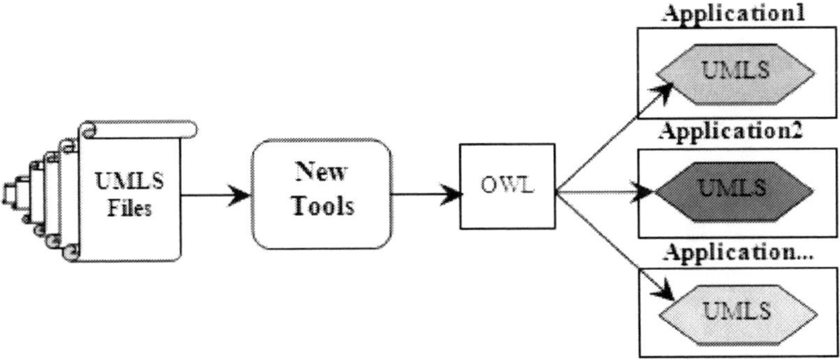


**Fig. 3.** New tools support for biomedical application(s)

New relations and circumstances will be continually discovered in huge medical and biological domains. Due to this dynamic nature biomedical categories or concepts cannot be totally defined, which have an effect on the content of Unified Medical Language System. Even, constantly enlarging biomedical necessities demand for a dynamic information representation model, UMLS has format limitations of information descriptions. Jimenez-Ruiz E. [15], for example, worked on these format limitations and manually translated the UMLS Semantic Network into OWL. However, such kind of manual static translations does not fully meet the needs of dynamically changing nature biomedical domain.

One way of integrating biomedical data is focusing on mapping specifications [18] which are also interested in the study of OBO ontologies [16]. As a result of differences in structure between the source and target terminologies, mapping relationships are many-to-many or one-to-many relations in general. Therefore, they are rarely one-to-one relations as stated by Oliver Bodenreider, et.al. [17].Therefore, even mappings are generated automatically or semi-automatically between vocabularies or ontologies, it is difficult to use them in an automated coding process.

It is seen that there is a call for tools that make UMLS content automatic representation suitable for dynamic semantic actions. Advances in the Semantic Web domain make it a suitable technology for works demanding semantic knowledge associated with biomedical systems. In this work, Web Ontology Language, which mainly offers strong syntax and machine interpretability, is suggested as a semantically convenient representation of UMLS that includes knowledge sources of concepts and categories used in the biomedical domain. New tools that transform whole or part of UMLS content into OWL will provide support for the above mentioned middle semantic layer to facilitate semantic interoperability between biomedical systems. (Fig. 4)

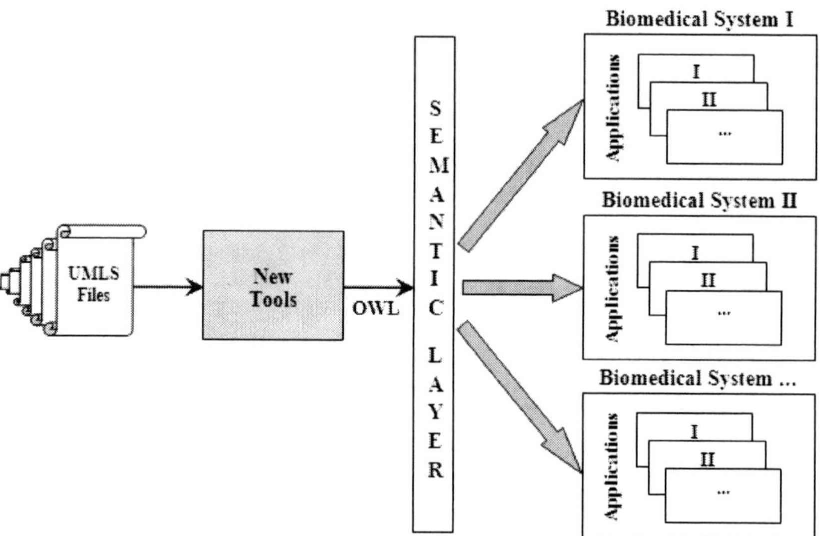


**Fig. 4.** New tools contributions for interoperability of biomedical systems

# 6 Conclusion

A contemporary pragmatic approach is proposed in this study to make use of a semantic middle layer with common standards for information exchange between biomedical systems. Since biomedical terminologies take vital part in the background of life sciences information systems and UMLS is an inclusive source for biomedical vocabulary, the need of new tools, which transform whole or part of UMLS content to a semantically sharable and reusable format, is also revealed for supporting knowledge-intensive biomedical actions. The tools transforming UMLS biomedical data in OWL that is the proposed standard offering strong syntax and machine interpretability for above mentioned semantic layer can help in upgrading existing systems without changing the current applications and technologies. Information retrieval, text mining, natural language processing, knowledge management, software agents and semantic web applications in biomedicine can take advantage of semantic representation of UMLS data in OWL for validating relations among entities, searching/aligning UMLS concepts or semantic types, and other related tasks.

# References

1. Humphreys, B.L., Lindberg, D.A.B., Schoolman, H.M., Barnett, G.O.: The Unified Medical Language System: An informatics research collaboration. Jamia 5(1), 1–11 (1998)
2. McCray, A.T.: An upper-level ontology for the biomedical domain. Comp. Func. Genom. 4, 80–84 (2003)
3. U.S. National Library of Medicine, The Future of the UMLS Semantic Network Workshop, Bethesda, Maryland (April 2005), http://mor.nlm.nih.gov/snw/

4. Schuyler, P.L., Hole, W.T., Tuttle, M.S., Sherertz, D.D.: The UMLS Metathesaurus: representing different views of biomedical concepts. Bull Med. Libr. Assoc. 81(2), 217–222 (1993)
5. Fan, J.W., Friedman, C.: Semantic classification of biomedical concepts using distributional similarity. Journal of the American Medical Informatics Association 14(4), 467–477 (2007)
6. Fa, J.W., Xu, H., Friedman, C.: Using contextual and lexical features to restructure and validate the classification of biomedical concepts. BMC Bioinformatics 8, 264 (2007)
7. U.S. National Library of Medicine, Unified Medical Language System (UMLS) 2011AA Release. National Library of Medicine Technical Bulletin November-December (2011), http://www.nlm.nih.gov/pubs/techbull/mj11/mj11_umls_2011aa_release.html
8. U.S. National Library of Medicine, The Unified Medical Language System, UMLS Reference Manual, http://www.ncbi.nlm.nih.gov/books/NBK9676/
9. Berners-Lee, T., Hendler, J., Lassila, O.: The Semantic Web - A new form of Web content that is meaningful to computers will unleash a revolution of new possibilities. Scientific American 284(5), 34–43 (2001)
10. Huang, S.L., Yang, C.W.: Designing a semantic bliki system to support different types of knowledge and adaptive learning. Computers & Education 53, 701–712 (2009)
11. Stroetmann, V.N., Rodrigues, J.M., et al.: Conceptual Framework for eHealth Interoperability. Semantic Health International Project (2007), http://www.semantichealth.org/DELIVERABLES/SemanticHEALTH_D1_1_finalC.pdf
12. Rubin, D.L., Lewis, S.E., Mungall, C.J., et al.: National Center for Biomedical Ontology: Advancing biomedicine through structured organization of scientific knowledge. OMICS-A Journal of Integrative Biology 10(2), 185–198 (2006)
13. Bechhofer, S., van Harmelen, F., Hendler, J., Horrocks, I., McGuinness, D.L., Patel-Schneider, P.F., Stein, L.A.: W3C, OWL Web Ontology Language Reference, http://www.w3.org/TR/owl-ref/
14. Lezcano, L., Sicilia, M.A., Rodrguez-Solano, C.: Integrating reasoning and clinical archetypes using OWL ontologies and SWRL rules. Journal of Biomedical Informatics 44(2), 343–353 (2011)
15. Jimenez-Ruiz, E.: The UMLS Semantic Network in OWL, http://krono.act.uji.es/people/Ernesto/UMLS_SN_OWL
16. Smith, B., Ashburner, M., Rosse, C., Bard, J., Bug, W., Ceusters, W., Goldberg, L.J., Eilbeck, K., Ireland, A., Mungall, C.J., OBI Consortium, Leontis, N., Rocca-Serra, P., Ruttenberg, A., Sansone, S.A., Scheuermann, R.H., Shah, N., Whetzel, P.L., Lewis, S.: The OBO Foundry: coordinated evolution of ontologies to support biomedical data integration. Nat. Biotechnol. 25(11), 1251–1255 (2007)
17. Bodenreider, O., Bean, C.A.: Relationships among knowledge structures: Vocabulary Integration within a subject domain. In: Bean, C.A., Green, R. (eds.) Relationships in the Organization of Knowledge, ch. 6. Kluwer Academic Publishers, Boston (2001)
18. Rebholz-Schuhmann, D., Nenadic, G.: Biomedical Semantics: the Hub for Biomedical Research 2.0. J. Biomed Semantics 1(1) (March 2010), http://www.ncbi.nlm.nih.gov/pmc/articles/PMC2895735/

# Harmonizing Software Standards with a Semantic Model

Elisabeth C.H. Verhelst[1], Huibert-Jan Lekkerkerk[2], Jandirk Bulens[1], and Sander J.C. Janssen[1]

[1] Alterra, Environmental Sciences Group, Wageningen University and Research Centre, Droevendaalsesteeg 3, 6708 PB Wageningen, The Netherlands
[2] Informatiehuis Water, Stationsplein 89, 3818 LE Amersfoort, The Netherlands

**Abstract.** The application of standards in the software development process supports interoperability between systems. Maintenance of standards must be guaranteed on the organisational and technical level. The use of semantic technologies can contribute to the standard maintenance process by providing a harmonizing bridge between standards of different knowledge domains and languages and by providing a single point of administration for standard domain concepts. This paper describes a case study of the creation of a semantic layer between software standards for water management systems in The Netherlands.

**Keywords:** interoperability, standard maintenance, semantic, ontology.

## 1 Introduction

Interoperability between software systems is supported by the adoption of standards in the software development process. Standards are available on different levels of abstraction: national and international, domain specific and software development methodology specific. It is essential to keep the standards aligned and coherent to avoid interoperability problems. Standard harmonisation activities can be observed at the organisational and technical level. On the organisational level this is arranged via organisational bodies such as advisory groups and forums of standard organisations. On the technical level a variety of instruments are necessary. We mention here syntactical and semantic measurements. This paper introduces a harmonisation method for standards on the semantic level. It is based on the creation of a semantic mediating layer between a set of existing software standards.

Semantic mediation has been proposed as a method to achieve integration in multi-disciplinary and multi-scale environments [1];[2]. Through semantic mediation an agreed conceptual schema can be built between parties for linking their systems [2]. The agreed conceptual schema serves as an additional knowledge layer. Semantic mediation is a crucial challenge for any integrated project

E. García-Barriocanal et al. (Eds.): MTSR 2011, CCIS 240, pp. 170–177, 2011.

[3]; [4]; [5], as it provides consistent and transparent building blocks in definitions and terms required for the methodological and technical linking of systems, data sources and indicators. Usually semantic mediation is based on an ontology, which is a specification of a conceptualization for a system [6]. An ontology consists of concepts and their relationships [7]. This paper focuses on the use of ontologies for semantic integration of standards.

## 2 Problem Statement

Standards are heavily interconnected. Progress in technological developments and changing requirements determine a continuous harmonisation necessity. It is a challenge to determine whether and how a change in a revised standard affects other standards. Examples of general proposed standard revision areas are: scope, normative reference, terms and provisions [8]. An analysis of standard interdependencies enables an in-depth understanding, necessary for standard maintenance. In addition, version control and change management procedures look after an reproducible, documented and structural standard maintenance process. On the syntactic level this means that standards have to remain consistent in the ability of exchanging information. Semantic harmonisation regulates the interconnection of standards between knowledge domains and languages. Ambiguity and redundancy in terms and definitions must be avoided. As an example of an organisational semantic harmonisation activity we mention here the designated maintenance agency at ISO, responsible for guarding the consistency of standard meta-metadata, such as country names, language names, character sets[1]. Our research aims to contribute to an automated manner of harmonisation of the terms and concepts within standards.

## 3 Software Development Methodology and Semantics

The importance of semantics in software development has been recognised for several stages in the development process. Research has been conducted in order to integrate semantics in the requirements specification phase [9]. Use of semantics has been proposed for process modeling [10]. In 2009 the organisation for development of enterprise integration standards Object Management Group (OMG) published the Ontology Definition Metamodel (ODM) [11]. The ODM aims to regulate 'the formal grounding for representation, management, interoperability, and application of business semantics...' by offering amongst others '...Options in the level of expressivity, complexity, and form available for designing and implementing conceptual models, ranging from familiar UML and ER methodologies to formal ontologies...' [11], p. 1). The ODM describes the relationship between the Model Driven Architecture (MDA) components such as UML and Common Logic (CL) and knowledge representation languages like OWL. Many research activities have contributed to documenting methods of

[1] `http://www.iso.org/iso/maintenance_agencies`

integrating semantic technologies with MDA[12]; [13]. In [14] the differences between OWL and UML have been unraveled. In our research we have concentrated on the meaning of terms and concepts used in the standards. In a case study we developed a semantic concept harmonisation layer between the Aquo standards in order to contribute to the interoperability of the standards on the semantic level.

## 4 Case Study

The Netherlands is a country that has a significant portion of its total area below sea level. The country has also many water bodies. Proper water management is therefore essential. For this purpose the Dutch water boards and the national government share several information systems. The Dutch 'Informatiehuis Water' (IHW) is responsible for setting and maintaining the standards for the development of these systems. This guarantees that water management organizations create systems that can interconnect and exchange information effectively. IHW maintains many standards, identified by a common name 'Aquo standard'. Every standard has its own purpose. The standards are related on the international level (ISO), national level (NEN3610) and via domain specific information models (IM) (IMRO, IMWA etcetera) (Fig. 1) The Aquo standard IMWA is part of the NEN3610 standard.

Details of the separate Aquo standards are given in Table 1. Though different in application, syntax format and maintenance software, all parts of the Aquo standard refer to the same domain concept origin. In Table 1 this is elaborated for the concept *'sluice'*. The objective of the research was to provide for a semantic

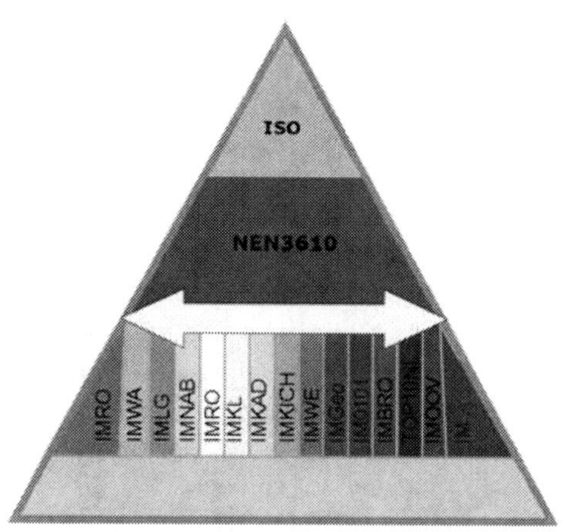


**Fig. 1.** Dutch standard NEN3610 and its dependencies

**Table 1.** Details of the separate standards in the Aquo Standard. The example column gives an example of the notation of the concept 'sluice' in the standard.

| Name | Application | Syntax Format | Example |
|---|---|---|---|
| Aquo-lex | Thesaurus | XML | <Code>KSL, 1009</Code><br><Term>sluice</Term><br><Definition>'An artificial, movable barrier that can open or close the connection between two waters'</Definition> |
| UM Aquo | Exchange of information | UML | FeatureType with Attribute TypeOfWatermanagementBody (retrieved from domain table list) |
| LM Aquo | Logical model | ER | Entity = sluice<br>Attribute = type of sluice; capacity; location |
| IMWA | Exchange of geo information | UML | Feature Type with Attribute geometryPoint |
| Domain tables | Keyword lists | XSD | <xs:enumeration value=" Watermanagement Body; sluice"/> |

layer that contains the concepts (e.g. 'sluice') and defines the relations of those concepts within the group of standards. At the same time a single point of concept maintenance was aimed for.

## 5 Results

The general approach was to first investigate which concepts in the standards were overlapping, how they were related and next to design a common semantic standard, or ontology. The central starting point was Aquo-lex, since it is by its origin a thesaurus containing concept definitions and relations. The Simple Knowledge Organization System (SKOS) [15] model was used to semantically disclose the concepts into a new version of Aquo-lex. SKOS is a powerful, well-ordered KOS model that can store semantic relations such as broader and narrower, as well as language specifications, definitions and synonyms. KOSses from different domains (schemes) can be connected together via the `skos:ConceptScheme` class. In order to link the concepts from the SKOS Aquo-lex to the other Aquo standards some extra OWL classes and properties needed to be defined. These properties explicitly describe the linking relation in a semantic way, e.g. `hasConcept` is the link of the ER diagram to the concept definition in SKOS (Fig. 2).

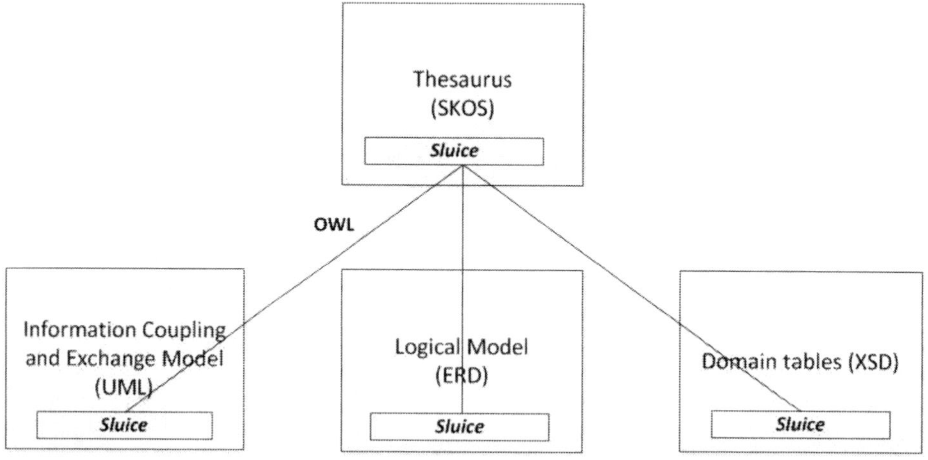


**Fig. 2.** The linked concepts in the separate standards, with the lines between the Sluice boxes identifying the semantic relation `hasConcept`

Once this SKOS-OWL data model was designed it had to be loaded with the actual values of the Aquo-lex standard (the SKOS part) and the values of the related linked standards (the OWL part). The conversion of Aquo-lex XML to SKOS was executed by means of a data conversion product[2]. In addition XSLT was used to reformat the conversion result to proper SKOS syntax. The conversion of ER and UML to OWL was done by hand on a small set of instances, as this was the best way to ensure proper match between concepts for the relatively small number of concepts. The values were given unique ids in order to make them independent of language and character coding limitations. For example: concepts were coded as `CONxxxx` and entities from the ER diagrams with `ENTxxx` (xxx referring to a number sequence). The result of the construction of the semantic layer was an OWL file that contained individuals of SKOS concepts originating from Aquo-lex and their relations with the rest of the Aquo standards. Below is listed a part of the OWL file, in which the concept with

[2] FME from SAFE software ( `http://www.safe.com`).

id `CON0001` is related via the **`hasConcept`** property to the EntityClass with id `ENT001`. The named value 'SLUICE' can be observed via the **`rdfs:label`** property.

```
<owl:Thing rdf:ID="ENT001">
<rdf:type rdf:resource="#EntityClass"/>
<rdfs:label rdf:datatype="http://www.w3.org/2001/XMLSchema#string"
>SLUICE</rdfs:label>
<catalogus:code
rdf:datatype="http://www.w3.org/2001/XMLSchema#string"
>KVP</catalogus:code>
<catalogus:hasConcept rdf:resource="#CON0001"/>
</owl:Thing>
```

The creation of this semantic layer has resulted in the conversion of the Aquo-lex standard from XML to SKOS. It can eventually lead to the elimination of the stored concepts in the other Aquo standards. In return these other standards are to be provided with linked concepts that are administered via the SKOS-OWL reository. The contents of the SKOS-OWL file was demonstrated via a graphical user interface (Fig. 3) that showed the concept definitions in the new SKOS Aquo-lex (central in the picture) together with the information of the related standards (surrounding frames).

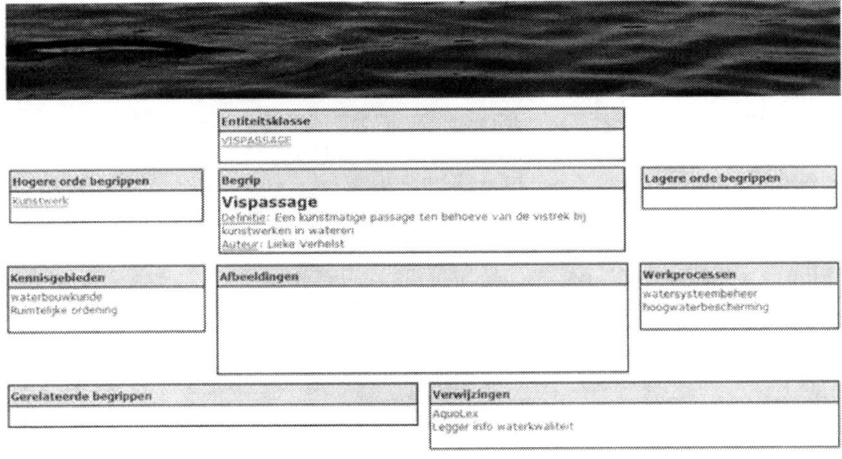


**Fig. 3.** A graphical user interface on top of the semantic layer (in Dutch)

## 6 Conclusion and Discussion

We have created a semantic model for harmonising concept definitions between software standards with SKOS-OWL. By itself SKOS does not provide sufficient properties to define concept relations beyond the thesaurus level. However the openness of the OWL syntax allows for addition of user-defined properties. We used this possibility in our SKOS-OWL design to add OWL properties that link concepts between standards. The developed semantic model is a first step in the harmonization of terms and concepts in the Aquo standards. While the GUI helps to make it accessible to a wider audience, further steps in the development of the Aquo standards are on integrating the syntax content of the different standards more tightly (see column 4 of Table 1). This would require clear rules for conversion between different programming paradigms, such as OWL-UML, OWL-ER (see, for example, [16]) and OWL-XSD translations and vice versa. In addition procedures need to be set up to ensure the consistency of such a set of linked standards and ontologies, as changes in one standard can affect other standards. Tools for easy editing need also be available, as ontology editors can be difficult to use for those not familiar with such tools or the concept of ontologies.

A semantic concept model can be used for two purposes. The first is to simplify standard concept definition and maintenance by means of a single concept administration point. The second purpose is to align standard concepts from different domains using the `skos:ConceptScheme` class. We suggest the development of a separate ontology specification per domain standard (similar to the result of our research), and then to link the separate ontologies into an umbrella ontology that combines all the concepts in the subsequent standards. The umbrella ontology can be used as a semantic bridge between the domain specific standards as depicted as the horizontal arrow in Fig. 1. The two purposes descibed contribute to the semantic interoperability of standards.

**Acknowledgments.** We gratefully acknowledge the financial support of Informatiehuis Water for carrying out this research. The authors would like to thank all colleagues who contributed to this study. We are grateful to the standard specialists Wilbert Vos, Hinne Reitsma and Willemien Joosse of IHW for their advice and the introduction to the Aquo standards. We would like to thank Arend Ligtenberg for his valuable feedback during preliminary investigations. We also thank our Software Engineers Steven Hoek, Inge la Riviere, Jappe Franke and Matthijs Danes for building the software implementation.

## References

1. Rizzoli, A.E., Donatelli, M., et al.: Semantic links in integrated modelling frameworks. Mathematics and Computers in Simulation 78(2-3), 412–423 (2008)
2. Villa, F., Athanasiadis, I.N., et al.: Modelling with knowledge: A review of emerging semantic approaches to environmental modelling. Environmental Modelling and Software 24(5), 577–587 (2009)
3. Jakobsen, C.H., McLaughlin, W.J.: Communication in Ecosystem Management: A Case Study of Cross-Disciplinary Integration in the Assessment Phase of the Interior Columbia Basin Ecosystem Management Project. Environmental Management 33(5), 591–605 (2004)

4. Bracken, L.J., Oughton, E.A.: 'What do you mean?' The importance of language in developing interdisciplinary research. Transactions of the Institute of British Geographers 31(3), 371–382 (2006)
5. Scholten, H.: Better Modelling Practice: an ontological perspective on multidisciplinary, model based problem solving, p. 313. Wageningen University. PhD, Wageningen (2008)
6. Gruber, T.R.: A translation approach to portable ontology specifications. Knowledge Acquisition 5, 199–220 (1993)
7. Antoniou, G., van Harmelen, F.: A semantic web primer. The MIT Press, Cambridge (2004)
8. Coetzee, S.: Results from a normative dependency analysis of geographic information standards. Computer Standards and Interfaces 33(5), 485–493 (2011)
9. Kaiya, H., Saeki, M.: Ontology based requirements analysis: lightweight semantic processing approach. In: Fifth International Conference on Quality Software (QSIC 2005), September 19-20, pp. 223–230 (2005), `http://ieeexplore.ieee.org/stamp/stamp.jsp?tp=&arnumber=1579139&isnumber=33358`
10. Rodrguez-Garca, D., Barriocanal, E.G., Alonso, S.S., Nuzzi, C.R.: Defining Software Process Model Constraints with Rules Using OWL and SWRL. International Journal of Software Engineering and Knowledge Engineering, 533–548 (2010)
11. OMG: Ontology Definition Metamodel Version 1.0 (2009)
12. Gašević, D., Devedžić, V., et al.: Model Driven Architecture and Ontology Development. Springer, Heidelberg (2006)
13. Álvarez, M.Á., G-Bustelo, B.C.P., Sanjuán-Martínez, O., Lovelle, J.M.C.: Bridging together Semantic Web and Model-Driven Engineering. In: de Leon F. de Carvalho, A.P., Rodríguez-González, S., De Paz Santana, J.F., Rodríguez, J.M.C. (eds.) Distributed Computing and Artificial Intelligence. AISC, vol. 79, pp. 601–604. Springer, Heidelberg (2010)
14. Hart, L., Emery, P., Colomb, B., Raymond, K., Taraporewalla, S., Chang, D., Ye, Y., Kendall, E., Dutra, M.: OWL full and UML 2.0 compared (2004)
15. Isaac, A., Summers, E.: SKOS Simple Knowledge Organization System Primer, W3C (2009)
16. Athanasiadis, I.N., Villa, F., Rizzoli, A.E.: Enabling knowledge-based software engineering through semantic-object-relational mappings. In: 3rd International Workshop on Semantic Web Enabled Software Engineering, 4th European Semantic Web Conference, Innsbruck, Austria (2007)

# Ontology Importance towards Enhancing Suggestions in a News Recommender System for a Financial Investor

Vasile Paul Bresfelean, Mihaela Bresfelean, and Ana-Maria Ghiran

Babeş-Bolyai University, Dept. of Business Information Systems,
Str. Theodor Mihali 58-60, 400599, Cluj-Napoca, Romania
{paul.bresfelean,anamaria.ghiran}@econ.ubbcluj.ro,
miha1580@yahoo.com

**Abstract.** In terms of the current information society, where news and information generally run at a remarkable speed, it is imperative that the events of capital importance for the world economy (and others fields) to be available and disseminated in real time. Thus, we refer to news that may influence in a large manner the stock quotes, forex, oil quotes and so on. In view of the fact that the Internet abounds in news, in our on-going research (under grant TE_316 "Intelligent methods for decision fundamentation on stock market transactions, based on public information") we seek to propose a solution for extracting news vital information, its interpretation in real time, followed by suggesting possible decisional alternatives to the representatives involved in the stock exchange market. We put forward in the present paper our starting ideas for a conceptual model for a semantic Web news search, which encompasses multi-agent technologies, semantic Web, text search and mining, RSS news feeds.

**Keywords:** News, splogs, multi-agent systems, ontology, Semantic Web, text mining.

## 1 Introduction

While the humankind, as we know it, encompasses a continuous transformation, one valid choice is to keep up the pace with these changes and take advantage from them, either by responding or perchance by predicting them [1]. For example, when a new matter transpires about Tsunami in Japan, or insurgency in a large Middle east oil producer, global news networks would swiftly report it. On the other hand, one cannot know when this surprising news will occur again, might be periodical, occasional or irregular. Consequently, it is important to have a fast reaction and interpretation to this news, being one step ahead of the others, and based on experience from the past events, to be able to foresee or anticipate the movements on the stock market, oil prices, raw materials, exchange rates and so on. For this reason, we appeal to a variety of news sources such as RSS feeds, sites of the news media, the television, in newspapers, new media (Twitter, blogs, etc.), weather forecast etc. Selection criteria for the news

E. García-Barriocanal et al. (Eds.): MTSR 2011, CCIS 240, pp. 178–185, 2011.

sources used in our research should take account of a high rank of international distribution, reliability of the news, the accessibility of RSS feeds [2].

Given that the Internet is bursting with news, in our on-going research, we seek to propose a solution to get relevant information from news, its interpretation in real time, followed by suggesting possible alternatives for decisions to the representatives involved in the stock exchange market. We present in this paper our starting ideas on a conceptual model for semantic Web news search, which includes multi-agent technologies, semantic web, text search and mining, RSS news feeds.

## 2 State of the Art in the Field

Recommender systems aim at addressing the problem of choice difficulty as they provide personalized suggestions that match the user needs [3]. Studies conducted for news recommender systems like MESH project [4] concluded that these have two different ways of usage: `pull` mode - the recommendations are driven by the user queries, and `push` mode - the recommendations can be offered even if the user did not express any request for a specific item. Because of this dual mode, the choice over the implementation of recommendation algorithm in the first case would be towards content based solutions, whilst in the second case towards collaborative filtering. The integration of these two in a hybrid approach can overcome the limitations of the algorithms applied alone.

With the intention of recommending news items, some researchers tried to model the users Web actions by analyzing the RSS blogs to which the user has subscribed [5], allowing to track down the individual profile from a specific user, by interpreting the users blog posts and identifying related similarities between the user generated content and information coming from other sources. The news items having good correlations with the user profile are finally recommended to the user [5]. Phelan et al. [6] consider a novel alternative to conventional recommendation approaches by harnessing a popular micro blogging service such as Twitter with a view to identifying emerging topics of interest, which can be matched against recent news coverage in an RSS feed, as the basis for story recommendation.

Other researchers [7] had implemented a prototype application to monitor RSS feeds of news stories, querying the Web for facts needed by OntoSem, translate them from OWL into OntoSems native KR language to understand the text, and exports the computed facts back to the Web in OWL. Yu et al. [8] proposed a Web portal which was owned and managed by the specific user, located on the users local desktop or local network, and constructed based on semantic Web technology for satisfying the requirements of a user to all his relevant information.

Some lines of work [9] represent a significant departure from previous keyword search models that assumed a static database by offering a query comprised by a collection of keywords, subsequently applied on multiple text streams (these can be RSS news feeds, TV closed captions, emails, etc). The authors use an incremental computation of the Jaccard coefficients between the participants and the content of the streams. There are frameworks [1] able to read news items and extract the valuable economic information from them, allowing the semiautomatic discovery of world changes of interest for particular users, and able to update its knowledge base to always represent the current state of the world.

## 3 Our Proposed Architecture

As a result of recent web developments and events, and based on up to date literature studies, we offer a conceptual model for a semantic Web news search system. It is intended for the users who require attention-grabbing events, arranged and interpreted by their own predilections.

Our model is differentiated from a simple news recommendation solution in sense that except from providing the news of interest for a user, it incorporates arguments in order to convince her/him about a certain course of action regarding financial investments. It acts in two phases: first as a recommender systems, gathering data about user profile and searched items, applying recommendation techniques and presenting suggestions; and second, interpreting the recommendations in order to support the user in decision making.

As can be seen in Fig. 1, the proposed model is formed by the 4 parts, as follows: the news data sources, the module based on a multi-agent environment, the resulting data collection module, and a final module for knowledge extracting.

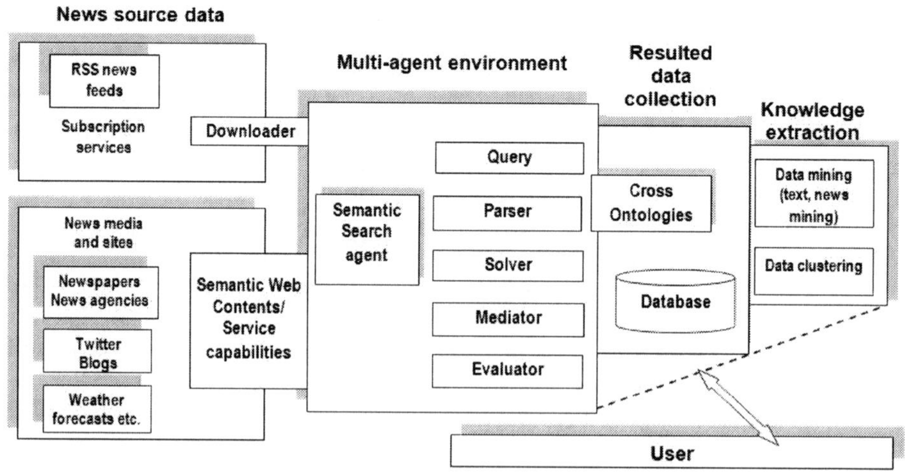


**Fig. 1.** Conceptual model for a semantic Web news search system, based on [15]

### 3.1 Data Sources

In our cross-media information extraction from different news agencies, blogs, and other sources from different areas, the lack of semantic interoperability results in massive assimilation work. The propagation of blogs and other new media consequences in an abundance of data, so that users are incapable of taking advantage of the collective knowledge.

The primary Internet sources used in our architecture are the relevant news sites and blogs selected according to the constructed user profile. As in other news recommender systems [4], our system is employing different recommendation algorithms, case the user is requiring or not for suggestions (pull-push mode). If the selection of data sources is directed by the user (it is subject to the pull mode), then the recommendation algorithm best suited is one finding items similar to those directly expressed by user (a content based approach). If the user does not manifest an explicit request, then the system is offering suggestions (in push mode) and in order to have more diversity, collaborative filtering methods are used.

Besides news, our users might be particularly interested in opinions expressed as blogs, and primary, those issued by financial analysts. As these analysts are the main processors of financial information, their opinion is very valuable for stock market investors which are contented if the system could make suggestions about them.

Blogs are some of the most interesting but difficult to process social media sources, nevertheless valuable for studying customers view on goods, services, whole companies. Our idea is to offer personalized information to users own interests, and also finding and matchmaking blogs/users which have common thoughts. But all together, blogs might contain ambiguous and confusing data due to various profits from advertising associations.

Splogs or spam blogs represent detrimental blogs intended to magnetize search engine traffic, and are used solely to support affiliate sites, degrade the quality of search engine results, and waste network resources [10].

As a consequence, some of the main directions from the research literature that we shall take in consideration when in fact building the news search system, include the following:

- Efficiently collect splogs by sampling blog homepages with common keywords on specific dates with most recurrent rate [11]; more than half of the collected splogs may possibly created by a very small number of spammers.
- Exploiting unique blog temporal dynamics so as to detect splogs [10]. The main idea behind it is to use temporal structural properties computed from self-similarity matrices across different attributes; these are combined with content based features.
- Using Support Vector Machines (SVMs) and Relaxed Online SVM (ROSVM) for content-based filtering and content-based splog detection [12].

- Analyzing the search results of a sequence of temporally ordered queries returned by a blog search engine [13], build and maintain blog profiles whose posts frequently appear in the top-ranked search results. Evaluation is based on using genuine data collected from blogs search engines.

The data sources are made of the RSS news feeds via subscription services, and also of news media and sites (newspapers, news agencies, Twitter, blogs, weather forecasts, etc.), by the use of the semantic web contents/service capabilities which semantically describe the general capabilities of the web site. An RSS document, so-called a feed or channel, includes full or summarized text, plus metadata such as publishing dates and authorship [16].

### 3.2 Multi Agent System and Ontologies

The multi-agent environment was developed starting from the most recent studies [15]. First, the RSS feeds collected from different source are aggregated and parsed.

Users requests are directed towards the search-agent responsible for complete processing, then the parser examines the call according to its field of knowledge, and sends its interpretation to a monitor, which has a central role collecting information from all parsers. In the case of contradictory versions from different parsers, the results are forwarded to a mediator who has access to ontologies, preceding assessments and probabilistic inspection to clarify meaning. The mediated outcome is then sent to a solver, in order to produce the final solution.

The evaluator component has the role to incorporate the user feedbacks in the system by storing in a specific database, thus permitting to approximate the system's standard of performance. As a consequence the model offers continuously updated data to allow discovering the users profile, identifying relevant similarities and correlations between the generated content and items coming from news data sources.

The ontology will hold the user preferences in order to have a semantic description over the interest topics. Ontologies have already proved their applicability in describing domain specific information and are increasingly employed in modeling the knowledge. An ontology is described in a specific language like OWL, and specially on a subset of it OWL DL derived from Description Logic. We would like to emphasize the role played by the semantic descriptions for the item search. In a simple recommender system, the returned results are for directly related concepts whilst in an ontology based can refer to indirect concepts. An example: A user could express an interest in companies that have low volatility shares. Without a semantic meaning associated with this, the returned results would include only a keyword match (that might be referring to information about shares of a company that deals with low volatility solvents). But if we express that low volatility shares are stock shares for which the price does not change often (equation 1), the results quality is clearly improved.

$$LowVolatilityShares \sqsubseteq StockShares \sqcap \exists haveLowFluctuationPrice.SharePrice \quad (1)$$

In order to construct the ontology based user profile, the agents will populate ontology with instances but also will be able to define new concepts or properties not currently included through using the specific ontology API (for instance in case of OWL language there are at least three APIs that could be used: Jena Ontology API [17], OWL API [18], Protégé OWL API [19]).

On the other hand, the ontology will be subject of the knowledge extraction module for inferring new information.

The ontology use enables us to uncover specific relationships, in addition to the general association linking terms. For example, the OWL based ontologies could be utilized for conceptualization of metadata but also as a knowledge base.

Applying automated reasoning capabilities like Pellet[1], Racer[2] over the knowledge base permit the verification of the ontology for inconsistencies. Using OWL as the ontology language allows us to define constraints over the ontology concepts or relationships. In case a constraint is violated, the reasoner will report it as an inconsistency. This is an important feature when employing a reasoning engine in order to differentiate or classify concepts in the ontology (the agent responsible for screening through all alerts and discarding the unnecessary feeds). Applying a reasoner to an ontology will enable:(i)`Check consistency of the ontology` - ensuring that there are no common instances of classes that have been declared disjunctive, e.g. an individual known as aggressive investor will not be found in category of preserving capital investors or conservative investors (ii)`Classify taxonomy` - inferring the subclasses or the subproperties, e.g. the user is interested in financial products that is in all subcategories of them: shares, bonds, treasury bills, annuities etc.(iii)`Compute inferred types` - if an individual is an instance of a class expression, e.g. if the user has opted towards buying stocks that means she/he has got a form of financial investment.

The whole system has been designed considering the full integration of the multi agent system with ontologies and the reasoning capabilities over the knowledge base. Agents are deployed inside an agent platform [20]. An agent platform is required in order to deliver functionalities like: surveillance and management of agents (creation, destruction, migration), agent communication, error notification and security mechanisms. The choice over the agent platform will have a certain influence over the proposed architecture depending on the capabilities offered by the agent platform (FIPA compliance, reasoning capabilities, support offered for documentation, up-to-date downloadable versions etc.) For instance, considering that we would like that our agents to be able to create OWL constructs, axioms, run inferences or load, save an entire ontology, we need them to include an OWL API. Currently all these are based on Java, therefore, it is better if our agents are also created using Java.

### 3.3 The Resulting Data and Presentation Module

With the intention of achieving a superior interpretation/deduction required for semantic awareness of texts, one needs to utilize a comprehensive ontology (such

[1] http://clarkparsia.com/pellet/

[2] http://www.racer-systems.com/

as WordNetG, GOLD, Dublin core, BMO etc.). Due to the fact that there is no definitive and consistent ontology, we propose to use Cross-Ontology similarity methods so as to evaluate terms from dissimilar ontologies. Another proposal is to incorporate a semantic bridge in the cross-media extraction models in order to assist information course and define mappings across diverse conceptualizations defined in different ontologies, which are not mutual but may possibly have an analogous connotation. Novel studies show how to integrate a semantic bridge [14] and to apply a rule language for defining it, encompassing two main advantages: increase the maintainability of the bridges and perform transformations between related concepts.

The RSS feeds used in our system might have large portions containing basically plain text and without any semantics. A solution would be by using Natural Language Processing (NLP) tools to convert natural language text into a structured representation thereby adding additional metadata in the RSS fields [7].

### 3.4 The Knowledge Extractor Module

The presence of structure and content information in ontologies enables us to plan several developments in the knowledge extraction module, based on data mining and data clustering technologies. It facilitates our model that besides the proposal of full articles, to also suggest only news fragments. Once an item is connected with a definite request, then its downloaded and stored in our database.

## 4 Conclusion

In this paper we included a part of our research in designing a model for a semantic Web news search system. We started by presenting some of the most recent studies in the field, then we introduced our model intended for the users who require attention-grabbing events, arranged and interpreted by their own predilections, reducing the time spent in finding by hand the news items. There were described the main modules and their components, and how we envision their functionality when the project will be finalized.

Future researches will be directed towards the completion of the model and its validation with factual data sources.

**Acknowledgments.** This work was supported by the CNCSIS TE_ 316 Grant.

## References

1. Schouten, K., Ruijgrok, P., Borsje, J., Frasincar, F., Levering, L., Hogenboom, F.: A semantic web-based approach for personalizing news. In: Proceedings of the 2010 ACM Symposium on Applied Computing, SAC 2010, pp. 854–861. ACM, New York (2010)
2. Arango, M.: Vanishing point. In: ACM Multimedia, pp. 1067–1068 (2005)

3. Ricci, F., Rokach, L., Shapira, B.: Introduction to recommender systems handbook. In: Recommender Systems Handbook, pp. 1–35 (2011)
4. Picault, J., Ribière, M., Bonnefoy, D., Mercer, K.: How to get the recommender out of the lab? In: Recommender Systems Handbook, pp. 333–365 (2011)
5. Di Massa, R., Montagnuolo, M., Messina, A.: Implicit news recommendation based on user interest models and multimodal content analysis. In: Proceedings of the 3rd International Workshop on Automated Information Extraction in Media Production, AIEMPro 2010, pp. 33–38. ACM, New York (2010)
6. Phelan, O., McCarthy, K., Smyth, B.: Using twitter to recommend real-time topical news. In: Proceedings of the Third ACM Conference on Recommender Systems, RecSys 2009, pp. 385–388. ACM, New York (2009)
7. Java, A., Finin, T., Nirenburg, S.: Text understanding agents and the semantic web. In: Proceedings of the 39th Annual Hawaii International Conference on System Sciences, vol. 03, p. 62.2. IEEE Computer Society, Washington, DC (2006)
8. Yu, H., Mine, T., Amamiya, M.: An architecture for personal semantic web information retrieval system. In: Special Interest Tracks and Posters of the 14th International Conference on World Wide Web, WWW 2005, pp. 974–975. ACM, New York (2005)
9. Hristidis, V., Valdivia, O., Vlachos, M., Yu, P.S.: Continuous keyword search on multiple text streams. In: Proceedings of the 15th ACM International Conference on Information and Knowledge Management, CIKM 2006, pp. 802–803. ACM, New York (2006)
10. Lin, Y.R., Sundaram, H., Chi, Y., Tatemura, J., Tseng, B.L.: Detecting splogs via temporal dynamics using self-similarity analysis. ACM Trans. Web 2, 4:1–4:35 (2008)
11. Sato, Y., Utsuro, T., Murakami, Y., Fukuhara, T., Nakagawa, H., Kawada, Y., Kando, N.: Analysing features of japanese splogs and characteristics of keywords. In: Proceedings of the 4th International Workshop on Adversarial Information Retrieval on the Web, AIRWeb 2008, pp. 33–40. ACM, New York (2008)
12. Sculley, D., Wachman, G.M.: Relaxed online svms for spam filtering. In: Proceedings of the 30th Annual International ACM SIGIR Conference on Research and Development in Information Retrieval, SIGIR 2007, pp. 415–422. ACM, New York (2007)
13. Zhu, L., Sun, A., Choi, B.: Online spam-blog detection through blog search. In: Proceeding of the 17th ACM Conference on Information and Knowledge Management, CIKM 2008, pp. 1347–1348. ACM, New York (2008)
14. Barnickel, N., Fluegge, M., Schmidt, K.-U.: Interoperability in egovernment through cross-ontology semantic web service composition. In: Workshop Semantic Web for eGovernment, 3rd European Semantic Web Conference (2006)
15. Weber, N., Braubach, L., Pokahr, A., Lamersdorf, W.: Agent-based semantic search at motoso.de. In: Braubach, L., van der Hoek, W., Petta, P., Pokahr, A. (eds.) MATES 2009. LNCS, vol. 5774, pp. 278–287. Springer, Heidelberg (2009)
16. Hu, C.L., Chou, C.K.: Rss watchdog: an instant event monitor on real online news streams. In: Proceeding of the 18th ACM Conference on Information and Knowledge Management, CIKM 2009, pp. 2097–2098. ACM, New York (2009)
17. Rajagopal, H.: JENA: A Java API for Ontology Management (2005)
18. Horridge, M., Bechhofer, S.: The owl api: A java api for working with owl 2 ontologies. In: OWLED (2009)
19. Knublauch, H.: Protg-owl API Programmer's Guide (2010)
20. Bellifemine, F.L., Caire, G., Greenwood, D.: Developing Multi-Agent Systems with JADE. Wiley, Chichester (2007)

# Detailed Presentation versus Ease of Search – Towards the Universal Format of Bibliographic Metadata. Case Study of Dealing with Different Metadata Kinds during Import to Virtual Library of Science

Jakub Jurkiewicz and Aleksander Nowiński

ICM University of Warsaw, ul. Pawińskiego 5a,
Block D, 5th floor 02-106 Warsaw, Poland
{j.jurkiewicz,a.nowinski}@icm.edu.pl

**Abstract.** We divide the existing scientific publication metadata types into two classes: field based and free-text based. The division is based on origins of data, however they correspond to the main functions of digital libraries: data presentation and search services. These classes are compared to present their advantages and disadvantages. Our own format, BWMeta, is presented and shown to have the advantages of both previously described classes. Another important quality of BWMeta is displayed, which is good preservation of information.

## 1 Need of Storing Both Strictly Defined Metadata and Free Form Text

In the middle ages books were very expensive and rare. They were protected by guards, and libraries were reserved for very few people. Invention of Guttenberg brought possibility of reading to much wider audience. In 20th century books were already very popular, and libraries could be found in every city. The biggest libraries contain millions of volumes. The real problem was to find desired information. This problem was solved by catalogues, which indexed books by specific metadata fields. In 70-ties, when computers were introduced into libraries, situation did not change that much. Number of indexed fields has increased and access to a whole catalogue was much easier and faster, but user still has to know a title, an author name or a topic of the book.

The introduction of text search was a great step forward. In the beginning it was limited only to some metadata fields, like abstract, title, author name or keywords. Now it is possible to use fulltext search - when complete article text may be used for searching. But despite the practical ability to perform powerful full text searches, metadata is still in wide use, because a precise query based on article abstract or keywords can give results more relevant than a full text search. Another reason for allowing metadata-only search, is a situation when

E. García-Barriocanal et al. (Eds.): MTSR 2011, CCIS 240, pp. 186–193, 2011.

the user is not allowed to access full text, however he or she may be interested in buying it.

Yadda[1] is a digital library system developed in ICM, which is used as a common platform for storing a number of different resources from different providers. Yadda offers a unified presentation and search over a number of a different collections of documents from different sources. It serves a huge collection of data directly from the publishers as well as a number of smaller bibliographical databases created and maintained in Poland. The system offers support for the classic publications (articles, books), but is being developed as a generic system capable of storing any other types of records (computer programs, multimedia files and others). This forced us to develop and maintain generic metadata format named BWMeta, suitable for multiple purposes. In this format we store all metadata in our library called Virtual Library of Science. This library is manged by Yadda software.

During our operation of the system an important part of the work was import and conversion of the metadata from various sources. It could be provider specific formats (like Elsevier format), some standard formats (Dublin Core) or direct import from relational database (Baztech - Polish bibliographical database).

Our experience is that in general metadata usually falls in one of two categories:

- Fulltext style data - additional tags are added to existing fulltext to identify metadata fields inside the object, usually it comes from publisher
- Metadata style data - data is organised as set of specific fields, typically it comes from libraries, or databases created by librarians

This is not a sharp distinction - in data acquired from both kinds of providers one can find elements specific to another class. Usually in fulltext data style one can find some metadata elements, and metadata style data sometimes formatted fulltext can appear.

What differs those two kinds of metadata classes is how they were created. Fulltext style metadata is created by adding additional tags to existing fulltext to improve data presentation to the user. Providers of this kind of data hardly care how it will look in another system, and data itself is not easy to translate into exact metadata form. This data is usually based on authors information added to articles, typically data in Latex format. In fulltext style data metadata are usually tags added to article content - data.

In metadata style class, which origins from normal library catalogues, data are created usually by librarians by filling up predefined set of fields. Fulltext may be added as additional part of the record. During import process to the Yadda system we often have to discuss with provider how this data should be presented to the user.

This format is usually more precise, but can be prone to data quality problems – for example, it is common to place initials in place of first name. This example shows essential difference between those types – dealing with unclear situations. In the fulltext style, unclear data can be represented as plain, untagged text, and the assumption is made that a human reader will interpret it properly. In metadata style class there are two possible solutions - either to break the constraints and put data into wrong field (hoping they will be displayed in understandable way), or discard it, because metadata schema has no place for it.

The metadata style formats used in daily Yadda operations are based on the relational databases: the Baztech database[7], BazEkon from Main Library of the Cracow University of Economics, CEJSH and AGRO databases. Another types of data in metadata style are the most important exchange formats like: RIS [4] or MARC [3]. The most important data in fulltext style for us is Tag by Tag format from Elsevier[6]. 9.2 millions of records are kept in Virtual Library of Science. Most of them have been imported from Elsevier Science (6.9 millions) and Springer (1.7 millions). Smaller sources like Baztech ( 230 thousand records) and Agro (360 thousand records) are very important for us because they cover Polish publication.

In Yadda system we have to deal with both kinds of metadata style as input in the best possible way. Our system has the following goals:

1. good data presentation - clear, precise and useful for scientific library user
2. display as much data as it possible
3. preserve all the original data, and do not lose information

To accomplish these goals we have developed our own metadata format - BWMeta. It has to be flexible - to support extensive number of information, and at the same time support free form content for displaying them for readers. Our format is based on XML - which is the best choice for this task, because allows storing free form text content, and organised metadata fields in the same document. The format has also been tested and has proven usability for librarians [2]. There is also additional future goal, as we are preparing to build libraries of multimedia, computer programs and experimental data, which require very flexible metadata system. BWMeta is such metadata format, which is suitable for describing various types of data and does not relay on typical publication bibliographic data.

All the previously mentioned 9.2 millions of records, originating from a number of various sources, have been converted to BWMeta - the storage format of the Virtual Library of Science. The format has evolved significantly during five years of operation - introduction of each of the previously mentioned data sources required us to make some changes to the BWMeta format. This was the driving force, motivating the BWMeta team to make it flexible enough to meet all the requirements.

## 2 Examples of Metadata Issues

A number of cases is provided below to illustrate various issues regarding scientific publication metadata. RIS and Tag by Tag formats are presented and compared to BWMeta. Our proposed solutions are described as well.

### 2.1 Title, Subtitle, Title Translation

Every, or nearly every article has a title. Sometimes a subtitle is added to the title, also translation of the title can be found in metadata.

Rarely, a general class of the article is added to the title, e.g. "Controversies in geriatric medicine". While defining the metadata format one could attempt to predict all the possible title types and describe them one by one. Unfortunately, this approach is very work consuming as the user interface has to support all of them. Another approach is based on semi-structured data – the XML "title" field with mixed content allowed and some parts of it tagged to denote specific meaning. This approach is clearly superior to the previous one, but some issues remain unresolved – for example articles with more than one title. Another issue is describing parts of title.

Example header could apply to both title, and title abbreviations. In BWMeta format we have decided that any element - like article or book may have many names of different types. It is also allowed in Dublin Core.

As a title we display name with no type. Interface allows displaying all names with types. Advantage of this solution is the fact, that parts of the software that do not work with specific title types do not need to understand it - it is passed to user, or to other modules.

Figures 1, 2 and 3 display real world examples of the "title" property, represented in various formats. The RIS format (as shown on figure 1) misses the "Controversies in geriatric medicine" string. The "Tag by tag" format (figure 2) does preserve this part of information, but it is difficult to interpret it or convert to another format – it is not clear which field it corresponds to. Overall, we

```
T1  - Commentary on frailty and vulnerability by Carlos Montes Paixao Junior
      and Alexandra Prufer de Queiroz Campos Araújo
```

**Fig. 1.** Example of titles section in RIS -format - metadata style

```
   <ce:dochead>
       <ce:textfn>Controversies in geriatric medicine</ce:textfn>
   </ce:dochead>
   <ce:title>
     Commentary on frailty and vulnerability by Carlos Montes Paixao Junior
     and Alexandra Prufer de Queiroz Campos Ara&uacute;jo
   </ce:title>
```

**Fig. 2.** Example of titles section in Tag by Tag -format - fulltext style

```
  <name lang="eng" type="abbreviation">Controversies in geriatric medicine</name>
  <name lang="eng">
     Commentary on frailty and vulnerability by Carlos Montes Paixao Junior
     and Alexandra Prufer de Queiroz Campos Araújo
  </name>
```

**Fig. 3.** Example of titles section in BWMeta format - our style mixed one

conclude that the BWMeta format meets all the requirements: all the information is kept, and the scope of each part of it is clear. The interoperability is good: it is easy to translate the information to another format, such as Dublin Core. Another important issue is possibility of describing provenance of title. It impossible to do in RIS format or Dublin Core. In our format BWMeta attribute type for element name could be used for storing this information.

## 2.2 Person and Affiliation Description

One of the most important issues to be solved by a publication metadata system is the management of information related to a person. In European culture it seems simple - a person has a name and a surname, as main describing elements. Apart from this person can have a degree, pedigree and other attributes. In the case of field-based metadata, the information about a person is often reduced to full name, sometimes written as "surname, initial" (Smith, J.). In fulltext style metadata, names are stored exactly in the way author wrote them, sometimes with additional tagging of text parts. But how to connect email with author,

```
AU  - Bauer, J.M.
AD  - Department of Internal Medicine and Geriatric Medicine,
      Friedrich-Alexander Universität Erlangen-Nürnberg,
      Prof.-Ernst-Nathan-Strasse 1, 90419 Nürnberg, Germany
N1  - doi: 10.1016/j.eurger.2010.05.011
N1  - Language of Original Document: English
N1  - Correspondence Address: Bauer, J. M.; Department of Internal Medicine and Geriatric
      Medicine, Friedrich-Alexander Universität Erlangen-Nürnberg, Prof.-Ernst-Nathan-Strasse 1,
      90419 Nürnberg, Germany; email: juergen.bauer@klinikum-nuernberg.de
```

**Fig. 4.** Example of author section in RIS -format - metadata style: some N1 are about an author, and some are not, but how to automatically distinguish their meaning?

```
<ce:author-group>
    <ce:author>
        <ce:given-name>J.M.</ce:given-name>
        <ce:surname>Bauer</ce:surname>
        <ce:cross-ref refid="cor1"><ce:sup>&#x204E;</ce:sup></ce:cross-ref>
        <ce:e-address type="email">juergen.bauer@klinikum-nuernberg.de</ce:e-address>
    </ce:author>
    <ce:affiliation>
        <ce:textfn>
             Department of Internal Medicine and Geriatric Medicine,
             Friedrich-Alexander Universit&auml;t Erlangen-N&uuml;rnberg,
             Prof.-Ernst-Nathan-Strasse 1, 90419 N&uuml;rnberg, Germany
        </ce:textfn>
    </ce:affiliation>
    <ce:correspondence id="cor1">
        <ce:label>&#x204E;</ce:label>
        <ce:text>Tel.: +499113985543; fax: +499113985542.</ce:text>
    </ce:correspondence>
</ce:author-group>
```

**Fig. 5.** Example of author section in Tag by Tag -format - fulltext style

```
<contributor institution="false" role="author">
   <attribute key="contact-email">
      <value>juergen.bauer@klinikum-nuernberg.de</value>
   </attribute>
   <attribute key="author-correspondence">
       <value>Tel.: +499113985543; fax: +499113985542.</value>
   </attribute>
   <name lang="zxx" type="canonical">J.M. Bauer</name>
   <name lang="zxx" type="surname">Bauer</name>
   <name lang="zxx" type="forenames">J.M.</name>
</contributor>
<affiliation id="affiliation-gen-id-0001">
   <text>
         Department of Internal Medicine and Geriatric Medicine,
         Friedrich-Alexander Universität Erlangen-Nürnberg,
         Prof.-Ernst-Nathan-Strasse 1, 90419 Nürnberg, Germany
   </text>
</affiliation>
```

**Fig. 6.** Example of author section in BWMeta format - our style mixed one

when email is added in affiliation? One option is to add additional links between information about the author and affiliation. This is often done but in this case we lose a lot of information if there is no link specified by data provider. In fulltext style metadata there is no problem, as typically order, placement and content of the affiliation text informs about connection between affiliation elements and authors. But this information can be present only in fulltext style metadata. In BWMeta we use also fulltext style with optional links which allows connecting the author with the affiliation if possible.

Figures 4, 5 and 6 present examples of the "author" On figure 4 in RIS format the part about the affiliation institute is missing. Some information about the author correspondence are marked by N1 type of field, but information about doi is put in the same type of field, so it is not clear which information refers to author, and which to the document.

On the figure 5 containing tag by tag and figure 6 containing BWMeta all information are present. We find BWMeta better one, because information about phone numbers and emails is kept in fields of the same type, while in Tag by Tag are split to different parts.

### 2.3 Bibliography

The part concerning bibliographic elements is a very complex one. It consists of a relatively big number of fields. Because of the great importance of bibliography in scientific work, all the information should be preserved. In some formats, the level of complication of bibliographic information is comparable to the complication of main record. In other cases, this information is reduced to a simple text field. Usually text field with some external links is enough for citation presentation to user, but sometimes user requires other forms, like bibtex citation, which can be constructed only if bibliographic information is splitted into fields. When we were constructing our format - BWMeta we decided that we do not want to lose information. For example we would like to know which world in the bibliography text means the first name of the second author, and which the first author middle name etc. It is very important, as this information may be used to locate cited work in the library resources and provide direct link to the work. This functionality is necessary in modern digital library. Of course there exist numerous standards for keeping metadata records like ISO 690 norm, but they do not provide enough space for keeping additional information. Such information could be internal an identifier of document within imported collection, on publisher site. This identifier often can not be resolved to our identifier during import process, and shouldn't be presented to users. However if we found it as identifier in other document we could be nearly 100% sure that we know where citation leads to.

On the other hand unified form is required for presentation purposes. This form should be exactly the same for every citation in whole system. It is possible to perform formatting in user interface, but this is expensive: whenever we add a new type of information in metadata (we import data from new source), we have to adopt the interface. If owners of the library do not want to modify system each time they add data from a new provider, then in metadata style they put data in the wrong field. For example they put initials in field that should be used for the first name. In fulltext style presentation is really nice - but how to keep information about elements that should not be displayed? For this reason in BWMeta format we kept both information: text that should be displayed - which is generated during data import, and classical metadata organised in XML tree. Metadata in this part is very flexible, as we use attributes with types and new types can be introduced easily. They could also be simplified to one text field. Usually, there is no modification required to any part of system, unless this part directly needs to use them. In Elsevier Tag by Tag format quite complicated structure is defined for storing bibliographic information. It is hard to parse this format, and during translation to other formats we usually lose a lot of information. In formats that kept bibliography element on single field it is sometimes hard to understand some small piece of information which could be important.

## 3 Conclusion and Future Work

None of two basic classes of metadata supports at once both basic functionality of digital libraries: preservation of original data and unified presentation of data. The first one demand possibility of storing as much different fields as it possible, while the second one requires only few fields, but with information in unified format. Fulltext class data are better suited for the presentation, while metadata style classes are better for the preservation.

In BWMeta both classes are mixed. Majority of fields is in metadata style class, but whenever required we support a free form description as in fulltext style class. Additionally it is possible to keep in BWMeta fields which are meaningless to the most of application, but can be used in future. This flexibility would allow in future fully automatic translation from other formats to BWMeta. This would allow storing data in out system in other formats and translate to BWMeta( or its software model) on demand. This is step in direction of harmonisation of metadata[8]. Currently storing data in other formats is impossible because some formats ( Tag by Tag, Springer format) do not keep information about journal and journal issues as a separate records, but as a part of record describing element. Storing journals, issues and volumes is requirement for Virtual Library of Science.

In future we need to merge attributes names with existing standards. Minimal set of standards that needs to be mapped to our names is: Dublin Core ( when used as rdf relation names), marc ontology. This would simplify exchanging data with existing metadata providers. It is possible to integrate with existing unification of such things MarcOnt[5].

**Acknowledgments.** This work is partially supported by the National Centre for Research and Development (NCBiR) under Grant No. SP/I/1/77065/10 by the Strategic scientific research and experimental development program: "Interdisciplinary System for Interactive Scientific and Scientific-Technical Information".

## References

1. The Virtual Library of Science, `http://yadda.icm.edu.pl`
2. Zamłyńska, K., Bolikowski, Ł., Rosiek, T.: Migration of the Mathematical Collection of Polish Virtual Library of Science to the YADDA platform. In: Proceedings of DML 2008 Towards Digital Mathematics Library, July 27. Masaryk University, Birmingham (2008)
3. Marc standarts, `http://www.loc.gov/marc/`
4. Thomson ResearchSoft, RIS format specification, `http://www.refman.com/support/risformat_intro.asp`
5. Kruk, S.R., Synak, M., Zimmermann, K.: Marcont - integration ontology for bibliographic description formats. In: Proc. DC 2005 (2005)
6. Pepping, S., Schrauwen, R.: Tag by tag. Elsevier Science, Amsterdam (2001)
7. Derfert-Wolf, L.: BAZTECH - Polish Technical Journal Contents. Slavic & East European Information Resources 5(1/2), 149–154 (2004)
8. Nilsson, M.: Harmonization of metadata, standards (2008), `http://ariadne.cs.kuleuven.be/lomi/images/5/52/D4.7-prolearn.pdf`

# Using Resource Description Format in Integrating and Querying Web Data Sources for Semantic Portals

Sabina-Cristiana Necula

Alexandru Ioan Cuza University of Iasi,
Faculty of Economics and Business Administration,
Carol I Blvd, no. 22, 700505, Iasi
sabina.mihalache@gmail.com

**Abstract.** This paper presents with examples the Resource Description Format use in integrating web data sources for semantic organizational portals. We define a vocabulary and describe data in RDF format in order to integrate available Linked data with local data from RDF triple store. We explore this data using SPARQL and we discuss the possible applications for semantic web portals. We conclude by presenting future work and general conclusions.

**Keywords:** semantic web portals, RDF, OWL, SPARQL, integration.

## 1 Introduction

The semantic web is based on the idea of a ***layered architecture***. Much like the ISO concept of layers in data communications, the semantic web architecture is composed of the following layers:

- URIs and Namespaces-the names of things
- XML and XMLS Data types-a means of communicating data
- RDF and RDF/XML-a basic language
- RDF Schema and Individuals-an ontological primitive
- Ontology languages, such as OWL-the logical layer
- Applications-the implementation layer.

The RDF data model [1] is similar to classic conceptual modeling approaches such as Entity-Relationship or Class diagrams, as it is based upon the idea of making statements about resources (in particular Web resources) in the form of subject-predicate-object expressions. These expressions are known as triples in RDF terminology. The subject denotes the resource, and the predicate denotes traits or aspects of the resource and expresses a relationship between the subject and the object.

This mechanism for describing resources is a major component in what is proposed by the W3C's Semantic Web activity: an evolutionary stage of the World Wide Web in which automated software can store, exchange, and use machine-readable information distributed throughout the Web, in turn enabling users to deal with the information with greater efficiency and certainty.

E. García-Barriocanal et al. (Eds.): MTSR 2011, CCIS 240, pp. 194–200, 2011.

RDF is a framework for describing Web resources, such as the title, author, modification date, content, and copyright information of a Web page.

## 2 Related Work

Ontologies have spawned fair degree of research interests, as evidenced in the literature. For instance, while Shum et al [2] utilize ontology principles for developing SholOnto, an ontology-based digital library server to support scholarly interpretation and discourse, Lai and Yang [3] have experimented utilization of ontology principles for deriving ontology-based metadata for Chinese information services in Chinese digital libraries. Yeh [4] designed an ontology-based portal for digital archive services. Uszkoreit et al [5] successfully employed Ontologies for creating a knowledge portal for the field of Language Technology. Kalfoglou et al [6] developed an ontology-driven web-based system for personalized news services.

Set in the backdrop of the importance of description logics, facet relations and ontologies in current digital information scenario, the present paper examines the use of description logics and facet relations for developing ontology-based knowledge management systems for digital libraries.

The RDF Book Mashup provided an early example of publishing Linked Data related to retail and commerce. The Book Mashup uses the Simple Commerce Vocabulary to represent and republish data about book offers retrieved from the Amazon.com and Google Base Web APIs.[7]

More recently, the GoodRelations ontology has provided a richer ontology for describing many aspects of e-commerce, such as businesses, products and services, offerings, opening hours, and prices. GoodRelations has seen significant uptake from retailers such as Best Buy and Overstock.comseeking to increase their visibility in search engines such as Yahoo! and Google, that recognise data published in RDFa using certain vocabularies and use this data to enhance search results.

## 3 Technologies Involved

Metadata (data about data) promises to overcome the problem of interoperability. An accepted standard for representing metadata should provide information about the syntax, the structure and the semantic context of data. Many standards have been proposed. For the moment the World Wide Web Consortium recommends RDF (S) (Resource Description Framework - Schema), and OWL (Ontology Web Language) for representing linked web data. RDF provides a general data model for Web resources, RDFS provides a vocabulary for describing properties and classes of RDF resource, and OWL provides richer semantics beyond RDF and RDFS. OWL represents data as instances of OWL classes related by properties and provides a powerful axiom language and it is used to represent ontologies. XML (eXtensible Markup Language) is the basis of OWL and RDF descriptions.

Ontology as a layer of building stack of the Semantic Web supports richer knowledge inference as well as knowledge representation [8]. Inference on the Semantic Web improves the quality of data integration on the Web, by discovering new relationships, automatically analyzing the content of the data, or managing knowledge on the Web in general. Inference based techniques are also important in discovering possible inconsistencies in the (integrated) data. The use of ontologies for knowledge sharing, heterogeneous database integration, and semantic interoperability has been long realized [9; 10; 11; 12].

A number of query languages have been developed to query RDF and OWL. SPARQL (SPARQL Protocol and RDF Query Language) is currently the de facto standard RDF query language. Since OWL can be serialized as RDF, SPARQL can be used to query it. There is thus a need for an expressive OWL query language that supports comprehensive querying of OWL [13]. These authors proposed SQWRL (Semantic Query-enhanced Web Rule Language) built on SWRL (Semantic Web Rule Language).

SPARQL makes it possible to send queries and receive results, e.g., through Hypertext Transfer Protocol (HTTP) or Service Oriented Architecture Protocol (SOAP).

# 4 Using RDF in Describing Web Data Sources

We used Security Exchange Commission data sets available in n3 format at http://www.rdfabout.com/demo/sec/. Besides this data about companies we used some financial data available in Excel files at http://pages.stern.nyu.edu/~adamodar/New_Home_Page/data.html. We wanted to show how we can integrate financial data by making use of Semantic Web technologies.

We developed a vocabulary/ontology for merging data from the two sources. The vocabulary and its namespaces is available at http://sandbox.metadataregistry.org/schemaprop/list/schema_id/49.html and depicted in Figure 1.

| Label | Type | URI |
|---|---|---|
| company | class | .../uri/schema/fin/company |
| Company name | property | .../uri/schema/fin/name |
| tickersSymbol | property | .../uri/schema/fin/tickersSymbol |
| industryName | property | .../uri/schema/fin/industryName |
| SICCode | property | .../uri/schema/fin/SICCode |
| xschangeCode | property | .../uri/schema/fin/xschangeCode |
| sizeClass | property | .../uri/schema/fin/sizeClass |
| StockPrice | property | .../uri/schema/fin/StockPrice |
| TradingVolume | property | .../uri/schema/fin/TradingVolume |
| MarketCap | property | .../uri/schema/fin/MarketCap |
| TotalDebt | property | .../uri/schema/fin/TotalDebt |
| FirmValue | property | .../uri/schema/fin/FirmValue |
| EnterpriseValue | property | .../schema/fin/EnterpriseValue |

**Fig. 1.** Financial data sets vocabulary

We created a turtle file in order to represent data from the Excel file. Sample content is presented in Figure 2.

```
@prefix dc:        <http://purl.org/dc/elements/1.1/> .
@prefix vcard:     <http://www.w3.org/2001/vcard-rdf/3.0#> .
@prefix ns:       <http://sandbox.metadataregistry.org/uri/schema/fin>
.

:company1
    ns:name    "China Infrastructure Invsmt" ;
    ns:tickersymbol  "CIIC" ;
        ns:industryname "Diversified Co" ;
        ns:SICCode "9913" ;
        ns:xschangeCode "NDQ" ;
        ns:SizeClass "4" ;
        ns:StockPrice "0.69" ;
        ns:TradingVolume "241738" ;
        ns:MarketCap "45.6" ;
        ns:TotalDebt "473" ;
        ns:FirmValue "519.1" ;
        ns:EnterpriseValue "517.8".

:company2
    ns:name    "AFP Imaging Corp" ;
    ns:tickersymbol  "IWKS" ;
        ns:industryname "Medical Services" ;
        ns:SICCode "8000" ;
        ns:xschangeCode "NDQ" ;
        ns:SizeClass "6" ;
        ns:StockPrice "18.5" ;
        ns:TradingVolume "364" ;
        ns:MarketCap "340.6" ;
        ns:TotalDebt "8.6" ;
        ns:FirmValue "349.2" ;
        ns:EnterpriseValue "348.4".
```

**Fig. 2.** Financial data sets available in Turtle format

We configured Joseki in order to query data. Services and datasets configuration are presented in Figure 3.

```
<#service2>
    rdf:type            joseki:Service ;
    rdfs:label          "SPARQL on the company model" ;
    joseki:serviceRef   "company" ;
    joseki:dataset      <#company> ;
    joseki:processor    joseki:ProcessorSPARQL_FixedDS ;
    .
<#service3>
    rdf:type            joseki:Service ;
    rdfs:label          "sec" ;
    joseki:serviceRef   "sec" ;
    joseki:dataset      <#sec> ;
    joseki:processor    joseki:ProcessorSPARQL_FixedDS ;
    .
## Datasets

<#company>   rdf:type ja:RDFDataset ;
    rdfs:label "company" ;
    ja:defaultGraph
      [ rdfs:label "company.ttl" ;
        a ja:MemoryModel ;
        ja:content [ja:externalContent <file:Data/company.ttl> ] ;
      ] ;
    .
<#sec>   rdf:type ja:RDFDataset ;
    rdfs:label "sec" ;
    ja:defaultGraph
      [ rdfs:label "sec.n3" ;
        a ja:MemoryModel ;
        ja:content [ja:externalContent <file:Data/sec.n3> ] ;
      ] ;
    .
```

**Fig. 3.** Services and datasets configuration in Joseki

Therefore we were able to address queried on financial data sets. An example of SPARQL query is presented in Figure 4.

```
prefix dc:          <http://purl.org/dc/elements/1.1/>
prefix vcard:       <http://www.w3.org/2001/vcard-rdf/3.0#>
prefix :            <http://example.org/company/>
prefix ns:          <http://sandbox.metadataregistry.org/uri/schema/fin>
PREFIX dc:        <http://purl.org/dc/elements/1.1/>
SELECT ?name
WHERE
  { ?company ns:name ?name }
```

**Fig. 4.** SPARQL query that returns the name of the companies

RDF is a general method to decompose any type of knowledge into small pieces, with some rules about the semantics, or meaning, of those pieces. The point is to have a method so simple that it can express any fact, and yet so structured that computer applications can do useful things with it.

While there are not many implementations in the field we benefit from the existing grounding technologies like ontology, RDF/OWL descriptions, SPARQL language.

## 5 Conclusions

The results of our initial research appear to be very promising. RDF descriptions, specifically when based on knowledge representation formalisms do add value and enhance the search experience. Taking lead from these initial findings, we intend to develop a full semantic portal, identifying different relationships that exist among the different web data sources. Further, the study would be expanded by adding more document instances.

We present in this paper the necessary configurations that are to be made in order to access a SPARQL endpoint, essentially in building semantic web mash-ups. Our future work will refer to scalability in order that web applications may work on big data sets.

**Acknowledgments.** This work was supported by CNCSIS-UEFISCSU, project number PN II-RU code 188/2010.

## References

1. http://www.w3.org/1999/02/22-rdf-syntax-ns
2. Shum, S.B., Motta, E., Domingue, J.: ScholOnto: An Ontology-based Digital Library Server for Research Documents and Discourse. International Journal of Digital Libraries 3(3), 237–248 (2000)
3. Mao-sheng, L., Xiu-dan, Y.: Ontology-based Metadata Schema for Chinese Digital Libraries. Department of Information Management, Peking University, Beijing (2001), http://www.cs.vu.nl/franksh/postscript/K-CAP01.pdf
4. Ching-Long, Y.: Development of an Ontology-Based Portal for Digital Archive Services. Department of Computer Science and Engineering, Tatung University (2002), http://www.iis.sinica.edu.tw/APE02/Program/chingyeh.pdf
5. Uszkoreit, H., Jorg, B., Erbach, G.: An Ontology-based Knowledge Portal for Language Technology. German Research Center for Artificial Intelligence and Saarland University, Germany (2003), http://www.mcgreg.net/pub/COLLATE-EnablerElsnet03.pdf
6. Kalfoglou, Y., Domingue, J., Motta, E., Vargas-Vera, M., Buckingham Shum, S.: myPlanet: An Ontology-driven Web-based Personalized News Service. Knowledge Media Institute (KMi), The Open University, Milton Keynes, U.K (2004)
7. Heath, T., Bizer, C.: Linked Data: Evolving the Web into a Global Data Space. In: Synthesis Lectures on the Semantic Web: Theory and Technology, 1st edn., vol. 1(1), pp. 1–136. Morgan & Claypool, San Francisco (2011)
8. Joo, J.: Adoption of Semantic Web from the perspective of technology innovation: A grounded theory approach. International Journal of Human-Computer Studies 69, 139–154 (2011)

9. Gruber, T.: The Role of Common Ontology in Achieving Sharable, Reusable Knowledge Bases. In: Allen, J., Fikes, R., Sandewall, E. (eds.) Principles of Knowledge Representation and Reasoning, pp. 601–602. Morgan Kaufman, San Mateo (1991)
10. Kashyap, V., Sheth, A.: Semantics-based Information Brokering. In: Proceedings of the Third International Conference on Information and Knowledge Management (CIKM), pp. 363–370 (1994)
11. Sheth, A.: Changing Focus on Interoperability in Information Systems: From System, Syntax, Structure to Semantics. In: Goodchild, M., Egenhofer, M., Fegeas, R., Kottman, C. (eds.) Interoperating Geographic Information Systems. Kluwer Publishers, Dordrecht (1998)
12. Wache, H., Vögele, T., Visser, U., Stucjenschmidt, H., Schuster, G., Neumann, H., Hübner, S.: Ontology-Based Integration of Information A Survey of Existing Approaches. In: Stuckenschmidt, H. (ed.) IJCAI 2001 Workshop: Ontologies and Information Sharing, pp. 108–117 (2001)
13. O'Connor, M., Das, A.: SQWRL: a Query Language for OWL. In: Fifth International Workshop OWL: Experiences and Directions (OWLED 2009), Chantilly, VA (2009)
14. Good Relations Ontology experiences, http://www.heppnetz.de/projects/goodrelations/

# Improving Metadata by Filtering Contextual Semantic Role Information

Diana Trandabăţ

University Al. I. Cuza Iaşi, Romania
dtrandabat@info.uaic.ro

**Abstract.** This paper proposes a method to automatically improve a web page's metadata using the semantic content of the page. Thus, using a semantic role labeling system, the web page content is parsed and the entities that frequently play core semantic roles are considered for addition to the web page's list of metadata. Semantic role analysis answers questions such as: "What role has an entity in a specific context?" or "When, why, where or how an event takes place?".

**Keywords:** natural language processing, semantic roles, metadata filtering.

## 1 Introduction

The motivation behind the work presented in this paper is the need for more accurate metadata for web pages created by non-specialists. If using a website template has became more and more frequent nowadays, few web paged developed starting from a generic template understand or respect the need for accurate metadata. Thus, users usually just copy-paste the metadata from the web site template, enriching the web with web pages having "html", "css" and "template" as metadata. Since metadata are information describing the web page's content, why can't they be derived from it? This paper argues for the use of semantic role labeling for extracting semantic metadata from web page content, in case the user provided metadata don't exist or are non-concluding.

The backbone of the proposed application is a semantic role labeling system. Using semantic role analysis, we are able to answer questions such as: "What roles do entities play in different contexts?" or "When, why, where or how an event takes place?" Thus, the system extracts semantic roles from the content of the web page (included in the paragraph tag), and created a map of roles and their instantiations in the document. If an entity appears frequently in the document having a core semantic role, it is then considered for as a metadata candidate for the current webpage.

The paper is structured in 6 sections. After an introduction in the field of semantic role analysis, we briefly present the current work in semantic roles in Section 2. Section 3 introduces the overall architecture of our application, describing the intermediary steps, while Section 4 presnets the backbone system, the semantic role labeling system. Section 5 presents some example of how metadata can be and filtered using semantic roles, and Section 6 draws the conclusions of this paper and discusses further envisaged developments.

E. García-Barriocanal et al. (Eds.): MTSR 2011, CCIS 240, pp. 201–208, 2011.

## 2 State of the Art

Natural language processing (NLP) is a key component of artificial intelligence. The NLP community has recently experienced a growth of interest in semantic roles, since they describe WHO did WHAT to WHOM, WHEN, WHERE, WHY, HOW etc. for a given situation, and contribute to the construction of meaning. Fillmore in [3] defined six semantic roles: *Agent*, *Instrument*, *Dative*, *Factive*, *Object* and *Location*, also called *deep cases*. His later work on lexical semantics led to the conviction that a small fixed set of deep case roles was not sufficient to characterize the complementation properties of lexical items, therefore he added *Experiencer*, *Comitative*, *Location*, *Path*, *Source*, *Goal* and *Temporal*, and then other cases. This ultimately led to the theory of Frame Semantics [4], which later evolved into the FrameNet project[1].

The semantic relations can be exemplified within the Commercial Transaction scenario, whose actors include a *buyer*, a *seller*, *goods*, and *money*. Among the large set of semantically related predicates, linked to this frame, we can mention buy, sell, pay, spend, cost, and charge, each of which indexes or evokes different aspects of the frame. The verb buy focuses on the *buyer* and the *goods*, backgrounding the *seller* and the *money*; sell focuses on the *seller* and the *goods*, backgrounding the *buyer* and the *money*; pay focuses on the *buyer*, the *money*, and the *seller*, backgrounding the *goods*; and so on. The idea is that knowing the meaning of any of these verbs requires knowing what takes place in a commercial transaction and, to some extent, knowing the meaning of all the predicates involved in the frame.

In the last decades, hand-tagged corpora that encode such information for the English language were developed (VerbNet[2] [7], FrameNet [1] and PropBank[3] [11]). For other languages, such as German, Spanish, and Japanese, semantic roles resources are being developed. For Romanian, [16] has started to automatically build such a resource[4].

For role semantics to become relevant for language technology, robust and accurate methods for automatic semantic role assignment are needed. With the SensEval-3 competition[5] and the CONLL Shared Tasks[6], Automatic Labeling of Semantic Roles, identifying frame elements within a sentence and tag them with appropriate semantic roles given a sentence, a target word and its frame [8], has become increasingly present among researchers worldwide. Most general formulation of the Semantic Role Labeling (SRL) problem supposed determining a labeling on (usually but not always contiguous) substrings (phrases) of the sentence *s*, given a predicate *p*, as in the following example:

---

[1] FrameNet web page: `http://framenet.icsi.berkeley.edu/`

[2] VerbNet web page: `http://verbs.colorado.edu/~mpalmer/projects/verbnet/downloads.html`

[3] PropBank web page: `http://verbs.colorado.edu/~mpalmer/projects/ace.html`

[4] `http://students.info.uaic.ro/~dtrandabat/RomanianFrameNet_v1.xml`

[5] SemEval web address: `http://www.senseval.org/`

[6] ConLL web address: `http://ifarm.nl/signll/conll/`

[The queen]$_{Agent}$ broke [the window]$_{Theme}$.
[You should see a doctor]$_{Theme}$ , [he]$_{Agent}$ said.

In recent years, a number of studies, such as [2] and [5], has investigated this task on the FrameNet corpus. Role assignment has generally been modeled as a classification task: A statistical model is trained on manually annotated data and later assigns a role label out of a fixed set to every constituent in new, unlabelled sentences. The work on SRL has included a broad spectrum of probabilistic and machine-learning approaches to the task, from probability estimation [5], through decision trees [14] and support vector machines [12], to memory-based learning [9]. While using different statistical frameworks, most studies have largely converged on a common set of features to base their decisions on, namely syntactic information (path from predicate to constituent, phrasal type of constituent) and lexical information (head word of the constituent, predicate).

An important drawback of the presented systems is that they don't treat nominal predicates, being only built for verbal predicates. Furthermore, they only consider one predicate per sentence, even if this is not always the case. For example, in the sentence `The awarding of the Nobel Prize to President Obama was largely debated`, we have two predicate words, the `awarding`, having as *Theme* the phrase `of the Nobel Prize`, and `debated`, having two arguments, an *Agent* `The awarding of the Nobel Prize to President Obama` and an *Manner* complement, `largely`. The semantic role labeling system presented in this paper considers both verbal and nominal predicates.

## 3 Proposed Architecture

The goal of our application is to help users add metadata to their web pages using the content on their page. Thus, the envisaged application contains a series of modules, as follow:

1. Extract the content of the paragraph tags of a web page and clean the obtained paragraphs (remove stopwords such as the functional prepositions "and", "to", "from", etc.);
2. Extract the page identifiers: title of the web page, text found in the headings (h1 to h6), anchor text for links;
3. Performing semantic role labeling on the text extracted at step 1, in order to identify the role played by each particular word;
4. Establish a map of entities which appeared frequently in the web page, with their respective semantic relations.
5. Compare the entities obtained in step 4 with the ones extracted at step 2.
6. Provide the user with a list of candidate metadata to select from.

The core module of this architecture is the semantic role labeling system. The system we used is based on the pre-trained semantic role modules of PASRL presented in [15]. PASRL (Platform for Adjustable Semantic Role Labeling) trains 12 classifiers form the Weka framework [6] with different feature sets, checks the performances of the different obtained models and select the best performing model. The 10 fold cross-validation results of all classifiers are also saved since they provide a confusion matrix that can be used to see which classes were correctly predicted by different classifiers. The output of PASRL is a Semantic Role Labeling System, a sequence of trained models which can be used to annotate new texts. PASRL offers pre-trained modules for different languages: English, German, Chinese, Czech and Japanese, trained on the corpus provided for research purposes by the CoNLL 2009 shared task. PASRL can also be trained for other languages, if an annotated training corpus is available.

We will further briefly describe each module of our system:

*Extract the content of the webpage.* This module gets the text inside paragraph tags and clean the obtained text by removing stopwords such as "and", "to", "from", etc. The html-to-text task is performed using a simple get method that extracts text from the web page, and selects only the content of a specific paragraph tag.

*Extract the page identifiers.* The title of the web page, the text found in the headings (h1 to h6) and anchor text from links are extracted. This information was considered important to create a sort of a "profile" of the web page, containing information that the user has considered important through formatting or by placing it in the title. The anchor of the text was considered also as important information since usually links relate to similar web pages, that may have their metadata correctly filled in, which turn them into a possible source for additional information;

*Performing semantic role* labeling on the extracted text, in order to identify the role each entity plays. Using the PASRL system (briefly described above), we semantically annotated the sentences in the web page;

*Establish a map of entities* which appeared frequently in the web page, with their respective semantic relations. The semantic roles identified in the previous step are attached to different entities. When one such entity has multiple semantic roles in the content of the page, it is considered to be of higher importance for the web page, since appearing in so many contexts makes it central to the text. A threshold is established to keep as metadata candidates only the entities that appeared at least in 5% of the content of the page, and that are annotated with the semantic roles of Patient, Agent, Beneficiary or Theme.

*Compare the entities:* The set of candidate metadata is further filtered by comparing it to the website profile created using the terms extracted using the second module above. The similarity is performed as simple perfect match or partial match for multi-word expressions.

Finally, the system will provide the user with a list of candidate metadata to select from: The resulted candidate list of metadata is presented to the user (limited to a maximum of 20 entries, ordered by a score given by the matching algorithm).

## 4 Description of the Semantic Role Labeling System

The training data used for the development of PASRL was the training and development resource from the ConLL 2009 Shared Task, consisting of manually annotated treebanks such as the Penn Treebank for English, the Prague Dependency Treebank for Czech and similar treebanks for Chinese, German, Japanese and Spanish languages, enriched with semantic relations (such as those captured in the Prop/Nombank and similar resources). The training data contains syntactic and dependency information, detailed in Table 1.

**Table 1.** Description of the input format for PASRL

| No. | Name | Description |
|---|---|---|
| 1 | ID | Token counter, starting at 1 for each new sentence. |
| 2 | FORM | Word form or punctuation symbol. |
| 3 | LEMMA | Lemma or stem (depending on particular data set) of word form, or an underscore, if not available. |
| 4 | POS | Part-of-speech tag, where the tagset depends on the language, or identical to the coarse-grained part-of-speech tag, if not available. |
| 5 | HEAD | Head of the current token, which is either a value of ID (meaning that the word with the ID n is the head-word of the current token) or zero ('0', if the head word is the ROOT _ctive node). |
| 6 | DEPREL | Dependency relation to the HEAD. The set of dependency relations depends on the particular language. |

PASRL is composed of two main sub-systems: a Predicate Prediction module and an Argument Prediction module (see Fig. 1). The Predicate Prediction module has two possible configurations, corresponding to the Predicate Identification and Predicate Sense Identification layers as marked on the left side of Fig 1. The first configuration involves a sequential identification of the predicates in a sentence (Predicate Identification modules), followed by the assignment of the predicate sense (Predicate Sense Identification modules on the right side of the figure). The second configuration allows for a joint learning of the predicates in sentence, together with their senses (the Sense Identification modules).

The second sub-system performs argument prediction, based on the dependency relations previously annotated with the MaltParser and the predicate senses (the Argument Identification modules).

For each level, the modules have three variants: *all*, *NP/VP* and *each*, related to the training set size. Training the whole system for a particular language requires running tens of classifiers, thus running the classifiers on the whole training size is a very time expensive task. Therefore, the training data has been filtered and, besides running the classifiers on the whole data size (*all* ), we offer the possibility to train, for each problem, different classifiers for the noun phrase or verb phrase predicates (*NP/VP* ) or even more refined, for each noun, respectively verb predicate in the training set (*each*).

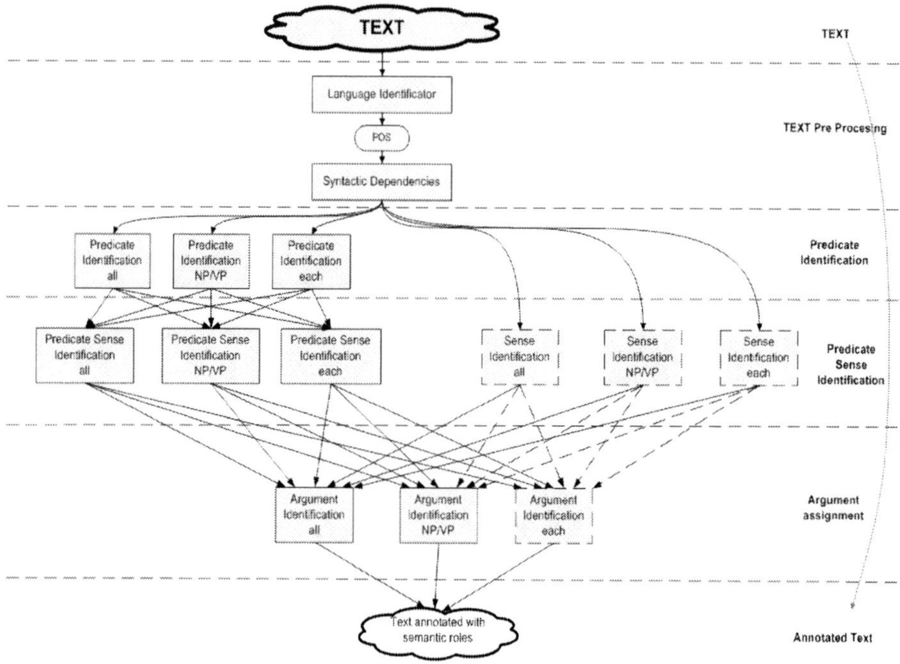


**Fig. 1.** Architecture of PASRL

For each module in Fig. 1, the set of 12 classifiers from Weka framework are trained. After running all the classifiers for all the modules, their performance is compared, and the path in Fig. 1 that obtains the highest summed performance is considered the best configuration. An example of such a best configuration can be: running the model created by the Predicate Identification – *all* subtask using the Decision Tree classifier, followed by the model created by the Predicate Sense Identification – *NP/VP* subtask created with the Decision Table classifier, and then the model of Argument Identification – *each* subtask created using the Naive Bayes classifier. The models for this best configuration are saved, and the best path is written to a configuration file. This configuration can then be used at a later time to annotate new texts with the developed SRL system. If all the created models are saved, and not just the best performing ones, the user can define, using the configuration file, the sequence of classifiers he wishes to run for each subtask in order to annotate new texts using the pre-trained models.

## 5 Example and Discussion

In order to exemplify the previous steps, consider the following sentences, annotated with semantic role information:

(1) [Welcome]$_{\text{TARGET}}$ [to my webpage]$_{\text{Theme}}$!
(2) [This webpage]$_{\text{Agent}}$ [contains]$_{\text{TARGET}}$ [information about the public transportation]$_{\text{Theme}}$.

(3) ...[after hearing about "ambient intelligence"]$_{Time}$, [I]$_{Agent}$ [enrolled]$_{TARGET}$ [at the Computational Linguistics Master programme]$_{direction}$.
(4) The best bookstore ever!.

Looking at the examples (1) and (2) above, one can tell that the most important piece of information in these sentences is the *Theme*. Usually, if when annotating a webpage with semantic roles obtained from PASRL the theme "my webpage" appears, with variations such as "my personal webpage", "my public profile", etc., then this theme is a perfect candidate for metadata describing the webpage.

In example (3), there is no group in the input sentence corresponding to a *Theme* semantic role. Therefore, we choose the sentence's semantic roles by eliminating the adjuncts (locative, temporal, modals, etc.) Thus, only "I" for example (3) can be kept as possible metadata candidate.

An interesting case is the one in example (4) above, where there is no predicate for the sentence. Since semantic roles are bided to the verb (or to a preciational noun) no verb implies no semantic roles assigned.

The examples above showed that in general the *Theme* semantic role specifies a metadata candidate, while adjunct semantic roles (Temporal, Location, Means, etc.) are usually discarded.

The list of semantic roles created at step 3 above will be ordered by the frequency of appearing in the pair specific-word-or-expression / semantic roles. These candidate metadata are compared to the website profile created using the webpage information.

The evaluation of the system depends on the performance of the semantic role module. As for the evaluation of the whole metadata creation module, the metrics of Ochoa and Duval for metadata quality are kept in mind: completeness, accuracy, provenance, conformance to expectations, logical consistency/coherence, timeliness and accessibility [10]. Similar evaluation metrics are used in [13], where a semi-automatic topic vocabulary extraction algorithm based on different techniques to extract metadata from the document content, structure and context are presented.

## 6 Conclusions

This papers has presented a semantic role labeling system applied to web texts, that represents the backbone of an application that improves the metadata keywords of a web page using its content. The semantic labeling system was developed using PASRL, a platform for supervised learning techniques. The developed platform tests several classifiers on different sub-problems of the SRL task (Predicate Identification, Predicate Sense Identification, Sense Identification, Argument Identification), chooses the ones with the greatest performance and returns a Semantic Role Labeling System (a sequence of trained models to run on new data). The training corpus used to develop the SRL system for Romanian is a semi-automatically derived semantic resource. Evaluations of the performances of the best algorithms are discussed.

The semantic role labeling system has performances varying from 54% to 75%, depending on their language, and performs well especially on short texts [15]. The overall accuracy of our system was not evaluated yet, but from our tests, the metadata candidate seem to be of major help in convincing the user to add metadata to their web pages.

**Acknowledgments.** The research presented in this paper was funded by the Sectoral Operational Programme for Human Resources Development through the project "Development of the innovation capacity and increasing of the research impact through post-doctoral programs" POSDRU/89/1.5/S/49944.

## References

1. Baker, G., Collin, F., Fillmore, C.J., Lowe, J.B.: The Berkeley FrameNet project. In: Proceedings of the COLING-ACL, Montreal, Canada (1998)
2. Chen, J., Rambow, O.: Use of deep linguistic features for the recognition and labeling of semantic arguments. In: Proceedings of the 2003 Conference on Empirical Methods in Natural Language Processing (2003)
3. Fillmore Charles, J.: The case for case. In: Bach, Harms (eds.) Universals in Linguistic Theory, pp. 1–88. Holt, Rinehart, and Winston, New York (1968)
4. Fillmore Charles, J.: Frame semantics. In: Linguistics in the Morning Calm, pp. 111–137. Hanshin Publishing, Seoul (1982)
5. Daniel, G., Jurafsky, D.: Automatic labeling of semantic roles. Computational Linguistics 28(3), 245–288 (2002)
6. Hall, M., Frank, E., Holmes, G., Pfahringer, B., Reutemann, P., Witten, I.H.: The WEKA Data Mining Software: An Update. SIGKDD Explorations 11(1) (2009)
7. Levin, B., Rappaport Hovav, M.: Argument Realization. Research Surveys in Linguistics Series. Cambridge University Press, Cambridge (2005)
8. Lluis, M., Carreras, X., Litkowski, K.C., Stevenson, S.: Semantic role labeling: An introduction to the Special Issue. Computational Linguistics 34(2), 145–159 (2008)
9. Roser, M., Daelemans, W., Van Asch, V.: A combined memory-based semantic role labeler of English. In: Proceedings of the Twelfth Conference on Computational Natural Language Learning, Manchester, UK, pp. 208–212 (2008)
10. Ochoa, X., Duval, E.: Times of Convergence. Technologies Across Learning Context, pp. 322–325 (2008)
11. Martha, P., Gildea, D., Kingsbury, P.: The proposition bank: An annotated corpus of semantic roles. Computational Linguistics 31(1), 71–106 (2005)
12. Sameer, P., Hacioglu, K., Krugler, V., Ward, W., Martin, J.H., Jurafsky, D.: Support vector learning for semantic argument classification. Machine Learning Journal 60(13), 11–39 (2005)
13. Sah, M., Wade, V.: Automatic metadata extraction from multilingual enterprise content. In: Proc. of the 19th ACM International Conference on Information and Knowledge Management (CIKM 2010), pp. 1665–1668. ACM, New York (2010)
14. Surdeanu, M., Harabagiu, S., Williams, J., Aarseth, P.: Using predicate-argument structures for information extraction. In: Proceedings of the 41th Annual Meeting of the Association for Computational Linguistics, Tokyo, pp. 8–15 (2003)
15. Trandabăț, D.: Natural Language Processing Using Semantic Frames, PhD Thesis, University Al. I. Cuza Iasi, Romania
16. Trandabăț, D.: Towards automatic cross-lingual transfer of semantic annotation. In: 6e Rencontres Jeunes Chercheurs en Recherche d'Information (RJCRI) 2011, Avignon, France, March 16-18 (2011)

# Towards a Knowledge Management Framework to Preserve Creativity in Small and Medium Enterprises

Fabio Sartori

Department of Computer Science, Systems and Communication (DISCo),
University of Milan - Bicocca,
viale Sarca, 336,
20126 - Milan (Italy)
Tel.: +39 02 64487913; Fax: +39 02 64487839
{sartori}@disco.unimib.it

**Abstract.** To support Small and Medium Enterprises (SMEs) in preserving their knowledge models and creativity is an important research topic and a challenge in the Knowledge Management field. SMEs are companies whose headcount or turnover falls below certain limits: they are typically characterized by high levels of creativity, which allow them to design and manufacture innovative products. However, SMEs are often characterized by significant technological gaps with respect to wider organizations and may suffer significantly due to the loss of key personnel and company experiential knowledge and competences. This paper introduces a methodological and computational framework for supporting SMEs in managing their creativity and overcome the problems above, based on storytelling and Case Based Reasoning (CBR) paradigms.

## 1 Introduction

The aim of the paper is to describe a conceptual and computational framework for the management of complex knowledge, that cannot always be captured exploiting traditional methodologies for the development of knowledge–based systems. Such knowledge typically concerns informal groups of people working and living within organizations, called Communities of Practice (CoPs) [1] as well as organizations with a low level of technicality, like e.g. Small and Medium enteprises.

Given the importance of SMEs in the global economy, especially in Europe, there has been a great deal of research in the Knowledge Management (KM) context, both from the theoretical and the practical standpoint to support SMEs in their day to day activities and to join SME networks. The main KM issues that have been recognized for SMEs are: their significant technological gaps with respect to wider organizations. Several technological solutions have been proposed in KM literature to enhance networking and knowledge sharing within collaborative communities (see KNOW-CONSTRUCT project [2] as an example); their

E. García-Barriocanal et al. (Eds.): MTSR 2011, CCIS 240, pp. 209–214, 2011.

vulnerability in terms of loss of key personnel, as a consequence of their small size [3]. The limited dimensions of SMEs, that undoubtedly is a benefit from the agility perspective, may in fact cause the lack of a shared structured framework for company experiential knowledge collection and management.

In this paper we mainly refer to the latter issue introducing a framework for the representation and management of collective creativity based on the integration of storytelling [4] and Case Based Reasoning (CBR [5]) methodologies. Storytelling [4] is a short narration through which an individual describes an experience on a specific theme [6]. Storytelling can be considered an effective way to capitalize the knowledge that is produced from the daily working activities within organizations [7]. Knowledge Management can profitably exploit storytelling as a way to make explicit the individual experiences, skills and competencies, to promote the negotiation processes through dialogues among people involved, to support the reification of new knowledge in order to make it available for the future and to help newcomers in the learning process about his/her job through the analysis of the problem–solving strategies and social context represented by the stories. In order to improve knowledge formalization and sharing within SMEs, CBR can be exploited to model problem–solving situations, solutions and outcomes into archives of stories that can be incrementally enriched with new experiences, building up a complete memory of innovative solutions adopted by a SME.

## 2 Creativity, Storytelling and Case Based Reasoning

### 2.1 Creativity and Knowledge Management

In [8] creativity is defined as *the tendency to generate or recognize ideas, alternatives, or possibilities that may be useful in solving problems, communicating with others, and entertaining ourselves and others.* People become creative due to their needs to tackle new, modified or complex situation, to exchange ideas or finding solutions to problems. Creativity arises from the capability to think in a different way with respect to the others.

In [9], the word *creative* can be referred to both novel products of value and the person(s) working on it, focusing on the fact that the novelty of a product is necessary but not sufficient to find creativity in it too, since it must also be of value or be appropriate to the demands of the current situation.

A very detailed analysis of creativity and creative people comes from Psychology. For example, in [10] it is observed that the term creativity is very often associated to people who think unusually, being interesting and stimulating for other groups of persons as well as people who study the world in novel and original ways or individuals capable to produce important changes in our culture. According to the concepts above, creativity is *any act, idea, or product that changes an existing domain, or that transforms an existing domain into a new one.* Thus, what is important in defining a creative person is to establish if the novelty of the ideas produced by him/her is accepted by other members of the group he/she is proposing them.

An interesting point of view is offered by the not academic research; for example, Linda Naimann[1] distinguishes between imagination and creativity: creativity is *the act of turning new and imaginative ideas into reality*, while innovation is *the production or implementation of an idea.* Thus, it is not sufficient to have ideas for being creative, it is necessary to put them into action too. This also the meaning of creativity for bigger companies, like IBM, for which [2] innovation means *using new ideas or applying current thinking in fundamentally different ways that result in significant change.*

The last definition is very close to how Knowledge Management has interpreted the term creativity; in [11], very clear definitions of knowledge, creativity, innovation and the relationship existing among them through the discipline of Knowledge Management are given by David Gurteen: *Creativity and innovation concern the process of creating and applying new knowledge. [...] Knowledge Management, however, is a new discipline and creativity and innovation need to be thought about in this new context.*". Also for Gurteen creativity and innovation are complementary: the generation of new ideas (i.e creativity) needs a way to implement them (i.e. innovation). On the other hand, no innovation is possible without a preexistent idea. As a consequence, existing knowledge is required to start the process (i.e. how is it possible to put into action an idea?) as well as the generation of new knowledge as the process output (i.e. what is the result of an idea application to a given context?). From the conceptual perspective, *dialogue* is indicated as the most suitable paradigm to take care of creativity in the Knowledge Management area. As a consequence, the natural technological support for this kind of creativity management is the adoption of *groupware* approach.

What's the reason of adopting a collaborative technological infrastructure, like groupware, to support creativity, that is typically associated to *singles* in the commonsense thinking? Indeed, Knowledge Management has (or should have) the scope to support *organizations* in their decision-making processes, and organizations are (or should be) *collaborative environments*, where decisions and results (both positive and negative) are taken and obtained by means of collective reasoning process, involving different actors according to their competencies and bounded by shared and common goals. Thus, Knowledge Management is (or should be) more interested to *collective creativity* [12] than *individual creativity*, that is the main subject of other disciplines.

According to [13], the importance of individual creativity is not diminished by collective creativity, since the latter cannot be considered as the mere sum of individual creativities. In other words, collective creativity can be thought as the composition of individual creativities: it is the result of many problem solving methods and many different competencies involved in the solution of a common problem. In order to obtain collective creativity from a set of individuals it is necessary to have *negotiation* among them.

---

[1] Available at `http://www.creativityatwork.com/articlesContent/whatis.htm`, last access 2011/05/06.

[2] Definition from the IBM Global Innovation Study, 2006.

This is the reason why collective creativity is typical of environments where Communities of Practice arise and act: in particular, it is interesting to focus on SMEs as environments where Communities of Practices are established to overcome typical problems affecting them. In [14] it is highlighted that within SMEs, due to the limited technology systems to store knowledge and information in a useful way, *it is often difficult to establish the common knowledge baseline, as knowledge and information is often located on individuals' personal computers, on Windows network file structures, in paper format or in tacit format.* SMEs are typically characterized by higher levels of creativity than wider organizations: this the reason why many SMEs are leaders in their sectors. Anyway, the absence of technological infrastructures to preserve and maintain knowledge and capability to innovate is one of the most critical limitations for them, since it prevents the creation of a shared archive of the overall knowledge owned by the SME. In this way, the collective creativity continues to be hidden in the minds of the individuals, with potentially negative consequences in case of their outage. For this reason, it is argued that Communities of Practices would be meant as *vehicles to manage knowledge of individuals on particular domains to ensure the creation, sharing, harvesting and leveraging of knowledge*, becoming active actors in the promotion of collective creativity within SMEs.

Finding methodological solution to discover CoPs inside SMEs and let them guide the decision–making processes involved in SMEs' activities is still an open and challenging problem in the Knowledge Management field: as suggested by Gurteen, it is necessary to chose both conceptual and computational tools that are able to fit with the negotiation and reification processes through which innovation and creativity are generated inside CoPs.

### 2.2 Towards a Framework for Collective Creativity Management

From the conceptual point of view, *storytelling* has been universally recognized as a very good starting point. As reported by S. Denning [15], storytelling build trust, unlock passion, overcomes hierarchies due to its intrinsically collaborative nature and its capability to flatten the communication among people (or communities). A story is a narrative account of a real or imagined event or events. Within the storytelling community, a story is more generally agreed to be a specific structure of narrative with a specific style and set of characters and which includes a sense of completeness. Through this sharing of experience stories allow to pass on accumulated wisdom, beliefs, and values. Stories explain how things are, why they are, and the different roles and purposes involved. Stories are the building blocks of knowledge, the foundation of memory and learning. In the Knowledge Management literature [16], stories are often considered as very profitable tools to make explicit tacit knowledge, with the possibility to exploit them in the process of generating collective creativity from individual creativity.

From the computational point of view *Case Based Reasoning* is one of the most suitable paradigm to deal with negotiation–reification processes within CoPs. The relationship between storytelling and CBR has been already analyzed in the past: in this approach, the story is the conceptual tool through which a tacit knowledge frame, for example a portion of a problem solving strategy, is made explicit. The story is structured as follows: 1. *observation and reflection*, the problem to be solved is analyzed, in order to identify its main characteristics, like e.g. the needed inputs and the expected outputs; *abstract concepts creation*, the problem is then generalized, in order to find possible similarities with one or more past problems; *experimentation in new situations*, the past problems are used as starting points for finding solutions to the current situation; *concrete experience*, the new problem with its solution is stored to be used in the future as a baseline for new problem solving activities.

This schema perfectly fits the 4R's cycle [17] for the development of CBR applications: the *problem description* as a case and the *retain* step of the CBR cycle correspond to the the point number 3 above; the *retrieved case* and the *new case with its solution* correspond to the point number 4 in the enumeration above; finally, *the retrieve, reuse and revise* steps of the CBR cycle correspond to points number 1 and 2 in the list. In this way, the case structure can be adopted as an approximation of a story representing the narration of a complete problem solving strategy adopted by an individual during an innovative solution generation (*individual creativity*); the collection of all the cases produced by an organization, i.e. the *case base*, can be thought as an approximation of all the stories produced by their members in their problem solving activities (*collective creativity*).

## 3 Concluding Remarks

In this paper we have presented a Knowledge Management framework for the development of systems to support work and learning in Small and Medium Enterprises. The approach is based on the integration of storytelling and Case Based Reasoning, and it can be exploited to support organizations where Communities of Practice are active too.

Case Based Reasoning has been successfully applied to support SMEs in the past: for example, the Symphony project [18] was funded by European Union to develop a KM system to support experts of human resources of SMEs in the employer selection process. In our framework, a story represents a decision making process about a problem to be solved. Stories can be archived and retrieved as cases according to the CBR paradigm: in this way, new cases (i.e. stories) can be continuously created and stored to be used in the future, building up a memory of all the experiences that can be then used as effective tools for supporting creativity management.

The framework has been recently applied in the context of the Flexibly Beyond project, funded by EU under the 6th Frame Program[3]. The main result obtained has been the validation of our approach in supporting SMEs: the typical problem solving strategy adopted by experts is based on reasoning by analogy method, which allows them to design new products by adapting older ones. The framework is very suitable to support them in this activity, providing a way to

[3] Grant number 2006-VP021-30140.

organize their archive of past experiences (i.e. stories about their work) into an organizational memory (i.e. the CKS–Base) that formalizes the knowledge patrimony produced by experts over the years, with significant benefits from the loss of key personnel point of view. In this way, experience and knowledge created by the SME and captured by the CBR system could be used as a very important training method alternative to the more traditional ones.

## References

1. Wenger, E.: Community of Practice: Learning, Meaning and Identity. Cambridge University Press, Cambridge (1998)
2. Soares, A.L., Simões, D., Silva, M., Madureira, R.: Developing enterprise sponsored virtual communities: The case of a sme's knowledge community. In: Meersman, R., Tari, Z., Herrero, P. (eds.) OTM 2006 Workshops. LNCS, vol. 4277, pp. 269–278. Springer, Heidelberg (2006)
3. Handzic, M.: Knowledge management in smes–practical guidelines. CACCI 1, 1–11 (2004)
4. Atkinson, R.: The life story interview. Sage University Papers Series on Qualitative Research Methods 44 (1998)
5. Kolodner, J.: Case–Base Reasoning. Morgan Kaufmann Publisher, Inc., San Francisco (1993); 2929 Campus Drive, Suite 260, San Mateo CA 94403
6. Bruner, J.: The narrative construction of reality. Critical Inquiry 18, 1–21 (1991)
7. Kleiner, A., Roth, G.: How to make experience your company's best teacher. Harvard Business Review 75(5), 172 (1997)
8. Franken, R.E.: Human motivation. Thomson/Wadsworth (2007)
9. Weisberg, R.W.: The study of creativity: from genius to cognitive science. International Journal of Cultural Policy 6(3), 235–253 (2010)
10. Csíkszentmihályi, M.: Creativity: flow and the psychology of discovery and invention. HarperPerennial (1997)
11. Gurteen, D.: Knowledge Management and Creativity. Journal of Knowledge Management 9(5), 97–107 (1998)
12. Sonicrim, L.S.: Collective Creativity. Design 6(3), 1–6 (2001)
13. Chaharbaghi, K., Cripps, S.: Collective creativity: wisdom or oxymoron? Journal of European Industrial Training 31(8), 626–638 (2007)
14. du Plessis, M.: The strategic drivers and objectives of communities of practice as vehicles for knowledge management in small and medium enterprises. International Journal of Information Management 28(1), 61–67 (2008)
15. Denning, S.: The Leader's Guide to Storytelling: Mastering the Art and Discipline of Business Narrative. John Wiley and Sons, Chichester (2011)
16. Bhardwaj, M., Monin, J.: Tacit to explicit: an interplay shaping organization knowledge. Journal of Knowledge Management 10(3), 72–85 (2006)
17. Aamodt, A., Plaza, E.: Case-based reasoning: Foundational issues, methodological variations, and system approaches. AI Communications 7(1), 39–59 (1994)
18. Bandini, S., Mereghetti, P., Merino, E., Sartori, F.: Case-based support to small-medium enterprises: The symphony project. In: Basili, R., Pazienza, M.T. (eds.) AI*IA 2007. LNCS (LNAI), vol. 4733, pp. 483–494. Springer, Heidelberg (2007)

# Semanticook: A Web Application for Nutrition Consultancy for Diabetics

Emin Akkoç and Nihan Kesim Cicekli

Middle East Technical University,
Department of Computer Engineering
emin.akkoc@gmail.com, nihan@ceng.metu.edu.tr

**Abstract.** This paper presents a web application whose purpose is to provide consulting for diabetic nutrution. The system has two main modules which are the carbohydrate counting module and the recipe module. The purpose is to make diabetic diets more flexible with the help of carbohydrate counting and also help patients cook meals with the materials they have at hand, which can be described as a reverse recipe and also informs them about the carbohydrate amounts of their meal and validates the suitability of those meals with their diets. The modules are implemented with JSP in the presentation layer, Java Servlet technology in the controller mechanism and data is stored in MySQL database and OWL ontology.

## 1 Introduction

One important phase of Diabetes medical treatment is nutrition management. Nowadays many of the diabetic patients are expecting miracles by applying diets written in a page, that are not flexible. Because those diets are not flexible, it is too hard to follow them and as a consequence the patients cannot continue the diets and quit after a while. That situation brings those patients into a desperate mood which effects their health and life quality.

Semanticook Project has two main modules which are the "CarbCounting Module" and "Recipe Module". The system mainly focuses on the nutrution management of type 1 Diabetes patients. Those two modules are run interactively with each other.

The CarbCounter module of the system handles carbohydrate counting of a selected meal by adding carbohydrate values of each of the ingredients with their amounts and then generates warning messages for the user about how many units of insulin should the user inject if he/she eats that meal. The CarbCounter module has another feature which is called the "Diet Validator". The Diet Validator has the purpose of comparing the meal with the corresponding slot in the users diet and generates warnings about the result. The carbohydrate counting is not a diet; it is an approach to achieve a goal. Patients who know carb counting can make their choices freely but not everybody knows about it. That freedom does not provide patients to eat at any amount but helps them to eat in the direction of their own choices.

The second module uses a food ontology to return a meal which can be prepared with the ingredients that the user has. It also gives carbohydrate information about the recipe with the help of the CarbCounting module.

E. García-Barriocanal et al. (Eds.): MTSR 2011, CCIS 240, pp. 215–224, 2011.

The rest of the paper is organized as follows. Section 2 gives some background information on type 1 diabetes and also summarizes the related work on diabetes control. Section 3 presents the overall system design and the used technologies. The implementation of the system is described in Section 4. Section 5 explains the usage of the system. Section 6 concludes the paper with some remarks about the future work.

## 2 Background Information

Type 1 diabetes is usually diagnosed in children and young adults. It was previously known as juvenile diabetes. In type 1 diabetes, the body does not produce insulin. Insulin is a hormone that is needed to convert sugar, starches and other food into energy needed for daily life. Only 5% of people with diabetes have this form of the disease. With the help of insulin therapy and other treatments, even young children with type 1 diabetes can learn to manage their condition and live long, healthy, happy lives [1].

If you have diabetes, your body cannot make or properly use insulin. This leads to high blood glucose, or sugar, levels in your blood. Healthy eating helps to reduce your blood sugar. It is a critical part of managing your diabetes, because controlling your blood sugar can prevent the complications of diabetes [2].

The patients should play an active role in their nutrition management and they have to know when and what to eat and also know the consequences when they have high and low blood sugar which are related directly to the carbohydrates and calories they get. The patients should also know which food contains how much carbohydrate and calory. It is a very difficult task to know every food and its calories. Only after the patients have that knowledge, they can arrange the dose of insulin which they have to inject. Carbohydrate counting is not a diet it is an approach to achieve a goal. Patients who know carb counting can make their choices freely but not everybody knows about it. That freedom does not provide patients to eat at any amount but helps them to eat in the direction of their own choices.

**Example:** One unit of insulin which Ali will inject handles 40 carbohydrates (This information should be given by the patient) and Ali wants to eat an apple pie. Total carbohydrate of the apple pie will be calculated by the system with the information available in the database and tells Ali how many units of insulin he should inject. The system also warns Ali if this meal is suitable with his diet.

There have been several projects on food ontologies and diabetes control [9,10,11]. In [9] the authors present the dietary ontology developed for the automated menu generator MenuGene. The menu generator uses Genetic Algorithms to prepare weekly menus for web users. In our work we do not aim to generate menus automatically. [10] describes the rapid prototyping of a food ontology oriented to the nutritional and health care domain that is used to share knowledge between the different stakeholders. They present the Food Ontology that they designed for the PIPS Project. We have been inspired by this ontology. However our aim is different in which we use the ontology to suggest alternative recepies for diabet 1 patients. The goal of the study in [11] is to propose an automated food ontology constructed for diabetes diet care. The methods include generating an ontology skeleton with hierarchical clustering algorithms (HCA), class naming by intersection naming, and instance ranking by granular ranking and positioning. Our system proposes substitutions based on the food at hand and the selected meal, while their system tries to generate a food ontology automatically by using all food characteristics.

## 3 System Design

The features are combined in a Web application with 3-Tier architecture. Inputs like personal and medical information are input by the user to the system and those information is stored into the database. Other inputs that user is going to send are about the foods that are available in the fridge. Those foods are matched with the ontology and the reasoning mechanism suggests foods according to those inputs. This event takes place in the application server. Other features like carbohydrate counting and diet warning mechanisms also take place in the application server.

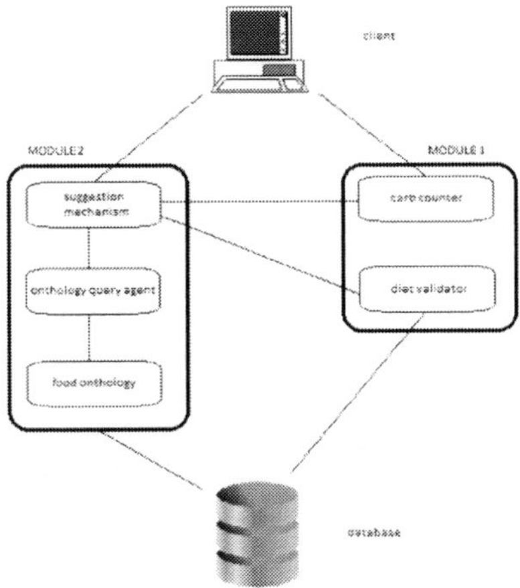


**Fig. 1.** System Architecture

The approach to solve the problem consists of two main parts:

1. Design and implement a module which helps diabetic type 1 people to arrange their nutrition according to their diets (given by a doctor or specialist) and do that by allowing people to eat variety of foods even desserts. (This is done with the help of carbohydrate counting mechanism which is mentioned earlier)
2. Design and implement a module which helps diabetic people to prepare their meal with the help of recipes.

There is a web site which gets inputs from the user which are:

- The diet which was written by a specialist.
- The carbohydrate and insulin ratio (Different for every patient).

The carbohydrate and insulin ratio is important for the system because it helps patients by using this ratio in carbohydrate counting. This ratio tells the system that how much carbohydrate taken can be handled by *a unit of insulin*.

With the system having those information about the patient, now it can tell the patient how many units of insulin he/she should get with a meal. Module 1 uses a database and a control mechanism and shows the result in a web page. Module 2 uses an ontology of foods to give effective recipes to the user. For example, the user can input the ingredients which are available in his/her fridge (lemon, flour, cream, eggs) and the system can suggest the user to cook lemon cheesecake with the calories calculated. The reason behind using ontology is to make effective suggestions. For example if the user has milk but not cream, the system should also know that the cheesecake can also be cooked by milk instead of cream.

The state of the art semantic web technologies, such as Java servlets, Ajax, Jena, SPARQL, RDF and OWL, are used in the implementation of the system. Servlets are the Java platform technology of choice for extending and enhancing Web servers. Servlets provide a component-based, platform-independent method for building Web-based applications, without the performance limitations of CGI programs. Unlike proprietary server extension mechanisms (such as the Netscape Server API or Apache modules), servlets are server- and platform-independent. This leaves you free to select a "best of breed" strategy for your servers, platforms, and tools [3].

AJAX stands for Asynchronous JavaScript and XML. It is a group of interrelated web development methods used on the client-side to create interactive web applications. With Ajax, web applications can send data to, and retrieve data from, a server asynchronously (in the background) without interfering with the display and behavior of the existing page. Data is usually retrieved using the XMLHttpRequest object. Despite the name, the use of XML is not needed (JSON is often used instead), and the requests need not be asynchronous [4].

Jena is a Java framework for building Semantic Web applications. It provides a programmatic environment for RDF, RDFS and OWL, SPARQL and includes a rule-based inference engine. The Jena Framework includes an OWL API, in-memory and persistent storage, SPARQL query engine.

RDF is a framework for describing Web resources. RDF is a directed, labeled graph data format for representing information in the Web. This specification defines the syntax and semantics of the SPARQL query language for RDF. SPARQL can be used to express queries across diverse data sources, whether the data is stored natively as RDF or viewed as RDF via middleware. SPARQL contains capabilities for querying required and optional graph patterns along with their conjunctions and disjunctions. SPARQL also supports extensible value testing and constraining queries by source RDF graph. The results of SPARQL queries can be results sets or RDF graphs [5].

The OWL Web Ontology Language is designed for use by applications that need to process the content of information instead of just presenting information to humans. OWL facilitates greater machine interpretability of Web content than that supported by XML, RDF, and RDF Schema (RDF-S) by providing additional vocabulary along with a formal semantics.

## 4 Implementation

Several food ontologies have been developed in the literature [7,8]. We have also developed a food ontology for the purposes of our system. Our food ontology is a subset of the existing proposals and it is used to store data and relations between different food categories. Foods, their carbohydrate values and relations between them are stored and defined in the ontology (see Fig. 2). The Food class is the super class of Meal class and has a relation with FoodCore which is “hasIngredient” which means, Food individual has 0 or more ingredients of FoodCore individuals. FoodCategory Class is needed for Dietetic validation. Diet specialists write diets with these categories. FoodCore class is needed to represent core elements of meals. Those elements are ingredients of meals. Mixture class is needed to contain meal ingredients. "BakedPotatoWithCreamMix” refers to the mixture of “BakedPotatoWithCream” which has mixture elements as “Potato,Cream, Cheese” (FoodCore). MixtureElements Class holds the elements of Mixture individuals.”CreamMix1” refers to ”Butter” and “CreamMix2” refers to ”Milk”.

Object properties are illustrated in Fig. 3. *belongsToCategory* property has FoodCore domain value and FoodCategory range value(Ex: Milk belongsToCategory MilkCategory). belongToCategory property has characteristics as “Transitive” and “Asymmetric”. Transitive means that if a belongsToCategory b and b belongsToCategory c then a belongsToCategory c. Asymmetric means a belongsToCategory b does not mean b belongsToCategory a. “belongsToCategory” has an inverse property “categoryIncludes” which means a belongsToCategory b, then b categoryIncludes a.

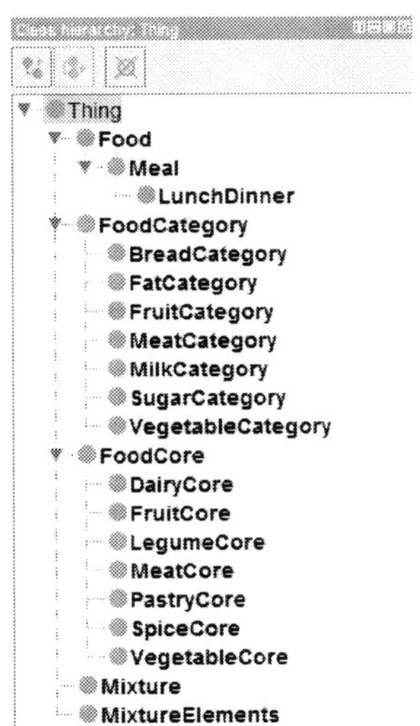


**Fig. 2.** Class Taxonomy

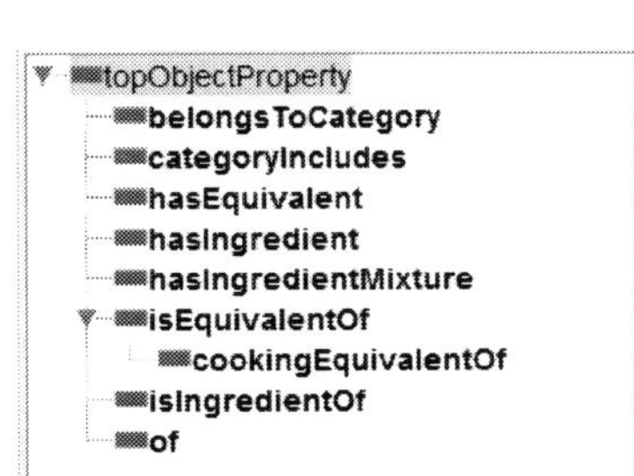


**Fig. 3.** Object Properties

*categoryIncludes* property has FoodCore domain value and FoodCategory range value (Ex: MilkCategory categoryIncludes Milk). categoryIncludes property has characteristics as "Transitive" and "Asymmetric". "categoryIncludes" has an inverse property "belongsToCategory" which means b categoryIncludes a, then a belongsToCategory b.

*hasEquivalent* property defines the relation between a mixture and its elements. For example Cream mixture has Milk and Butter elements.

*hasIngredient* property has Food domain value and FoodCore range value (Ex: BakedPotatoWithCream hasIngredient Potato). hasIngredient property has characteristics as "Transitive" and "Asymmetric". "hasIngredient" has an inverse property "isIngredientOf" which means b hasIngredient a, then a isIngredientOf b.

*hasIngredientMixture* property defines the relation between a meal and its mixture. For example BakedPotatoWithCream has BakedPotatoWithCreamMix.

*isEquivalentOf* property defines the relation between a foodCore and its mixture. For example Cream has CreamMix. This property is critical when finding the substitutable ingredients. isEquivalentOf property has characteristics as "Transitive" and "Symmetric". Symmetric means that if a isEquivalentOf b then b isEquivalentOf a.

*isIngredientOf* property has FoodCore domain value and Food range value (Ex: Potato isIngredientOf BakedPotatoWithCream). isIngredientOf property has characteristics as "Transitive" and "Asymmetric". "isIngredientOf" has an inverse property "hasIngredient" which means a isIngredientOf b, then b hasIngredient a. *of* property defines the relation between the MixtureElements and FoodCore (Ex: CreamMix has elements CreamMix1 of Milk and CreamMix2 of Butter).

To find the ingredients of a meal, first, the mixture of the meal, then the elements of that mixture should be found. The structure is shown in Fig. 4.

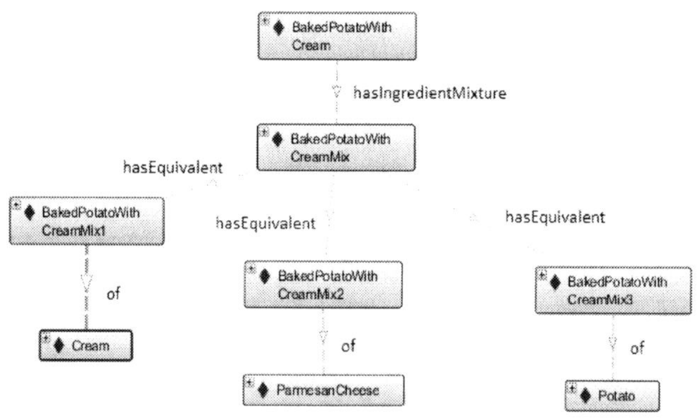


**Fig. 4.** Meal Structure

During the suggestion process there exist an operation which adds value to the search operation. Suppose that our fridge contains all the ingredients of Baked Potato With Cream except Cream, but we have Butter and Milk. If we search with traditional methods, the suggestion mechanism will not return that meal just because we do not have cream. But substitution mechanism knows to obtain cream by mixing butter and milk which are available in the fridge. So it returns the meal with an extra note like "combine milk and butter in order to get cream". This operation is achieved with a function called "expandFridge()". The function expands the fridge by looking at the ingredients and adds mixtures that are made with these ingredients to the search parameters (Ex: expand fridge adds cream by mixing milk and butter). The ontology also includes individuals such as butter, cream, milk, parmesan cheese which are stored as members of core classes such as the DairyCore class.

The individuals have their own object and data properties. For instance object property assertions of Cream individual of DairyCore Class include `hasEquivalent`. Cream individual belongs to the Fat Category of Diet Categories and has an equivalent of Cream Mixture. Data Properties of Cream are "gramsPerUnit" and "carbohydrate" which indicates how much amount of carbohydrate does that individual has in how many grams of that individual. The object property assertions of Cream Mixture indicate that there are two elements in that mixture which are "Butter" and "Milk".

The properties of an individual of Meal class include "hasIngredient" object. "hasIngredientMixture" property indicates that this meal has a mixture. Data Property "recipe" stores the recipe of that meal. Fig. 5 illustrates the properties of a Meal.

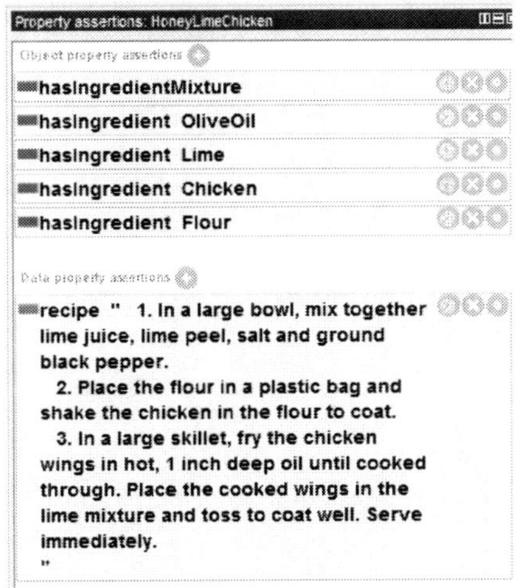


**Fig. 5.** HoneyLimeChicken Meal properties

## 5 User Interfaces

There are mainly three user interfaces. The first one is for entering the diet of the patient as prescribed by the doctor (see Fig. 6) The purpose of the diet screen is to input the diet which is written by a diet specialist to the system. There are slots (cells) in the table which has coordinates as days of week and periods. Those cells can be filled with the upper panel. The user chooses which cell to fill first and then adds the food category and amount.

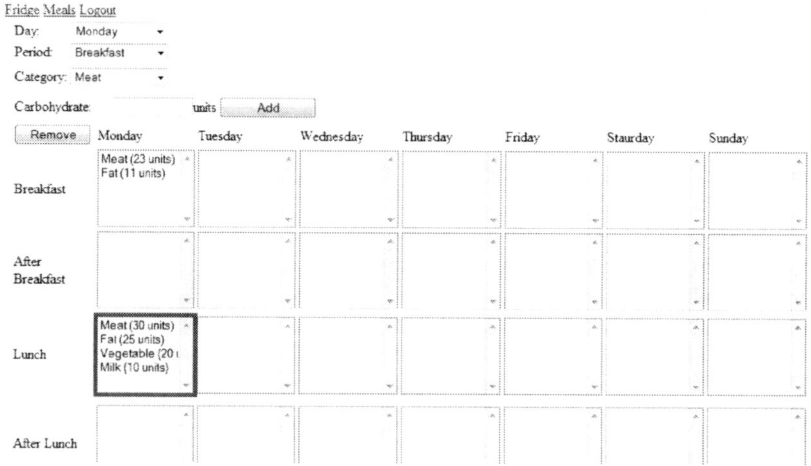


**Fig. 6.** Diet screen

The second user interface is for showing the contents of the fridge, i.e. the ingredients at hand. The user inputs foods which are available at home by first selecting the category, and then food. This list is used in the meal suggestion mechanism (Fig. 7).

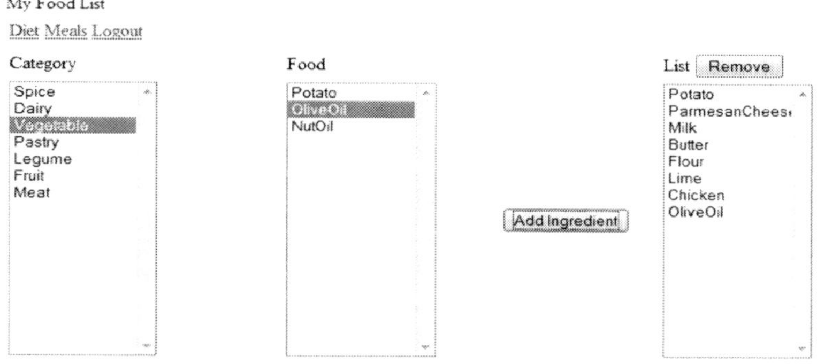


**Fig. 7.** Fridge screen

Fig. 8 shows the third user interface which is used for meal suggestion. The list on the left side of the screen shows meals which can be prepared using the ingredients which were input by the user (fridge). Before selecting the meal from the list, the user can choose the day and period from the comboboxes to get information about the diet program. The Result area shows amounts of ingredients for the meal. The Recipe area outputs the recipe of the selected meal.

The Info area gives information about the diet slot. The system warns the user that the diet indicates that patient should get 30 units of carbohydrate of meat category food, 25 units of carbohydrate of Fat category of food and 20 units of carbohydrate of Vegetable Category. This area also gives information about the total carbohydrate of the selected meal and how many units of insulin should be injected if that meal is eaten.

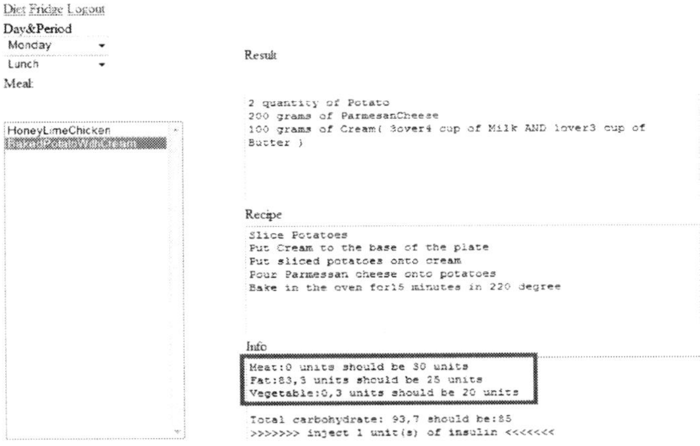


**Fig. 8.** Meal Suggestion

## 6 Conclusion

Semanticook Project is implemented to help diabetic type 1 patients with their nutrition management and prepare meals that are suitable to their diets. With this project it will be easier for the users to break the constraints of non-flexible diets.

Presentation layer of Semanticook Project is implemented in JSP and server requests are handled by AJAX. Ontology is created and populated in Protege 4.1 tool, all screenshots that are present section 5. A research is done with Diet Specialist Harika Akkoç (Sami Ulus Childrens Hospital) and Dr. Mehlika Işıldak (Baskent University Hospital) to get information about diabetic diets and carbohydrate counting. Querying ontology is handled by SPARQL querying language and used in java servlets with the help of JENA framework. MySQL database is used to store information about users and diets of users, and database querying is handled with SQL with the help of MySQL JConnector (JDBC Connector) in servlets. Java code is written in Eclipse J2EE Helios IDE.

Web based health care technologies are spread through Worldwide and many software companies started researches about these projects. Personal Health Record systems standing at the base of those kinds of projects to store medical information about patients and some extensions are made such as observing measurements and diet modules. Those kinds of systems are used by patients as well as doctors. If implemented in SOA, Semanticook can be easily integrated with such systems (Google PHR) and can be used effectively for nutrition management of diabetic patients and non-diabetic patients.

## References

1. American Diabetes Association, Official Pocket Guide to Diabetic Exchanges, 2nd edn., http://www.diabetes.org/diabetes-basics/type-1/
2. Vern Cherewatenko, M.D.: Diabetes – A Nutrition Update (Winter 2005), http://www.douglaslabs.com/pdf/nutrinews/Diabetes%20II%20NN%20%28Winter-05%29.pdf
3. Java Servlet Technology Overview, http://www.oracle.com/technetwork/java/overview-137084.html
4. Ullman, C.: Beginning Ajax. wrox (March 2007) ISBN 978-0-470-10675-4, http://www.wrox.com/WileyCDA/Section/id-303217.html
5. Prud'hommeaux, E., Seaborne, A.: SPARQL Query Language for RDF (January 2008), http://www.w3.org/TR/rdf-sparql-query/
6. Blanchette, K.: RD, CDE, Carbohydrate Counting (1996), http://www.endocrinologist.com/Carbohydrate-Counting.html
7. Dougherty, A.J.: Food Ontology, http://frdcsa.org/~andrewdo/WebWiki/FoodOntology.html
8. Genomic Standards Consortium (GSC), Food Ontology Project, http://gensc.org/gc_wiki/index.php/Food_Ontology_Project
9. Gaál, B., et al.: Dietary Ontology for Nutrition Counseling Expert System. In: 10th Intl. Protégé Conference, Budapest, Hungary, July 15-18 (2007)
10. Cantais, J., et al.: An example of food ontology for diabetes control. In: Proceedings of the International Semantic Web Conference 2005 Workshop on Ontology Patterns for the Semantic Web (2005)
11. Li, H.C., et al.: Automated Food Ontology Construction Mechanism for Diabetes Diet Care. In: International Conference on Machine Learning and Cybernetics, Hong Kong, August 19-22, pp. 2953–2958 (2007)

# From Relational Databases to Linked Data in Epigraphy: Hispania Epigraphica Online

Fernando-Luis Álvarez[1], Joaquín-L. Gómez-Pantoja[2], and Elena García-Barriocanal[1]

[1] Information Engineering Research Unit, Computer Science Dept., University of Alcalá, Ctra. Barcelona km. 33.6 – 28871 Alcalá de Henares (Madrid), Spain
{fernandol.alvarez,elena.garciab}@uah.es
[2] Dep. Of History.- University of Alcalá, C/ Colegios 2, E-28801 Alcalá de Henares (Madrid), Spain
gomez.pantoja@uah.es

**Abstract.** Epigraphic databases store metadata and digital representations of inscriptions for information purposes, heritage conservation or scientific use. At present, there are several of such databases available, but our focus is on those that are part of the EAGLE consortium, which aims to make available the epigraphy from the ancient classical civilization. Right now, the EAGLE partners share a basic data schema and an agreement on workload and responsibilities, but each repository has it own storage structure, data identification system and even its different idea of what an epigraphic database is or should be. Any of these aspects may lead to redundancy and hampers search and linking. This paper describes a system implementation for epigraphic data sharing as linked data. Although the described system was tested on a specific database, i.e. Hispania Epigraphica Online, it could be easily tailored to other systems, enabling the advantage of semantic search on several disparate databases.

**Keywords:** Linked Data, Semantic Web, Epigraphy.

## 1 Introduction

In a previous paper (Álvarez-García et al., 2010), we detailed the historical and convenience reasons for which epigraphers are partial to use digital databases. First, there is a long tradition to collect and edit inscriptions in *corpora* or collections, put together according several criteria: geographical, chronological or by subject, like divinities, military, freedman, magistrates and so on. Since information provided by epigraphs is rarely out of the ordinary, time and experience have shown that universal catalogues are more useful than smaller ones, which means that those projects are always expensive both in terms of the scholarly work needed and of the cost of publishing hefty volumes which target a restricted group of readers and are in the need to be brought up to date very often. Finally, "the bigger is better" axiom has pernicious side effects when searching and recovering data from books that catalogue

E. García-Barriocanal et al. (Eds.): MTSR 2011, CCIS 240, pp. 225–233, 2011.

as many as 10000 monuments, because hand-made indexes (always a cumbersome and ungrateful task) constrain the number of available search-keys.

A digital database solves all of those inconveniences, since there is not theoretical limit to the number of records it could held; every DB managing software has some kind of built-in data searching and recovering functions[1]; and digital publishing allows for an almost cost-free publication, including something scholars had been claiming for since the first *corpus titulorum*: the ability to add as many pictures of monuments (and other media) as you feel fit and without the constraints money and space force on printed copy (Gómez-Pantoja, 2011).

These explain the proliferation of epigraphic databases during past years and the strengthening of some of them as acceptable research tools in the field, specially since the Web has became a recognized powerful information provider and the adoption of the UTF standard allows printing and searching non-Latin scripts and any other outdated or weird characters as, for instance, the *litterae Claudianae* or some Roman fraction numbers[2]. However, epigraphic databases need to give an additional step in becoming interlinked and accessible for machine processing. Schemas as EPIDOC allow for a uniform representation of epigraphic metadata (Cayless, 2003) and several systems using this proposed standard have already been deployed (Bodard, 2008). But they still do not solve the issues of linking to other epigraphic collections through the Web. The paradigm of linked data (Bizer, 2009) brings the required ingredients for such interlinking. However there is a need to provide a cost-effective and generic solution for current epigraphic databases that do not require full conversion or maintenance of RDF representations in triple stores, as this would represent a too drastic change from the current deployments that are based on relational databases. In that direction, this paper reports the practical implementation of a relational-to-linked data solution for the Hispania Epigraphica database. This represents a first experience aimed at gaining experience for other databases to undergo the same process in the future.

The rest of this paper is structured as follows. Section 2 provides background information on the state of epigraphic collections inside the EAGLE consortium.

---

1 In fact, the most visible feature of any epigraphic database is its utility as an *index universalis* for the printed copy. This is specially true in the case of the Epigraphik Datenbank Clauss-Slaby (`http://oracle-vm.ku-eichstaett.de:8888/epigr/epigraphik_en`) which, as of the date of this writing, stores digital texts for 404.465 inscriptions from more than 19.500 places and taken from over 1.140 books and papers.

2 Until few years ago, scholars objected epigraphic databases on the ground of its poor graphic ability to represent non-Latin text and diacritics. This led to the use of non-standard fonts in some databases, which probed to be a very bad move since it compromised overall compatibility and system upgrading. On the contrary, those who tried ways to solve the problem at a deeper level, got useful and smart solutions, as it happens with the EpiDoc initiative (http://wiki.digitalclassicist.org/EpiDoc), a XML encoding tool initially designed for epigraphers to print non-standard characters that could be also used to write structured documents compliant with the TEI standard. EpiDoc allows to publish epigraphic corpora, suitable for semantic use and exchange in the Web, as in these two instances: `http://irt.kcl.ac.uk/irt2009/IRT256a.html` and `http://www.steinheim-institut.de/cgi-bin/epidat?function=Ins&sel=bay&inv=0001`).

Section 3 reports on the use of D2RQ for the exposure of Hispania Epigraphica and how the identification problem can be handled. Then, Section 4 reports on the type of configurations needed. Finally, conclusions and outlook are provided in Section 5.

## 2 Background

The abovementioned advantages of digital collections in epigraphy were so obvious even before some of them were generally available, that the *Association Internationale d'Epigraphie Grecque et Latine*, a main scholar's society, officially appointed a Commission to define the guidelines a digital collection must comply to be persistent, sustainable and useful for scientific research[3]. The Electronic Archive of Greek and Latin Epigraphy (EAGLE) was launched in the late 1990s, with the aim to build up a universal digital repository for all the Classical Greek and Latin inscriptions. In essence, EAGLE was built following the project Prof. Alföldy had imagined and launched in Heidelberg in the mid-1980s (Feraudi-Gruénais, 2010), which later on became an EAGLE founding partner, the Epigraphische Datenbank Heidelberg or EDH for short; but, since the number of Italian monuments is roughly a half of all known Latin epigraphs, the responsibility to digitalize those inscriptions (and a several thousands of seeding records) were transferred to Rome and thus, a second database, EDR (short for Epigraphic Datebase Rome) was started.

The mimetic acronyms reflect aptly the common identity of both databases, which were originally designed to run on a local environment, under close editorial control and with restricted access. But the idea of an universal and exclusive database was deemed unfeasible on two counts: first, the WWW remarkable grow and popular acceptation, which made open access an unavoidable alternative, initially for dissemination, then for entering and editing data. Then, personnel and resources were clearly insufficient to digitalize all the known Greek and Latin inscriptions in a short time. Help was then sought, opening EAGLE even from those not fulfilling the guidelines. This happened first with the Epigraphic Database Bari (EDB), which resulted from what some Univerità di Bari's scholars started in the early '80 aiming to supersede the chronic shortage of money and work force affecting another great epigraphic *corpus*, the Christian inscriptions from Rome. Due to the project's early start, EDB has some structural issues (database architecture tied to extinct or outdated operative systems, proprietary encoding for some special characters and so on), which partially compromises online search and consult and, more important, compatibility with other EAGLE partners. The fourth member of the consortium, is *Hispania Epigraphica Online*, which aims to digitalize the rich Portuguese and Spanish ancient epigraphic heritage. The database is built around the use of Web pages for both data administration and dissemination and takes advantage of all the Web 2.0 service, including total content transparency for Web search services. The core of HEpOl are tables in which every input could be re-used in other records to avoid repetitive operations and to allow data coherence in queries.

For convenience reasons, each EAGLE repository is financed, managed and even hosted independently, but there is a mutual agreement to further cooperation. Until

---

[3] http://www.eagle-eagle.it/Italiano/Documenti/Document2_it.html

now, this resolve has led to allocate specific responsibilities to each partners, based along geographical lines for EDH, EDR and HEpOl and on subject and geography in the case of EDB. Also, there is a commitment to unify criteria on some data fields to allow global queries, for which a test site is already done and working[4]. Thus, pairing up equivalent fields in each database seems a sensible strategy since all of them are built around a relational model, in which the inscription is the record's main subject, which is surrounded by other data like the monument's description and measures, its places of finding and conservation, paleographic information, bibliography, and so on.

Although the procedure greatly shortens old days searches on a bookshelf full of weighty volumes[5], this kind of query is hardly precise, as it was shown on our prior paper (Álvarez-García et al., 2010). First, it is basically a string search, which means that, at the end, it rests on the user cooperation, either to refine the searching terms through the usual querying tools or to filter the returned data by hand; and secondly, no effort was made to avoid data redundancy, since every database was designed to be a stand-alone system. The main victim of these flaws is the most sought item of every record, the inscription, which is a text and thus, full of nuanced meanings; and any particular epigraph could exist in the field bibliography as two or more distinct entities, at it often happens with fragmentary or unclear texts.

As an interim solution, some databases opted to multiply the number of record fields to single out every distinctive text meaning, which led to waste computer resources and, more important, converted data entry into a repeated and boring task. And in regard of duplicate texts, some projects are tagging their records with a supplementary ID, obtained from M. Depauw's *Trimegistos,* which previously need to check texts and corresponding metadata to assure uniqueness[6].

## 3 Using D2RQ with Hispania Epigraphica Resources

As a case study for an automated solution of the abovementioned problems, we have undertaken the conversion of one of the EAGLE databases, HEpOl.[7] Since individual conversion of each resource to RDF would require a full schema conversion and associated investment in technology for maintaining a triplestore, the alternative approach is to resort to D2RQ, which converts a relational database and to RDF exposure in a straightforward way and provides an SPARQL access point to query resources. Thus, requesting information on any particular property or attribute, it would call up all the related resources, both in the target database and in all of those repositories linked with our projected system. D2RQ can run under Windows (as a service) or as Java application on Linux; and could query either a MySQL database (as it is the case of HEpOl) or Oracle, PostgreSQL and Microsoft

[4] `http://www.eagle-eagle.it/Italiano/index_it.htm`

[5] As David Noy aptly described it at `http://ccat.sas.upenn.edu/bmcr/2007/2007-08-46.html`

[6] `http://www.trismegistos.org/about.php`: the process used to generate that unique ID is described there.

[7] `http://www.eda-bea.es`

SQL servers. It also includes a scripting function to generate a basic mapping of the target database. Concretely, by running this utility:

```
generate-mapping [-u username][-p password][-d driverclass] [-o outfile.n3][-b base uri] jdbcURL
```

D2RQ returns a file in N3 format, which will act as a mapping between the target database and its RDF representation. This .n3 file follows this schema: (a) Prefixes and Namespaces, (b) Server, (c) Database and (d) Structure of Tables.

The configuration started by inspecting the system to be represented in RDF and looking for an individual ID for each resource. For inscriptions, the simplest solution is to use the existing traditional IDs, which could be taken from either printed or digital corpora. For instance, *CIL* II 3042 = *IRAl* 27 = *HEpOl* 23[8]. Most of the time, alone or in concordance, those serials account for a valid and unambiguous identification; but sometimes the system is unreliable, as when the stone was successively engraved on both sides, as it happens in this case. Thus, the correct identification needs to be represented as (*CIL* II 3042 = *IRAl* 29; *CIL* II 3043 = *IRAl* 27) = *HEpOl* 23), which is an obviously long and verbose procedure.

As it was already discussed, the most notable attempt to assure the uniqueness of ancient is Trismegistus, which treats every written document on papyrus, regardless of the source from which it emanates or its condition, as a unique entity, generating from the involved records a mixed version, to which a single and unrepeatable ID is given. The system has been already adopted by a papyrological portal and aggregator[9], as the only way to guarantee the parts of the same document —very often owned by different Musuems and collections—are always treated as the same entity. During a meeting held in Rome in November 2010, the EAGLE Consortium agrees to adopt the same procedure for inscriptions and Trimegistos IDs been already issued for some databases. This feature would become a keystone in our attempt to export HEpOl data structure and content into a Linked Data system, since it guaranties information could be exchanged more quickly and efficiently after the other database are exported to RDF and their resources are correctly paired to the corresponding ones in HEpOl. However, in the interim the approach to be followed is having several URIs for the same resource, each using one of the traditional versions for the same inscription, and relying on some internal ID conversion. This offers maximum flexibility and allows different ID schemas to be used in an interoperable way. Backlinks with the alternative IDs for a resource allow for accessing the mappings. This is coherent with integration of different reference systems as discussed by Doerr, Schaller and Theodoridou (2004).

---

8 Meaning the number a particular inscription received in *CIL* II = *Corpus Inscriptionum Latinarum. Vol.II, Inscriptionum Hispaniae Latinae*, ed. By E. Hübner Berlin 1869, with a supplement printed in 1892; *IRAl*: R. Lázaro Pérez, *Inscripciones Romanas de Almería*, Almería 1980. And http://www.eda-bea.es/pub/record_card_1.php?rec=23.

9 `http://www.papyri.info/ddbdp/p.oxy;4;744#` shows how one record stores two parts of the same document, which are now hold by two institutions, one European, the other American.

## 4 Mapping HEpOl and Basic Linking

Each epigraphic database relies on a different relational schema, even though the descriptive elements are to a large extent the same. To illustrate a concrete example, let's start with a concrete HEpOl record (Fig. 1):

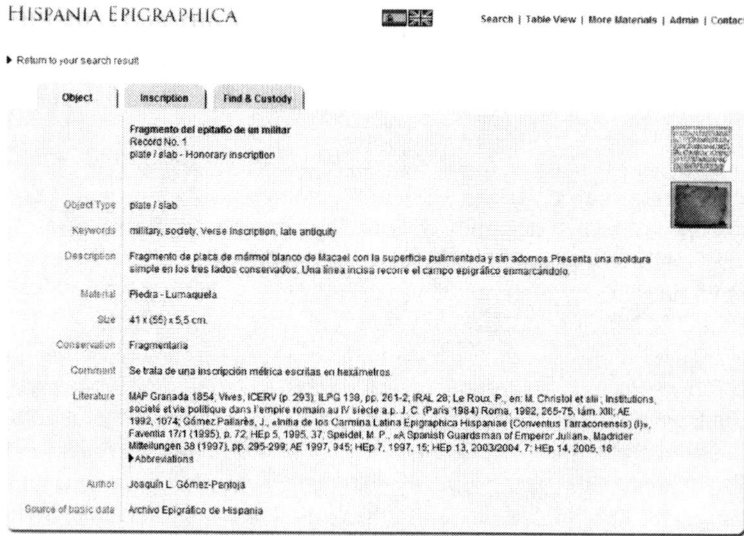


**Fig. 1.** HEpOl record number 1

Data is stored in tables, linked as usual by keys fields and unique identifiers, which allows for SQL querying. The fields considered for the exposure were the following:

| Title | Type | Object Type |
|---|---|---|
| KeyWords | Description | Material |
| Size | Bibliography | Text |
| Translate | Found Place | Name |
| Conventus | Roman Province | Actual Place |

The table shows several fields whose content could be linked to other linked data systems, concretely *Material* and *Actual Place* were linked to DBPedia and *found place* and *actual place* to Geonames.

To specify the URI of each of those resources linked to HEpOl, the following namespaces are used:

```
@prefix dbowl: <http://dbpedia.org/ontology/>.
@prefix dbPed: <http://dbpedia.org/resource/>.
@prefix dbprop: <http://dbpedia.org/property/>.
@prefix geoNames: <http://www.geonames.org/ontology#>.
@prefix
GnSearch: http://www.geonames.org/search.html?q=>.
```

```
@prefix
geoNamesMap:<http://www.geonames.org/maps/showOnMap?q=>.
@prefix hEp: <http://oe.dynalias.net:2020/resource/> .
```

This instructs the server on the prefixes that have to be used when they access the required elements in the dataset. The last prefix point to the address hosting our own Linked data (in the temporary, preproduction phase).

In our example, the inscription with number "1" has as value for *Place of Finding*: "Abla, Almería, Andalucía, España". Using GnSearch, GeoNames will provide information about this place, such as Latitude and Longitude, which could be placed on a map using *geoNamesMap*.

In addition to the above, it is necessary to declare each of the fields, their relationship and links to other datasets. For instance, the table Inscriptions would be described as:

```
map:INSCRIPTIONS a d2rq:ClassMap;
      d2rq:dataStorage map:database;
      d2rq:uriPattern "INSCRIPTIONS/@@INSCRIPTIONS.ID@@";
      d2rq:class hEp:Inscriptions;
      d2rq:classDefinitionLabel "Inscriptions";
      .
```

That is, the declaration includes the table name, the URI of each resource referenced in the database, the class within our system and the labels for the class. According with this scheme, all fields must be set up for the correct description of the inscriptions, as in this instance:

```
map:INSCRIPTIONS_placename_in_antiquity a d2rq:PropertyBridge;
      d2rq:belongsToClassMap map:INSCRIPTIONS;
      d2rq:property hEp:placename_in_antiquity ;
      d2rq:join  "INSCRIPTIONS.placename_in_antiquity_ID     =>
placename_in_antiquity_conn.ID";
      d2rq:propertyDefinitionLabel "placename_in_antiquity ";
      d2rq:column
"placename_in_antiquity_conn.placename_in_antiquity";
.
```

Being inscriptions the main resource, it features a property named `placename_in_antiquity`, which refers to the column with the same name in the table.

Other properties are configured in the same way, as well as the various correspondences with tables and datasets from which information is coming. Thus, if geographic information is obtained from Geonames, the declaration will look like this :

```
map:INSCRIPTIONS_f_place a d2rq:PropertyBridge;
      d2rq:belongsToClassMap map:INSCRIPTIONS;
      d2rq:property hEp:f_place;
      #d2rq:propertyDefinitionLabel "INSCRIPTIONS f_place";
      #d2rq:column "INSCRIPTIONS.f_place";
      d2rq:property hEp:Found_Place;
      d2rq:uriPattern
"http://www.geonames.org/search.html?q=@@INSCRIPTIONS.f_place@@"
;
.
```

An example of a linked configuration is the `Found_Place` property. Using the search of GeoNames Uri, String resource adding to seek, create a link to the DataSet. This step is need to configure the link as follows:

```
map:INSCRIPTIONS_material_refined_definition_conn_ID a
d2rq:PropertyBridge;
      d2rq:belongsToClassMap map:INSCRIPTIONS;
      d2rq:join    "INSCRIPTIONS.material_refined_definition_ID
=> material_refined_definition_conn.ID";
      d2rq:uriPattern
"http://dbpedia.org/resource/@@material_refined_definition_conn.
material_refined_definition_en@@";
      d2rq:property hEp:Material_refined_definition;
      #d2rq:property     owl:sameAs;
      #d2rq:uriPattern
"http://www.freebase.com/view/en/@@material_refined_definition_c
onn.material_refined_definition_en@@";
      .
```

This way, after configuring each property, D2RQ will recall a Content Negotiation routine to allow sharing between servers or to serve as RDF frontend to display data according to a RDF schema.

In summary, the process of exposure requires the following steps: (a) determining which elements (table attributes) in the relational database are to be exposed, then (b) configuring D2RQ for them, (c) identifying and configuring external sources to provide links and finally (d) implementing if required multiple IDs for the same resource using different traditional schemas. All these can be done with virtually no re-coding or maintenance of the legacy Web applications providing access to the database. However, this approach is still not making homogeneous the schemas of the different databases, which would require using EPIDOC as a vocabulary for exposing linked data. This can be achieved in a second step using some reconfigurations that reuse that vocabulary, when it is eventually made available.

## 5 Conclusions and Outlook

The linked data approach provides the required ingredients to make epigraphic databases evolve into a consistent and interlinked distributed database, improving its usefulness for researchers and the public. However, epigraphic databases are currently stored in relational databases and a cost-effective solution on legacy deployments appears as the best strategy for a gradual transition to schemas with richer semantics. This paper has reported the lessons learned in the creation of the linked data exposure of Hispania Epigraphica using D2RQ. A number of configurations allow for a coherent sharing and some basic links to other linked data systems. This combined with URIs allowing for different traditional reference systems provides a baseline for a cost-effective solution to exposing current epigraphic databases without a need of changing their data stores.

Future works will attempt to repeat the experience with other EAGLE databases and build interlinking based on elements that are relevant for researchers and common vocabularies reusing the effort done in the EPIDOC initiative.

**Acknowledgements.** This paper benefited from financial help from the Spanish Ministry of Innovation and Technology (project HAR2008-04820-C04-03/HIST).

## References

1. Álvarez-García, F.L., García-Barriocanal, E., Gómez-Pantoja, J.L.: Sharing Epigraphic Information as Linked Data Epidoc. In: Sánchez-Alonso, S., Athanasiadis, I.N. (eds.) MTSR 2010. CCIS, vol. 108, pp. 222–234. Springer, Heidelberg (2010)
2. Bizer, C.: The Emerging Web of Linked Data. IEEE Intelligent Systems 24(5), 87–92 (2009)
3. Bodard, G.: The Inscriptions of Aphrodisias as electronic publication: A user's perspective and a proposed paradigm. Digital Medievalist 4 (2008)
4. Cayless, H.: Tools for Digital Epigraphy. In: Proc. of the Association for Computing in the Humanities. Association for Literary and Linguistic Computing, Athens (2003)
5. Doerr, M., Schaller, K., Theodoridou, M.: Integration of complementary archaeological sources. In: Computer Applications and Quantitative Methods in Archaeology Conference (CAA 2004), Prato, Italy, April 13-17 (2004)
6. Feraudi-Gruénais, F.: Latin on Stone: Epigraphy and Databases. In: Feraudi-Gruénais, F. (ed.) Latin on Stone: Epigraphic Research and Electronic Archives, pp. 1–17. Lexington Book, Lanham (2010)
7. Gómez-Pantoja, J.L.: Amor Virtual o por qué se llevan tan bien Inscripciones y Ordenadores. In: Iglesias Gil, J.M. (ed.) Actas de los X Cursos sobre sl Patrimonio Histórico, Reinosa, Julio 2009, pp. 67–96. Universidad de Cantabria, Santander (2011)

# Content Models for Enhancement and Sustainability: Creating a Generic Framework for Digital Resources in the Arts and Humanities

David Stuart[1], Brian Aitken[2], Daisy Abbott[3], Alex Chassanoff[4], Mark Hedges[1], and Andrew McHugh[2]

[1] Centre for e-Research, King's College London, London, UK
{david.stuart,mark.hedges}@kcl.ac.uk
[2] HATII, University of Glasgow, Glasgow, UK
{B.Aitken,A.McHugh}@glasgow.ac.uk
[3] Digital Design Studio, Glasgow School of Art, Glasgow, UK
D.Abbott@gsa.ac.uk
[4] School of Information & Library Science, University of North Carolina, Chapel Hill, USA
achass@email.unc.edu

**Abstract.** In this paper we describe a framework to increase the accessibility and reuse of objects stored in digital repositories. In many cases digital repositories are created with little consideration to the wider information environment, or the extension of the repository to items beyond the initial collection. The CMES framework emphasizes providing a structure for creating appropriate content models, selecting appropriate metadata schemas, and the need for a modular approach to the creation of a user interface.

**Keywords:** Fedora, Islandora, content modelling, digital repositories.

## 1 Introduction

Digitizing analogue collections is an increasingly important part of many cultural institutions, primarily to facilitate access to the collections, but also as a way of digitally preserving the objects (Poll, 2010). Many of the collections that have been digitized so far have been made available online, although too often in isolation, focusing on making one specific collection available, rather than taking into consideration the wider information environment and the ability of a repository to incorporate as yet unthought-of content. Such a disjointed approach has resulted in data silos (Nichols, 2009), where the true value of the digitization is lost as there is less scope for inter-collection and intra-collection discovery, with additional silos being created when new collections are digitized. What is required is a framework that both allows digital objects to be as widely accessible as possible and that is extensible to include additional digital collections that may be very different from the original digital objects. Creating a framework that makes digital objects as widely accessible as possible not only means being able to search and retrieve objects in isolation, but also enabling

E. García-Barriocanal et al. (Eds.): MTSR 2011, CCIS 240, pp. 234–244, 2011.

connections to be drawn between objects, whether this is between objects within the same collection, within different collections in the same repository, or with objects outside a repository. Such a framework will both increase the accessibility and reuse of objects in digital repositories.

The aim of the Content Models for Enhancement and Sustainability (CMES) project is to create such a generic framework for digital resources in the arts and humanities, specifically addressing the issues of sustainability and the innovative re-use of resources in a broad range of collections that can be extended to support additional digital resources. This paper describes the proposed framework developed as part of the CMES project, as well as the approach taken to the design of the user interface. As such it reflects the constraints and the potential of the software it is built upon, in this case the Fedora digital repository software (http://fedora-commons.org) and its enhanced content models.

## 2 Background

The Fedora Commons digital repository software is designed for modelling complex objects and the relationships between them. These complex objects can be the aggregate of different types of content, with the relationships between objects expressed in RDF, not only allowing hierarchical relationships but also graph-like relationships (Lagoze et al., 2006). Representations of digital objects within Fedora are formalised as 'content models', which explicitly describe the digital objects. Content models can describe the number and types of file that form datastreams within the digital objects adhering to a particular content model, enabling the simple reuse and management of digital objects. All data objects automatically contain the RELS-EXT datastream which stores the RDF triples associated with a particular object. More recently the Enhanced Content Model framework has been incorporated, enabling the specification of relationships between objects and the declaration of schemas for any XML datastreams (Blekinge-Rasmussen & Christiansen, 2009).

When a repository is to contain a collection of known objects it is not necessary to create an elaborate framework to deal with all eventualities, instead content models can be designed to deal with very specific objects. However digital collections are often heterogeneous, with significant differences both within and between genres of objects. For example, whereas multiple collections may contain images, these images may come in different formats (e.g., JPEG, TIFF), or multiple formats, or multiple images may be associated with the same object. Alternatively there may be multiple different types of file associated with the same object, for example, an object may have both images and text documents relating to its provenance. Content models can be created at different levels of abstraction, and a number of different approaches have been taken depending on the type of collection and the purpose of the repository, although as the University of Wisconsin Digital Collections Centre have argued: "New kinds of objects should not necessarily mean new content models" (University of Wisconsin, 2009). It is necessary to have a balance between extremely abstract content models that provide little indication of the type of information it may contain and how it may be deployed, and too rigid content models which become overly clunky as they try to contain every piece of information.

The complexities of creating a framework can be seen most clearly when dealing with metadata, as collections with same types of object often have very different metadata, in different formats and at different levels of specificity. Whilst the draft content models of the RepoMMan Project include the metadata streams within the data object content models (Green, 2006), most projects have seperated all or part of the metadata from the digital object, both to allow the referencing of objects that might not have been digitized, and to reduce duplication where multiple objects may have the same metadata.

The NSDL/N-Core approach (Krafft, Birkland, Cramer, 2008) separates metadata completely from the resources. The framework is based around five types of object: resource objects which contain the actual resources; metadata objects which contain the metadata associated with an object; aggregation objects which aggregate resources and other aggregations; metadata provider objects which aggregate and provide provenance data for metadata objects; and agent objects that specify the source for metadata and aggregations. The idea of a trinity of resource objects, metadata objects, and aggregation objects form the basis of many Fedora collections, although the boundaries between such objects are not always the same.

Many repositories have taken a middle-way between the extremes, keeping some metadata with the resource object, and separating other metadata out as separate metadata objects. For example, the WGBH Library chose to separate the descriptive metadata from resources, allowing multiple resources to share a common set of metadata, whilst rights metadata and instantiation metadata was included in the resource object (Beer, Pinch, & Cariani, 2009). The ICPSR Content Models for Social Science data are formed of three objects, a research data data object, a research document data object, and an aggregating data object. The research document data object contains descriptive data for the associated research data data object, whilst the aggregating object contains both descriptive metadata and PREMIS metadata (ICPSR, 2010).

The difficulty of cross-collection searching has been recognized for a number of years, with difficulties in combining disparate resources caused by differing metadata and vocabularies (Schreibman et al. 2008). In their survey of approaches to metadata interoperability Haslhofer & Klas (2010) identify three approaches: model agreement; meta-model agreement; and model reconciliation. The first, model agreement, refers to adhering to an agreed standardization, for example a generic metadata standard such as Dublin Core. Widespread adherence to any single scheme is not only difficult to achieve, but such solutions are often based on adhering to the lowest common denominator (Blanke et al., 2008) and are often unsuitable because organizations already have metadata in their own proprietary formats. Meta-model agreement refers to achieving interoperability through the use of a common meta-model in which existing standards are placed. For example, METS provides a wrapper in which a user can place any metadata element they wish. Increased flexibility, however, comes at a price of increased complexity in the creation of a user interface to deal with multiple standards, and the content of the metadata is not standardized (McDonough, 2006). Model reconciliation refers to cross-walking between different schemas. A common metadata standard both simplifies the development and ingest processes (Beer, Pinch, & Cariani, 2009), although cross-walking is not necessarily a simple process, and can result in the loss of subtleties in the information.

## 3 Initial CMES Collections

The initial CMES framework is based on an analysis of five digital collections in the areas of digital texts and performing arts: The Stormont Papers, 84 volumes of the Stormont parliamentary papers; the Inscriptions of Aphrodisias, a collection of files relating to 2,000 inscriptions from Aphrodisias in Turkey; Scottish Traditions of Dance, a collection of interviews, photographs and video relating to Scottish dance; Adolphe Appia, a collection of images, video, and 3D models relating to the theatre innovator Adolphe Appia; and digitised audio from the King's Sound Archive. The resources were selected due to their perceived value to the research community and the complementary nature of the content. Moreover, usage is currently restricted by the custom interfaces and data models that were defined at creation. They will benefit from enhanced interfaces and models that support cross-analysis of content. Analysis of the collection highlights a number of the issues that will need to be addressed by any proposed framework.

The collections contain a wide variety of content types, in various formats, and at different quality. For example, images which appear in each of the five collections, can be found as both TIFFs and JPEGs. Within the Stormont collection each page consists of a single TIFF file, whilst the King's Sound Archive contains a single JPEG of each record label. The images within the Scottish Traditions of Dance collection and the Adolphe Appia collection have both an archival quality TIFF file and a lower resolution JPEG, whilst the Inscriptions of Aphrodisias collection has both a low resolution thumbnail JPEG and a high resolution JPEG. Unless there are to be multiple image content models reflecting the different existing file structures, it is necessary to create an all-encompassing image content model with the steps necessary to make collections adhere to it without the loss of information.

Despite the diversity of images in the different collections, an image content model is relatively simple to create as the files may be changed from one format to another, or resized. It is more difficult to model the 3D Models, despite them only appearing in one of the collections. Within the Adolphe Appia collection 3D models exist in three formats: Turn Tool Scenes; Virtual Reality Modelling Language files; and Quest educational files. Turn Tool Scenes not only have a TNT file, but also an associated HTML page. Where a model has been created using Virtual Reality Modelling Language there may be multiple WRL files in a single folder whilst the Quest 3D Educational files are standalone executables. There are few tools for converting files from one format to another, and greater scope for loss of functionality.

Within the King's Sound Archive the audio files are not only in different formats, but also different versions. As well as multiple versions of the edited music files designed for public consumption, there are also unedited WAV files for preservation purposes.

Within the collections there are both a number of compound and aggregate objects. Compound objects are objects made up of different file types, for example, the interviews within the Scottish Traditions of Dance combine both audio files and transcriptions of these audio files, whilst each side of a digital record in the King's Sound Archive comprises both images and audio files. Aggregates combine more than one object of the same type, for example the record is an aggregate of the two sides. However not all files necessarily fall into easily distinguishable aggregate or compound objects. For example volumes in the Stormont collection may be thought of as a compound of the xml files and an aggregate of all the pages.

The collections have varying quality and quantity of metadata, provided at different levels, and often in their own idiosyncratic format. The Stormont collection contains an XML file with descriptive, rights, event, and technical metadata about the collection as a whole in a TEI format. The Scottish Traditions of Dance collection provides primarily descriptive metadata as a single XML file including all the records in the collection. The Adolphe Appia collection contains descriptive and rights metadata in a single record associated with each object, currently held within an Excel spreadsheet. The Inscriptions of Aphrodisias collection contains rights metadata for each of the files in the collection. There is a single METS record for each record in the King's Sound Archive, providing information about the struture and contents of the record and its associated files.

## 4 CMES Framework

It is important that a framework is not only suitable for the creation of content models to deal with the initial collections, but can incorporate additional collections. Such collections may include data types that have not been included previously, or in a format that has not been included previously. For example, whilst in the initial collections images are in .jpg and .tif formats, there may be future collections which have .gif images. Alternatively a collection may include data sets in Excel spreadsheets, for which no content model has as yet been created. Nonetheless analysis of the initial collections, as well as the literature, demonstrates a number of necessary criteria for the framework.

- Content models must enable the constituent parts of composite objects to exist and to be manipulated independently.
- Metadata should be included at different levels of abstraction.
- Metadata should be applied consistently across different collections to facilitate inter-collection retrieval.
- Use of generic metadata schemes should not result in the loss of metadata or the subtleties of metadata, as the data is transferred into one generic format.
- The framework needs to cope with new types of content that may not have been put in Fedora before.

The proposed CMES framework will make use of four layers of content models:

1. Collection content model – Identifying the resources within a collection, it is not only important for the current collection information, but also for the enabling additional collections to be created based on existing resources.
2. Composite content models – Content models designed to bring together more than one resource content model. For example, an interview may be a composite of an audio and a text file. By having composite models that reflect specific type of content the resources can then be manipulated in different ways.

3. Resource-Metadata content models – Whilst two different collections may both contain the same resource types, e.g., images, they do not necessarily contain the same metadata. To preserve both the fullness of the metadata and to ease cross-collection analysis, metadata will be kept in its original state as well as being cross-walked to a generic format to facilitate inter-collection connections.
4. Resource content models – Content models designed to reflect basic objects, e.g., image, video, audio, document. These are not associated with any particular file type, but rather have data streams reflecting the most appropriate file types.

On occasion there will be original files that need to be included, but do not conform to the existing content model. For example a GIF image would not fit easily into an image content model that only included TIFFs and JPEGs, and a content model that could include all file types would be extremely large, cumbersome, and quickly become dated. The proposed solution is to create an additional simple content model for any new file types, which then expresses a relationship with the main content model for that particular resource type. The relationship between GIF model and the Image file can be expressed in RDF. This approach will also be applied to the 3D models for which there is no one definitive format. Each model will be exported into a flash version which will form the basis of the main 3D model digital object, with the original files stored in to a simple content model.

## 5 Metadata

Metadata may be variously categorized in different ways, for example, descriptive, structural, and administrative (NISO, 2004), or administrative, descriptive, preservation, technical, and use (Gillard, 2008). Some of these types of metadata are more suitable than others at different levels of the collection. For example, where different rights are associated with different resources in the same collection, it would be difficult to reduce this to a single set of rights for the whole collection. At the same time, whilst there has traditionally been greater emphasis on the item-level activities, rather than collection-level descriptions, collection-level descriptions have become more important with digital resources as numerous collections are brought together in one place the collection metadata plays an increasingly important part in information retrieval (Macgregor, 2003), as demonstrated by the development of schemas such as void (Vocabulary of Interlinked Datasets) (Alexander et al., 2009). It is important that the metadata that is captured or created is useful, and not purely captured because it can be. As Foulonneau & Riley (2008, p.124) note "It is generally good practice to share, in addition to any other desired formats, the most complete descriptive metadata records available...A good solution is to offer a simple format such as Dublin Core, and supplement this with a more robust one appropriate to your environment, such as MARC or MODS in a library."

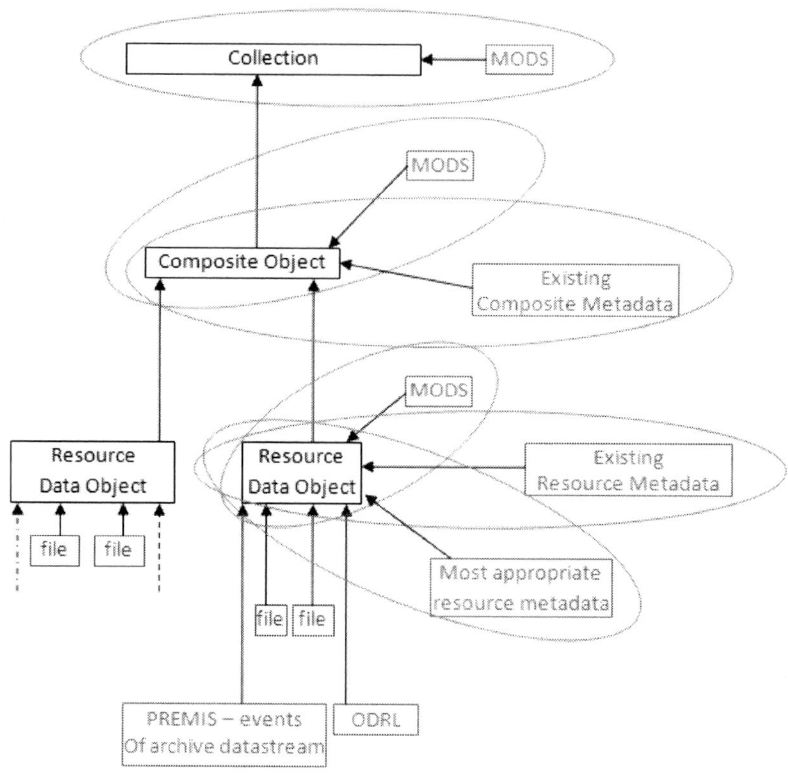


**Fig. 1.** Proposed CMES Framework

The proposed CMES framework needs to include descriptive, preservation, structural and rights metadata, although different types of metadata will be required at each of the four levels.

**Collection objects** – Description metadata is required at a collection level to aid with resource discovery. As there is a great variety in the types of content, both between collections, and within collections, a generic metadata schema is required. Where collection data is available, for example, as it is with the Stormont papers, that data should be cross-walked into MODS. Both Dublin Core and MODS are widely used in the digital library sector, and either may be deemed suitable (Gartner, 2008).

**Composite objects** – Descriptive metadata and structural metadata are required to aid with resource discovery and to identify resources forming the composite object. Structural metadata can be kept in the RELS-EXT datastream. Where descriptive metadata exists at the composite level, for example, the Scottish Traditions of Dance Trust has metadata records at the level of the interview, that metadata should be cross-walked into MODS. Where there is benefit from keeping the original file it should be stored in a simple content model, with the relationship to the composite model stored in its RELS-EXT datastream.

**Resource-metadata** – Descriptive data is also important at the resource data level, as well as at the composite object level, both for those objects that are not part of a composite object and for those that are not. As well as descriptive metadata in the form of existing metadata, and that metadata cross-walked into MODS, metadata will also be cross-walked into the most appropriate file metadata type (i.e. MIX for images), to make the resources within the repository as accessible as possible to external users.

**Resource data object –** As well the descriptive metadata in the level of resource-metadata objects, it is necessary to include two additional datastreams to include events metadata (i.e. how the archive version of the information was created and processed) and rights metadata with the resource object.

## 6 User Interface

Each of the collections that has been analysed and incorporated into the framework currently has its own legacy web interface that connects to and is driven by an existing underlying data model. The fundamental elements and interconnections within each data model are being extracted, abstracted and redefined through the content models that form the core of the CMES approach. A comparable approach is also being taken with the user interfaces associated with the collections.

Merely generating updated versions of the legacy web interfaces, although perfectly feasible, would not be a worthwhile approach for two important reasons. Firstly, creating modern equivalents of the legacy interfaces would simply result in another set of 'time capsule' interfaces that will likely become as outdated as the original interfaces as web standards and best practice guidelines continue to evolve. Secondly, one of the principal aims of the project is for the content models to be generic enough and suitably flexible to be applied to other collections that share a broadly similar data structure. Therefore the user interfaces through which the content may be accessed must also be suitably flexible and provide adequate facilities for customisation.

For these reasons a modular approach to the development of user interfaces within the project is being adopted. Loosely coupled user interface components are being developed based on the structure defined in the underlying content models, enabling future developers and users to extend or replace an individual module without affecting other modules used by the same or other content models. This will ensure that future extensions of the content models can be reflected in the user interface without impacting on existing content held within the repository.

The Fedora Commons digital software repository provides mechanisms for making content available through its web service API (application programming interface). Using the API, developers can build user interfaces that connect to and retrieve data from the repository via web service requests. Although this approach enables repository developers to create user interfaces that are suited specifically to their applications, it also means that a web service based client application has to be developed by each project that makes use of Fedora Commons, as no such client application is provided by default.

As outlined by the developers of Fedora Commons, the available web service API has "grown in an ad-hoc fashion as the software has matured", resulting in a large number of complex methods that "obscure the elegance of Fedora's underlying architecture" (Fedora Commons, 2008). Rather than develop a user interface from the ground up based on this overly complicated API, CMES decided to investigate pre-existing projects that have attempted to provide a generic user interface management framework for Fedora Commons.

Two potential solutions were identified: Hydra (http://projecthydra.org) and Islandora (http://islandora.ca). Each of these open source development projects share similar goals and objectives, namely to provide a flexible and modular set of user interface tools that can be employed to provide a web-based front-end to a Fedora based repository.

The Hydra project is a collaboration that started in 2008 between the universities of Hull, Stanford and Virginia and as its name implies, if offers a "one body, many heads" approach, with the repository acting as the "body" upon which one or more feature-rich applications can be attached as the "heads". Hydra is based on the Ruby on Rails web application framework with search and retrieval facilities provided by Blacklight, a Ruby based library catalogue web application. Blacklight in turn utilises Apache Solr to provide powerful full-text searching.

Islandora is developed by the University of Prince Edward Island's Robertson Library. The approach taken by Islandora is to develop a configurable module for the PHP based Drupal content management platform that enables Drupal to interact with a Fedora based repository. Content from the repository can then be formatted and repurposed for display using the huge selection of standard Drupal modules that are available. As with Hydra, Islandora offers powerful full-text searching by means of the Apache Solr search platform.

Both Hydra and Islandora provide feature-rich and flexible approaches to developing front-ends for a Fedora repository but Islandora has proven to be slightly better suited to the requirements of the CMES project. The Islandora project is currently at a greater level of maturity than Hydra, which is important for a short-term project such as CMES. Islandora provides a more comprehensive set of documentation and there is an active user community posting problems and solutions through the Islandora Google Group.

A further deciding factor in the decision to adopt Islandora is the technology stack that it utilises. The Islandora approach of integrating Fedora with Drupal provides a higher level entry point to developing user interfaces than Hydra's lower level Ruby on Rails approach. Many user interface issues such as providing themes, site structures and configuring navigation routes and menus can be handled through Drupal's web-form based administrative interface without requiring a single line of code to be written, which is not the case with Hydra.

By utilising Drupal, Islandora also provides access to more than 8,000 user interface modules that are currently freely available to customise Drupal's core functionality, many of which offer an off the shelf solution to the challenge of presenting data extracted from the Fedora repository. The modular approach of Drupal will also ensure that the CMES project, or any subsequent user of the data contained within the CMES repository, is able to create reusable user interface components to meet any requirements that cannot be met by any existing modules.

In addition to providing a set of flexible web-based user interfaces to the data stored within the repository as described above, the CMES project will also offer direct access to the Fedora web service API, thus ensuring that future users of the collections have the facility to further develop and integrate their own tools to process the resources in ways that have yet to be considered.

## 7 Conclusions

Analysis of the five arts and humanities collections demonstrates the wide variety of files in ostensibly similar collections, and the limitations of the current repositories in enabling the digital objects within them to reach their full potential. There is a need for a more flexible framework, both in terms of the underlying content models and in the creation of user interfaces. CMES has provided such a framework, both with its levels of content models, its incorporating multiple types of metadata, and a modular approach to a user interface.

This paper has discussed the development of the CMES framework; however there are many potential pitfalls between the theoretical and the practical, and future research will necessarily focus on the implementation and evaluation of the framework. Evaluation of the framework needs to take into consideration the accessing and reuse of objects, surveying of researchers and teaching staff regarding resource use for research and teaching, and the take-up of the enhanced content models for new collections.

## References


1. Alexander, K., Cyganiak, R., Hausenblas, M., Zhao, J.: Describing Linked Datasets – On the Design and Usage of void. In: The "Vocabulary Of Interlinked Datasets". Linked Data on the Web Workshop 2009 (2009)
2. Beer, C.A., Pinch, P.D., Cariani, K.: Developing a flexible content model for media repositories: a case study. In: Proceedings of the 9th ACM/IEEE-CS Joint Conference on Digital Libraries, JCDL 2009, pp. 97–100 (2009)
3. Blanke, T., Aschenbrenner, A., Küster, M., Ludvig, C.: No Claims for Universal Solutions - Possible Lessons from Current e-HumanitiesPractices in Germany and the UK. In: Proceedings of the 4th IEEE e-Science Conference, Indianapolis (2008)
4. Blekinge-Rasmussen, A., Christiansen, K.F.: Enhanced Content Models (2009), https://wiki.duraspace.org/download/attachments/13770428/FedoraOR09.pdf?version=1&modificationDate=1267451847631
5. Fedora Commons (2008), https://wiki.duraspace.org/display/DEV/RESTful+Fedora+Proposal
6. Foulonneau, M., Riley, J.: Metadata for Digital Resources, Chandos, Oxford (2008)
7. Gartner, R.: Metadata for digital libraries: state of the art and future directions. JISC Technology & Standards Watch (2008), http://www.jisc.ac.uk/media/documents/techwatch/tsw_0801pdf.pdf
8. Gilliand, A.J.: Setting the Stage. In: Introduction to Metadata. J. Paul Getty Trust (2008), http://www.getty.edu/research/publications/electronic_publications/intrometadata/setting.pdf
9. Green, R.: RepoMMan Project: Experiences with Fedora during the project's first year (2006), http://www.hull.ac.uk/esig/repomman/downloads/D-D8-fedora-exp-v10.pdf

10. Haslhofer, B., Klas, W.: A Survey of Techniques for Achieving Metadata Interoperability. ACM Computing Surveys 42(2), Article 7 (2010)
11. ICPSR (2010), `https://wiki.duraspace.org/display/DEV/ICPSR+Content+Models+for+Social+Science+data`
12. Krafft, D.B., Birkland, A., Cramer, E.J.: NCore: Architecture and Implementation of a Flexible Collaborative Digital Library. In: JCDL 2008 Joint Conference on Digital Libraries, Pittsburgh, PA, USA, June 16-20, pp. 313–322 (2008)
13. Lagoze, C., Payette, S., Shin, E., Wilper, C.: Fedora: an architecture for complex objects and their relationships. International Journal on Digital Libraries 6(2), 124–138 (2006)
14. Library of Congress, `http://www.loc.gov/standards/metadata.html`
15. Macgregor, G.: Collection-level descriptions: metadata of the future? Library Review 52(6), 247–250 (2003)
16. McDonough, J.P.: METS: standardized encoding for digital library objects. International Journal on Digital Libraries 6(2), 148–158 (2006)
17. Nichols, S.: Time to Change Our Thinking: Dismantling the Silo Model of Digital Scholarship. Ariadne 58 (2009), `http://www.ariadne.ac.uk/issue58/nichols`
18. NISO. Understanding Metadata (2004), `http://www.niso.org/publications/press/UnderstandingMetadata.pdf`
19. Poll, R.: NUMERIC: statistics for the digitization of European cultural heritage. Program: Electronic Library and Information Systems 44(2), 122–131 (2010)
20. University of Wisconsin (2009), `https://wiki.duraspace.org/display/DEV/University+of+Wisconsin+Digital+Collections+Center+-+Complex+Content+models`
21. Schreibman, S., Roper, J.O., Guegen, G.: Cross-collection Searching: A Pandora's Box or the Holy Grail? Literacry and Linguistic Computing 23(1), 13–24 (2008)

# Integration of CERIF, ESB and Cloud in the Development of a National Global Research and Administrative Management System

Yi Wang[1], Simon Kerridge[1], Peter Smith[1], Simon Foster[2], Steve Coppin[3], and Stephen Trowell[2]

[1] University of Sunderland, Academic research, Edinburgh Building, City Campus, Chester Road, UK
[2] University of Exeter, Research and Knowledge Transfer, the Innovation Centre, Rennes Drive, Exeter, Devon, UK
[3] University of Kent, Learning and Research Development, Information Services, Canterbury, UK

**Abstract.** In the UK, no single supplier produced IT System has been identified which supports the management and administration of a University's research program from pre-award through to post award. The requirements of UK higher education research management and administration include program management, costing and pricing, application development and submission, contract and project administration, monitoring and reporting. UK universities commonly use a mixture of spreadsheets and databases, working alongside existing finance and administration systems to manage these activities.

This paper presents the scope and design of a modular Research Management and Administration System for the UK higher education sector which allows for the management and administration of research projects from 'cradle to grave'. The system should embody best practice processes for research management and administration, whilst maintaining the capability [through open systems architecture] to interface with a University's existing systems such as Finance, HR and Student systems.

**Keywords:** ESB, Cloud, Higher Education, integrated system, Interface, Research management and administration system [RMAS], CERIF.

## 1 Introduction

The usage of information technology at UK universities is characterized by historically grown system platforms, little integration and an incomplete support of business processes in university administration. such systems mainly implement tasks in ways that are specific to certain organizations [1]. From the study made for the RMAS[1] project, Binge [2] no single system was identified that could provide a fully scoped integrated research management and administration solution for use by UK Higher Education Institutions [HEIs].

---

[1] See `http://www.exeter.ac.uk/research/rmas/` (accessed 29th June 2011).

E. García-Barriocanal et al. (Eds.): MTSR 2011, CCIS 240, pp. 245–255, 2011.

A very small number of HEIs had developed systems with a view to scaling them up for wider use. It was also clear that no commercial supplier had a cost effective and complete solution. The lack of a comprehensive software system has resulted in many difficulties in turning research management data into useful information, in the right place, at the right time.

RMAS will provide, for the first time, a transferrable integration of the administrative management of research activities. The present position is that there are a variety of corporate and other systems that together hold, sometimes in multiple formats, the data that are required to manage research activities effectively. There are however many difficulties in turning these data into useful information that can be used and shared effectively.

The RMAS project aims to accomplish the following:

1. The development of a shared services solution, based on the shared requirements of a group of HEIs that will specify an RMAS. The consortium will manage a tendering framework such that suppliers can provide components of the overall RMAS. The consortium will also form the first tranche of users, contributing to the development of an RMAS and benefiting from early adoption.

2. The development of an RMAS should seek to take advantage of a Service Oriented Architecture [SOA] so far as it does not compromise the timely development and implementation of a system, and to also explore options for delivery via Software as a Service [SaaS].

3. The expansion of the broad set of user requirements into a more detailed set of functional requirements which would enable suppliers to respond to an invitation to tender. The full cost of development and a technical solution could then be considered and a financial plan developed. At this point it would be possible to quantify any requirements for funding and propose possible sources.

4. Each institution has own existing services and systems. Some of these modules need to be replaced, others will be retained and entirely new functionality might also be added. An RMAS should enable changes within different institutions; a promising architecture to facilitate the interchanging of modules is an Enterprise Service Bus [ESB]. An ESB can provide flexible and reusable *infrastructure* services that may be configured and augmented easily and quickly.

This paper first gives an overview of existing systems and their drawbacks and then identifies the new characteristics of the RMAS. After that the detailed design of RMAS is discussed and a set of conclusions are presented.

## 2 Previous Work

The UK Joint Information Systems Committee [JISC] e-learning Framework is based on a service-oriented approach ["SOA"] with the aims of achieving interoperability, reusability of services, more rapid development times, and the ability to combine services to create new composite services. A quote from a key document expresses the aim to "*...facilitate the integration of commercial, home-grown, and open source components and applications within institutions and regional federations, by agreeing common service definitions, data models, and protocols*" [3].

Binge [2] indicated that in the absence of an off the shelf integrated RMAS some universities have resorted to writing their own in-house systems. Those universities without the expertise or dedicated resource to write such a system are forced into using a mix of systems not designed to work with each other. Such approaches are inefficient, incurring a significant overhead through double handling of data and the opportunity for errors is very prevalent. HEIs are seeking a robust solution for their research management and administration processes that fully interfaces with existing core systems and will result in increased accuracy, efficiencies and economies.

Welland [4] argued that a few institutions had already made progress toward integration of part of their research support. The University of Glasgow has developed a Research System [25] to replace one which had originated from the MAC initiative in 1994. The development commenced in 2004, building on the previous system and adding a web front end. The development project ended in March 2008, although further developments were anticipated in respect of automation of award calculations, electronic document management, linkage to Je-S, extensive data cleansing and further access roles.

The University of Liverpool has developed an Integrated Research Information System [IRIS] [5] to enable research activity to be administered and reported on at all levels. IRIS combined an Oracle Portal front end with approval workflows, email notifications and management information reports. It allowed research administrators, departmental staff and individual academics to manage their research activity on a daily basis, aided internal research collaboration and ensured timely and up to date management information and reporting.

One of the preferred design methodologies for new implementations of a SOA is the Enterprise Service Bus [ESB] [6 and 7]. A service bus acted as a conduit along which information passes. The ESB should connect services together, mediate the data that are transferred and control the Web service network. A user sends a message which will be loaded into the bus, then traverses through any necessary security checks and data transformations on the way before being routed to its final destination.

Application Service Provision [ASP], and Software as a Service [SaaS] have all influenced the more recent emergence of cloud computing [8] and academia was particularly influential in driving grid computing forward. The approach where many individual physical servers are accommodated onto a much smaller number of virtual servers [9] has also proved to be a key enabler in the emergence of cloud computing. Cloud computing allows huge storage and processing capacity to be concentrated in server farms and has also allowed for the sharing of the physical resources. Individual users of cloud-based services may share physical servers but use different virtual servers or virtual machines, thus facilitating the keeping of their details separate and private [10]. There is already significant interest within HE and FE regarding the potential of cloud computing to support education, research and administrative activities [11].

RMAS should have a data model that allows for the collecting, maintaining and exchanging of good quality, comprehensive and current research information. In particular, the use of a standard exchange format, the Common European Research Information Format [CERIF], to improve interoperability of data between the different stakeholders [Funding Councils, Research Councils and other funders, HESA, Higher Education Institutions and other].

Researchers have examined the suitability of CERIF versus other possible standards [12, 13, 14 and 15] and recommended that CERIF as a standard exchange format between the stakeholders. CERIF provides a canonical reference data model at both data and metadata levels. CERIF originated in 1988 but was based on work by several national Research Councils especially in projects IDEAS [16] and EXIRPTS [17]. A Group was formed to formulate CERIF91 [18 and 19].

## 3 RMAS Characteristics

An RMAS will enable the information to be accessible within and outside of HEIs in a way that developments by individual HEIs would be unlikely to achieve alone. The new administrative systems will focus on meeting the needs of end users, particularly academic faculty; a group that are not well supported by many current administrative systems. The use of rules engines and workflow will make it practical to implement more processes and deliver more services online, directly to end users. The following are the four characteristics of the RMAS.

1. **Efficient:** An RMAS would free up research active staff by providing an efficient tool to develop, monitor and progress awards, allowing staff to maximize their outputs and spend more of their time on doing research.
2. **Flexible:** An RMAS should be a flexible system would allow institutions to cope with growth in research activity without the need for increased staffing levels. A flexible RMAS combined with imaginative business policies and processes can be a source of comparative advantage for an institution.
3. **Interoperable:** The system should interoperate with other systems to support processes that cross departments, systems, and institutions.
4. **Scalable:** The ability to support additional users and processes at low incremental cost requires automated, self-service processes that serve end users directly, applying rules accurately, consistently, and with minimum human intervention. Rules engines should be used to ensure that the system is highly scalable and that changes in rules can be implemented in a cost-effective and timely manner.

It is anticipated that an integrated RMAS solution would produce both cashable and non-cashable savings. An estimate of potential savings falls in the range of 10 - 20% of administration costs, which in the case of an institution such as the University of Exeter, equates to £250K – £500K p.a. [20]. A business case developed by [Bolton, 2010] suggested that higher saving of 25-30% could be achieved by adopting CERIF with an integrated RMAS. If 20 HEIs of a similar size to Exeter adopted RMAS, the more conservative estimated efficiency savings over a 5 year period would be in the range of £25m - £50m.

## 4 RMAS Design

A process map was created at the outset of the RMAS project to indicate the scope and broad areas of activity and how these related to interfaces and other activities. This was updated following the responses to a survey and is reproduced below [shown in Appendix 1]. Information from a number of HEIs was gathered after the survey and the diagrams they provided varied from straightforward organisation charts to very complex process maps which showed the interactions of processes across the entire university.

Figure 1 illustrates the system architecture of a UK university. The major challenge for this system is the heterogeneous interfaces and expensive maintenance due to a piecemeal implementation over 15 years. Whilst tightly coupled in some areas the administrator/user has to re-enter the same information manually into different systems and applications in other areas. It is a more or less complete stack of functionality but with different architectures; full integration is still elusive.

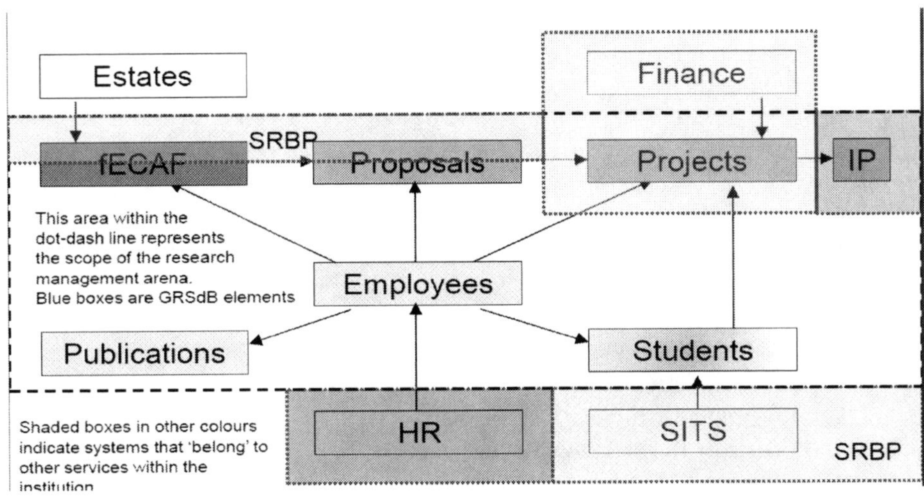


**Fig. 1.** A system architecture of a UK university [based on [26]]

RMAS is currently moved towards a system to reduce development costs. By exposing existing functionality as services, which can be reused in current and future applications, universities can achieve significant cost reduction. The movement towards XML and open Web services standards opened the floor for reuse within and across universities. The initial overhead in creating reusable services can be quickly recouped as soon as other applications are assembled out of these reusable services.

A standards-based, loosely coupled modular design will make it possible to add other applications that either enhance included functionality or add functionality that is not included. For example, a portal will provide a user interface that can be configured to suit a wide variety of users and can present information from other systems. An SOA will make the core functionality on which higher level processes depend available to multiple applications and to other systems. This will ensure that processes that cross traditional departments [and systems] can be supported. A modular design, together with an open source licensing model and the publication of service interface information, will allow both open source and commercial modules to be used with the system.

A demand survey was developed with input from the Steering Group and was undertaken between 21 November and 24 December 2008 using the Bristol Online survey system. All HEIs in England, Scotland, Wales and Ireland were invited to take part. Use was made of the BUFDG, ARMA and UCISA organisations to make HEIs aware of the survey. Follow up communications to the Directors of Finance of the Russell and 1994 Group Universities were sent out, to encourage participation.

The response to the survey was excellent, with 60 replies from 58 institutions. This represented approximately 30% of HEIs by number, but was almost 60% in terms of research turnover based on 2006/07 data. The survey requested information about 15 different elements or modules of research administration, which broadly covered the entire lifecycle of a grant, and might be included in an RMAS. The result of the survey is 8 core modules decided for RMAS, which are connected by an ESB. There are:

1. Academic Expertise
2. Funding Sourcing Tool
3. Proposal Management
4. Costing and Financial Management of Projects Post Award
5. Process Workflow
6. Outcomes & Outputs
7. Electronic Document Management
8. Customer Relationship Management

The RMAS modules are to be connected by an enterprise service bus [ESB] as illustrated in Figure 2, providing services to different types of clients [6]. Those services can be used in specific applications. The physical location of the data is immaterial to the service users. The data exchange between the hub and the local applications uses CERIF XML. Note that many of the existing applications do not provide a data export in CERIF XML format. An important task of data exchange is the translation to CERIF XML document standards for research information.

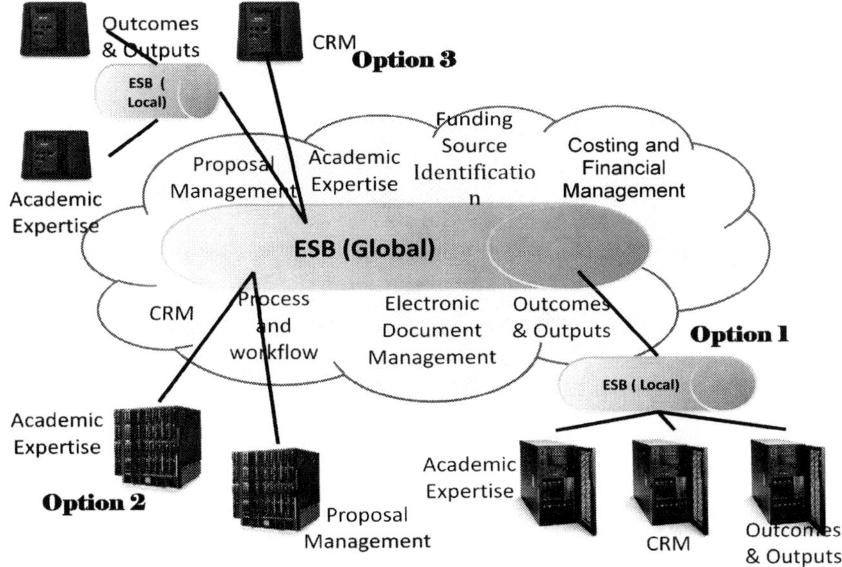


**Fig. 2.** The technical framework

Three strategies have been identified for how to connect local [existing] modules to the global ESB.

1. **Option 1:** Each local function has a local connector which exchanges information with the global ESB
2. **Option 2:** All function is connected to a local ESB which exchanges information with global ESB
3. **Option 3:** A mix of Option 1 and 2.

In order to enable the cooperation of heterogeneous applications an additional software layer, called middleware, is needed. Messages or events are exchanged among systems to integrate data or functions. Those ideas are continued in the enterprise application integration [EAI] approach [22], i.e., independent applications are loosely coupled via middleware.

The middleware layer establishes connections to existing applications using adaptors or connectors, converts different data formats, maps schemas and supports data exchange among different applications [routing, queuing, transaction management]. Enterprise Information Integration [EII] is an integration approach at enterprise level and provides a basis for the implementation of portals or analytical applications [23 and 24]. Integration requires a common understanding of the data shared by different users and applications.

Although so far semantics and ontology technology have not yet been implemented in RMAS, it is this future approach which will enable semantic interoperability among modules, both locally and globally. Ontologies will be applied to determine how to map the schemas and values used in one database, or by one application system, into schemas and values used in other databases, or by other application systems.

The same paradigm applies to unstructured data. A search for grants proposals in one set of documents may return the names of parties who have responded to application or contacts, whether or not they have succeeded in the application. A search for applicants in a different set of documents may return only names of parties who have made an applied for a certain funder in the last five years. When comparing the two lists of names, the enterprise ontology might either resolve the semantic discrepancies, or else make those discrepancies known, as part of the result set.

## 5 Conclusion

This paper describes the major component in the development of a shared services solution, where a group of HEIs are working together to specify an RMAS. The development of this RMAS will seek to take advantage of Service Oriented Architecture [SOA], cloud computing and ESB. The system should make it easier for end users to do the things they need to do by anticipating their needs; helping them make choices, set goals, and track their progress and reducing the time it takes them to complete administrative tasks.

The RMAS will allow systems to share data, eliminating redundant data entry and storage. A study [27] has shown that academic staff and research managers and administrators believe that an RMAS can increase both the quality and quantity of research undertaken. Staff will have more time to support people and better tools for providing that support. The project will in addition use CERIF [28] as a prerequisite for efficient data sharing.

This system will also reduce the number of enquiries made to staff. Because it applies the relevant rules accurately and consistently, its use will increase the number of staff able to provide accurate up to date information about research. Further, this research information can then be better used for management, planning and strategy decision by university senior management.

**Acknowledgement.** The authors thank JISC and HEFCE for providing funding and support for this project and everyone who supported us during the development of this paper.

## References

[1] JISC, Research information management Developing tools to inform the management of research and translating existing good practice (2010), http://www.researchdatatools.com/downloads/2010-research-information-management-2.pdf (last viewed June 28, 2011)

[2] Binge, A.: Research Management and Administration System-feasibility study. University of Exeter (2009)

[3] JISC (2011), http://www.jisc.ac.uk/whatwedo/programmes/elearningframework.aspx (last viewed June 24, 2011)

[4] Welland, D.: RMAS-UCISA CISG Research Management Systems Seminar, Research Management Systems Seminar. University of Liverpool (2010), http://www.ucisa.ac.uk/en/groups/cisg/Events/2010/researchmanage/~/media/groups/cisg/events/2010/RMS/RMS05_Deborah_Welland.ashx (last viewed June 27, 2011)

[5] Keggin, N.: Research Management as an Integrated Research System, Research Management Systems Seminar, University of Liverpool (2010), http://www.ucisa.ac.uk/en/groups/cisg/Events/2010/researchmanage/~/media/groups/cisg/events/2010/RMS/RMS06_Nikki_Keggin_Kathy_May.ashx (last viewed June 27, 2011)

[6] Chappell, D.A.: Enterprise Service Bus. O'Reilly & Associates, Sebastopol (2004)

[7] Gilpin, M., Vollmer, K.: The Forrester Wave: Enterprise Service Bus, Q4 2005, PDF document (2005), http://www.capeclear.com/esbwave/ (retrieved online February 20, 2006)

[8] Gruman, G.: What cloud computing really means. InfoWorld (2008), http://www.infoworld.com/d/cloud-computing/what-cloud-computing-really-means-031 (retrieved June 02, 2011)

[9] Hein, T.: BEING GREEN Comes Naturally. Baseline 91, 40–42 (2008)

[10] Armbrust, M., Fox, A., Griffith, R., Joseph, A., Katz, R., Konwinski, A., Lee, G., Patterson, D., Rabkin, A., Zaharia: A view of cloud computing. Communication of the ACM 53(4), 50–58 (2010)

[11] JISC, Invitation to Tender: Technical Review of Cloud Computing for Research, p. 1–14 (2009)

[12] Natchetoi, Y., Wu, H., Babin, G., Dagtas, S.: EXEM: Efficient XML data exchange management for mobile applications. Information Systems Frontiers 9, 439–448 (2007), http://portal.acm.org/citation.cfm?id=1285888 (last viewed June 28, 2011)

[13] Price, D., Caddick, S.: How to stay on top. Times Higher (August 5, 2010), http://www.timeshighereducation.co.uk/story.asp?storycode=412909 (last viewed June 28, 2011)

[14] Rogers, N., Ferguson, N.: Exchanging Research Information in the UK. EXRI-UK: A study funded by JISC (2009), http://ie-repository.jisc.ac.uk/448/1/exri_final_v2.pdf (last viewed June 28, 2011)

[15] Joerg, B., van Grootel, G., Jeffery, K.: CERIF 2008 1-1 XML Data Exchange Format Specification (2008), http://www.eurocris.org/fileadmin/cerif-2008/CERIF2008_1.1_ (last viewed June 28, 2011)

[16] Jeffery, K., Lay, J., Miquel, J.-F., Zardan, S., Naldi, F., Vannini-Parenti, I.: IDEAS: A System for International Data Exchange and Access for Science. Information Processing and Management 25(6), 703–711 (1989)

[17] Naldi, F., Jeffery, K., Bordogna, G., Lay, J., Vannini-Parenti, I.: A Distributed Architecture to Provide Uniform Access to Pre-Existing Independent, Heterogeneous Information Systems RAL Report 92-003

[18] van Woensel, L.: Towards harmonisation of databases on research in progress – Final report of the European Working Group on Research Databases (November 1988b); Published by the Liaison Committee of Rectors' Conferences of Member States of the European Communities and Directorate General for Science, Research and Development of the Commission of the European Communities; financed by the Commission of the E.C., contract PSS*0058/B, compiled by Dr. L. Van Woensel

[19] Skjellaug, B.: Temporal Data: Time and Relational Databases. University of Oslo (1997)
[20] Welland, D.: Research Management and Administration System, University of Exeter (2009), http://www.exeter.ac.uk/media/universityofexeter/research/rmas/documents/RMAS_Part_2_eeport_v2_1_070110.doc (last viewed June 22, 2011)
[21] Bolton, S.: The Business Case for the Adoption of a UK Standard for Research Information Interchange. Report to JISC (2010), http://www.jisc.ac.uk/media/documents/publications/reports/2010/Businesscasefinalreport.pdf (accessed July 31, 2011)
[22] Linthicum, D.S.: Enterprise Application Integration. Addison Wesley, Reading (1999)
[23] Deßloch, S., Maier, A., Mattos, N., Wolfson, D.: Information Integration – Goals and Challenges. In: Datenbank-Spektrum, vol. 6. dpunkt-Verlag (2003)
[24] Sheth, A.P., Larson, J.A.: Federated Database Systems for Managing Distributed, Heterogeneous and Autonomous Systems. ACM Comp. Surveys 22(3) (1990)
[25] McCutcheon, V.: A Working Research Management and Administration System, Research Management Systems Seminar, University of Liverpool (2010), http://www.ucisa.ac.uk/en/groups/cisg/Events/2010/researchmanage/~/media/groups/cisg/events/2010/RMS/RMS04_Valerie_McCutcheon.ashx (last viewed June 28, 2011)
[26] Kerridge, S.: Electronic Research Administration: A case study from a non research intensive university in the UK. International Network of Research Managment Societies [INORMS] Congress, Cape Town, INORMS / SARIMA / ACU (2010)
[27] Kerridge, S.: What do Academic Staff think of Research Management and Administration? In: ARMA 2011 Poster, Glasgow (2011)
[28] Joerg, B.: CERIF 2008 v1.2 (2011), http://www.eurocris.org/Index.php?page=CERIF2008&t=1 (retrieved April 22, 2011)

# Appendix1: System Overview

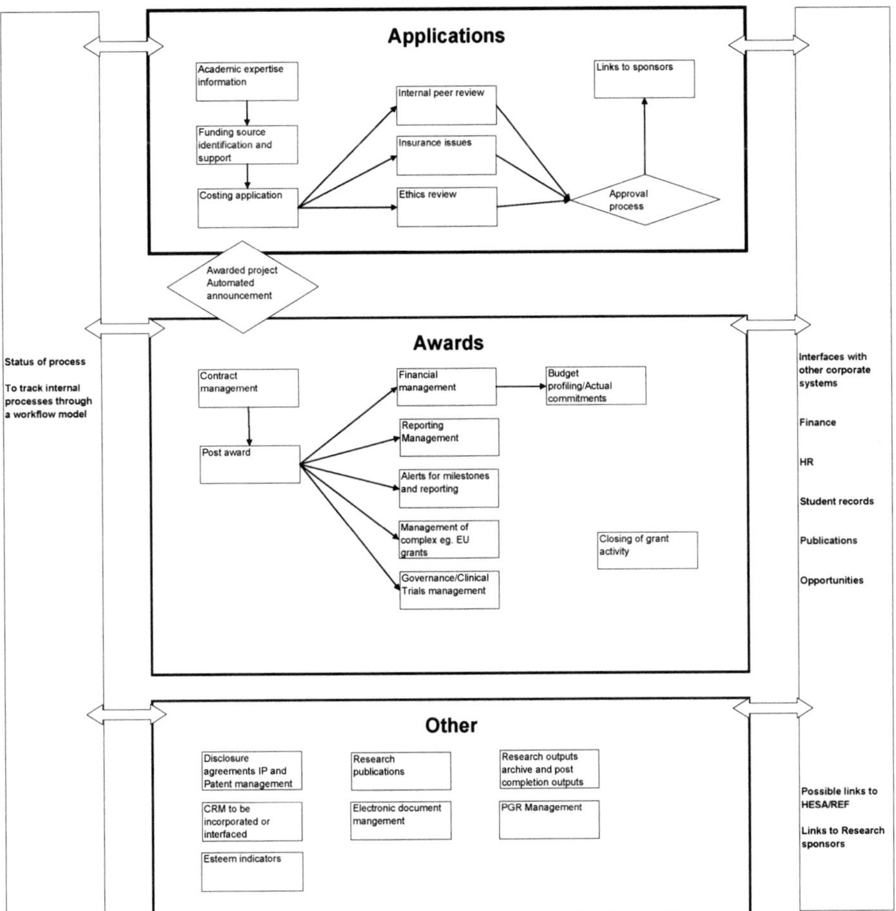

# Towards a Sharable Research Vocabulary (SRV) – A Model-Driven Approach –

Brigitte Jörg[1], Keith Jeffery[2], and Geert van Grootel[3]

[1] Deutsches Forschungszentrum für Künstliche Intelligenz – Bereich Sprachtechnologie (DFKI GmbH, LT), Alt Moabit 91c, 10559 Berlin, Germany
brigitte.joerg@dfki.de

[2] Science and Technology Facilities Council (STFC), Rutherford Appleton Laboratory, Didcot, Oxfordshire OX11 0QX, UK
keith.jeffery@stfc.ac.uk

[3] Vlaamse Overheid – Department Economie, Wetenshap en Innovatie, Koning Albert II-laan 35, bus 10, B-1030 Brussel, Belgium
Geert.VanGrootel@ewi.vlaanderen.be

**Abstract.** It is known, that research is a key driver of innovation. Not only are governments interested in ongoing activities and outcomes, but also managers, researchers, the media, and increasingly society as such. Research Information is naturally stored in information systems by means of an underlying conceptual model to describe the system entities by their attributes and relationships. So-called 'conceptual models' do usually not employ the vocabularies or rules required at application levels. Where research information systems are gaining ground, they are recognized as valuable for maintaining multiple vocabularies, in support of inter-system operation, information integration and re-use, and for the better understanding of Research as a domain and thus its ontological foundation. We propose a model-driven framework for developing a sharable Research vocabulary based on CERIF[1], which scales to serving multiple stakeholders.

**Keywords:** Research Information, CERIF, Meaningful Metadata, Model-driven Metadata, Metadata and Semantics, Information Systems.

## 1 Introduction

Research Information is naturally stored in information systems by means of an underlying conceptual model [32, 33, 34] to describe the system entities by their attributes and relationships. So-called 'conceptual models' do usually not employ the vocabularies or rules required at application levels. Where research information systems are gaining ground, they are recognized as valuable [5, 7, 11, 12, 15, 20, 23, 28, 29] for maintaining multiple vocabularies, in support of inter-system operation, information integration, re-use and access, and for the better understanding of Research as a domain

[1] CERIF – the Common European Research Information Format – a EU Recommendation to Member States: http://cordis.europa.eu/cerif/, http://www.euroCRIS.org/

E. García-Barriocanal et al. (Eds.): MTSR 2011, CCIS 240, pp. 256–268, 2011.

and thus, its ontological foundation [9, 10, 31]. The method we present for vocabulary development is inline with the paradigm of Model-Driven Architectures (MDA) towards system-specifications and interoperability based on formal models. It specifies three viewpoints on a system, where the initial one is computer-independent (CIM), followed by an expression in a platform-independent model (PIM) language that is translated into platform-specific models (PSMs) [27]. We propose CERIF [19] as the underlying (PIM) model towards a sharable Research vocabulary (SRV), which scales to serving multiple stakeholders. CERIF maintains (a) a holistic concept of Research and (b) supplies the means to manage the semantics in the Research context.

Various vocabularies have been developed, and some have been in use for more than a century, and still are. The history clearly refers to the 'Library and Information Science' (LIS) area maintaining classification systems of science and knowledge organization systems. The Web and its technologies enabling networked systems proliferate the production of knowledge organization systems beyond the traditional library environment into research, markets and society, where inter-linking calls for formalization and standardization [1, 4, 6, 8, 12, 16, 21, 24, 29], and hopefully conceptualization, before agreement [3, 26] and interoperation. We think, that a SRV essentially requires two things: A formal syntax and a declared semantics. With CERIF, the formal syntax is supplied by the conceptual model entities and the structure inherent in their representation and inter-linkage, the declared semantics is supplied by its conceptual sub model – the 'semantic layer' – first introduced in [17].

We present the Research domain by introducing the CERIF model in section 2 and present the 'semantic layer' as a host and guide to multiple vocabularies in section 3, before we map three well-established and known formats, SKOS, SBVR, Dublin Core to CERIF in section 4, to validate the CERIF model's intension by extension. Within section 5 we demonstrate a real-life SBVR-driven CERIF vocabulary development implementation by Collibra, and conclude by directing to ontologies and linked data.

## 2 The CERIF Model and Some History

The Common European Research Information Format (CERIF) is a conceptual model of the Research domain; it identifies participating entities and describes them by their basic attributes and via relationships they maintain with each other. In CERIF there exist 'base entities' like person, organization, project; without them research is hardly possible. The base entities produce 'result entities' like publication, patent, or product. Results can only be achieved in an appropriate environment, which in CERIF terms is '2$^{nd}$ level entities', like equipment, funding, or facility. On top, CERIF allows for multilingual field descriptions of names, keywords, or abstract. Figure 1 (CIM) shows base and result entities surrounded by 2$^{nd}$ level entities and indicates relationships as lines. In running CERIF implementations (PSM) or Current Research Information Systems (CRISs), relationships are maintained in 'link entities'.

So-called 'link entities' in CERIF are underspecified binary relationships, where the linking names or roles are not part of the model as such, but maintained as concepts or vocabulary terms within the 'semantic layer' [17, 18, 19]. The highly normalized structure of the semantic layer (Figure 3) allows for the maintenance of multiple vocabularies, at best driven through CERIF 'link entities', and is very flexible. Figure 2 presents a typical link entity structure in column 1 and some

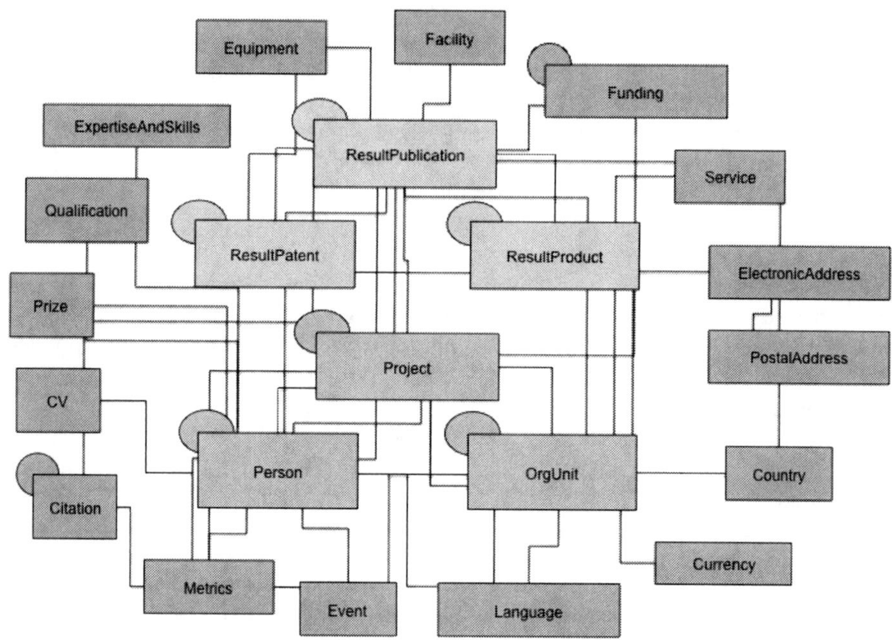


**Fig. 1.** CERIF Entities and their Relationships

populated link entity examples in columns 2-4. A CERIF link entity is underspecified in the conceptual model because it only indicates the two linked entities in its name without an explicit role or label, but refers to a relationship term via cfClassID/cfClassSchemeID instead. Each role in a link entity belongs to a classification scheme and is enhanced by time-stamps. This linking mechanism allows for multiple roles/labels, classes or terms to be assigned with multiple classification schemes and is indicated in columns 2-4.

| **cfEntity1Name_Entity2Name** | cfPers_Proj | cfPers_OrgUnit | cfPers_OrgUnit |
|---|---|---|---|
| **cfEntity1ID**<br>**cfEntity2ID** | cfPersID<br>cfProjID | cfPersID<br>cfOrgUnitID | cfPersID<br>cfOrgUnitID |
| **cfClassID**<br>**cfClassSchemeID** | coordinator-uuid<br>example-scheme-uuid | affiliation-uuid<br>example-scheme-uuid | cerif-tg-leader-uuid<br>example-scheme-uuid |
| **cfStartDate**<br>**cfEndDate** | 2005-04-01<br>2007-11-31 | 2000-08-15<br>2038-12-31 | 2005-05-01<br>2038-12-31 |

**Fig. 2.** Generic CERIF link entity structure in column 1, and some CERIF model link entities in column 2-4

CERIF is maintained as an Entity-Relationship-Model (ERM); in MDA terms, a PIM, transformable to multiple database languages (PSMs), where table names are prefixed with 'cf' and shortened to ensure cross-database support; e.g. cfPers is the short name of a person entity. CERIF also specifies a model-driven interoperation format through CERIF XML according to W3C recommendation [4].

CERIF and CRIS initiatives are not new. The first serious efforts for international cooperation among research information systems date back to the Seventies, where a UNESCO/ICSU[2] committee was set up to investigate the situation and UNISIST had published a "Study report on the feasibility of a world science information system" [22]. CERIF as a standard format and EU recommendation to Member States arose from a Conference of European University Rectors and in parallel, heads of research funding organizations of G7 countries [16]. Since 2002, CERIF has been entrusted to the care of euroCRIS[3], and improved and updated continuously. In the meanwhile, it is adopted in national systems of eight European countries, widely used in others, and considered internationally. As an example, in UK there are several ongoing publicly-funded projects promoting the utilization of CERIF for interoperation between universities, research funders, innovators and others with an emphasis on research evaluation.

## 3 The CERIF 'Semantic Layer'

Figure 3 shows the 'semantic layer' entities (PIM). The heart of the semantic layer is the classification entity (cfClass) tied to a scheme (cfClassScheme) via foreign key (PFK) to preserve a system unique identification. It additionally requires dates and allows for a URI (cfURI), should class requirements be time-constraints or uniform identification. The cfClassId identifies a class or concept to which multiple terms, descriptions, definitions, examples can refer (cfClassTerm, cfClassDescr, cfClassDef, cfClassEx), in many languages.

A recursive classification entity, the cfClass_Class link entity, allows for conceptual mappings or relationship kind variations within and across multiple classification systems or schemes, such as synonym; broader term. A class scheme is identified by its cfClassSchemeId, has a name and allows for a description,

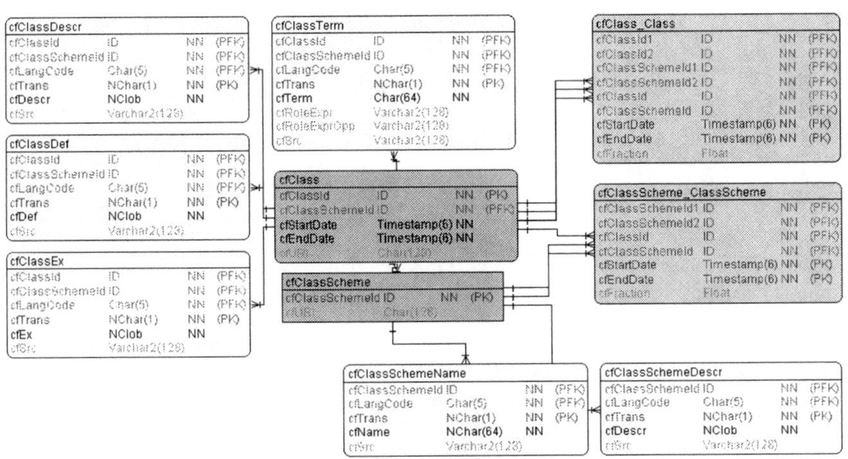


**Fig. 3.** CERIF Semantic Layer Entities

[2] UNESCO/ICSU: United Nations Educational, Scientific and Cultural Organization/ Internationals Council for Science

[3] euroCRIS: `http://www.eurocris.org/`

cfClassSchemeDescr. The presented 'semantic layer' in figure 3 incorporates CERIF task group-agreed features next to being officially released; each therein contained multilingual entity allows for a cfSrc attribute to inform about the source of the term, description, definition, example or name. The term entity has been enhanced by the addition of two essential attributes: role and role-expression-opposite. We understand, that a term describes the function of a role or the properties of a role associated with an entity as indicated in table 1, where in a person-publication relationship the role "authoring" leads to a role expression (cfRoleExpr) "is author of", and inversely (cfRoleExprOpp) "is authored by", and where the property in the person entity is "author", and in the publication entity it is "authored".

**Table 1.** CERIF term, role and expression examples

| CERIF Link Entity | Term | Role | Relationship 1 | Relationship 2 |
|---|---|---|---|---|
| cfPers_ResPubl | **Author** | authoring | is author of | is authored by |
| cfPers_OrgUnit | **Commissioner** | commissioning | is commissioner at | is commissioned by |
| cfPers_OrgUnit | **Manager** | managing | is manager of | is managed by |
| cfProj_Pers | **Manager** | managing | is manager of | is managed by |
| cfProj_Fund | **Pending** | pending | is pending | is pending for |
| cfOrgUnit_OrgUnit | **Part** | (is-a) | is part of | has part |
| cfOrgUnit_OrgUnit | **Funder** | funding | is funded by | is funder of |
| cfOrgUnit_OrgUnit | **Member** | membership | is member of | has member |
| cfOrgUnit_OrgUnit | **Acquisition** | acquiring | was acquired by | has acquired |

In CERIF, roles are always inaugurated from functional or operational relationships in link entities, as indicated in column 1 of the above table, introduced in figure 2, and demonstrated in figure 4 through an ERM extract of some CERIF entities.

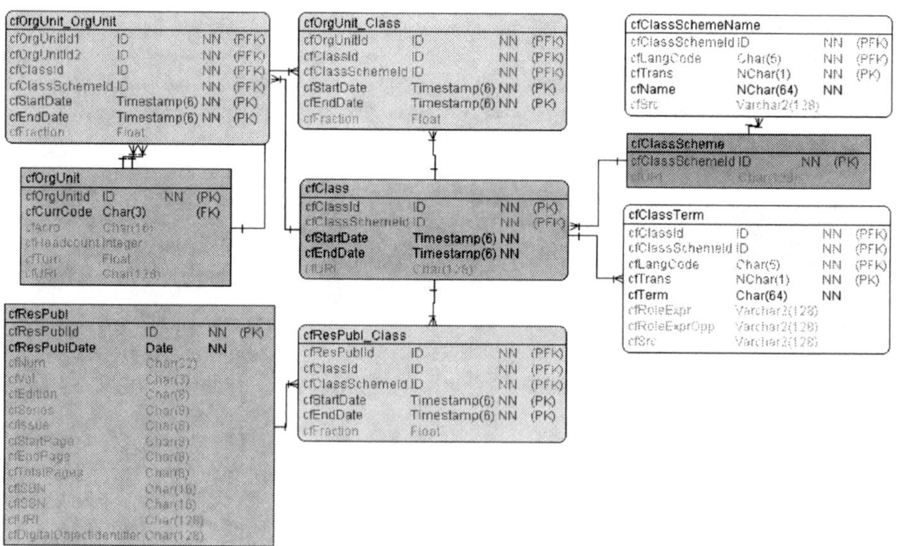


**Fig. 4.** Some CERIF Entities and Link Entities – an ERM Extract

The latest CERIF release 2008–1.2 maintains vocabularies such as *organizational-types* through cfOrgUnit_Class: 'Academic Institution', 'University', 'Government', or *organizational-structure* through cfOrgUnit_OrgUnit: 'Member', 'Acquisition', 'Funder', 'Spin-Off'. A CERIF XML example representation for organizational types is available with figure 5, showing the semantically declared xml organization type concepts for "Academic Institution" and "University". Each concept (cfClass) has its identifier (cfClassId), refers to its corresponding scheme (cfClassSchemeId), allows for a URI (which includes the scheme name as a prefix in the example), and is valid during a time period. The label or name of a concept (cfClassTerm) refers to the concept identifier and classification scheme, and can be multilingual (cfTerm); the same holds for a description (cfClassDescr) and (cfDescr).

```
<?xml version="1.0" encoding="UTF-8"?>
<CERIF release="2008-1.2 Organisation Types (Semantics)"
    date="2011-06-13"
    sourceDatabase="euroCRIS: CERIF Release 2008 - 1.2">
   .....
   <cfClass>
    <cfClassId>cfacademicinstitution-uuid</cfClassId>
    <cfClassSchemeId>cforganisationtypes-2008-1.2-uuid</cfClassSchemeId>
    <cfStartDate>2010-11-02</cfStartDate>
    <cfEndDate>2038-12-31</cfEndDate>
    <cfURI>CERIF-2008-1.2-cfOrganisationTypes#Academic_Institution</cfURI>
   </cfClass>
   <cfClassTerm>
    <cfClassId>cfacademicinstitution-uuid</cfClassId>
    <cfClassSchemeId>cforganisationtypes-2008-1.2-uuid</cfClassSchemeId>
    <cfTerm cfLangCode="en" cfTrans="o">Academic Institution</cfTerm>
   </cfClassTerm>
   <cfClassDescr>
    <cfClassId>cfacademicinstitution-uuid</cfClassId>
    <cfClassSchemeId>cforganisationtypes-2008-1.2-uuid</cfClassSchemeId>
    <cfDescr cfLangCode="en" cfTrans="o">Academic institution is an educational institution dedicated
        to education and research, which grants academic degrees.</cfDescr>
   </cfClassDescr>
   <cfClass>
    <cfClassId>cfuniversity-uuid</cfClassId>
    <cfClassSchemeId>cforganisationtypes-2008-1.2-uuid</cfClassSchemeId>
    <cfStartDate>2010-11-02</cfStartDate>
    <cfEndDate>2038-12-31</cfEndDate>
    <cfURI>CERIF-2008-1.2-cfOrganisationTypes#University</cfURI>
   </cfClass>
   <cfClassTerm>
    <cfClassId>cfuniversity-uuid</cfClassId>
    <cfClassSchemeId>cforganisationtypes-2008-1.2-uuid</cfClassSchemeId>
    <cfTerm cfLangCode="en" cfTrans="o">University</cfTerm>
   </cfClassTerm>
   <cfClassDescr>
    <cfClassId>cfuniversity-uuid</cfClassId>
    <cfClassSchemeId>cforganisationtypes-2008-1.2-uuid</cfClassSchemeId>
    <cfDescr cfLangCode="en" cfTrans="o">A university is an institution of higher education and
        research, which grants academic degrees in a variety of subjects. A university is a
        corporation that provides both undergraduate education and postgraduate education.</cfDescr>
   </cfClassDescr>
   ...
</CERIF>
```

**Fig. 5.** CERIF XML Vocabulary Extract from Organizational Types

## 4 Related Work and Mapping

Available classification systems such as thesauri, taxonomies, classification schemes, topic maps, ontologies and subject headings share similar elements. Knowledge organization formats like SKOS – the Simple Knowledge Organization System, are recommended by W3C and increasingly used. The Open Management Group (OMG) specified a vocabulary format, allowing for preservation of semantics: SBVR – the Semantics of Business Vocabulary and Rules, which is currently used for CERIF vocabulary developments.

SKOS aims to overcome legacy structures by defining a common data model for sharing and linking knowledge organization systems via the Web, based on RDF [21] (PIM). We investigate the declared classes and properties under the SKOS namespace [24] (PIM) and map them to CERIF 'semantic layer" entities (PIM) in table 2. SBVR constructs aim at enhancing vocabulary descriptions by terminology entries, definitions, and additional specifications such as notes and examples [25]. The SBVR 'Business Vocabulary" distinguishes so-called semantic communities (CIM) by their shared understanding of concepts, while speech communities use language expressions in vocabularies stored as a term dictionary (PIM) inline with the concepts of the semantic community (CIM). In SBVR terms, Research is a semantic community (CIM); the CERIF entities, types and roles are speech communities (PIM), each maintaining vocabularies that can be structured inline with the CERIF model, like cfOrgUnit-OrgUnit-2008-1.2-Roles, cfPerson-Person-2008-1.2-Roles, and where a conceptual term (PIM) corresponds to the global dictionary (PIM).

By investigating widespread and elaborated formats of comparable dimensions, we aim at validating the CERIF model's intension by extensions. Because of its formal syntax and declared semantics (PIM), CERIF can store – in the 'semantic layer' – terms in structural relationships emulating dictionaries, lexicons, thesauri and domain ontologies. Moreover, since CERIF exhibits at entity instance level a triple structure (e.g. Person A is author of Publication X; Term A is synonym of Term B), all the usual logical processing operations are sustainable. Thus, CERIF becomes a superset over semantic stores such as dictionaries, thesauri, or ontologies and a mapping to CERIF allows resolution of conflicts in term representation and meaning.

With table 2 we present a (CIM) mapping from SKOS (PIM) to CERIF (PIM), but will not go into details towards mapping additional constraints like range and domain, or inverse property definitions in SKOS.

**Table 2.** Basic SKOS to CERIF Mapping

| SKOS Types | SKOS Entity | SKOS-CERIF-Mapping (CIM) | CERIF Entity |
|---|---|---|---|
| Class | Collection | is type of | cfClassScheme |
| Class | Concept | is a | cfClass |
| Class | ConceptScheme | is a | cfClassScheme |
| Class | OrderedCollection | is type of | cfClassScheme |
| lexical | altLabel | is term of class in lexical scheme | cfTerm |
| mapping | broadMatch | is term of class in mapping scheme | cfTerm |

**Table 2.** (*Continued*)

| | | | |
|---|---|---|---|
| semantic rel. | broader | is term of class in semantic rel. scheme | cfTerm |
| semantic rel. | broaderTransitive | is term of class in semantic rel. scheme | cfTerm |
| docu | changeNote | is time-stamped descr. in new/old class | cfClassDescr |
| mapping | closeMatch | is term of class in mapping scheme | cfTerm |
| docu | definition | is a | cfDef |
| map prop | editorialNote | is time-stamped descr. in new/old class | cfClassDescr |
| | | is descr. in mapping scheme | cfClSchDescr |
| | | is cerif publication record reference | cfResPubl |
| docu | example | is a | cfEx |
| conc schemes | hasTopConcept | is term of class in concept scheme | cfTerm |
| lex label | hiddenLabel | is term of class in lexical scheme | cfTerm |
| docu | historyNote | is time-stamped descr. in new/old class | cfClassDescr |
| | | is descr. in docu scheme | cfClSchDescr |
| | | is cerif publication record reference | cfResPubl |
| conc schemes | inScheme | inherent cerif linkage (mandatory) | cfClSchID |
| map prop | mappingRelation | is term of class in mapping scheme | cfTerm |
| conc coll. | member | is term of class in conc coll. scheme | cfTerm |
| conc coll. | memberList | is term of class in conc coll. scheme | cfTerm |
| map prop | narrowMatch | is term of class in mapping scheme | cfTerm |
| semantic rel | narrower | is term of class in semantic rel. scheme | cfTerm |
| semantic rel | narrowerTransitive | is term of class in semantic rel. scheme | cfTerm |
| notations | notation | is | cfClassScheme |
| docu | note | is time-stamped descr. in new/old class | cfClassDescr |
| | | is descr. in docu scheme | cfClSchDescr |
| | | is cerif publication record reference | cfResPubl |
| lex label | prefLabel | is tem of class in lexical scheme | cfTerm |
| semantic rel | related | is term of class in semantic rel. scheme | cfTerm |
| map prop | relatedMatch | is term of class in mapping scheme | cfTerm |
| docu | scopeNote | is time-stamped descr. in new/old class | cfClassDescr |
| | | is descr. in docu scheme | cfClSchDescr |
| | | is cerif publication record reference | cfResPubl |
| semantic rel | semanticRelation | is term of class in semantic rel. scheme | cfTerm |
| conc schemes | topConceptOf | is term of class in concept schemes | cfTerm |

SKOS and SBVR refer to the semantic layer of CERIF. Subsequently, we want to demonstrate the power of the entire CERIF model for vocabulary developments by taking into account the Research context. We consider the Dublin Core format (PIM) an appropriate candidate for demonstrating a (CIM) mapping towards the CERIF publication entity (PIM) in context. The Dublin Core Metadata Initiative developed first a 13 and subsequently a 15 element set of metadata elements to describe WWW resources based on XML. It has found wide adoption, especially in digital repositories and particularly for resource discovery. The Dublin Core elements are presented in the table 3 through a CERIF mapping; here we will not investigate qualified Dublin Core.

**Table 3.** Dublin Core – CERIF Publication Entity Mapping

| DC Element | DC Resource - CERIF Publication Entity Mapping (CIM) | CERIF Entity |
|---|---|---|
| Title | is attribute in cerif multilingual entity cfResPubTitle | cfTitle |
| Creator | is pers role in DC scheme of cerif link entity cfPers_ResPubl | cfTerm |
| | is org role in DC scheme of cerif link entity cfOrgUnit_ResPubl | cfTerm |
| Subject | is subject role in DC scheme of cerif link entity cfResPubl_Class | cfTerm |
| Description | is attribute in cerif multilingual entity cfResPublDescr | cfDescr |
| Publisher | is org role in DC scheme of cerif link entity cfOrgUnit_ResPubl | cfTerm |
| Contributor | is pers role in DC scheme of cerif link entity cfPers_ResPubl | cfTerm |
| | is org role in DC scheme of cerif link entity cfOrgUnit_ResPubl | cfTerm |
| Date | is attribute in cerif publication entity cfResPubl | cfResPublDate |
| Type | is type role in DC scheme of cerif link entity cfResPubl_Class | cfTerm |
| Format | is format role in DC scheme of cerif link entity cfResPubl_Class | cfTerm |
| Identifier | is attribute in cerif publication entity cfResPubl | cfResPublId |
| | it may be a | cfURI |
| | it may be a | cfISBN |
| | it may be a | cfISSN |
| Source | is an attribute within multilingual semantic layer entities | cfSrc |
| Language | is a language code identifier in cerif multilingual entities cfResPublTitle, cfResPublSubtitle, cfResPublAbstr | cfLangCode |
| Relation | in cerif, a publication has relationships with entities: - publication | cfResPubl_ResPubl |
| | - person (relationship roles maintained as cfTerms via) | cfPers_ResPubl |
| | - organization ( " ) | cfOrgUnit_ResPubl |
| | - project ( " ) | cfProj_ResPubl |
| | - event ( " ) | cfResPubl_Event |
| | - facility ( " ) | cfResPubl_Facil |
| | - equipment ( " ) | cfResPubl_Equip |
| | - product ( " ) | cfResPubl_ResProd |
| | - patent ( " ) | cfResPubl_ResPat |
| | - classification ( " ) | cfResPubl_Class |
| | - citation ( " ) | cfResPubl_Cite |
| | - metrics ( " ) | cfResPubl_Metrics |
| Coverage | a geographic coverage is country code reference in cfCountry | cfCountryCode |
| | a temporal coverage is managed by cfStartDate/cfEndDate in | *cfLinked_Entities* |
| Rights | a description of access restrictions is discussed as a | cfScheme/Layer |
| | a claimed rights of creator is discussed as a | cfClassScheme |

Dublin Core as a metadata standard for research information suffers some severe disadvantages. It has concepts such as 'creator' and 'contributor', which are better considered as roles of 'person' or 'organization', and the very abstract concept of 'relation(ship)'. A criticism of Dublin Core e.g. [2, 14] concerned mainly the lack of formal syntax and declared semantics, and led to (a) the adoption of qualified Dublin Core by some implementers utilizing namespaces to remove ambiguity of synonym terms under different elements; (b) more recent attempts to utilize RDF (Resource Description Framework) for Dublin Core so coding the elements as triples and thus amenable to logic processing. Over a period of 10 years or more, the Dublin Core experts have approached the structure of CERIF.

With the latest release CERIF 2008–1.2 [19], several vocabularies have been published. Related approaches to standardizing research information have recently been started by CASRAI[4] where the primary entities of interest are currently person and activity (comparable to CERIF project), with organization and infrastructure to be added. An ontology to enable networking of scientists, VIVO, is in use among seven partner institutions at Cornell University, to the benefit of scientists; and a controlled vocabulary is used to standardize and enhance information retrieval for a wider range of users [5]. VIVO is grounded in the Semantic Web world and employs formats like SKOS, BIBO, FOAF, Dublin Core, Geopolitical ontology and Event ontology.

## 5 Model-Driven CERIF Vocabulary Development with BSM

In order to organize and facilitate collaborative development of sharable vocabularies, the CERIF task group has transferred the model-driven vocabulary development to the Business Semantics Glossary (BSG), an advanced implementation of the OMG standard SBVR by Collibra[5], inline with the Business Semantics Methodology (BSM) in figure 6. The BSM is essentially constituted of a set of complementary cycles aimed at creating a set of consolidated language neutral semantic patterns for application in a variety of semantic environments. The BSM allows for a strong community driven development of a shared conceptual model.

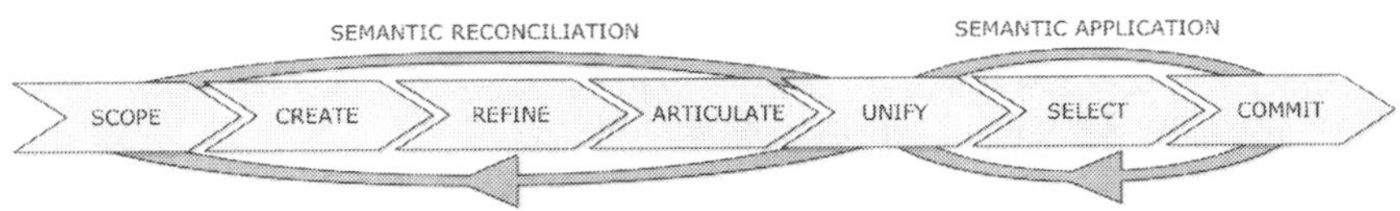


**Fig. 6.** Business Semantics Management Overview: Semantic Reconciliation and Application.

EWI[6] currently co-develops a set of end user services focused around import and export of vocabularies, classification and categorization management, and complex rule building facilities, for exchange of thesauri in SKOS or ISO5964 [13] or as CERIF sub model types upon request by end users, but with coherent semantic mapping of relation expressions between thesauri terms. The classification and categorizations services are primarily designed for the handling of coded classification schemes, including history and version management, as well as mapping across versions. The classification schemes and semantic mappings can be exported into above mentioned formats, RDF or OWL. EWI chooses this approach in order to overcome the direct coupling that very often exists between the physical and logical representation of business vocabularies and the absence of a coherent and well-modeled conceptual layer. The entire semantics management is thus decoupled

---

[4] The Consortia Advancing Standards in Research Administration Information (CASRAI): `http://casrai.org/`

[5] Collibra: `http://www.collibra.com/`

[6] Department of Economy, Science and Innovation, Flemish Government (EWI) `http://www.ewi-vlaanderen.be/en/`

from the physical representations, where the latter can be generated on demand by the customer. EWI is currently focusing on the creation of a CERIF meta model within the BSG, that will form the foundation for the semantic mapping of different existing models in the Flanders Research Information Space. It is the aim to have a CERIF meta model as the central point of reference for all semantic mappings in the RI domain in Flanders

## 6 Conclusion

The presented framework, vocabulary examples and mappings show the complexity and ambiguity inherent in vocabularies. A model-driven approach to supply a formal syntax and declared semantics, supports the understanding of Research as a domain, typical processes and the entities concerned, and supports development of appropriate role and type vocabularies for multiple application contexts and stakeholders. We consider a model-driven approach the necessary underlying steps towards sustainable ontological foundations of the Research domain and thus valuable exposure of linked data.

**Acknowledgements.** We wish to thank Stijn Christaens from Collibra for supplying the Business Semantic Glossary application to the CERIF TG and thus supporting further CERIF vocabulary development. Additionally, we wish to thank DFKI for support and enabling activities within euroCRIS currently via META-NET through the T4ME grant agreement no. 249119. Last but not least, we wish to thank all CERIF TG actives for highly valuable discussions and results.

## References

1. Alexander, K., Cyganiak, R., Hausenblas, M.: Describing Linked Datasets – On the Design and Usage of void, the "Vocabulary of Interlinked Datasets". In: Proceedings of Linked Data on the Web Workshop LDOW 2009, Int. WWW Conference (2009)
2. Asserson, A., Jeffery, K.: Research Output Publications and CRIS. In: Nase, A., Van Grootel, G. (eds.) Proceedings of CRIS 2004, pp. 29–40. Leuven University Press, Leuven (2004)
3. Batini, C., Lenzerini, M., Navathe, S.B.: A Comparative Analysis of Methodologies for Database Schema Integration. ACM Computing Surveys 18(4), 323–364 (1986)
4. Bray, T., Paoli, J., Sperberg-McQueen, C.M., Maler, E., Yergeau, F.: Extensible Markup Language (XML) 1.0, 5th edn. W3C Recommendation (November 2008)
5. Carey, J.: Faculty of 1000 and VIVO: Invisible Colleges and Team Science. Issues in Science and Technology Librarianship (Spring 2011)
6. Dublin Core Metadata Element Set: [DCMI Recommendation] standardized as ISO 15836: 2009. Dublin Core Metadata Initiative (2009)
7. DG General Expert Group: Assessing Europe's University-Based Research. DG General for Research, European Commission (2010)
8. Fugazza, C., Dupke, S., Vaccari, L.: Matching SKOS Thesauri for Spatial Data Infrastructures. In: Sánchez-Alonso, S., Athanasiadis, I.N. (eds.) MTSR 2010. CCIS, vol. 108, pp. 211–221. Springer, Heidelberg (2010)

9. Guarino, N.: Formal Ontologies and Information Systems. In: Guarino, N. (ed.) Proceedings of FOIS 1998, pp. 3–15. IOS Press, Amsterdam (1998)
10. Guarino, N., Welty, C.: An Overview of OntoClean. In: Staab, S., Studer, R. (eds.) Handbook on Ontologies, International Handbook on Information Systems, pp. 151–172. Springer, Heidelberg (2003/2004)
11. Hornbostel, H.: From CRIS to CRIS: Integration and Interoperability. In: Asserson, A., Simons, E.J. (eds.) 8th CRIS Confernce on Enabling Interaction and Quality: Beyond the Hanseatic League, pp. 29–38. Leuven University Press, Leuven (2006)
12. Ivanovic, D.: Data Exchange between CRIS UNS, Institutional Repositories and Library Information Systems. In: Proceedings of 5th International Quality Conference (2011)
13. ISO 5964: Documentation – Guidelines for the establishment and development of multilingual thesauri. International Organization for Standardization (ISO) TC46/SC9 (2008)
14. Jeffery, K.: An Architecture for Grey Literature in a R&D Context. International Journal on Grey Literature 1(2), 64–72 (1999)
15. Jeffery, K., Lopatenko, A., Asserson, A.: Comparative Study of Metadata for Scientific Information: The Place of CERIF in CRISs and Scientific Repositories. In: Proceedings of CRIS 2002, Kassel, Germany (2002)
16. Jeffery, K.: The CERIF Model as the Core of a Research Organisation. Data Science Journal 9 (2010)
17. Jörg, B., Jeffery, K., Asserson, A., Van Grootel, G., Grabczewski, E.: CERIF 2006-1.1 Model Introduction and Specification. euroCRIS (October 2007)
18. Jörg, B.: CERIF: Formal Contextual Relations to guide through the maze of Research Information. In: Int. Conf. RIS in the EU, Bratislava, Slovakia (2009)
19. Jörg, B., Jeffery, K., van Grootel, G., Asserson, A., Dvorak, J., Rasmussen, H.: CERIF 2008–1.2 Model Introduction and Specification. euroCRIS (November 2010)
20. Joint, N.: Current Research Information Systems, Open Access Repositories and Libraries: ANTAEUS. Library Review 57(8), 570–575 (2008)
21. Manola, F., Miller, E.: RDF Primer, W3C Recommendation (February 2004)
22. Martin, M.D.: Reference Manual for Machine-Readable Bibliographic Descriptions, Unesco 1974, 71 p., compiler, Paris (1974)
23. McNie, E.C.: Reconciling the supply of scientific information with user demands: an analysis of the problem and a review of the literature. Environmental Science & Policy 10(1), 17–38 (2007)
24. Miles, A., Bechhofer, S.: SKOS Simple Knowledge Organization System. W3C Recommendation (August 18, 2009)
25. OMG available Specification: Semantics of Business Vocabulary and Business Rules (SBVR), v1.0, Open Management Group (January 2008)
26. Oei, J.L.H., ven Hemmen, L.J.G.T., Falkenberg, E.D., Brinkkemper, S.: The Meta Model Hierarchy: A Framework for Information Systems Concepts and Techniques. Technical Report No. 92-17, Department of Information Systems, University of Nijmegen, The Netherlands (1992)
27. Poole, J.D.: Model-Driven Architecture: Vision, Standards and Emerging Technologies. In: Lee, S.H. (ed.) ECOOP 2001. LNCS, vol. 2072. Springer, Heidelberg (2001)
28. Sicilia, M.A.: On Modeling Research Work for Describing and Filtering Scientific Information. In: Sánchez-Alonso, S., Athanasiadis, I.N. (eds.) MTSR 2010. CCIS, vol. 108, pp. 247–254. Springer, Heidelberg (2010)

29. Stracke, C.M.: The Benefits and Future of Standards: Metadata and Beyond. In: Sánchez-Alonso, S., Athanasiadis, I.N. (eds.) MTSR 2010. CCIS, vol. 108, pp. 354–361. Springer, Heidelberg (2010)
30. Van Grootel, G., Spyns, P., Christiaens, S., Jörg, B.: Business Semantics Management Supports Government Innovation Information Portal. In: Meersman, R., Herrero, P., Dillon, T., et al. (eds.) OTM 2009 Workshops. LNCS, vol. 5872, pp. 757–766. Springer, Heidelberg (2009)
31. Wand, Y., Weber, R.: An Ontological Model of an Information System. Transactions on Software Engineering 16(11), 1282–1292 (1990)
32. Wand, Y., Weber, R.: Research Commentary: Information Systems and Conceptual Modeling – A Research Agenda. Information Systems Research 13(4), 363–376 (2002)
33. Wand, Y., Weber, R.: On Ontological Foundations of Conceptual Modelling: A Response to Wyssusek 2006. Skandinavian Journal of Information Systems 18(1), 127–138 (2006)
34. Wyssusek, B.: On Ontological Foundations of Conceptual Modelling. Skandinavian Journal of Information Systems 18(1), 63–80 (2006)
35. Wintraecken, J.V.R.: NIAM Information Analysis Method: Theory and Practice. Kluwer Academic Publishers, Norwell (1990)
36. Zimmermann, E.: CRIS-Cross: Current Research Information Systems at a Crossroads. In: Simons, E., Asserson, A. (eds.) CRIS 2002. Leuven University Press, Leuven (2002)

# INSPIRE: Managing Metadata in a Global Digital Library for High-Energy Physics

Javier Martin Montull

CERN, 1211 Geneve 23, Switzerland
javier.martin.montull@cern.ch

**Abstract.** Four leading laboratories in the High-Energy Physics (HEP) field are collaborating to roll-out the next-generation scientific information portal: INSPIRE. The goal of this project is to replace the popular 40 year-old SPIRES database. INSPIRE already provides access to about 1 million records and includes services such as fulltext search, automatic keyword assignment, ingestion and automatic display of LaTeX, citation analysis, automatic author disambiguation, metadata harvesting, extraction of figures from fulltext and search in figure captions. In order to achieve high quality metadata both automatic processing and manual curation are needed. The different tools available in the system use modern web technologies to provide the curators of the maximum efficiency, while dealing with the MARC standard format. The project is under heavy development in order to provide new features including semantic analysis, crowdsourcing of metadata curation, user tagging, recommender systems, integration of OAIS standards and innovative metrics.

**Keywords:** Digital library, high-energy physics, metadata curation.

## 1 Introduction

The High-Energy Physics (HEP) community, which comprises around 30,000 researchers worldwide [1], has been using the SPIRES database [2] as one of their main information resources over the last 40 years. CERN, DESY, Fermilab and SLAC are undertaking the ambitious project of replacing the system by a more modern one that uses state-of-the-art technologies to help researchers access documentation in innovative and flexible ways.

The system is built on top of an Open Source digital library software developed at CERN, Invenio [3], currently used in over 25 institutions worldwide.

INSPIRE combines the curated content present in the SPIRES database with the materials harvested from the main information providers in the field, with arXiv.org as the primary source. The latter currently contains more than 850k bibliographic records.

This paper describes the functionality currently available in the INSPIRE system [4] and the development that will be incorporated in the coming years. The main focus of this article is the treatment of metadata in the system, from the initial ingestion to the automatic and manual curation that takes place in order to offer high quality content to scholars in the field.

E. García-Barriocanal et al. (Eds.): MTSR 2011, CCIS 240, pp. 269–274, 2011.

## 2 INSPIRE Overview

The INSPIRE system takes advantage of the features provided by the Invenio software, first deployed at CERN in 2002.

This software is based on open standards (MARCXML, MARC21, OAI-PMH) and is capable of managing any type of bibliographic entity, such as an article, paper, thesis, book, picture, video using the Machine-Readable-Cataloging (MARC) standard [5] for the representation of bibliographic information.

This digital library software is highly modular allowing the INSPIRE collaboration to develop new functionality in an agile manner. Its strength resides in the fast and powerful search engine, which uses Google-like syntax, the flexible document type submission and approval workflow, the flexible metadata treatment and the user personalization (with Web 2.0 features like baskets, notifications, user groups and user comments).

Since the INSPIRE project began, new functionalities have constantly been introduced to Invenio's code base. For instance, it is currently possible to perform searches inside the fulltexts, displaying snippets of the matches in the search page. The INSPIRE collaboration has been negotiating with publishers in order to extend this functionality to restricted articles. In April 2010, the first agreement was signed with Springer in that direction.

INSPIRE actively participates in the ORCID initiative [6], which aims to solve the author ambiguity problem that affects many fields of research. In a big step towards achieving this goal, INSPIRE deployed its own author disambiguation algorithm in the year 2011 that makes use of data such as affiliation history, coauthor information, citation patterns and more, in order to disambiguate authors with a high success rate. In Fig. 1, an author page is shown where all the name variants for a given author are displayed, together with all the important information related to the author.

In the same research track, an interface for the authors to claim their papers, which is shown in Fig. 2, has been developed in order to increase the quality of the author pages. This interface uses metadata information to group research papers that have a high probability of having been written by the given author, allowing the user to confirm/disclaim the proposals and add new papers.

Finally, the joint effort among CERN, the APS [7] and the experimental collaborations has resulted in the definition of the authors.xml format [8] that allows collaborations to present their author lists with each author being uniquely identified.

Other important features that have been put in place include the extraction of plots from arXiv.org tarballs, which allow scholars to view all plots from a record in the detailed page as well as search inside the captions (which are extracted from the LaTeX files), automatic keyword assignment and citation analysis.

Some of the citation analysis tools include citation count graphs and recommendations based on the citation network [9].

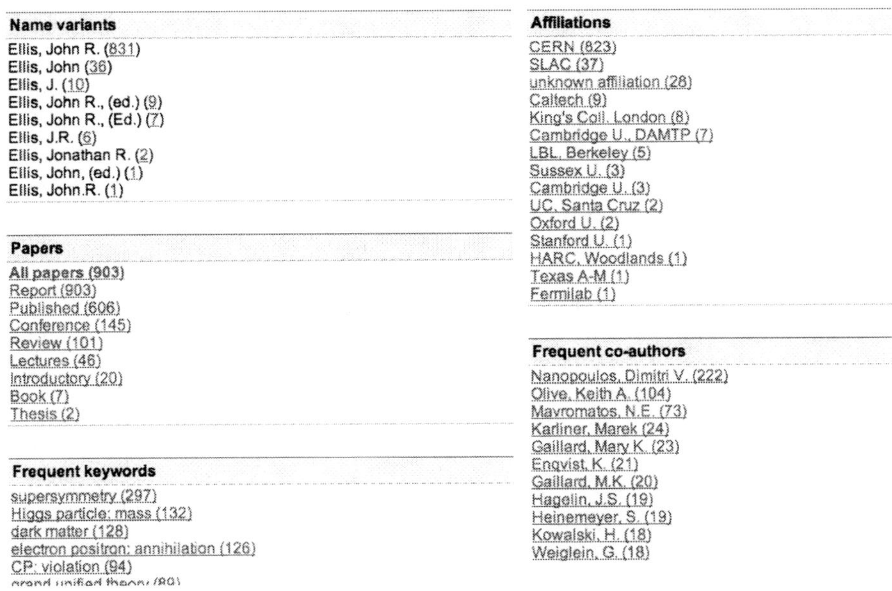


**Fig. 1.** Author page with author disambiguation information

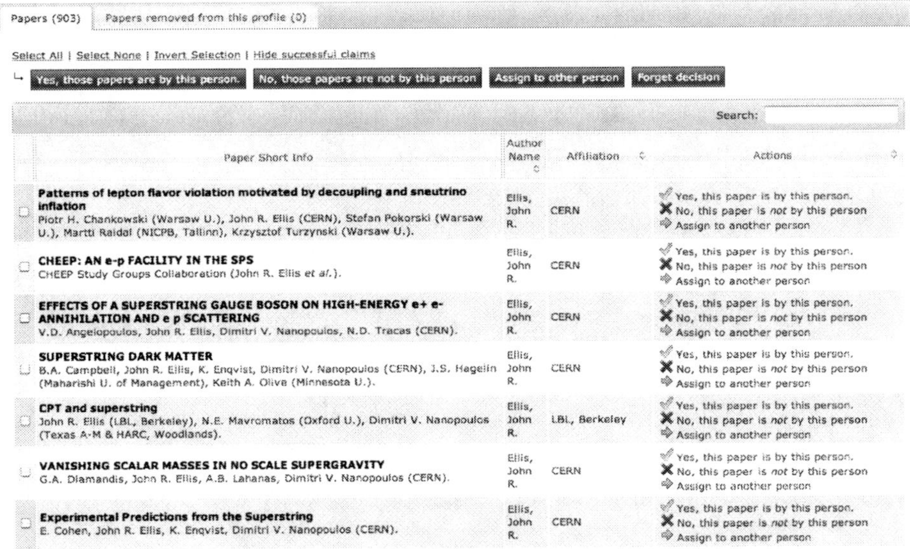


**Fig. 2.** Interface to claim papers for a given author

## 3 Managing Metadata in INSPIRE

The SPIRES database has remained popular over the course of decades thanks to its trusted content curated by SLAC, Fermilab and DESY. One of the main challenges that the INSPIRE system faces is providing the appropriate curation tools necessary to efficiently maintain all the metadata ingested by the system.

For INSPIRE to become independent from SPIRES, which is the final goal, it is necessary to mimic SPIRES's harvesting scheme. In addition, some filters have been designed to perform more automated classification and data-mining of incoming content.

The process starts with the harvest from external repositories and publisher feeds using common protocols (OAI-PMH) and the mapping of the harvested data to be compliant with INSPIRE's data representation, MARC21.

If the external repository provides additional files alongside the metadata, like figures or the original LaTeX source, these are extracted and indexed for user search.

To help minimize manual curation work, an automatic analysis of the fulltext files is made to extract all references and make them searchable.

Before the final insertion of the harvested record, checks are performed to avoid duplication in the database and to allow manual intervention in case of conflicts.

Once the records are part of the system, some manual curation is needed to achieve the highest possible quality in the metadata information.

All tools developed for metadata processing are Rich Internet Applications (RIAs), which allow seamless access from any given location. The technologies used include AJAX, the Python programming language and MySQL databases.

The main metadata editing tool build into the system, shown in Fig. 3, is called BibEdit and allows the editing of a single record. The record is displayed with its internal representation (MARC21) and once opened a cataloger can edit any field/subfield in a convenient way.

Every modification sent to the system creates a unique object so that any error introduced can be reverted. These revisions are shown in the interface as well, so that the cataloger can check the history of modifications, revert or merge a previous version.

In addition, useful functions such as search for records, undo/redo changes, keyboard shortcuts and preview of the record are provided.

In line with the overall goal of increasing productivity and reducing errors, new features are under development and will be available in the future iterations of INSPIRE. These include autocompletion of fields based on knowledge bases, special modes to manage citations and authors (some collaborations have thousands), and a module devoted to checking that all combinations of fields/subfields are valid.

Once the user submits the desired changes to the record, the server produces a MARCXML file, checks its compliance with the standard and sends the modification to be included in the database.

In addition to BibEdit, which only allows catalogers to correct one record at a time, a different tool named MultiEdit can apply arbitrary transformations across an entire set of records.

MultiEdit, shown in Fig. 4, first performs a search to retrieve the set of records that the cataloger wants to correct. Once this is done, it is possible to filter the MARC tags displayed in the interface and the collection the records belong to.

After having selected the records to be modified, the interface allows the cataloger to define a set of actions that will be applied to them. The actions include adding fields/subfields, deleting them and replacing some strings among others. A preview of the modifications is then shown and all of them can be sent into the system at once.

As mentioned at the beginning of this section, the automatic harvesting will sometimes detect that the record to be added to the system is a potential duplicate. At other times a preprint is published and a new record has to be introduced with just few changes. Without an appropriate tool it is challenging for a cataloger to spot the differences between similar records and actions taken can be error prone.

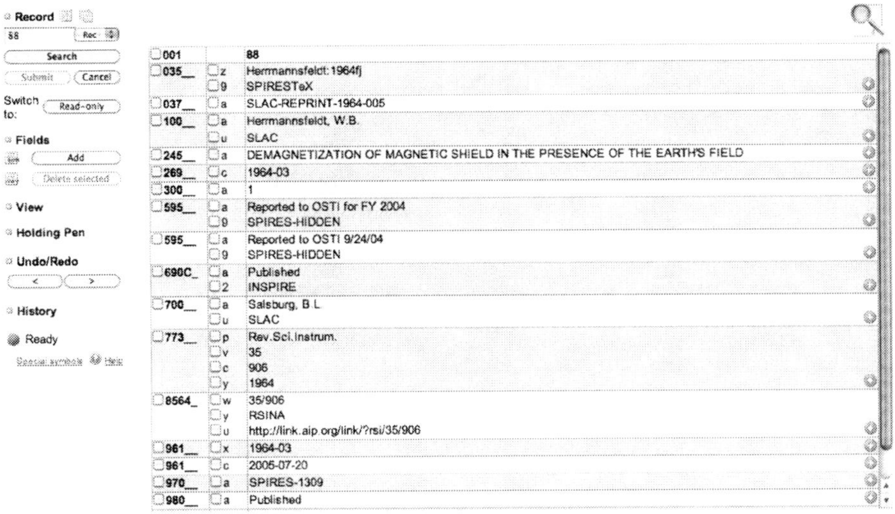

**Fig. 3.** BibEdit, tool for editing a metadata record

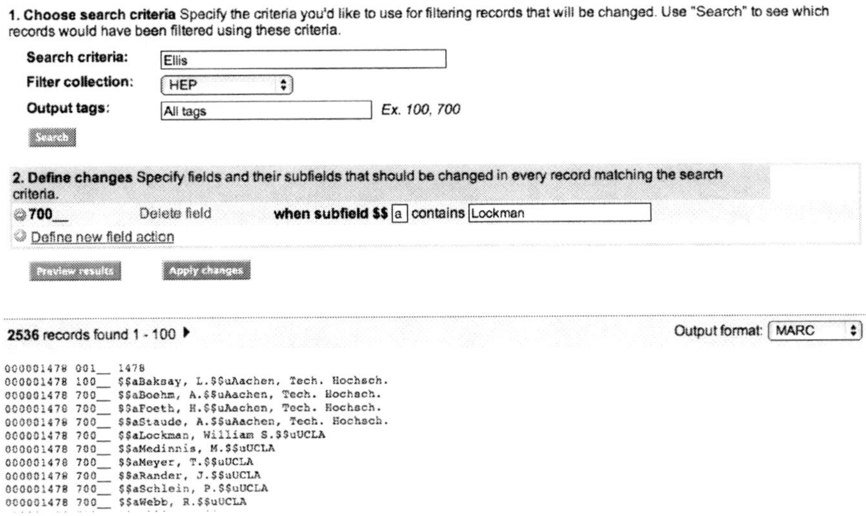

**Fig. 4.** MultiEdit, tool for editing multiple metadata records

Thus, INSPIRE also contains a tool named BibMerge which consists in a side-by-side comparison of MARC records so that the cataloger can easily find the differences (comparison techniques such as the Levenshtein distance are used) and merge both records once the appropriate fields have been added or removed.

## 4 Conclusion

The management of metadata in a global High-Energy Physics (HEP) digital library such as INSPIRE leads to multiple challenges, given the objective of producing highly trusted information used for research.

In order to increase the productivity of the catalogers in charge of metadata curation, all metadata harvested into the system from external sources must be preprocessed, so that as much information as possible is automatically ingested.

In cases where manual curation is required, the INSPIRE system provides a variety of tools that allow catalogers from the four participating laboratories (CERN, SLAC, DESY and Fermilab) to edit the metadata using cutting-edge technologies.

The use of these techniques allows INSPIRE to maintain high quality content and maximize the productivity of catalogers involved in the collaboration.

## References

1. Travis Brooks (May 22, 2011), `http://www.projecthepinspire.net/ape09.pdf`
2. `http://www.slac.stanford.edu/spires/`
3. `http://www.invenio-software.org`
4. `http://inspirebeta.net`
5. `http://www.loc.gov/standards/marcxml`
6. `http://www.orcid.org`
7. `http://www.aps.org`
8. `http://www.slac.stanford.edu/spires/hepnames/authors_xml/`
9. Holtkamp, A., Mele, S., Simko, T., Smith, T.: Realizing the dream of a global digital library in High-Energy Physics, CERN-OPEN-2010-019

# Metadata Management and Sharing in Multimedia Open Learning Environment (MOLE)

Manolis Mylonakis, Polyxeni Arapi, Nikos Pappas, Nektarios Moumoutzis, and Stavros Christodoulakis

Laboratory of Distributed Multimedia Information Systems and Applications, Technical University of Crete (TUC/MUSIC), 73100 Chania, Greece
{manolis,xenia,nikos,nektar,stavros}@ced.tuc.gr

**Abstract.** This paper presents a framework and an architecture for learning resource management and sharing aiming at facilitating the implementation of such functionality on top of existing Learning Management Systems. It also presents the implementation of this framework and its integration with the MOLE (Multimedia Open Learning Environment – http://www.moleportal.eu/) system. Main components of this architecture are: (a) the LOM Editor, an intuitive web based tool that is able to accommodate different Application Profiles, while getting adapted accordingly; (b) The LOM Repository that stores the metadata generated by the LOM Editor and implements the common repository services (search/expose, submit/store, request/deliver) for the management of metadata records; (c) the user interfaces that exploit those services to expose the metadata management functionality to end-users; and (d) an OAI-PMH interface that allows for harvesting of the repository metadata from large repositories/federations (e.g. ARIADNE, Organic.Edunet etc.) on top of the repository.

**Keywords:** eLearning infrastructures, metadata management and sharing, LOM Editor, Application Profiles, OAI-PMH.

## 1 Introduction

One of the major challenges for learning organizations today is sharing learning resources by facilitating discovery and retrieval of the learning content they develop and store in their repositories [1]. While learning content repositories already cater for their local users, there are no agreed profiles that address the needs of the learning domain, and no established practices for combining existing specifications into complete solutions [2]. Individual organizations are developing their own ad hoc solutions, following different technical approaches, mechanisms, and metadata models. As a result, the opportunity to establish broader interoperability is limited.

Establishing practices that combine existing specifications into integrated solutions in the form of agreed profiles, in order to address the needs of specific learning communities in terms of learning object sharing, discovery, and exchange, will be beneficial to all stakeholders [2]:

E. García-Barriocanal et al. (Eds.): MTSR 2011, CCIS 240, pp. 275–286, 2011.

- *Educators* will become more productive by being able to easily discover learning content that addresses the needs of their students, thus maximizing re-use and reducing the cost of reproducing new resources.
- *Students* will gain access to the highest-quality learning resources available, making a significant impact on the quality of their learning experience and their learning outcomes.
- *Content providers* will have the opportunity to advertise their products making them globally discoverable.
- *System vendors* will be able to make their systems compliant with major federations of learning resources, by only supporting a minimum set of specifications, and
- *Federation builders* will secure their investment by developing durable infrastructures based on standard specifications.

A concrete evidence of the significance of this challenge is the fact that the IMS Global Learning Consortium, has established a group on Learning Object Discovery and Exchange (LODE) [2] aiming to facilitate the discovery and retrieval of learning content. The group examines and adapts specifications that are being applied to digital libraries, generic repositories and learning repositories to address the current lack of agreed profiles addressing the needs of the learning domain and practices for combining existing specifications into complete solutions. The project group will study search mechanisms, meta-data harvesting, learning content exchange, content identification as well as collection and service description. It will also define a small number of scenarios for the discovery and exchange of learning resources (e.g., federated searching, harvesting, etc.) with the objective to develop:

- A set of specification profiles to support the scenarios
- Sample implementation(s)
- A conformance domain-profile

The ultimate goal is interoperability among systems involved in those scenarios. Interoperability is achieved when a system (e.g., a LMS) end user is able to discover a compatible learning object hosted on a separate system (e.g., a learning object repository) using a LODE-compliant discovery service. Addressing federated discovery (through either federated search or harvest-driven centralized search) presents the greatest interoperability challenge. Establishing interoperability in terms of LODE requires that the federations should be based on LODE search and LODE registry specifications. However, federation is not a requirement for compliance. The term "federated" is used in a loose sense to refer to a group of distributed, independently managed and potentially heterogeneous repositories, whether or not any agreements, trust relationships etc. exist between them.

In this paper we report on the design and implementation issues related to the support of learning content sharing requirements on top of an existing Learning Management System (LMS). We focus on those aspects that are related with the use of learning metadata standards and their Application Profiles (APs) for the description of learning resources, as well as the implementation of harvesting protocols that will make them available to large repositories/federations, technical issues that should be addressed in eLearning infrastructures.

Our learning infrastructure is based on the MOLE (Multimedia Open Learning Environment – http://www.moleportal.eu/) system. MOLE is a multilingual multimedia information system for managing courses, supporting learning processes and learning communities through the Web. The MOLE multi-tenant architecture supports multiple instances of MOLE using the same core to serve the needs of multiple communities. Each particular community has specific needs for the descriptions of the learning material that are more or less different from the needs of other communities. In this sense, applying the framework described in this paper to MOLE was appropriate in order to support different Application Profiles to suit those community-specific needs and overcome the problems of semantic interoperability.

We have adopted the Learning Object Metadata standard (LOM). Our annotation tool is an intuitive web-based LOM editor that is not strictly bound to a specific AP, as it is the case for most LOM metadata editors. The LOM editor can use multiple APs, thus supporting efficient metadata specifications in any application-specific context. The metadata records are managed by a repository that implements the common repository services (search/expose, submit/store, request/deliver). Appropriate user interfaces have been implemented exploiting those services to expose the metadata management functionality to end-users. Finally, an OAI-PMH interface has been implemented on top of the repository. This interface supports metadata harvesting from large repositories/federations (e.g. ARIADNE, Organic.Edunet etc.). This way, the learning content created by an organization could be made known and exploited by other organizations that are connected to those federations.

The paper focuses on the learning resources metadata management and sharing in MOLE and implements the general architectural framework we propose. Section 2 presents the proposed architectural framework for supporting metadata management and sharing of learning resources on top of a LMS. Section 3 provides details regarding the implementation of this framework in the case of MOLE and considers various technical issues that should be considered for implementing the framework in other Learning Management Systems. Related work is presented in section 4, while section 5 concludes and presents directions of future research and developments.

## 2 Metadata Management and Sharing Framework and Architecture

A typical LMS can be conceptually represented with the architecture on the left part of Fig. 1. In this architecture, a LMS consists of the following:

- A set of Learning/Collaboration services
- a set of Course Creation/Management Services, and
- a Learning Resources Repository for the storage of the learning resources (physically of by reference) and their metadata.

In this section we present a framework and an architecture for learning resource management and sharing aiming at facilitating the implementation of such functionality on top of existing Learning Management Systems. The main component of this architecture that allows for the extension of an LMS for effective resource management and sharing is called the *Metadata Management and Sharing System*.

The *Metadata Management and Sharing System* allows for the creation of LOM metadata descriptions based on different APs, supporting the needs of different communities in different educational contexts. The technical experts can take advantage of the system, in order to develop an appropriate AP by using the *Application Profile Builder* that the users can later use in order to create the corresponding LOM metadata descriptions through the *LOM editor*. The LOM XML documents that are produced can be searched and edited through appropriate user interfaces. Moreover, the *OAI-PMH Interface* implementing the OAI-PMH protocol on top of the *LOM Metadata Repository* allows for the exposure of the metadata to *Learning Resources Federations/Consumers*.

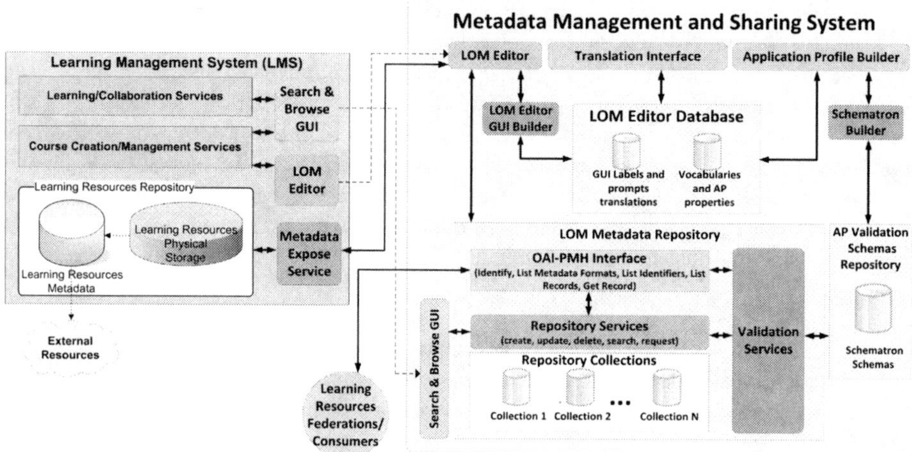


**Fig. 1.** The Metadata Management and Sharing Architecture

Fig. 1 illustrates in detail the architecture of the *Metadata Management and Sharing System (MMSS)* and its connection points (interfaces) with the LMS. The *MMSS* consists of the following components:

- The *Application Profile Builder* is a web based utility (user interface) that offers a graphical representation of the LOM metadata elements and their structure. Through that utility, the AP expert can navigate through the LOM elements, setting their multiplicity and the vocabularies, taxonomies or ontologies that will be used as value spaces for each of them. Moreover, the expert can state which elements are mandatory, recommended or optional. Furthermore, the expert can create custom vocabularies using list of values, taxonomies or domain ontologies, that can be later used in specific metadata elements. That information is saved in the *LOM Editor Database*.
- The *LOM Editor Database* is responsible for the storage of the system properties, the vocabularies, the translations of the GUI and the help descriptions. It also stores the relevant information that is necessary for the system in order to support the APs.

- The *Translation Interface* is a web based wizard utility that can be used by the experts in order to translate the LOM editor's user interface, labels, prompt messages and help information related to each metadata element into different languages. It takes advantage of the "Google Translate API", suggesting translations to the user simplifying the translation procedure. The information related to translations is also kept in the *LOM Editor Database*.
- The *LOM Editor* is a web application that offers to the users a formal way to create or edit LOM metadata descriptions. The elements that are presented on the editor's user interface are each time adapted to the AP that is being used. The elements associated with vocabularies are supported by auto complete services. For each element a short description (help) is available, explaining the usage of the corresponding element for each AP. The LOM metadata description is finally saved in the *LOM Metadata Repository*.
- The *LOM Editor GUI Builder* utilizes the information that is kept in the *LOM Editor Database* and builds the *LOM Editor's* user interface.
- The *Schematron Builder* is a component invoked by the *Application Profile Builder* as soon as an AP is saved. It is responsible for the creation of schematron schemas [3] according to the corresponding APs. These schemas are stored in the *AP Validation Schemas Repository*" and are used for the validation of the XML documents.
- The *AP Validation Schemas Repository* is the responsible component for the storage of the Application Profile validation schemas (schematron).
- The *LOM Metadata Repository* is responsible for the storage and management of the generated LOM XML descriptions. It consists of the following parts:
  - The *Validation Services* that are used to validate the XML documents that are generated by the *LOM Editor*, utilizing the information that is kept in the *AP Validation Schemas Repository*. The documents that are invalid are also kept in the repository in order to be later completed by the user, but will not be published through the *OAI-PMH interface* until they reach their final state.
  - *Repository Services* is a set of services over the repository collections that conform to the IMS Digital Repositories Interoperability (IMS DRI) Specification [4]. The IMS DRI specification provides recommendations for the interoperation of the most common repository functions enabling diverse components to communicate with one another: search/expose, submit/store, gather/expose and request/deliver.
  - The *OAI-PMH interface* is the responsible part for the exposure of the metadata descriptions. It is an implementation of the OAI-PMH protocol, a widely accepted protocol for the exposure of repositories information. The LOM XML documents that are exposed by this protocol have been previously validated based on the corresponding Application Profile and the *Validation Services*.
- The *Search & Browse GUI* is a web based graphical user interface that can be used to search and browse the content of the *LOM Metadata Repository*, utilizing the *Repository Services*.

The connection points between the Learning Management System and the Metadata Management and Sharing System are the following, as illustrated in Fig. 1:

- The *Metadata Expose Service* that passes to the *LOM Editor* the learning resource metadata values stored in the LMS that can be mapped to the current LOM AP. This way, the LOM Editor automatically fills in the values of the corresponding elements in the LOM AP metadata description each time a new metadata description for a learning resource is being created.
- The *LOM Editor box* on the side of the LMS corresponds to the integration of the *LOM Editor* in the LMS. This requires no effort, since the LOM editor integration is easily done using technologies like iframes without any changes needed to be done on the side of the LMS. The web interface of the LOM editor can be customized using the *LOM Editor GUI Builder* in order to fit the web interface of the LMS.
- The *Search & Browse GUI box* on the side of the LMS. As in the case of LOM Editor, the Search & Browse GUI can be easily integrated using iframes.

## 3 Implementation and Usage

The proposed architecture has been successfully implemented on top of MOLE (Multimedia Open Learning Environment – http://www.moleportal.eu/), developed by the Laboratory of Distributed Multimedia Information Systems and Applications (TUC/MUSIC) of the Technical University of Crete. In the following sections we introduce MOLE and its main characteristics and we describe its integration with the metadata management and sharing architecture proposed in this paper.

### 3.1 Multimedia Open Learning Environment (MOLE)

MOLE is a multilingual multimedia information system for managing courses, supporting learning processes and learning communities through the Web. More specifically, MOLE fosters distance-learning by enabling communication between tutors/trainers and students, cooperation among students and access to coursework information and learning resources. In doing this, MOLE platform also supports the combination of traditional classroom-based lessons and practical sessions, with self-study and e-learning. This, so called, "hybrid" or "blended" approach provides a significant learning opportunity as it combines the immediacy of communication among the instructor and the learners and the irreplaceable practical training in laboratories and the convenience, flexibility and self-regulation of education without the time and space constraints. This hybrid organization aims to exploit the strengths of both approaches (traditional and tele-education).

An important characteristic of MOLE is its multi-tenant architecture (Fig. 2) that can support multiple instances of it using the same core to serve the needs of different projects or communities (Fig. 3).

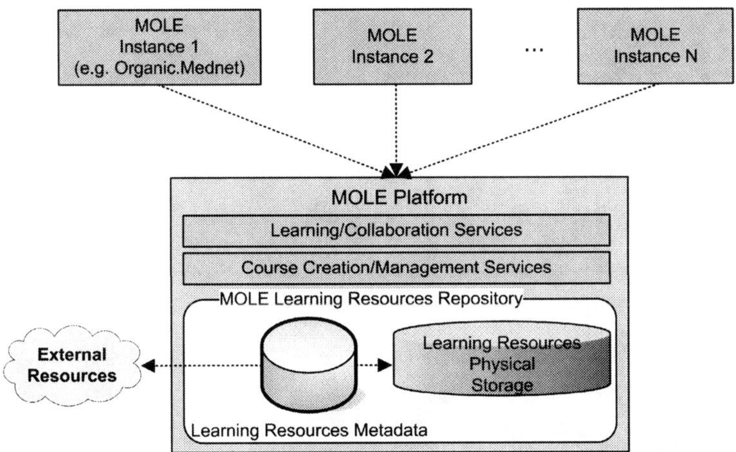


**Fig. 2.** MOLE has a multi-tenant architecture able to support different communities and projects

MOLE offers a set of services for

- the organization and management of digital educational content (e.g. lectures, notes, exercises, technical lab material, literature, FAQs etc.)
- informing learners (e.g. announcements, mail, calendar, personal rating etc.)
- assisting learning communities (e.g. mailing lists, live chat rooms, forums, personal messaging, instant messaging, annotation tools, video conference etc.)
- supporting educational activities (e.g. registration to courses, formation of lab teams, exercise uploads and deadline management, assessment tests, multimedia presentations etc.)
- monitoring services (course usage statistics, class performance indicators)

Special emphasis in MOLE system is given in the use of multimedia as a powerful learning means and for that reason it is continuously enriched with new multimedia features. In order to support learning applications MOLE meets the following technological requirements:

- Effective support of multimedia and video/audio data streams management.
- Mechanisms for multimedia presentations synchronization
- Support of synchronous and asynchronous learning activities.
- Support of live synchronized multimedia transmission through the system, and access to the recorded multimedia content.
- Support of learners' intervention in live sessions.
- Demonstrations and presentations of software by recording the video presentation and speaker combined with video and slides.
- Educational content creation in two ways

- Through the Web based interface without the need for specialized software installations
- Through an autonomous application for the creation of high quality video and editing tools for offline content creation and uploading in the system at different times.

- Communication tools to support educational communities
  - Discussions with video and audio in real time
  - Video Conferencing Services with collaboration tools
  - Asynchronous multimedia communication messages
- Advanced multimedia collaborative annotation tools on educational materials using multimedia

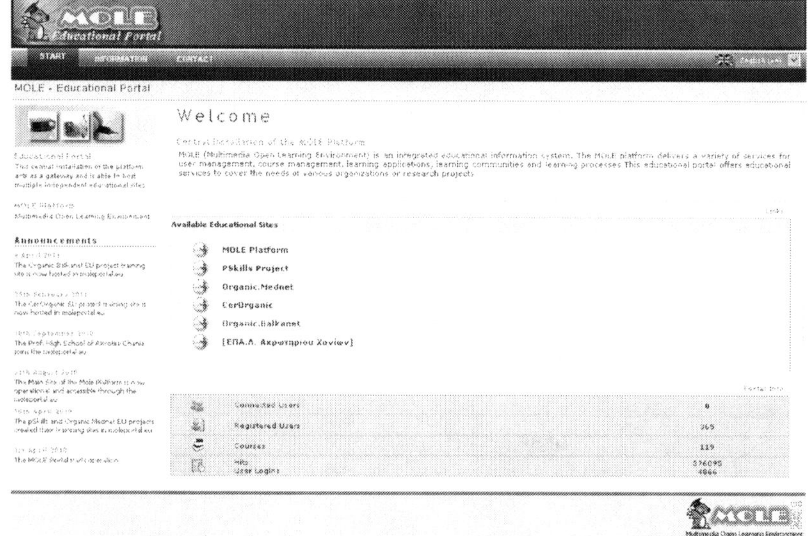


**Fig. 3.** Communities and projects currently supported by MOLE

## 3.2 Integration of the Proposed Architecture with MOLE

The *MMSS* has been integrated with MOLE following the approach presented in Fig.1. The communication between MOLE and the *MMSS* is done over HTTP. Each time a new metadata description is created in the *LOM Editor* for a learning resource residing in LMS, the LOM Editor calls the *Metadata Expose Service* to pass via a POST request the metadata of the learning resource stored in the LMS. This is done according to the mapping between the LMS metadata elements and the LOM AP metadata elements known to the *Metadata Expose Service*. Afterwards, the metadata author can further enrich the learning resource metadata following the rules imposed by the current AP that affect the presentation of the LOM Editor GUI. The information that is passed to the *MMSS* must at least contain the following metadata:

- the *organization identifier* that will be used in order to organize the resources in collections. Each collection represents an instance of the MOLE platform for a specific community.
- the *resource identifier* that will be used in order to identify a resource and the corresponding LOM XML description in the system.
- the *title* of the resource, which will be used as a title of the LOM description.

The richness of the LOM metadata for a learning resource depends a) on the number and type of the metadata elements used in the LMS that are able to be mapped to LOM elements depending on the current AP and b) on the metadata author who may further enrich or not the metadata following the rules expressed in the AP.

In the case of MOLE and using the Organic.Edunet AP [5], the metadata elements of MOLE that are mapped to the elements Organic.Edunet LOM AP (OE LOM AP) are the following (beyond the basic info described above):

- the *description*, that is related to the resource that is kept in the platform (mapped to the General/Description element of the OE LOM AP)
- the *keywords*, relevant to the learning resource being described (mapped to the General/Keyword element of the OE LOM AP)
- the *contributor*, who creates the LOM metadata description (mapped to the Meta-Metadata/Contribute element of the OE LOM AP)
- the *resource location*, which represents the location that the corresponding resource will be accessible (mapped to the Technical/Location element of the OE LOM AP)
- the *resource type* (mapped to the Educational/Learning Resource Type element of the OE LOM AP)
- the *format* of the resource as mime type (mapped to the Technical/Format element of the OE LOM AP), and
- the *file size* representing the size of the resource in bytes (mapped to the Technical/Size element of the OE LOM AP).

The *MMSS* has been developed as a web based application that takes advantages of Java, Java Server Pages and Java Script Markup Language. The core services are implemented in Java and served through the Apache Tomcat servlet container [6]. The *LOM Editor Database* utilizes a MySql database and the services for the management of the information (store/retrieve) are also implemented with Java technologies. The ontologies that are used as value spaces in metadata elements are processed using the Jena OWL API [8] in order to get thereafter presented through an appropriate graphical interface. The communication between the related graphical user interfaces is done through HTTP requests and the exchange of information is done through JSON or XML messages. The *Validation Schemas* that are produced by the Schematron Builder are kept in the local storage. The LOM metadata descriptions in the LOM metadata repository are stored in a native XML database (eXist-db [8]) and the *Services* provided by the LOM metadata repository are also implemented in Java. The *Validation Services* utilize the Xalan Java XSLT processor [9].

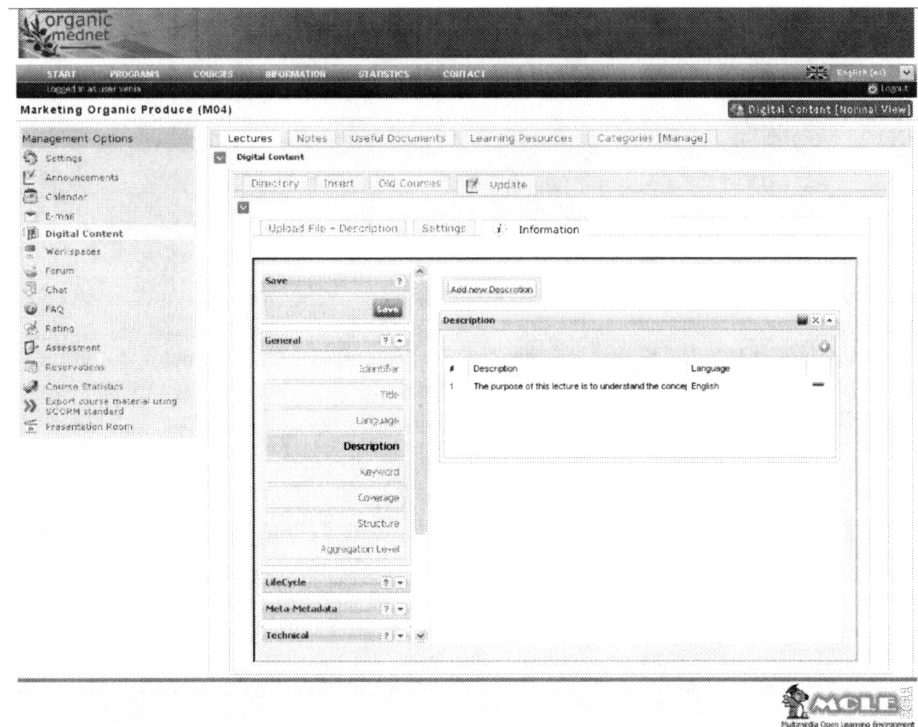


**Fig. 4.** The LOM Editor integrated in MOLE

As already mentioned, one of the advantages of the web implementation of the *LOM Editor* and the *Search & Browse GUI*, as well as the communication through HTTP requests, is that they can be easily encapsulated in different LMSs by using for example iframes, as in the case of MOLE. Fig. 4 shows the integration of the LOM Editor in MOLE using iframes.

## 4 Related Work

Different learning communities have different needs to support different educational contexts. Each community has its own needs for the description of the learning material and needs support for semantic interoperability. Those needs can be supported by the use of APs of LOM standard. To support the implementation of APs as well as the creation of metadata descriptions related to a specific AP, different software applications have been implemented. Two of the most known applications are the LomPad [10] and the ASK-LOMAP v1.0 [11].

The LomPad is a meta-tagging tool for learning objects. It is bilingual (French, English) and supports three predefined different applications of the IEEE-LOM standard (LOM strict, SCORM, and CANCORE APs). It is a java desktop application and does not provide facilities neither for metadata management nor for metadata exposure. Moreover, it does not support the creation of different APs.

On the other hand, ASK-LOMAP v1.0 is a web based tool for the creation of LOM APs and metadata authoring tool. Although it supports the creation of LOM APs, it does not support the combination of different vocabularies for the same element. Additionally, it does not support the use of taxonomies or ontologies as an element value space through its interfaces. The metadata authoring tool is not multilingual and it seems that multiplicity of some elements presenting appropriate graphical interfaces is not fully supported, although defined in the corresponding APs. Moreover, no search/browse facilities in the content of the repositories are provided and there is no support for metadata exposure to federations.

Compared to the above mentioned approaches, the advantage of the approach presented in this paper is that it provides a complete framework for metadata management and sharing according to different APs to support different communities that can be easily integrated in an existing eLearning system.

The closest approach to our approach in terms of its objectives is [21], where a Web-based tool is presented that has been developed to facilitate learning object annotation in agricultural learning repositories with IEEE LOM-compliant metadata and its prototype development within Organic.Edunet Project. Learning objects are organized in a portfolio-like user interface in a learning repository, which can be connected with Organic.Edunet federation or external federations as ARIADNE and LRE. The repository back-end is resource-oriented and stores its metadata according to an RDF representation of the Organic.Edunet AP. An interface exposes the repository closest to the internal representation and is based on REST. Several existing tools have been used and adapted: 1) A configurable metadata editor called Annotation Tool, built upon the code-library SHAME (http://shame.sf.net), 2) The electronic portfolio system Confolio (http://www.confolio.org) which allows flexible management of folder-based repository interfaces, and 3) The SCAM (http://scam.sf.net) semantic metadata and resource repository.

## 5 Conclusions and Future Work

Educational resource sharing is emerging as a viable means to improve the quality of and access to education [1]. Towards this end, the use of learning metadata standards for the description of learning resources, as well as the implementation of harvesting protocols that will make them available to large repositories/federations are technical issues that should be addressed in eLearning infrastructures. The approach discussed in this paper addresses the requirements set by the eLearning community and reflected in the work in progress from the LODE group of the IMS Global Consortium. The solutions proposed here on the design and implementation issues enable an LMS to share and exchange learning content with other systems leveraging the advances in related standards and ensuring the widest possible exploitation of learning content and related investment made by learning organizations today.

Currently, our system supports four European Projects (Organic.Mednet [12][13], CerOrganic [14][15], Organic.Balkanet [16][17], and pSkills [18][19]). Using the infrastructure described in this paper, the first three projects related with organic agriculture describe their learning resources using the Organic.Edunet AP and make them available to the Organic.Edunet federation [20]. Although currently used inside

MOLE, the LOM editor and repository design and implementation allows for its use in other applications with minimum effort.

Future developments under consideration include: a) The investigation of mappings between APs, their implementation and use to facilitate exchange and content between learning organizations/communities, and b) Automating as much as possible the creation of LOM metadata by using information stored in the learning object repositories, user profiles as well as forums and other communication facilities provided by LMSs.

**Acknowledgments.** The work presented in this paper is partially funded in the scope of the Organic.Mednet Project (LdV ES/09/LLP-LdV-TOI-149061) "Developing the Skills of Organic Agriculture Trainers for the Mediterranean" and the pSkills Project (502843-2009-LLP-GR-COMENIUS-CMP) "Programming Skills Development in Secondary Education by means of Modern Educational Programming Languages".

## References

1. Elearnspace: Why We Should Share Learning Resources, http://www.elearnspace.org/Articles/why_we_should_share.htm (retrieved on May 29, 2003)
2. IMS Global Learning Consortium Learning Object Discovery and Exchange (LODE) Project Group, http://www.imsglobal.org/lode.html
3. Schematron, http://www.schematron.com/
4. IMS DRI. IMS Digital Repositories specification V1.0 (2003), http://www.imsglobal.org/digitalrepositories/
5. Kastrantas, K., Palavitsinis, N., Manouselis, N., Palmer, M., Ebner, H., Sanchez, S.: D5.1.2. Educational Metadata for Organic.Edunet Learning Objects (Multilingual). Organic.Edunet Project Deliverable, Final version (May 20, 2009)
6. Tomcat Servlet Container, http://tomcat.apache.org/
7. Jena – A Semantic Web Framework for Java, http://jena.sourceforge.net/
8. eXist-db, http://exist.sourceforge.net/9
9. Xalan, http://xml.apache.org/xalan-j/
10. LomPad, http://sourceforge.net/projects/lompad/
11. ASK-LOMAP v1.0, http://www.ask4research.info/asklomap/
12. Organic.Mednet Project Official Website, http://www.organic-mednet.eu/
13. Organic.Mednet Project MOLE instance, http://om.moleportal.eu/
14. CerOrganic Project Official Website, http://www.cerorganic.eu/
15. CerOrganic Project MOLE instance, http://cerorganic.moleportal.eu/
16. Organic.Balkanet Official Website, http://www.organic-balkanet.eu/
17. Organic.Balkanet project MOLE instance, http://ob.moleportal.eu/
18. pSkills Official Website, http://pskills.ced.tuc.gr/
19. pSkills Project MOLE instance, http://pskills.moleportal.eu/
20. Organic.Edunet federation, http://www.organic-edunet.eu
21. Ebner, H., Manouselis, M., Palmér, M., Enoksson, F., Palavitsinis, N., Kastrantas, K., Naeve, A.: Learning Object Annotation for Agricultural Learning Repositories. In: IEEE ICALT 2009 Conference, Riga, Latvia (2009)

# From Concept to Sharing, to Delivery: Modeling a Quality Controlled Lifecycle for Learning Resources within a Federation of Repositories

Anastasios Koutoumanos[1], Vassilios Protonotarios[1], Alexios Dimitropoulos[1], Konstantinos Kastrantas[1], John Stoitsis[2], Salvador Sánchez-Alonso[2], and Cleo Sgouropoulou[3]

[1] Agro-Know Technologies, 15 Grammou str., 15235 Vrilissia, Athens, Greece
{tkout,vprot,adimitropoulos,kkastrad}@agroknow.gr
[2] Universidad de Alcalá, 28871 Alcalá de Henares, Spain
jstoitsis@ieee.org, salvador.sanchez@uah.es
[3] Technological Educational Institute of Athens, GR-12210, Greece
csgouro@teiath.gr

**Abstract.** An ever increasing number of members of educational communities engage in creating new digital resources within all fields of science and practice at all educational level within Learning Object Repositories, which, in terms, allows for even more efficient identification and delivery of existing material. This has in effect shaped a new type of activities related to the creation, sharing and re-use of learning objects through distributed repositories, associated within a federation with common policies, protocols and formats, which gave rise to an exponential growth of available material. However, this great success brings along a set of challenges, the most important being the demand for a quality controlled process, eliminating the information overload and safeguarding the added-value of learning object repositories, through models such as peer review and validation of learning objects and associated metadata. The authors of this paper provide insight on the approach for capitalizing on the European and international demand for quality-controlled educational content on Organic Agriculture and Agroecology, describing a framework and the associated quality issues for modeling the publication lifecycle of digital learning objects within a federation of learning object repositories.

**Keywords:** Multilinguality, multilingual, internationalization, organic.lingua, organic.edunet, repository, workflow, BPMN.

## 1 Introduction

The sheer number of 'Learning Objects', in the form of digital, multimedia material suitable for educational purposes, has experienced a constant growth during the last few years. In fact, an ever increasing number of members of educational communities engage in creating new digital resources within all fields of science and practice at all educational levels. This trend has led to the need of effectively indexing and storing

E. García-Barriocanal et al. (Eds.): MTSR 2011, CCIS 240, pp. 287–299, 2011.

learning objects and associated metadata in Learning Object Repositories (LORs), which, in terms, allows for even more efficient identification and delivery of existing material.

A recent development in this field, which stems from the need of all stakeholders to share, re-use and realize the full potential of the wealth and breadth of learning objects, is the uptake of the open access movement, in order to safeguard interoperability between LORs and facilitate access that allows for immediate use of the learning resources by end users. This has in effect shaped a new type of activities related to the creation, sharing and re-use of LORs through distributed repositories, associated within a federation with common policies, protocols and formats, which gave rise to an exponential growth of available material.

However, this great success brings along a set of challenges, the most important being the demand for a quality control process, eliminating the information overload and safeguarding the added-value of learning object repositories, through models such as peer review and validation of learning objects and associated metadata.

Within this paper we provide insight on the approach for capitalizing on the European and international demand for quality-controlled educational content on Organic Agriculture and Agroecology, describing a framework and the associated quality issues for modeling the publication lifecycle of digital learning objects within a federation of learning object repositories.

In fact, Multilingual Language Technologies play a key factor in protecting and increasing the usability of the large group of European languages [1]. Forty-four percent (44%) of European users report a perceived loss of interesting information because of lack of online content in a language that they can understand [3]. It is vital that affordable and robust language technology is developed and integrated into software and websites in order for them to be of more value to their users and to trigger the sharing of knowledge. Building systems for cross-lingual information and knowledge management will minimize language barriers among others for education boosting international cooperation and sharing of knowledge [11].

Organic.Lingua (www.organic-lingua.eu) is an EU-funded project that aims to build automated multilingual services and tools that facilitate the discovery, retrieval, exploitation and extension of digital educational content related to organic agriculture and agro-ecology. The project builds upon the established Organic.Edunet Web portal which has adopted a federated, standards-based approach that facilitates the incremental growth of the network along with the integration of more sophisticated services for its end users [4].

## 2 The Organic.Edunet Federation of Repositories

Organic.Edunet (www.organic-edunet.eu) is a learning portal that provides access to quality digital learning resources on organic agriculture and agroecology, appropriate for school and university level, thus targeting pupils, students, teachers and researchers, as well as general learners [6]. Learning portals are Web portals that provide access to digital learning objects, usually stored in digital repositories [5, 9]. The scope of learning portals exceeds indexing and providing access to this content, focusing on the facilitation of content reuse and sharing [2]. The Organic.Edunet Web portal was developed in the context of the "Organic.Edunet: A Multilingual Federation of Learning Repositories with Quality Content for the Awareness and Education of European Youth about Organic Agriculture and Agroecology", in order to facilitate end-users search, retrieval, access and use of the content in the connected learning repositories.

When it comes to searching for learning objects Organic.Edunet, as it is usually the case with learning portals, provides a set of advantages to its users compared to traditional search engines. Its focus on a specific area makes it easier for users to locate resources that are mostly relevant to their topic of interest. Also, resources are validated by experts on their field assuring they are of a standard quality. Finally a number of search mechanisms are set on the portal to facilitate retrieval of relevant resources. As a result, access to high-quality, specialized content that is appropriate for educational and learning purposes is facilitated [8].

The Web portal is currently available in seventeen languages, thus already facilitating its use by users from around the globe. However, the portal's features and the facilitated open access to an ever expanding quality controlled catalogue of resources, result in user communities from even more countries becoming increasingly interested in using the portal. Importantly 23.73% of the unique visitors per day came from countries not having content in their language (for example France, Italy and China).

The Organic.Lingua project aims to capitalize on this international demand, by transforming Organic.Edunet into a truly multilingual platform. This will be achieved by re-engineering the architecture and enhancing the existing services, in order to eliminate the identified shortcomings in multilingual support and cross-language resource organization, search and retrieval. Indeed, the project is targeting three main goals: (i) to enhance the existing Organic.Edunet portal, introducing automated multilingual services that will further support its uptake from the targeted audiences, (ii) facilitate multilingual features of the portal and (iii) further extend its geographical and linguistic coverage. In effect, the successful deployment of the Organic.Lingua services and methodology will transform Organic.Edunet into a truly international platform with multilingual services, tools and content.

The following paragraphs provide details on the key statistics of the Organic.Edunet portal and some insight on the multilingual aspects of the provided services, user interface and content made available through the federated repositories.

### 2.1 Key Statistics of the Organic.Edunet Portal

Since its official launch in January 1st, 2010 until April 30th, 2011 Organic.Edunet has received around 41,000 visits. In total, the various pages of the portal have been accessed almost 174,000 times. At the same time there were more than 2,000 registered portal users with their number rising quickly. The table below summarizes some key portal-level statistics for each of the countries participating in the project.

**Table 1.** Overview of key statistics for the Organic.Edunet portal [6]

| Country | Visits / day | Page views / day | Unique visitors / | Bounces / day |
|---|---|---|---|---|
| Austria | 2.79 | 15.79 | 1.82 | 1.10 |
| Estonia | 7.41 | 23.86 | 6.38 | 4.32 |
| Germany | 3.02 | 11.67 | 2.17 | 1.60 |
| Greece | 18.87 | 101.64 | 11.74 | 8.22 |
| Hungary | 6.25 | 24.31 | 4.27 | 3.53 |
| Norway | 2.77 | 11.46 | 1.77 | 1.52 |
| Romania | 4.14 | 22.18 | 2.79 | 2.08 |
| Spain (excl. Alcala) | 3.01 | 9.46 | 2.24 | 1.15 |
| Sweden | 0.41 | 1.58 | 0.29 | 0.20 |
| United Kingdom | 1.87 | 8.94 | 1.19 | 1.02 |

The resources accessed through the Organic.Edunet Web portal cover a wide range of topics, ranging from organic agriculture and agroecology, to ecology, nature, green, biodiversity, environment, energy, food security and climate change. The resources also cover a wide audience, ranging from professors and teachers to students and lifelong learners and exist in various file formats, such as documents, presentations, videos and lesson plans.

The portal facilitates four search mechanisms for locating resources, i.e. text-based search, browse through subjects, tag-based search and semantic search as well as a special search function for educational scenarios.

Most of the resources available in the Organic.Edunet portal are provided from established educational institutes, where they were used for either research or educational purposes. A survey conducted has shown that 93.7% of the portal users believe that it has useful material that can be used for educating young people on organic agriculture and agroecology, thus serving its original goal [7]. Furthermore, this survey has shown that the majority of users (88.8%) were satisfied after using the portal. The degree of user satisfaction is crucial for those involved in the development and operation of a Web portal [10] and thus this last information is of extra significance.

## 2.2 Multilingual Needs of Multilingual Users

The user interface of the Organic.Edunet Web portal has already been translated into several (17) languages and has been designed in a way to further facilitate both multilingual and multicultural usage and navigation. Moreover, the knowledge representations recorded in the federated repositories (ontologies and metadata schemas) have been manually translated by human experts to several European languages. Figures 1 and 2 highlight the impressive availability of content and associated metadata descriptions in different languages, both in native form and translations , as of June 2011.

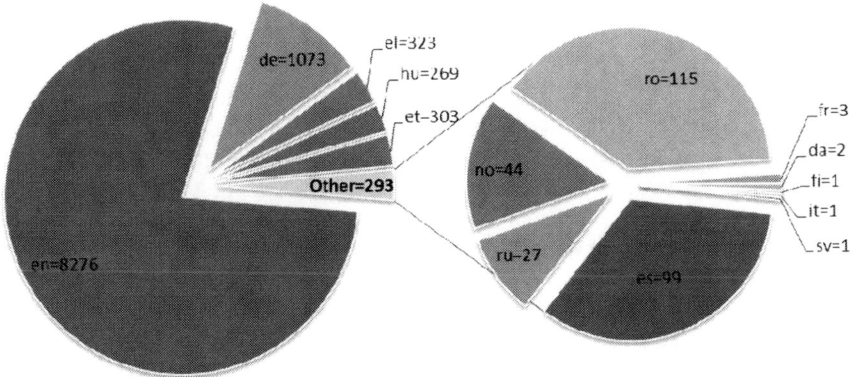


**Fig. 1.** Distribution of native language of resources in the Organic.Edunet federation

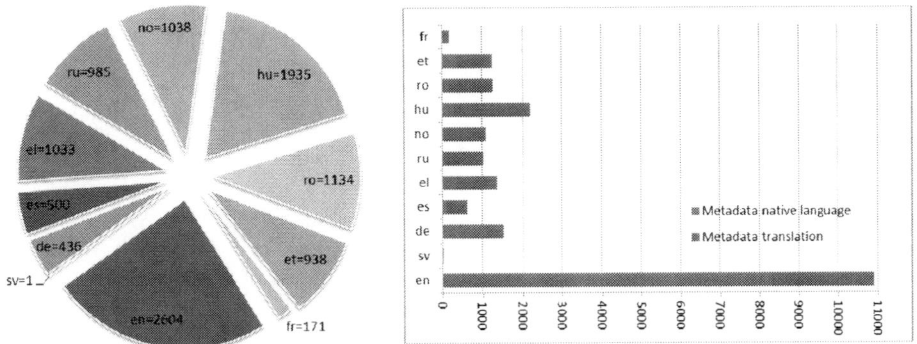


**Fig. 2.** Number of metadata translations (per language)

Apparently though, the portal's features and the facilitated open access to an ever expanding quality controlled catalogue of resources, result in user communities from even more countries becoming increasingly interested in using the portal. In fact, from the portal's official launch in January 2010 until April 30$^{th}$, 2011, almost 30.500 unique visitors from 159 countries requested access to the portal's content! However, the established localization process makes translation error prone and time consuming, and fails to provide quality and efficient cross-language functions and services. Indeed, the current solution relies completely on human effort in terms of translating and does not use existing linguistic resources and tools to help make resources and descriptions available in different languages and therefore facilitate the allocation, retrieval and sharing of knowledge. Moreover the different kinds of knowledge organization schemes, (namely, keywords -as expressed in metadata-, ontology terms and tags provided openly by users) are presented in several languages leading to restricted retrieval capabilities and fragmentation of the provided knowledge.

## 3 The Organic.Lingua Project

The Organic.Lingua project (www.organic-lingua.eu) aims to capitalize on this international demand, by transforming Organic.Edunet into a truly multilingual platform. This will be achieved by re-engineering the architecture and enhancing the existing services, in order to eliminate the identified shortcomings in multilingual support and cross-language resource organization, search and retrieval. Indeed, the project is targeting three main goals: (i) to enhance the existing Organic.Edunet portal, introducing automated multi-lingual services that will further support its uptake from the targeted audiences, (ii) facilitate multilingual features of the portal and (iii) further extend its geographical and linguistic coverage. In effect, the successful deployment of the Organic.Lingua's services and methodology will transform Organic.Edunet into a truly international platform with multilingual services, tools and content.

The main targeted audiences of Organic.Lingua fall in the following categories:

- **School teachers**, involved in teaching of Organic Agriculture (OA) & Agro-Ecology (AE) related topics, either directly (e.g. through environmental education activities) or indirectly (e.g. through educational activities on cross-disciplinary topics like biology, chemistry, economics or history).
- **Academics and researchers**, involved into teaching, tutoring or researching OA & AE related topics.
- **Content providers and other stakeholders**, involved in the creation, production, organization and/or publication of content around OA & AE. This group includes a wide variety of institutions and initiatives, such as: academic and research institutions, EU-funded initiatives, public organizations, private not-for-profit organizations, agricultural libraries or publishers, existing portals and repositories, as well as end-users themselves (communities of practice).

With regards to the Organic.Edunet Web portal, new multilinguality features are expected to further facilitate users who search for educational content, making it easier for them to access all resources relevant to their topic of interest. Fragmentation of knowledge as a result of language barriers is expected to be significantly reduced. To this end cross-lingual search will act as the key language technology developed. Also, the cost-effectiveness of the translation work will be increased by providing support in the process of labels, metadata and resource descriptions. Furthermore the generation of suggested descriptive metadata from text and translations is going to increase the completeness of metadata.

In essence, Organic.Lingua aims to provide a wider and more efficient multilingual access and effective exploitation of the OA & AE learning material available through the federated repositories. Our approach to this end is to analyze and re-engineer the service infrastructure, the related facilities and associated business models in order to support internationalization, multilingual access, effective localization and automatic translation services for the aggregated metadata descriptions and learning objects. In particular the re-engineering of the system architecture will integrate machine translation modules to enable on the fly translation of user interface elements and support for human translators (in the form of suggestions for manual translation), as well as cross-language search and retrieval services, contemplating the existing text-based search, multilingual keyword browsing, semantic search and tag-based (cloud) search.

# 4 Modeling the Content Lifecycle

In order to effectively tackle the challenging tasks of the Organic.Lingua project an analysis of the lifecycle of content within the Organic.Edunet repositories has been carried out, focusing in multilingualism and exploring the various ways different user categories interact with the system. The outcomes of this analysis will make the case for concrete language technology requirements, which will be the drivers for selecting the appropriate linguistic techniques and tools that will be integrated in the current architecture of the system. In doing so, the existing lifecycle of contents in Organic.Edunet will be revised, exhaustively identifying opportunities to apply language technologies for better multilingual support.

Before we carry on with details on this analysis, it is interesting to look into the definition of repository itself. A definition which perfectly accounts for the Organic.Edunet approach is indeed the following:

> "A university-based institutional repository is a set of services that a university offers to the members of its community for the management and dissemination of digital materials created by the institution and its community members. It is most essentially an organizational commitment to the stewardship of these digital materials, including long-term preservation where appropriate, as well as organization and access or distribution."[1] [12]

## 4.1 Main Actors and Roles

In order to better understand the content publication lifecycle within the Organic.Edunet federation of repositories of learning resources, it would be useful to describe the different roles of the people engaged and the affiliated organizations. The main roles operating within the Organic.Edunet platform and the affiliated organizations are presented in Table 2 and described below. Please note that we do not investigate roles and procedures outside the Organic.Edunet platform. In that sense, for example, the creation of digital learning resources is not analyzed, since it happens outside the context of the Organic.Edunet platform.

- **Content provider:** A content provider is a formal community or an institution that deposit digital resources, to be publicly available, in an identified repository. The repository can be thematic or institutional. The content provider is responsible for (a) the quality of the digital resources, and (b) the comprehensive description of the resource with appropriate and valid metadata.
- **Content Collector:** The content collector is the person responsible within each content provider for deciding on the learning objects that should be shared through the Organic.Edunet Web portal and makes sure that they are compliant to the portal's scope, policies and practices. For new content providers connecting to the Organic.Edunet Network or for current content providers wanting to contribute more learning objects, the role of the Content Collector is vital, as he is the person that initiates the process.

[1] Note that this definition refers to university-based repositories, but it suits equally well for any institutional repository with content related to education and research.

**Table 2.** Overview of the roles involved in the content lifecycle

| Role | Responsibility | Other terms used to describe the role |
|---|---|---|
| **Content Provider** | | |
| Content collector | Selects the content that will be available on the Organic.Edunet platform and pre-checks it against the Core Criteria | Content Curator |
| Metadata Editor | Uploads of the content to the repositories and annotates them with metadata. | Metadata Curator, Metadata Author |
| **Organic.Edunet Federation** | | |
| Technical Expert | Is responsible for arranging the way in which each content provider is connected to the Organic.Edunet. | Repository Curator |
| Peer reviewer | Reviews the metadata records to ensure the high quality of the metadata production. | Domain Expert, Metadata Curator |
| Validator | Takes the final decision for the acceptance of an entry as a LO on the Organic.Edunet platform. | |

- **Metadata Editor:** The role of a metadata editor is to check the lack of information and to enrich the records before being integrated into Organic.Edunet. This process can be highly time consuming for metadata editors and makes the new multilingual and innovative components to be introduced more important. Expertise on the topics of OA and AE is needed, since at this point the collected content will have to be evaluated and processed in order to prepare the Organic.Edunet metadata fields. Though some of the fields do not require special expertise on OA and AE issues (such as the date of creation of the content or authors), important fields such as key words and ontologies have to be completed by experts of OA and AE or at least from people that have knowledge of the related issues beyond the basic level. Furthermore the metadata authors are also responsible for translating the metadata in English. This means that they should possess a good knowledge of English and the OA and AE terms in this language to be able to correctly attribute the meaning of the metadata elements (such as title, descriptions, etc.).
- **Technical Expert/ Repository Curator:** The role of the Technical Expert or Repository Curator is to facilitate the process by providing the technical knowledge and the solutions needed for a Content Provider to be able to contribute content to the Organic.Edunet Web portal. He will serve as a contact point for all technical aspects of the procedure and as such he is expected to have expertise in issues related to databases, digital repositories and metadata. Also a deep knowledge of the Organic.Edunet architecture and technical characteristics is required.

- **Peer Reviewer:** In order to ensure a high quality of the metadata, this meaning relevance of the metadata with the content, completeness of the metadata and enrichment of the metadata a peer review process is necessary. This process will be done by Peer Reviewers within the Organic.Edunet Network which should be experts in the fields of OA & AE. The Peer Reviewers will evaluate content against set criteria in terms of quality and appropriateness of the contained information and completeness of metadata records. Furthermore, the peer reviewers should not be involved in the process of production of metadata (in terms of creation in the local language or translation) of the LOs that they are reviewing, so as to make sure that they can provide an objective, bias-free opinion and critic.
- **Validator:** The Validator is the person within the Organic.Edunet Network responsible for taking the final decision for the acceptance of an entry as a learning object on the Organic.Edunet platform. The Validator should be experienced in the production of digital content for the internet and web content platforms and possess good knowledge of issues related to digital repositories. It is very important that he will be capable of evaluating content in terms of quality of the contained information and usability for educational purposes as well as having good understanding of Organic.Edunet scope, practices and policies

### 4.2 Quality Assurance Policy

The Organic.Edunet quality strategy has the basic aim to assure quality of the learning materials (criteria and indicators for judgment and evaluation), quality of the workflow processes of the participating repositories (procedures, rules and processes) and give recommendations for the quality of educational scenarios which make use of learning materials. The challenge is therefore threefold, equally set in the field of defining processes for the handling of learning materials within the repositories, finding suitable criteria, indicators and guiding principles to evaluate the learning materials themselves, and a set of methodologies and recommendations for educational scenarios in the field of education for organic agriculture.

Organizations contribute through the aggregation of their institutional collections within the Organic.Edunet federation. These collections may use the Organic.Edunet Repository Tool to organize their resources and set up a compliant repository, or use their own repository management system and make their metadata available through the Open Access Initiative – Protocol for Metadata Harvesting (OAI-PMH). Examples of existing institutional collections already connected to Organic.Edunet are the Food & Agriculture Organization (FAO) and of the ECOLOGICA Association.

In order for an institution to make its collection of resources available through the Organic.Edunet portal, a publicly available Quality Assurance procedure that is compliant to the Organic.Edunet requirements must be adhered and employed. To assess this compliance the following process would have to be undertaken:

1. An institutional contributor registers to the portal and indicates the interest to contribute learning resources.
2. The Quality Centre starts an accreditation process in which the evaluation and quality procedures of the institutional contributor are identified and checked against the quality charter of Organic.Edunet.

3. For the accreditation the institutional provider has to complete a self-assessment form with details about their own quality strategies, this will be audited by a peer-review team of two peer-reviewers and further inquiries will be done if necessary.
4. The institutional provider will receive the "Organic.Edunet Quality Certificate" for three years which permits the contribution of learning repositories, either through joining the federation or through repeatedly or mass uploading of content.
5. After three years a re-audit has to take place.
6. All contributed resources will be checked as "quality certified" and user evaluations/ ranking mechanisms will be available.

### 4.3 Business Process Modeling

The analysis of the content lifecycle, including the interactions and the tasks engaged by the different roles within the Organic.Edunet platform, as they were described in the previous paragraphs, has been modeled using BPMN diagrams. Business Process Modeling Notation (BPMN) is a standard notation that is easily understandable by business users. The business users include "the business analysts that create the initial drafts of the processes, the technical developers responsible for implementing the technology that will perform those processes, in addition to the business people who will manage and monitor those processes" [13]. The lifecycle modeling is in fact an ongoing task and will be validated by project partners within the next period. BPMN models have proven both effective for capturing the interactions among all players and efficient for documenting the processes and facilitating discussion and consensus among all involved parties. The following diagram presents an overview of the key phase of Organic.Edunet's content lifecycle, that of publication of new content in an associated repository.

## 5 Next Steps – The Road Ahead

The Organic.Lingua Vision, as it was defined in [14] is to *"Boost the usage, exploitation and extension of knowledge on Organic Agriculture and Agroecology by using new / innovative ways to facilitate searching and to minimize the segmentation of knowledge due to lingual barriers through the introduction of automated multilingual services, supported by a quality-controlled collaborative and interactive approach while developing a sustainable low-cost and re-usable solution to the problem of multilinguality across Europe and the whole globe."*

Indeed, the newly developed services and tools in the Organic.Lingua framework will try and study the ways in which the newly introduced multilingual components can be integrated in the existing content population procedures of Organic.Edunet as well as the ways in which the revised content population procedure can be adopted by new Organic.Edunet federation providers. The project partners will engage in research and work-tasks focused on enhancing the approach to linguistic support and coverage of the existing Organic.Edunet infrastructure and content. More specifically, work within the project will concentrate on the following objectives:

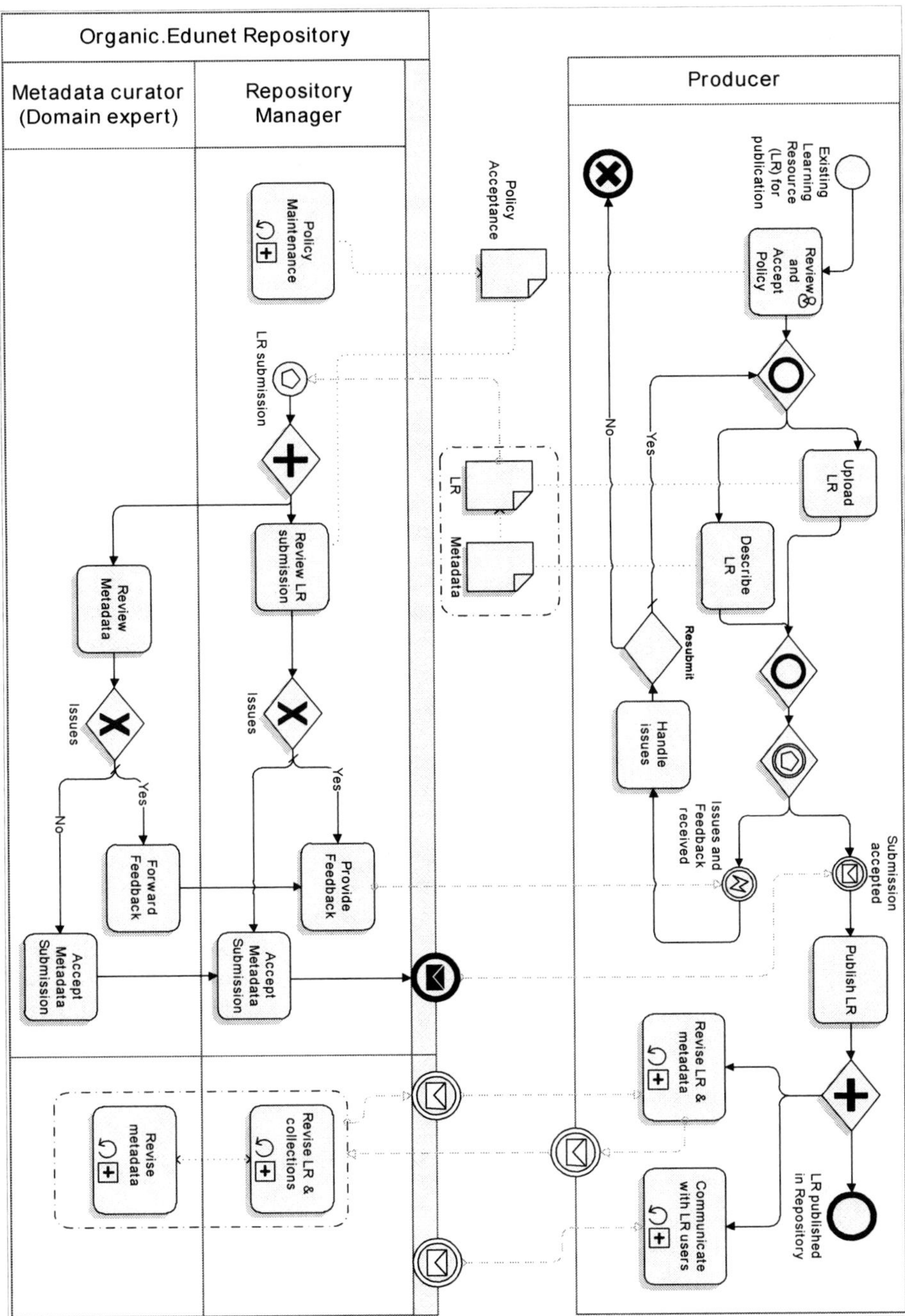


**Fig. 3.** BPMN overview of the content lifecycle

1. Identification of user needs regarding linguistic support in on-line information services, using the existing features of the Organic.Edunet Web portal.
2. Identification and exploration of new or anticipated user interface, functional and technical requirements related to linguistic technologies, analyzing established services and approaches, perceived gaps and weaknesses of existing solutions.
3. Development and validation of innovative and effective methods, processes and workflows for linguistic support in Organic.Edunet, over and beyond mere integration of linguistic technology.
4. Enhancement, extension and re-engineering of the existing Organic.Edunet Web portal towards a wider and cost-effective multilingual support, broadening in this way the current audience and impact of the service.
5. Deployment of large pilot trials that will demonstrate a cost-effective approach for delivering language technologies in existing Web portals based on open standards and open source software, which can be easily transferred to other domains.
6. Development of guides of good-practice and awareness on the practical application of language technologies, fostering the adoption of tools and techniques with predictable and validated results.

Open linguistic infrastructures will be given a special relevance in the technology assessment tasks of Organic.Lingua, without excluding other technologies that are currently not open. This enables potential synergies with projects addressing methodological foundations as FlaReNet and T4ME, and also with other initiatives or projects with similar objectives.

In specific terms, the Organic.Edunet service will benefit from the significant broadening of its current user base, covering content in more languages, integrating a large set of innovative aspects provided by the Organic.Lingua framework. In particular, support for the full learning resource creation lifecycle, will open opportunities for innovating both in digital description and preservation practices and also in finding specificities of the use of language technologies. Moreover, the thorough analysis of the user interaction needs for translation of metadata and ontologies in a practical and real-world setting. This represents an opportunity to advance in the knowledge about the application of language technologies with an emphasis on effectiveness and cost-effectiveness that would make multilingual support more effective. The planned exploration, evaluation and measurement of the impact of cross-lingual search in a real world setting, will enable the achievement of conclusions and metrics that relate to the effects of multilingualism in on-line information services. Conscious selection of language technologies according to clearly defined user requirements along with controlled piloting and evaluation will enable a comparison of different technological applications and the current support of different language pair combinations. Furthermore, machine translation used as an aid in the translation of the different elements of the system (metadata, ontologies, descriptions, etc.), will enable opportunities for combining some of these elements to improve the overall process. Automated summarizing and keyword extraction, used as aids for the generation of meta-descriptions, will allow for a comparison of fully manual translations provided in the existing version of the system. Last, but not least, cross-lingual facilities in information retrieval, will make resources more accessible by breaking language barriers.

**Acknowledgments:** The work presented in this paper has been carried along the Organic.Lingua project, co-funded by the European Commission under the ICT Policy Support Programme (ICT PSP).

## References

1. Burchardt, A., Rehm, G., Sasaki, F. (eds.): The Future European Multilingual Information Society. Vision Paper for a Strategic Research Agenda (2001)
2. Duncan, C.: Digital Repositories: the 'back-office of e-Learning or all e-Learning?'. In: Proc. of ALT-C 2002, Sunderland, September 9-11 (2002)
3. European Commission, Directorate-General for Translation: Size of the language industry in the EU, Kingston Upon Thames (2009)
4. Manouselis, N., Kastrantas, K., Sanchez-Alonso, S., Cáceres, J., Ebner, H., Palmer, M., Naeve, A.: Architecture of the Organic.Edunet Web Portal. Int. J. of Web Portals (IJWB) 1(1), 71–91 (2009)
5. Holden, C.: From Local Challenges to a Global Community: Learning Repositories and the Global Learning Repositories Summit, Academic ADL Co-Lab (November 2011)
6. Palavitsinis, N., Protonotarios, V., Manouselis, N.: Applying Analytics for a Learning Portal: the Organic.Edunet Case Study. In: Proc. of the 1st International Conference Learning Analytics and Knowledge (LAK 2011), Banff, Alberta, Canada (March 2011)
7. Protonotarios, V., Palavitsinis, N., Manouselis, N.: Organic.Edunet Web Portal: User Satisfaction Analysis. In: Proc. of the Joint Congress of EFITA 2011 and WCCA 2011, Prague, Czech Republic, July 11-14 (2011)
8. Protonotarios, V., Palavitsinis, N., Ungur, M., Manouselis, N.: A learning portal for organic education: A reflection on the user experience (2011)
9. Richards, G., McGreal, R., Hatala, M., Friesen, N.: The evolution of learning object repository technologies: Portals for On-line Objects for Learning (Electronic Version). Journal of Distance Education 17(3) (2003)
10. Sampson, D., Manouselis, N.: A Flexible Evaluation Framework for Web Portals based on Multi-criteria Analysis. In: Tatnall, A. (ed.) Web Portals - the New Gateways to Internet Information and Services. Idea Group Publishing, Hershey (2005)
11. Uszkoreit, H.: Language Technology A First Overview. German Research Center for Artificial Intelligence, 1–4 (2000)
12. Lynch, C.A.: Institutional Repositories: Essential Infrastructure for Scholarship in the Digital Age. ARL (226), 1–7 (2003)
13. Rius Gavidia, A., Sicilia Urbán, M.A., García-Barriocanal, E., Macarro Palazuelos, G.: Towards Automated Specifications of Scenarios in Enhanced Learning Technology. Intl Journal of Web-Based Learning and Teaching Technologies 3(1), 68 (2008)
14. Organic.Lingua deliverable, D2.1 White Paper on Organic.Lingua Vision (2011)

# A Survey on the Importance of Learning Object Metadata for Relevance Judgment

Panos Balatsoukas[1], Emmanouel Garoufallou[2], Stella Asderi[3], and Rania Siatri[2]

[1] Department of Computer and Information Sciences, University of Strathclyde, UK
Panagiotis.Balatsoukas@cis.strath.ac.uk
[2] Department of Library Science and Information Systems, ATEI of Thessaloniki, Greece
mgarou@libd.teithe.gr, rsiatri@yahoo.gr
[3]American College of Thessaloniki, Library, Greece
sasderi@act.edu

**Abstract.** The purpose of this study was to investigate university students' perceptions about the usefulness - importance of learning object metadata elements of the UK LOM Core application profile when judging the relevance of learning objects. In order to address this objective an online questionnaire survey was conducted among university students using a clustered sampling technique. A total of 252 students respondent to the questionnaire. Participants represented different disciplines (for example, from Humanities & Social Sciences to Sciences & Engineering), levels of familiarization with the web and educational level (postgraduates and undergraduates). The results showed that participants preferred content related and educational metadata elements as well as metadata elements about the date and the language of a learning object as opposed to technical metadata elements or metadata elements about the version of the learning object. The conclusions show that these findings can have implications in the design and use of learning object metadata schemas for students in the UK Higher Education.

**Keywords:** Learning object metadata, LOM, educational metadata, usefulness, relevance judgment.

## 1 Introduction

Metadata can support multiple roles in the provision of e-learning, including the description, management, retrieval and re-usability of learning objects and learning designs, as well as the delivery of on-demand and personalised learning experiences to users. For this purpose, a variety of educational or learning object metadata standards and schemas have been developed, including the Dublin Core Education (DCEd) (DCMI Education Working Group 2006) and the IEEE Learning Object Metadata (LOM) standard (IEEE Learning Technology Standards Committee 2002). These standards and schemas cover many of the educational, technical and content-related characteristics of learning objects or learning content in general.

In the United Kingdom, the use of learning object metadata, such as the IEEE's LOM standard, is an important predicate for the successful implementation of

E. García-Barriocanal et al. (Eds.): MTSR 2011, CCIS 240, pp. 300–311, 2011.

e-learning systems in higher and further education. For example, the JISC's e-learning programme (`http://www.elearning.ac.uk/`) has funded several UK projects on e-learning, some of which have investigated the creation and metadata annotation of learning objects in higher and further education, the creation of auto-generated learning object metadata, and the establishment of learning object metadata creation workflows. Although the LOM standard has been an important component of all these efforts, little is known about users' perceptions regarding its usefulness for finding and evaluating information.

The purpose of this study was to investigate university students' perceptions about the usefulness - importance of learning object metadata elements of the UK LOM Core application profile when judging the relevance of learning objects. In particular, the objectives of this study were:

- To identify the learning object metadata elements of UK LOM Core perceived by university students as being important when judging the relevance of learning objects.
- To examine the effects, if any, of educational level (undergraduates versus postgraduates), subject knowledge (Science & Engineering versus Humanities & Social sciences), and use of the web (expert versus novice users) on respondents' perceptions of the importance of learning object metadata elements.

## 2 Literature Review

There is no formal definition of learning object metadata or metadata for learning objects. Some authors, including Karampiperis and Sampson (2003), argue that metadata for learning objects should support the retrieval of learning objects 'in an educationally efficient and effective way'. In the context of this paper the term *learning object* is used to denote an electronic resource that is created from smaller components (such as data and information objects) and that can support one or more learning objectives (Balatsoukas *et al.* 2008). A learning object can be re-used in multiple educational contexts as part of larger modules and courses (Polsani 2003).

### 2.1 The Learning Object Metadata (LOM) Standard

The draft IEEE standard (IEEE LTSC 2002) proposes the structure and semantics of metadata elements used for the description of learning objects. The standard describes various characteristics of a learning object by defining an approximate number of 80 metadata elements, grouped under nine general categories. These are: 1. General (Descriptive information of the learning object as a whole, such as: identifier, title, language, description, keyword, coverage, structure, aggregation level); 2. Life-cycle (Elements related to the creation or revision history of the learning object as well as information about those who have contributed to the development, creation, revision of the learning object); 3. Meta-metadata (Information regarding the creation of the LOM metadata record); 4. Technical (The category consists of elements that describe the technical characteristics of the learning object such as the format, size, location and technical requirements); 5. Educational (The educational category aggregates elements regarding pedagogical and educational information about the use of a learning object. Such elements include: Interactivity Type, Learning Resource Type, Interactivity Level, Semantic density, Audience/Intended End User Role, Context, Typical Age Range, Difficulty, Typical Learning Time, Description, Language); 6. Rights (The rights category includes information regarding the intellectual property rights and conditions of use of the learning object); 7. Relation (The relation category presents information about the relationship of the described learning object with other objects); 8. Annotation (The annotation category provides a comment regarding the use of learning objects); 9. Classification (The classification category classifies the content of the learning object based on an appropriate classification system).

The LOM standard formed the basis for the development of many nation-wide and project-specific LOM application profiles. Some of the most prominent LOM application profiles include the CanCore project in Canada, the UK LOM Core in the UK and the SCORM LOM metadata application profile developed by the US Advanced Distributed Learning (ADL) project.

## 2.2 The UK LOM Core

The UK LOM Core is an application profile of the IEEE LOM. It does not propose new elements but specifies a basic minimum set of mandatory elements and includes recommendations and guidelines about the use of these elements in UK higher and further education. In particular, the UK LOM Core Working draft 0.3 defines twenty-seven data elements as mandatory (Campbell 2004).

It is worth noting that although the purpose of the UK LOM Core is to support the description of learning objects, elements related uniquely to the structure, granularity, pedagogical context and use of a learning object have not been defined as mandatory elements. The rationale behind this decision was based on the fact that these elements and their associated vocabularies have not been widely implemented yet and their use is not well understood.

## 2.3 User Studies on the Usefulness of Learning Object Metadata Elements

Some empirical research has been conducted for assessing the usefulness of learning object metadata elements. For example, the *MetaTest* project in U.S. employed a user study examining what metadata elements were most frequently used by users to search for and evaluate learning objects (Liddy, et al, 2003). The metadata schema used in this study was based on GEM (a Dublin Core application profile). The results of the study showed that users preferred viewing the descriptions of learning objects rather than the other metadata elements of GEM. When the description was not present then users tended to focus on other metadata elements and specifically the *Title*, *Subject*, *Rights* and *Creator* elements.

Plodzien et al (2006) investigated the perceived importance of the type of learning object characteristics. Forty seven university students and teachers evaluated 56 learning objects (including manuals, e-books and educational multimedia resources) against 18 criteria of quality developed specifically for the purpose of this study. The results showed that both teachers and students perceived the content, presentation of new knowledge and the use of examples of applying new knowledge as the most important criteria for evaluating the quality of learning objects. In addition, teachers also perceived the criterion of 'self-evaluation' as an important instructional component, while students perceived the existence of a dictionary of key concepts and feedback as important.

In a similar study, Small et al (1998) found a variety of criteria that school teachers employed to judge relevance of learning sources, such as: topic, subject, content description, materials, forms, grade level, purpose, lesson outline, summary, assessment, prerequisite, audience and title of lesson plan among others.

In summary, the review of the literature showed a lack of studies examining users' perceptions of the importance of learning object metadata elements for evaluating the relevance of learning material on the web. The studies by Small et al (1998) and Plodzien et al (2006) were not focused on specific learning object metadata elements, while the MetaTest study did not show an in-depth view of users' perceptions.

## 3 Methodology

The method of an online questionnaire survey was selected as the most appropriate for the needs of this study. The online questionnaire consisted of two sections:

- Section A – This section collected demographic data about the respondents' 1. level of study (undergraduate or postgraduate level), 2. subject area (*Science and Engineering* or *Humanities and Social Sciences*) and 3. level of familiarity with electronic information and learning resources and the WWW. These constitute the independent variables that were employed for statistical testing.
- Section B – This section lists the questions that investigated respondents' perception about the importance of each UK LOM Core element under examination (one question per metadata element). This section lists a total of 23 items (i.e. metadata elements) and a Likert scale was used as the response format for each item (7-point scale, where 1= not at all important and 7=very important) (Table 1). Likert scores have been successfully employed by other researchers in the past to measure users' perceived usefulness or importance of specific learning object characteristics for relevance judgment (see for example, Small et al. 1998; Plodzien et al 2006). Similarly, the selection of a 7-point Likert scale was built on previous experience gained by researchers in the field (Choi and Rasmussen 2002). It was thought that a larger rating scale would confuse respondents and reduce the reliability rate of the response (Nunnally 1978).

The selection of the 23 metadata element items was based on three factors:

1. These elements were defined as mandatory by the developing UK LOM Core.
2. These elements were proposed by the CanCore, the Canadian application profile of LOM, to be more appropriate for presentation in search interfaces that help users to make decisions about the relevance of a learning object.
3. In addition to the mandatory elements of the UK LOM Core few more elements were included in the questionnaire. These were:
   a. the *Difficulty* element. This element was not proposed for mandatory use by the UK LOM Core. However, the decision to include the *Difficulty* element in this study was made in order to enhance the number of educational metadata elements, as well as because the particular element was identified as an important criterion for relevance judgment by previous studies (Small et al. 1998).

**Table 1.** The 23 metadata elements

| | |
|---|---|
| 1. Costs | 13. Format |
| 2. Summary – Description | 14. Tutor's comments |
| 3. Keywords – Subject terms | 15. Author |
| 4. Title | 16. Installation |
| 5. Language | 17. Copyright |
| 6. Learning resource type | 18. Relation |
| 7. Date | 19. Learning time |
| 8. Interactivity | 20. Technical requirements |
| 9. Audience | 21. Structure |
| 10. Difficulty | 22. Version |
| 11. Coverage | 23. Size |
| 12. Location (Identifier) | |
| Likert score 1 = Not at all important – 7 = Very important | |

b. the *Relation* metadata element which is common both in the UK LOM Core and Dublin Core but it is not perceived to be a mandatory element in the UK LOM Core,

c. the *Comments* element that represents the *Annotation* category of LOM, as well as,

d. the *Installation* and *Structure* elements that characterize the re-usability and technical characteristics of a learning object.

Following the results of a pilot study, some items (i.e. metadata elements) in the second section of the questionnaire were accompanied with examples or short definitions of their scope. This action minimised the risk of presenting respondents with ambiguous metadata terms.

Finally, the selection of metadata elements from UK LOM Core was based on the fact that the particular schema provides a richer coverage of metadata elements for learning objects than any other educational metadata standard or schema (such as ARIADNE or Dublin Core-Education).

### 3.1 Sample

The population of this study consisted of Loughborough University students. The study examined the perceptions of undergraduate and postgraduate students as well as students from the Sciences – Engineering and Humanities – Social sciences.

Given the large size of the student population (17,601 registered students) there is a need for sampling techniques that can balance the requirements of this study. For example, the selected sample should be inclusive enough so that it does not exclude representatives from the two educational levels (undergraduates – postgraduates) and subject areas (Science – Engineering and Humanities – Social sciences). For the needs of this study a cluster sampling technique was implemented.

In the context of this research, cluster analysis has been selected as there is no sample frame for the whole population of Loughborough University available to the researcher. Based on this technique, two lists of all the academic departments within the university were compiled. Each department represented a cluster. The first list included all the departments from Science and Engineering faculties (14 departments), while the second listed all departments within the Faculty of Humanities and Social Sciences (10 departments). Seven departments were randomly selected from the first list and five departments from the second list. Therefore, a 50% of the total number of departments in each cluster was selected. All the students within each of the 12 departments formed the sample units of this research. The 12 departments selected for this study are shown in Table 2:

**Table 2.** Departments selected for participation

| *Science and Engineering* | *Humanities and Social Sciences* |
|---|---|
| Electronic and electrical engineering | Design and Technology |
| Chemical Engineering | English & Drama |
| Human Sciences | Geography |
| Information Sciences | Politics, International relations and European studies |
| Chemistry | Social Sciences |
| School of Mathematics | |
| Aeronautical and Automotive engineering | |

### 3.2 Questionnaire Distribution and Return Rate

An online version of the questionnaire was circulated via e-mail to the selected sample. Finally, a total of 287 students participated in the study. However, 252 responses qualified for inclusion in the analysis. This corresponds to a response rate of approximately 4.2%. Thirty-five questionnaires were excluded from the analysis because these were not completed by participants. From a more pragmatic perspective a response rate of approximately 5% is common in questionnaire surveys and many researchers report their findings on this basis (Smith 2009).

## 4 Results

### 4.1 Reliability and Validity of the Questionnaire

Two tests were performed to investigate the reliability and construct validity of the questionnaire developed for the purpose of this study. In particular, Construct validity involved the correlation between individual scales and the total mean of the questionnaire. The findings of the correlation tests revealed that all questionnaire items were significantly correlated with the overall mean score of the questionnaire at the 0.01 level revealing a good level of construct validity. The Cronbach's Alpha value for reliability was calculated after the data collection. Cronbach's Alpha for the reliability of the questionnaire from the current sample was 0.849 well above the 0.70 threshold.

Except for the two measures of construct validity and internal reliability, the items of the questionnaire were also tested for content or face validity. This type of validity examined whether a particular instrument covers the main content components or constructs of inquiry.

## 4.2 Participants' Profile

A total of 207 participants in the study were undergraduate students (first, second, third and final year students). However, only 45 postgraduate students (taught and research postgraduates) provided input to this questionnaire. Table 3 presents the frequencies and percentages of undergraduate and postgraduate participants.

**Table 3.** Participants' level of study

| **Level of Study** | **Frequency** | **%** |
|---|---|---|
| First Year | 82 | 32.5 |
| Second Year | 59 | 23.4 |
| Third Year | 53 | 21 |
| Forth Year | 13 | 5.2 |
| **Sub-total for Undergraduates** | 207 | 82.1 |
| Taught Postgraduates | 12 | 4.8 |
| Research Postgraduates | 33 | 13.1 |
| **Sub-total for Postgraduates** | 45 | 17.9 |
| **Total** | 252 | 100 |

Responses were obtained from seven out of 12 departments selected for participation in the study. In particular, the departments that respondent to this call for participation were: Chemistry (n=30, 11.9%); Design and Technology (n=10, 4%); English and Drama (n=18, 7.1%); Geography (n=59, 23.4%); Information Science (n=24, 9.5%); Politics (n=49, 19.4%); Human Sciences (n=62, 24,6%). In total, 119 participants (47,2%) were from a 'Science & Engineering' department, while 133 respondents (52,8%) were students of a 'Humanities & Social Sciences' department. All participants in the study were very familiar with the use of WWW search engines, the library OPAC, the *Learn* (institutional e-learning system), e-journals, e-reading lists and the *MetaLib* portal. However, they were less familiar with the use of e-book collections and the institutional repository.

Finally, the majority of the respondents in the survey were frequent users of the WWW which was used for more than five hours per week (n=185, 73.4%). However, a total of 67 participants used the WWW less than four hours per week (26.6%) (these were characterised as non frequent users of the WWW).

## 4.3 Descriptive Statistics

Table 4 summarises the Likert scores of each metadata element. The five most useful elements for judging the relevance of learning objects were: *Costs* (6.08), *Summary – Description* (5.67), *Keywords* (5.24), *Title* (5.02) and *Language* (4.89). Based on these results students valued more the content-related metadata as well as metadata about the language of a learning object. The cost of learning objects was the most important criterion for selecting a particular resource. The five least important metadata elements, however, were: the *Size* of the learning object (3.29), the *Version* of the learning object (3.70), the *Structure* of a learning object (3.80), the *Technical requirements* for using the learning object (4.09) and finally, the *Learning time* (i.e. the time needed for a learner to interact with the learning object) (4.17). It appears that many of the metadata elements that underpin the creation and technical manifestation of a learning object (such as *Version*, *Size*, *Technical requirements* and *Structure*) were considered less useful for relevance judgment than the metadata elements about the content, authorship, and educational characteristics. In addition, the *Learning time* was the least useful educational metadata element (when compared to other elements of this category, such as *Interactivity*, *Learning resource type*, *Audience* and *Difficulty*). Finally, the Likert scores for most of the educational metadata elements ranged between 4.19 and 4.83 which represent a positive behaviour towards this type of elements.

### 4.4 Differences between Learning Object Metadata Elements

A Friedman test was conducted to investigate statistically significant differences between the metadata elements under examination. The Friedman test showed that the level of importance varied significantly across the 23 learning object metadata elements: $\chi^2$ (22, N=252) = 1001.553, p<0.0005, two-tailed. In order to identify which metadata elements differed significantly a set of Wilcoxon tests were performed that compared several pairs of metadata elements. A close examination of the results showed that the three most important metadata elements (*Costs, Summary – Description* and *Keywords- Subject terms*) differed significantly from the remaining metadata elements either at the 0.01 or 0.05 level. Also, the educational metadata elements (*Audience, Learning resource type* and *Difficulty*) and three general purpose metadata elements (*Title, Interactivity* and *Coverage*) differed significantly from the *Structure*, *Relation* and *Copyright* metadata elements as well as from the category of metadata that represented the technical characteristics of a learning object (e.g. *Technical requirements*, *Installation*, *Format*, *Location*). These differences were significant at the 0.5 level. Non significant differences were observed between the metadata elements that represented technical characteristics (*Technical requirements*, *Installation*, *Format*, *Location*) and the *Learning time*, *Structure*, *Copyright* and *Author* metadata elements. The interpretation of the findings from the Wilcoxon tests and the results shown in Table 4 suggest that the availability (i.,e. *Cost*) and the topicality (i.e. *Summary – Description* and *Keywords*) of the learning object are the most important characteristics of a learning object for relevance judgment. However, it becomes evident that educational and content related metadata elements were significantly more important for relevance judgment than the metadata elements that describe the technical characteristics, the structure, the author and the copyright of a learning object as well as the relation of a learning object with other learning objects.

**Table 4.** The Usefulness of Learning Object Metadata Elements

| Metadata Elements | Likert Scores | Metadata Elements | Likert Scores |
|---|---|---|---|
| Costs | 6.08 | Format | 4.48 |
| Summary – Description | 5.67 | Tutor's Comments | 4.31 |
| Keywords – Subject Terms | 5.24 | Author | 4.30 |
| Title | 5.02 | Installation | 4.27 |
| Language | 4.89 | Copyright | 4.23 |
| Learning Resource Type | 4.83 | Relation | 4.19 |
| Date | 4.82 | Learning Time | 4.17 |
| Interactivity | 4.81 | Technical Requirements | 4.09 |
| Audience | 4.77 | Structure | 3.80 |
| Difficulty | 4.77 | Version | 3.70 |
| Coverage | 4.70 | Size | 3.29 |
| Location (Identifier) | 4.49 | | |

## 4.5 Impact of the Educational Level on the Usefulness of Learning Object Metadata Elements

The results of the Mann-Whitney test revealed that there were statistically significant differences between undergraduates and postgraduates in the importance they attach to the usefulness of three learning object metadata elements. These were the *Author*, *Technical requirements* and the *Learning time* elements. In particular, postgraduate students tended to perceive the *Author* (U=3624.5, N1=207, N2=45, p=0.028, two tailed) and *Technical requirements* (U=3526.5, N1=207, N2=45, p=0.015, two tailed) elements as more important than undergraduates. Undergraduate students, however, found the *Learning time* (U=3594.5, N1=207, N2=45, p=0.022, two tailed) element to be more important.

## 4.6 Impact of Subject Discipline on the Usefulness of Learning Object Metadata Elements

Also, the results of the Mann-Whitney test showed that there were no significant differences between students studying a Science or Engineering subject and students studying a Humanitites or Social Sciences subject in the importance they attach to the usefulness of most learning object metadata elements. This was not the case for two metadata elements where statistically significant differences were observed at the 0.05 level. These were: the *Requirements* element (U=6779.500, N1= 119, N2=133, p=0.046, two tailed), and the *Installation* (U=6565.000, N1=119, N2=133, p=0.018, two tailed). In particular, students from the *Sciences or Engineering* perceived these two elements as more important for judging the relevance of a learning object when compared to students from the *Humanities or Social sciences*.

### 4.7 Impact of Frequency of Use of the WWW on the Usefulness of Learning Object Metadata Elements

Finally, the results of the Mann-Whitney test revealed that there were no significant differences between the two levels of use of the WWW (frequent versus non frequent users) in the importance they attach to the usefulness of most learning object metadata elements. Significant differences at the 0.05 level were observed only in the case of the *Summary - Description* element (U= 5093.500, N1=67, N2=185, p=0.024, two tailed) which frequent users of the Web found it more important for judging relevance than the less frequent users of the Web.

## 5 Conclusions

The survey on users' perceptions of the importance of learning object metadata was focused on specific metadata elements already included in the UK LOM Core application profile. The results of the survey showed that participants preferred content-related and educational metadata elements as well as metadata elements about the date and the language of a learning object as opposed to technical metadata elements or metadata elements about the version of the learning object. In particular, participants perceived important the use of the *Title*, *Summary* (Description) and *Keyword* metadata elements when judging the relevance of a learning object according to its topical – subject relatedness. This is in accordance with the results of other studies that showed that different types of participants (including academics, undergraduate and postgraduate students) perceived the *Description*, *Title* and *Keywords* to be the most useful metadata elements for judging relevance in scholar IR systems (e.g. Wang and Soergel, 1998). These metadata elements were also perceived to be among the most useful elements by K-12 teachers (Small et al., 1998) university teachers and higher education students (Plodzien et al. 2006; Liddy, et al. 2003).

The findings also showed that participants perceived some educational metadata to be among the most important metadata elements for judging the relevance of learning objects. These were the *Interactivity* (includes both level and type)[1], *Learning resource type*, *Difficulty* and *Audience*. These metadata elements had a mean Likert score greater than 4.7 and differed significantly from the median (=4.49). The same educational metadata elements were identified as highly useful for searching and evaluating learning objects in different contexts. For example, Small et all (1998) found that K-12 teachers tended to value the same elements in the case of the GEM metadata schema. These were: the *Material type* (=*Learning Resource Type*), *Audience* and *Prerequisite knowledge* (= *Difficulty*). Other metadata elements that participants in the survey perceived to be among the most useful were the *Date*, *Language* and *Cost*. These metadata elements were defined as useful in other studies investigating the interaction of K-12 teachers and academics with LORs (Small et al., 1998;

---

[1] The Interactivity Type and Interactivity Level elements were included into a single questionnaire item. The reason for combining these two elements in a single question was related to the fact that participants in the pilot study perceived both elements to be related.

Plodzien, et al. 2006), thus suggesting a common set of highly important elements that developers and designers of learning object repositories and digital libraries should take into account in order to meet the needs of university students.

Metadata elements that participants in this study perceived to be of moderate importance, were: the *Location* (identifier), *Format*, *Tutor's comments*, *Author*, *Installation* and *Copyright*. The Likert scores for these metadata elements ranged between 4.49 and 4.23 and did not differ from the median (4.49). The remaining metadata elements (*Relation*, *Typical learning time*, *Technical requirements*, *Structure*, *Version* and *Size*) were considered to be the least useful metadata elements. Based on the Wilcoxon tests these metadata elements were significantly less important than the *Location* metadata element (i.e. the median). Also, it appears that the impact of individual differences (level of familiarity with the web, educational level and subject background) becomes significant in the case of the least and moderate important elements rather than among the highly important ones. For example, the Technical requirements metadata element appeared to be significantly more important for postgraduate students studying Sciences or Engineering rather than undergraduates or students from a humanities or social sciences discipline. Therefore, it becomes evident that the use of these elements should be contextualised according to the needs of specific communities of users and makes necessary the study of users' relevance judgment and information searching behaviour in order to establish the reasons why individual differences exist.

## References


1. Balatsoukas, P., Morris, A., O'Brien, A.: Learning objects update: review and critical approach to content aggregation. Journal of Educational Technology and Society 11(2), 119–130 (2008)
2. Campbell, L.: UK Learning object metadata core: working draft, version 0.3_ 12042. CETIS (2004), `http://metadata.cetis.ac.uk/profiles/uklomcore/wip/uklomcor_v0p3_1204.doc` (accessed July 27, 2011)
3. Choi, Y., Rasmussen, E.: Users' relevance criteria in image retrieval in American History. Information Processing and Management 38(5), 695–726 (2002)
4. DCMI Education Community (2006), `http://dublincore.org/groups/education/` (accessed July 27, 2011)
5. IEEE LTSC, Draft standard for learning object metadata (2002), `http://ltsc.ieee.org/wg12/files/LOM_1484_12_1_v_Final_Draft.pdf` (accessed July 27, 2011)
6. Karampiperis, P., Sampson, D.: Enhancing educational metadata management systems to support interoperable learning object repositories. In: IEEE International Conference on Advanced Learning Technologies (ICALT 2003), July 9-11, pp. 214–218. IEEE Computer Society, Athens (2003)
7. Liddy, E., Allen, E., Finneran, C., Gay, G., Hembrooke, H., Granka, L.: MetaTest: evaluation of metadata from generation to use. In: Proceedings of the 3rd ACM/IEEE-CS Joint Conference on Digital Libraries, p. 398. IEEE Computer Society, Washington, DC (2003)
8. Nunnally, J.: Psychometric theory. McGraw-Hill, New York (1978)

9. Polsani, P.R.: Use and abuse of reusable learning objects. Journal of Digital Information 3(4) (2003), http://journals.tdl.org/jodi/article/viewArticle/89/88 (accessed July 27, 2011)
10. Plodzien, J., Stemposz, E., Stasiecka, A.: An approach to the quality and reusability of metadata specifications for e-learning objects. Online Information Review 30(3), 238–251 (2006)
11. Sicilia, M., Garcia, E., Pages, C., Martinez, J., Gutierrez, J.: Complete metadata records in learning object repositories: some evidence and requirements. Int. J. Learning Technology 1(4), 411–424 (2005)
12. Small, R., Sutton, S., Miwa, S., Urfels, C., Eisenberg, M.: Information seeking for instructional planning: an exploratory study. Journal of Research on Computing in Education 31(2), 204–219 (1998)
13. Smith, R.: Strategic planning for public relations. Taylor and Francis, New York (2009)
14. Wang, P., Soergel, D.: A cognitive model of document use during a research project. Study I. Document selection. Journal of the American Society for Information Science and Technology 49(2), 115–133 (1998)

# A Novel Approach towards Skill-Based Search and Services of Open Educational Resources

Kyung-Hun Ha[1], Katja Niemann[2], Uta Schwertel[3], Philipp Holtkamp[4], Henri Pirkkalainen[4], Dirk Boerner[5], Marco Kalz[5], Vassilis Pitsilis[6], Ares Vidalis[6], Dimitra Pappa[6], Markus Bick[1], Jan Pawlowski[4], and Martin Wolpers[2]

[1] ESCP Europe Campus Berlin, BIS, Heubnerweg 8-10, 14059 Berlin, Germany
[2] Fraunhofer FIT, Schloß Birlinghoven, 53754 Sankt Augustin, Germany
[3] imc AG, Altenkesseler Strasse 17/D3, 66115 Saarbrücken, Germany
[4] University of Jyväskyla, Mattilanniemi 2, Agora Building, Jyväskylä, Finland
[5] Open Universiteit Nederland / CELSTEC, Valkenburgerweg 177, 6419 AT Heerlen, The Netherlands
[6] Division of Applied Technologies - NCSR DEMOKRITOS, Patriarchou Gregoriou and Neapoleos str, 153 10 Aghia Paraskevi, Greece

{kyung-hun.ha,markus.bick}@escpeurope.eu,
{katja.niemann,martin.wolpers}@fit.fraunhofer.de
uta.schwertel@im-c.de
{philipp.holtkamp,jan.pawlowski,henri.j.pirkkalainen}@jyu.fi
{dirk.boerner,marco.kalz}@ou.nl
{avidal,vpitsilis,dimitra}@dat.demokritos.gr

**Abstract.** Open educational resources (OER) have a high potential to address the growing need for training materials in management education and training. Today, a high number of OER in management are already available in a large number of repositories. However, users face barriers as they have to search repository by repository with different interfaces to retrieve the appropriate learning content. In addition, the use of search criteria related to skills, such as learning objectives and skill-levels is not generally supported. The European co-funded project OpenScout addresses these barriers by intelligently connecting leading European OER repositories and providing federated, skill-based search and retrieval web services. On top of this content federation the project supports users with easy-to-apply tools that will accelerate the (re-) use of open content.

**Keywords:** Open educational resources, content reuse, competences, federated search, management education.

## 1 Introduction

Management is a large education and training business field in Europe. Training topics range from general management and leadership to very specific issues like IT-Governance in the banking industry. Many enterprises, especially small and medium sized enterprises (SME), lack the financial resources and necessary flexibility to meet specific training needs of their employees. Therefore, the flexible access to a large

E. García-Barriocanal et al. (Eds.): MTSR 2011, CCIS 240, pp. 312–323, 2011.

amount of open learning resources can foster self-directed competence development much better and faster than formal courses and study programs. The term "Open Content" refers to online accessible digital assets (texts, images, graphics or multimedia) that are published under a licence that explicitly allows an end-user to access, copy, modify or redistribute the content with no or minimum costs [1]. In educational contexts, the term Open Educational Resources (OER) is often used and denotes digital objects (courses, course materials, modules, textbooks, videos, tests, simulations, tools etc.) which can be freely used, adapted, and distributed amongst interested stakeholders [2]. As it can expand access to learning and bridge the gap of formal and informal learning OER is of interest for both educational learning and corporate training.

There is already a large amount of OER on management topics available in learning object repositories (LOR). However, most business education organizations and SMEs do not use these contents or integrate them into their technical solutions (e.g. Learning Content Management Systems) [3]. Therefore, end-users face barriers when utilizing the content for learning in management. They have to search repository by repository with heterogeneous interfaces and different content categorization in order to retrieve the relevant content. Additionally, SMEs need approaches to training which match the existing and required staff competencies to solve ad-hoc problems. But the usage of search criteria related to competences and skills, such as learning objectives or competence-level is generally not supported by open repositories. Moreover, the systems do not support users with easy-to-apply tools and mechanisms that enable reuse, adaptation and internationalization of the content. Within the OpenScout project[1] these barriers are addressed.

OpenScout is an EU co-funded project which aims at providing skill-and-competence based search and retrieval web services that enable users to easily find, access, use and exchange open content for management education and training. The project not only connects leading European OER repositories but integrates its search services into existing learning suites like CLIX[2] and LearneXact[3], social networks and personal portals, to enable access from environments familiar to a wide range of users. The paper is organized as follow. In chapter 2, we describe the OpenScout system focusing on the interfaces and the infrastructure. In chapter 3 we will discuss the OpenScout application profile. This includes the competence based approach and the cultural classification for supporting the reuse with context and cultural specific information. Finally, we conclude and give an outlook of future work in chapter 4.

## 2 OpenScout System

### 2.1 User Interfaces

**OpenScout Portal.** The OpenScout web portal provides to end users all services and tools employed in the project from a single access point. Users can access open content and can use integrated tools (authoring tools as well as tools for

---

[1] www.openscout.net

[2] `http://www.im-c.de/en/products/learning-management-system/product-overview/what-is-clix/`

[3] `http://www.learnexact.com/`

communication and collaboration) which enable them to adapt and localize materials to their needs. Moreover, users have access to social network functionalities as well as to data social actions and annotations like tagging, rating, and commenting.

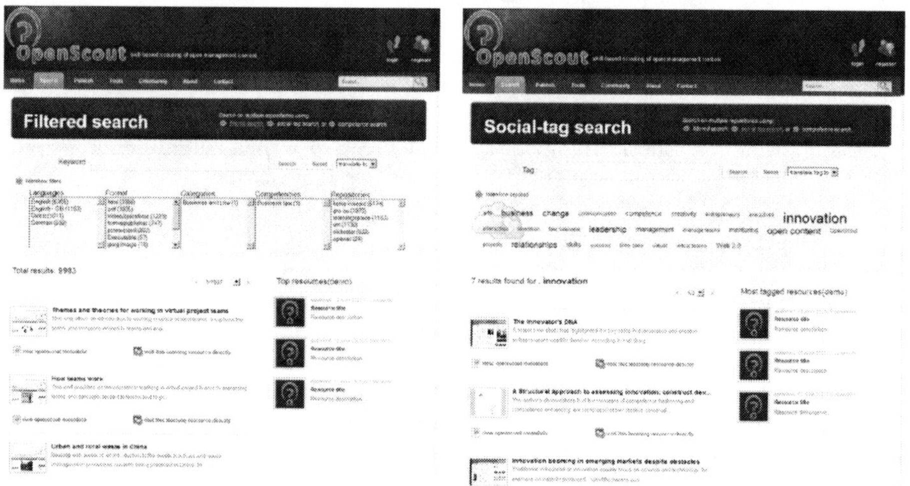


**Fig. 1.** OpenScout keyword search interface **Fig. 2.** OpenScout tag search interface

For supporting the learner while searching for the most suitable learning objects (LO), OpenScout offers a facetted search (see Fig. 1) and social or tag-based search (see Fig. 2) in addition to the general keyword search. The facetted search functionality allows the user to filter the search results according to the properties of the retrieved LOs, e.g. content type, competences, language, and repository. In the result list, all retrieved LOs are briefly described with their title, the beginning of their description as well as their URL and accompanied by a screenshot.

Users can choose to directly access a resource by clicking on the link or to first experience more information about it. In the detailed object view (see Fig. 3), users are presented with general information about the LO, such as its description(s) and learning resource type(s) as well as its social metadata (tags, comments, rating) and users can add social metadata if they are logged in. The users are also enabled to recommend learning resources by sharing them to popular social networks such as Facebook, LinkedIn and Twitter and hence increase the popularity of these resources and of the portal as a whole. Furthermore, registered users can interact with the OpenScout community: see their friends, their added tags, their interests and their recommendations.

**OpenScout Widget(s).** In addition to web portal access, the OpenScout system also aims to offer widgets that can be embedded in external systems like Social Networks, Learning Management Systems or other platforms. Currently, OpenScout offers a web-based search Widget, which allows a user to directly search from familiar learning environment.

**Fig. 3.** OpenScout learning resource presentation

The Widgets offers a search-relevant subset of the above listed portal services in a condensed way. The functionalities include search for content, check content metadata details, access the query results by clicking direct links or recommend and share learning resources to friends in existing social networks. Moreover, the Widget isn't limited to the OpenScout content federation but allows for optionally searching in other databases such as YouTube, Slideshare or Wikipedia. The system calculates a ranking of the search results of all repositories to support users in selecting relevant resources. Users not only have the option to select the repositories to search in, but to restrict the search results to certain content types or languages, choose to search in all fields, or just in the author or tag fields. Finally the Widget code supports localization of the interface language to different languages (currently German and English are instantiated).

The currently available Widget is already integrated into the Learning Management Systems CLIX and eXact LMS, and into the personal portal iGoogle. It can be integrated into any other OpenSocial[4] compatible container, or can be embedded through an iFrame into any Web page. Integrating OpenScout through Widgets in existing applications will not only achieve a wide distribution and sustainable use of its services but also accelerate the (re-)use of open content massively.

## 2.2 Infrastructure

The OpenScout system copies metadata describing learning resources that are stored in several learning object repositories (LOR) into the central OpenScout Repository Federation. The harvesting approach was preferred over direct search for performance and reliability reasons [4]. To enable the harvesting, an increasing number of repositories provide access to their metadata instances by offering an OAI-PMH [5] target. So far, the OpenScout system integrates the repositories OpenLearn[5],

[4] http://www.opensocial.org/
[5] http://openlearn.open.ac.uk

OpenER[6], SlideStar[7] and INSEAD[8] amongst others. The OpenScout Repository Federation offers an OAI-PMH target as well, to enable the content providers to re-harvest enriched metadata instances of their learning resources.

Furthermore, learning resources can be uploaded by OpenScout users. These include learning resources that are modifications or aggregations of resources that are already described through metadata stored in the OpenScout Repository Federation, but also independent learning resources. A Simple Publishing Interface (SPI) [6] is used to publish the learning resources and the respective metadata in the OpenScout Content Repository.

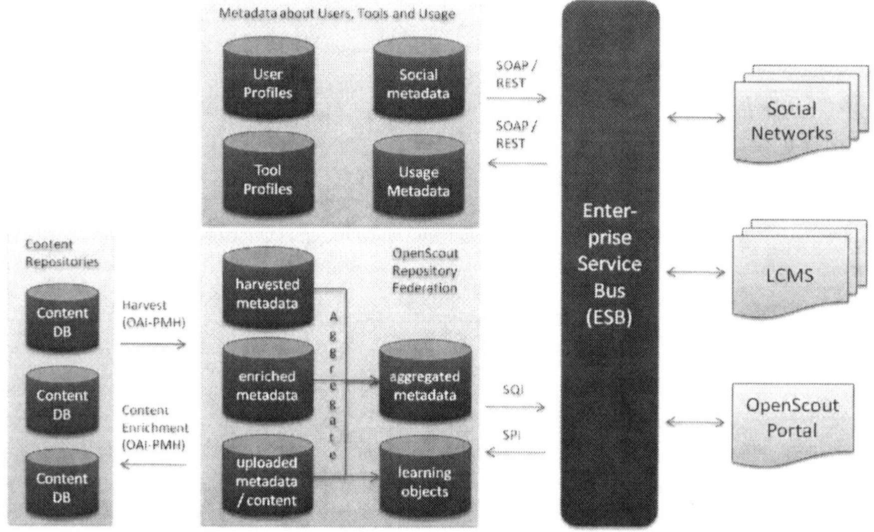


**Fig. 4.** The OpenScout architecture

Once stored in the OpenScout Repository Federation, the learning resources are available online using an Apache web server[9] while the metadata instances are indexed using Apache Lucene[10] to enable fast queries. The stored and indexed metadata instances can be queried through a Simple Query Interface (SQI) [7] which provides interoperability between search applications and LORs as it is neutral in terms of query languages and result formats. Currently, the OpenScout system allows Lucene, PLQL [8] level 0 and PLQL level 1 queries. As query result, statistics about the LOs, that match the query, can be returned to enable facetted search or the metadata instances are returned directly. The first one is enabled using SOLR[11] which is a search server based on Lucene.

---

[6] http://www.opener.ou.nl
[7] http://slidestar.de
[8] http://www.insead.edu
[9] http://httpd.apache.org/
[10] http://lucene.apache.org/
[11] http://lucene.apache.org/solr/

OpenScout also maintains user profiles (containing information about the users, e.g. interests and competences), tool profiles (containing data about tools that can be recommended for adapting learning resources), social metadata (data added by users, e.g. tags and ratings), and usage metadata (data about the user's actions and the usage of objects). This data is stored in separate databases and is accessible through REST and / or SOAP services.

Since the OpenScout user interfaces (e.g. the web portal) need to access and combine data from different sources, the backend layer has to allow scalability, and an easy way to extend services with a plug-in based architecture without affecting the client layer. In order to address these issues, the enterprise service bus (ESB) technology, namely the open source ESB Apache ServiceMix[12] is used. This means, all service communication takes place via the ESB. Furthermore, whenever an event occurs (e.g. a user conducts a search or opens a document), the ESB calls a web service that stores the event as CAM (Contextualized Attention Metadata) [9] instance in the usage metadata database to enable further services (e.g. for recommending LOs).

## 3 OpenScout Application Profile

The IEEE Learning Object Metadata (LOM) standard is a metadata standard to describe educational resources. Its purpose is to facilitate search, acquisition, use and exchange of learning objects by creating a standardized way to describe learning resources while taking into account the diversity of contexts in which the learning objects and their metadata are reused[13]. A metadata instance comprises information about the LO that is grouped in nine categories: general, life cycle, meta-metadata, educational, technical, educational, rights, relations, annotation, and classification categories. In contrast to the LOM standard, where all data elements are optional, the OpenScout application profile (AP) defines some mandatory and recommended data elements to enable the functionalities the project aims for. However, the set of mandatory elements was kept small so that further repositories can be more easily integrated[14]. In the following, we describe how we adapted and extended the LOM standard to enable skill- and competence based search as well as re-use, adaption and localizing of existing learning materials to the needs and contexts of learners and teachers.

### 3.1 Cultural Classification

OpenScout aims to support stakeholders in management education to re-use, adapt and localize existing learning materials to their own needs and context. One part of the solution is to provide a set of integrated tools and support mechanisms (best practices and user guides) [10]. Additionally OpenScout wants to support the re-use and adaptation by storing and presenting cultural and context specific information regarding the learning resources. This is done by the means of the cultural specification which we discuss in this paper.

[12] http://servicemix.apache.org/home.html

[13] http://ltsc.ieee.org/wg12/20020612-Final-LOM-Draft.html

[14] For further details see D1.1 Analysis Report on Federated Infrastructure and Application Profile available at http://www.openscout.net/downloads/deliverables

In recent literature on OER, there have been several indications of challenges towards utilization of OER in different contexts. Deimann et al. [11] argue that OER should provide context information and give indications where the material fits. OER should be "culturally portable" and support users to evaluate if the information is suitable for their context [11]. The role of cultural specification of OpenScout is to support this portability of OER.

Development and provision of such a specification should take into account the end-users of OER and capture possible contexts the materials can be applied in. OpenScout organized an interview and a brainstorming session with domain experts ranging from members of Higher Education to Business Schools and SME clusters in order to capture requirements for the specification. The results were strengthened with relevant literature after accomplishing a basic set of requirements. Following insights could be captured regarding the specification:

- The specification should provide means for a registered end-user as well as for content providers to add information to the LOs that isn't included typically in the LOM
- Users should be able to add information to the LOs when accessing certain OER in OpenScout portal or when uploading material to the OpenScout repository
- The specification should support and explain clearly: **what type of learning material it is** (is it a full course, case study, presentation etc.), **where the material has been used / usage context** (including information on the type of domain, location and language), and **how the material was applied** (including information on learning style, the focus and roles of stakeholders).
- Specification should give users the option to point out adaptation needs that indicate how the material **could be adapted by others** (should indicate if the material is culture and context specific and if some parts are general in a sense that they might fit different domains and countries etc.)

The requirements were also emphasizing the need to build a comprehensive classification with clear attributes for each of the aspects. Additionally the application of OER sets its own requirements, since OER might be used not just in Higher Education and Business Schools but also on other levels of education and in vocational training for SMEs and large enterprises etc. The cultural specification should take these aspects into consideration.

To get to this comprehensive classification with clear attributes, the brainstorming session mentioned above identified an initial list of important aspects which were necessary for the classification. Based on the initial list, a literature review focusing mainly on research in the fields of cultural differences and barriers of OER was conducted to build the first version of the cultural classification.

In the next step the Learning Object Metadata standard was analyzed in order to see to which extent the requirements for the cultural specification were covered. Several fields of the LOM standard were fitting to the categories of the cultural classification but the possible values were set too narrow. For other categories of the cultural classification no fitting field could be found. Therefore, we adjusted the OpenScout AP by extending the values of several entries and adding some additional entries to the AP. The first version of the cultural classification including the value extension and the AP integration and changes can be seen in (see Table 1). During the course of the project the presented cultural specification will be further developed.

**Table 1.** Initial OpenScout Cultural Classification

| Category | Source | Explanation | Value Extension | AP Integration |
|---|---|---|---|---|
| Language | Richter &Pawlowski [12], Expert Interviews | Language of the learning object | | LOM 1.3 |
| Learning Resource Type | Expert Interviews | Type of content | Presentation, Research Paper, Checklist, Course module, Full course, Video, Case Study, Modelling tools, Games, Serious games, Virtual worlds, Role-playing games, MUD | Extension of the value space of LOM 5.2 |
| Usage Context | Expert Interviews | Intended context of usage of the learning object | Conference, Vocational Training – SME, Vocational Training – Large Company | Extension of the value space of LOM 5.6 |
| Industry Sector | Expert Interviews | Industry sector covered by the learning object | Aerospace & Defence, Automobiles, Banking &Insurance, Chemicals, Construction & Building Materials, Electronic & Electrical Equipment, Energy & Oil & Gas, Engineering & Machinery, Forestry & Paper, Household Goods & Textiles, Information Technology, Leisure, Entertainment & Hotels, Media & Photography, Mining, Pharmaceuticals & Health, Real Estate, Retailing, Telecommunication Services, Transport, | Extension of the value space of LOM 9.1 and definition of a new classification that to be used for LOM 9.2 |
| Region | Hofstede [13], Henderson [14], Richter &Pawlowski [12] | Intended region for the usage of the learning resource | According to Country ISO Char 3 Code | New field LOM 5.12 |
| Adaptation needs | Expert Interviews | Cultural and context specific adaptation needs identified for the learning resource | Free text field | Extension of the definition of LOM 8 |
| Learning theory | Henderson [14] | Which learning theory does the learning resource belong to? | Behaviourist, Cognitivist, Constructivist | New field LOM 5.13 |

**Table 1.** (Continued)

| | | | | |
|---|---|---|---|---|
| Focus | Henderson [14] | Abstraction level of the learning resource | Abstract, Concrete | New field LOM 5.14 |
| Teacher / Learner Role | Henderson [14] | Relation between Teacher and Learner | Strict control, uncontrolled | New field LOM 5.15 |
| Learning Strategy | Henderson [14] | Strategies expected from the learners | Individualism, Collectivism | New field LOM 5.16 |
| Religion | Richter &Pawlowski [12] | Religious aspects influencing the learning resource | Free text field | Extension of the definition of LOM 1.6 |
| History | Richter &Pawlowski [12] | Historical aspects influencing the learning resource | Free text field | LOM 1.6 |
| Politics | Richter &Pawlowski [12] | Political aspects influencing the learning resource | Free text field | Extension of the definition of LOM 1.6 |

As mentioned above, OpenScout follows a user driven approach to add these additional information to the existing LOs either at the time of publishing (uploading) or when accessing LOs in the OpenScout portal. Users are able to enrich the existing metadata of LOs with this information according to their personal impression.

The cultural specification will be used in different ways within the OpenScout System: Personal recommendations of LOs and support of adaptation processes. For personal recommendations the information provided by the cultural specification will be matched with information provided by the user profiles and with user actions. For adaptation of learning materials, the cultural specification can support the user to judge whether modifications are needed to fit the material to own context and whether specific challenges in the adaptation are pointed out through the specification.

### 3.2 Competences

The OpenScout projects implements a skill- and competence based search functionality for OER in the field of business and management education [15]. Taken the paradigm of lifelong learning, competence-based learning addresses directly the need of individuals to upgrade their knowledge, skills and competence in a discipline throughout their lives as required for a lifelong competence development.

Competence-based approaches in the field of formal and/or non-formal education are becoming more common and offer the opportunity to develop flexible services that meet the needs of learners, trainers and potential employers. With the implementation of competence based education more tailored and personalized approaches are possible that allow very specific training and learning activities without the need to follow complete education programs. In order to support and use effectively this link between competence and education, there is also the need to provide reusable definitions of competences, across the different systems. To realize the competence-based search and browsing of OER in business and management education the project uses a competence service library that offers top-down and bottom-up metadata enrichment and support for individual competence development in the future [16]. The competence service library contains services for top-down implementation of a competence taxonomy and tagging services that allow to build bottom-up folksonomies.

```
<classification>
	<purpose>
		<source>LOMv1.0</source>
		<value>competency</value>
	</purpose>
	<taxonPath>
		<source>
			<string language="en">Classification System</string>
		</source>
		<taxon>
			<id>Domain Identifier</id>
			<entry>
				<string language="en">Domain Title</string>
			</entry>
		</taxon>
		<taxon>
			<id>Competence Identifier</id>
			<entry>
				<string language="en">Competence Title</string>
			</entry>
			<mineqf>Minimum EQF Level</mineqf>
			<maxeqf>Maximum EQF Level</maxeqf>
		</taxon>
	</taxonPath>
</classification>
```

**Fig. 5.** Competences in the application profile

The project consortium has produced a competence taxonomy that consists of competence categories and descriptions of competences in the domain. Learning resources in the OpenScout Repository Federation can be enriched with competences and levels according to the European Qualification Framework (EQF).

In compliance with the IEEE Standard for LOM, we store the metadata related to competencies for each LO in the classification section of the LOM (see Fig. 5). Each LO can have any number of competences associated to it. For each competence of a LO, the minimum and maximum proficiency scale values are included.

To collect, catalogue, manage, and maintain the competence metadata a toolset has been created. The core of the toolset is the competence catalogue that stores the competence taxonomy. It contains competence domains and their related competences as well as external resources, and a proficiency scale description based on EQF. For example, the competence domain “economics” comprises the competences “Being able to apply economic and business forecasting" and "Being able to conduct capital budgeting and investment decisions" amongst others.

Domain experts from the project consortium will enrich learning resources in the future with specific information about related competences and proficiency scales. Main target of the implementation is to support users with individual competence profiles to find learning resources that fit to their competence development goals.

## 4 Conclusion / Outlook

This paper describes technologies and strategies of the project OpenScout to make OER in the area of business and management education more accessible and appealing to learners. OpenScout not only focuses on a federated search, but stresses that open content can only be successful and sustainable if end-users collaboratively work on the resources. Therefore, OpenScout offers tools for re-use and adaptation and puts a strong focus on processes enabling users to localize existing learning materials to their own needs and context with a specific cultural specification of the material. By introducing a competence based approach to learning resources OpenScout supports effectively life-long learning of the individual learner in SMEs and training/education institutions

Currently, we work on fully integrating all components into the prototype. A competence tagging pilot is running where domain experts annotate federated learning resources according to competences. The integration of tools into the OpenScout portal is improved so that a resource specific recommendation of suitable tools can be provided. Furthermore, mechanisms for recommending tools and learning resources based on the users' context are currently deployed using the collected usage data. The search Widget is re-designed and a tighter integration into existing Social Networks is implemented.

As part of a continuous evaluation the portal has been presented to target users from the management domain at various events and has received positive feedback and interest. In order to achieve a long term success OpenScout is continuously working on integrating additional management related repositories. Content providers collaborating with OpenScout not only have functional benefits (e.g. access to enriched metadata) but can also expect an increased visibility, international recognition and an innovative image within their community which in turn can increase enrolment to their professional offerings.

## References

[1] Open Content (Wikipedia), `http://en.wikipedia.org/wiki/Open_content` (accessed March 23, 2010)

[2] Open Educational Resources (UNESCO), `http://www.unesco.org/new/en/communication-and-information/access-to-knowledge/open-educational-resources/`

[3] Atkins, D.E., Brown, J.S., Hammond, A.L.: A Review of the Open Educational Resources (OER) Movement: Achievements, Challenges, and New Opportunities. The William and Flora Hewlett Foundation (2007)

[4] Ternier, S., Verbert, K., Parra, G., Vandeputte, B., Klerkx, J., Duval, E., Ordóñez, V., Ochoa, X.: The Ariadne Infrastructure for Managing and Storing Metadata. Emerging E-Learning Technologies. IEEE Internet Computing 13(4), 18–25 (2009)
[5] OAI: Open Archives Initiative Protocol for Metadata Harvesting, Protocol Version 2.0 (June 14, 2002)
[6] Ternier, S., Massart, D., Van Assche, F., Smith, N., Simon, B., Duval, E.: A Simple Publishing Interface For Learning Object Repositories. In: Luca, J., Weippl, E. (eds.) Proceedings of World Conference on Educational Multimedia, Hypermedia and Telecommunications 2008, pp. 1840–1845. AACE, Chesapeake (2008)
[7] Simon, B., Massart, D., van Assche, F., Ternier, S., Duval, E., Brantner, S., Olmedilla, D., Miklos, Z.: A Simple QueryInterface for Interoperable Learning Repositories. In: Proceedings of the 1st Workshop on Interoperability of Web-based Educational Systems, pp. 11–18 (2005)
[8] Ternier, S., Massart, D., Campi, A., Guinea, S., Ceri, S., Duval, E.: Interoperability for searching learning object repositories, the prolearn query language. D-Lib Magazine 14(1/2) (January/February 2008)
[9] Schmitz, H.-C., Wolpers, M., Kirschenmann, U., Niemann, K.: Contextualized Attention Metadata. In: Roda, C. (ed.) Human Attention in Digital Environments. Cambridge University Press, Cambridge (2011), http://www.cup.es/catalogue/catalogue.asp?isbn=9780521765657
[10] Pirkkalainen, H., Pawlowski, J.: Open Educational Resources and Social Software in Global E-learning Settings. In: Yliluoma, P. (ed.) Sosiaalinen Verkko-oppiminen. International Multimedia & Distance Learning Oy. Ltd., IMDL (2010)
[11] Deimann, M., Bastiaens, T.: Special Session OER: Integrating OER and Instructional Design – Towards a more holistic view. In: Conference ICL 2007, Villach, Austria, September 26-28 (2007)
[12] Richter, T., Pawlowski, J.M.: The Need for Standardization of Context Metadata for e-Learning Environments. In: Proc. of e-ASEM Conference, Seoul, Korea (October 2007)
[13] Hofstede, G.H.: Cultures and organizations: Software of the mind, 2nd edn. McGraw-Hill, New York (1997)
[14] Henderson, L., Cook, J.: Theorizing Multiple Cultures Instructional Design Model for E-learning and E-teaching. In: Edmundson, A. (ed.) Globalized e-learning Cultural Challenges. Idea Group Inc., USA (2007)
[15] Kalz, M., Specht, M., Nadolski, R., Bastiaens, Y., Leirs, N., Pawlowski, J.: OpenScout: Competence based management education with community-improved open educational resources. In: Halley, et al. (eds.) Proceedings of the 17th EDINEB Conference.Crossing Borders in Education and Work-Based Learning, pp. 137–146. Feba Erd Press, Maastricht (2010)
[16] Klemke, R., Ternier, S., Kalz, M., Specht, M.: Implementing infrastructures for managing learning objects. British Journal of Educational Technology 41(6), 873–882 (2010)

# How Organizing Knowledge in Meaningful Ways Enhances Search and Supports Independent Learning

Karen Vignare[1], Christine Geith[1], Brian Collins[2], and Prabode Weebadde[2]

[1] Michigan State University, MSUglobal,
East Lansing, MI, USA
vignare@msu.edu, cgeith@gmail.com
[2] Venturit, Inc.
Okemos, MI, USA
{bp.collins,prabode}@gmail.com

**Abstract.** Focused on creating and finding OERs for food safety, food security, and agriculture curriculum improvement, each MSUglobal project has progressively delivered better opportunities for learners to discover and learn using content that has been gathered and organized by content experts. The partnerships and technology choices have produced a series of platforms, which have focused on the needs of specific tasks. The projects have built techniques for improving search and discoverability through more automated metadata. The projects have also created opportunities to include rich multimedia.

**Keywords:** Open Education Resources (OER), metadata tagging, OER search, discoverability, food safety, agriculture OER.

## 1 Introduction

Open Educational Resources (OER) need to be found to be useful. Once resources have been found they need to be organized in meaningful ways so that learners might use them to develop a deep understanding of content.

Focused on creating and finding OERs for food safety, food security, and agriculture curriculum improvement, each MSUglobal project has progressively delivered better opportunities for learners to discover and learn using content that has been gathered and organized by content experts. The partnerships and technology choices have produced a series of platforms, which have focused on the needs of specific tasks. This work has led to a cutting edge media platform that is designed to support learning across domains. The media platform, designed and developed by Venturit Inc., is based on Cognitive Flexibility Theory, a constructivist learning theory that argues that new media, because it supports nonlinear, multi-perspective learning, can support a deep understanding of content by enabling learners to work with a large number of examples to see the different ways that conceptual knowledge is combined and applied in real world contexts [1].

The Hewlett Foundation funded the initial project, which mapped food safety guidance from Global Food Safety Initiative (GFSI), an internationally recognized corporate association. GFSI and Michigan State University faculty created

E. García-Barriocanal et al. (Eds.): MTSR 2011, CCIS 240, pp. 324–336, 2011.

competencies, workshops, and assessments related to food safety topics. Building a searchable environment on Drupal, an open source tool, and DiscoverEd, the leading search code for OER resources, soon led to limitations that required the code to be built directly in Nutch [2]. The expansion of metadata characteristics, which can be either pulled or added by a curator was needed for the food safety project and AgShare, a Bill & Melinda Gates Foundation project to initiate, create, and use OERs in Master's of Agriculture curriculums in universities throughout Africa [3]. Hiring a content expert who was able to discover, organize, and manage appropriate materials optimized the gathering and arrangement of appropriate OER materials.

While technically effective, the metadata and semantic search still lacked a visible structure that supported the deep understanding of content. The inclusion of multimedia and social community features in a platform that is grounded in learning theory presented an opportunity for the some facets of the newest projects to build a global online learning community related to the study of food systems for veterinary students funded by United State Department of Agriculture.

The new media platform enables curators to manage metadata, semantic data, and gather and organize open educational resources. Content experts create posts that feature different types of media uploaded from a computer, the cloud, or grabbed from online sites, such as YouTube and Vimeo. Related content, such as OERs, can be linked to posts. In order to make curation more manageable, metadata related to embedded sources are brought in when available (e.g., title, author, description, date published, closed captioning, creative commons license, and source). The platform enables learners to easily search and compare and contrast content. Because content experts have mapped the domain, learners are able to work with content in linear and nonlinear fashion in an educationally friendly manner.

In order to take advantage of the strengths of different technologies the platform integrates multiple open source technologies. These technologies include: Kaltura's Community Edition; MSUglobal's search indexing tool; Nutch framework; DiscoverEd Nutch plugin; the Solr enterprise search platform; Ruby on Rails framework; MySQL database; and Apache web server.

## 2 Background

MSUglobal has four concurrent projects all using different applications and software to achieve the goals of each grant or contract. The projects are all related in that they require or plan to use open education resources, use some form of search whether pulled through scraping, RSS, DiscoverEd or metadata, extend the resources through discoverability- essentially good use of meta tagging, and build on the richness of each project to provide new layers of capacity. The projects also required building for both non-linear learners as well as linear learning paths.

### 2.1 Project One: Food Safety Knowledge Network Description

Prior to the start of the project, MSU faculty met with the GFSI Technical Working Group entitled, Global Markets Working Group and created a subgroup called the FSKN Working Group. The purpose of the FSKN Working Group was to further the goals of the GFSI Global Markets Working Group by guiding the development of training processes, curriculum, and assessment directed at the food safety managers of suppliers in countries whose food safety systems are less developed than those typically in place in North American, Western European, and other developed markets. The GFSI Global Markets Working Group and FSKN effort focused initially on Basic Level Requirements for the Food Manufacturing sector. UNIDO also contributed to the development and use of these resources through several training programs. Ongoing work has led to development of Intermediate Level requirements for the food manufacturing sector, as well as Basic and Intermediate Level requirements for primary production of fruit and vegetable products through project funding for USAID and India Horticulture Development Association.

The Basic Level for Food Manufacturing protocol finalized by the GFSI Global Markets Working Group encompasses 13 key areas of company requirements (Figure 1). The Basic Level Requirements represent the first 30 percent of the total company requirements that the GFSI Global Markets Working Group is defining. Working from the Basic Level requirements for companies, the FSKN Working Group defined 89 competencies that articulate the knowledge and skill sets an individual food safety manager should have upon completion of training at the Basic Level. The remaining requirements are being released for Intermediate Level Requirements in cooperation with GFSI. GFSI and its Technical Working Groups have continued to provide guidance for corporations and for the individuals, MSU faculty and experts who have developed the competencies and resources.

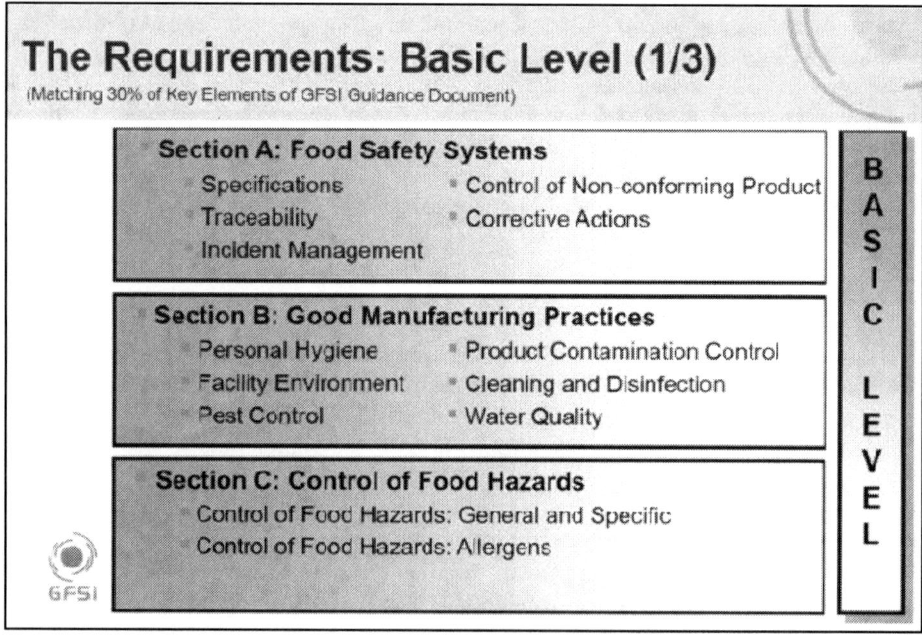


**Fig. 1.** GFSI Basic Level Requirements (Food Manufacturing Scope) for Less Developed Businesses and Small Suppliers

**Creating Processes for OER Development: The Benefits of Resources, Not Course Packages**. While the FSKN Working Group began creating competencies, MSU began creating instructional documentation and materials to support the development of OER aligned with the competencies. To accomplish the goals of publishing OER, the MSU team designed OER processes and procedures for collecting and creating content. For creating content, it was integral to think strategically about instructional use of materials. This meant moving beyond the concept of a whole course and into the level of individual learning resources. The goal was to enable learning resources to be combined to create a full course for Basic Level for Food Manufacture, or to be used separately if a learner, or trainer, desired to work on only a certain sub-set of competencies.

To create the individual resources at low cost, the MSU team determined that capturing existing trainings would provide the basis for the individual OER. Proprietary capture software Camtasia and Relay were used to record presenters during training programs. The capture process allowed for four immediate file output types—Power Point, flash video, mp4 and audio mp3. The audio files were immediately transcribed through automated technologies to produce text transcripts. In addition, the Power Point slide presentations were made available as PDFs and Open Office documents. Providing a wide variety of file formats, particularly those that are editable using common as well as open non-proprietary software, is an important enabler of OER remixing [4].

The learning resources were first produced in English. Training manuals were also produced. All formats of these resources (except, so far, for the transcripts and training manuals) were translated into Chinese and Arabic by pilot partners. These translated resources were also openly licensed and shared back to the FSKN as OER. In addition, ten stand-alone online learning modules are in review and seven modules have been created from their storyboards in the Connexions OER Repository (http://cnx.org).

**Identifying Existing Resources.** While capturing presentations was important for providing a base of OERs aligned with competencies, the MSU team also identified existing resources from training providers, universities, and non-governmental organizations. The FSKN pilot team reviewed the potential resources for alignment with the competency framework and agreed on content that could be added to the website. The MSU team prepared a policy document to explain OER and also an FSKN Toolkit document to explain how to contribute content.

**Platform Development.** At MSU, MSUglobal and its internal partner, Virtual University Design and Technology (vuDAT) as well as its external partner, Venturit Inc., used open source technology for all aspects of the initiative. This included creating the FSKN web site (http://foodsafetyknowledgenetwork.org), creating an open source tool to map the OER to the competency frameworks, and to integrating the open source Creative Commons DiscoverEd search tool to improve the discoverability of FSKN resources. Open source platforms were immediately investigated because we anticipated that solutions we would develop for FSKN would be useful to a much broader audience and we wanted to encourage its use in developing countries.

For content management, Drupal was chosen because of its large community of developers and extensive list of modules developed for various website features and functionalities. Drupal served as the base to build from for accomplishing the tasks described in the planning phase. Drupal, however, did not provide a competencies database module nor did it provide a method for searching and pulling content to the website. vuDAT programmers created an open source module that provides a way of creating a competency database and correlating resources to individual competencies. The current website demonstrates only the top level of competencies, but the module was built so that it could go five layers into a hierarchy. This allows for much deeper detail within competencies and eventually a learner could choose only a sub category of a competency to learn.

To improve the discoverability the resources in the FSKN as it grows, we decided to integrate the Creative Commons' DiscoverEd open source search tool into the competency database. DiscoverEd provides an enhanced search experience for users looking for educational resources by looking for materials from selected curators and displaying valuable metadata on the resources. The competency database only includes competencies while learning resources are culled to match a competency. This required writing code in Nutch by Venturit Inc., to customize DiscoverEd so it would work with Drupal. It also required that both an RSS feed and Open Architecture Index (OAI) be incorporated. These tools pull with an RSS or OAI standard from our own and other websites materials to support learning of the competencies. In addition, by adding OAI especially for MSU materials, we have begun an education process which demonstrates the use of OAI for content repositories. If a provider does not want to or cannot implement the OAI standard, they have the option of uploading to a website that does use OAI. Due to the innovative code from Venturit Inc. the competency module written for Drupal was no longer necessary.

### 2.2 Project Two: AgShare- Open Education Resources for Masters in Agriculture Curriculums in African Universities

Africa requires bold and innovative agricultural development to help the continent's smallholder farming families move out of poverty and hunger. Higher education institutions play a major role in developing the agriculture sector, for better or worse, because it is the graduates of agricultural programs who will go on to become the architects of future agricultural policies at continental, regional, national, and district levels. The products of postgraduate agricultural programs typically become future government leaders, directors and senior staff in non-governmental organizations and donor agencies, and captains of the agribusiness industry. For those responsible for the advancement of higher education, therefore, the challenge is to build research, training, and outreach programs to produce graduates that can respond to the demands for innovation.

Michigan State University, led by MSUglobal and the College of Veterinary Medicine, has partnered with OER Africa to lead four university pilot projects. They each have a different focus which allows the core team to pilot test different approaches for authoring, sharing, customizing and using OER to strengthen MSc agriculture curriculum. Each pilot has completed their development of open education resources. OER Africa held workshops on OER development and OER production technologies. They provided each pilot with continual assistance through the development and now are assisting with the monitoring and evaluation. Each pilot was also assisted by a skilled researcher or information specialist, who identified numerous open content materials, but not necessarily creative commons licensed materials, to support the development of modules and courses. An information specialist as part of the search meta-tagging enhancement is an increasingly important role in usage and dissemination of OER [5]. The resources are currently available at OER Africa, AgShare portal page, http://www.oerafrica.org/agshare/AgShare Home/tabid/1290/Default.aspx

Makerere University of Uganda is focused on improving dairy production while improving the use of graduate students in the processes of working with farmers and providing science-based, timely OER which can be used in MSc course modules in both the veterinary school and the agriculture school. Involving students directly in the research process, capturing the work with and for farmers, will provide replicable model for many other important commodities. Haramaya University of Ethiopia is developing a master's level course that directly impacts Extension. It is a course in the RUFORUM Agriculture Information Communication Management master's program and has potential for network-wide localization and replication, as well as a model for subsequent courses.

United States International University of Kenya is developing an OER module that utilizes a practical case method and active learning to build an ICT applications course for mid-sized farmers and farmer organizations. This OER module will also meet the needs of MSc agriculture faculty in at least one partner university. The pilot will also develop an AgShare Fellows program that will embed expertise within the faculty and students to create and use OER, which are aimed at assisting farmers to transform their traditional practice of farming as a livelihood to farming as a business enterprise. The collaborative content generation is through partnerships between USIU business school, agriculture universities and community-wide partners. The pedagogical model and the OER development process have the potential for an entirely new agribusiness curriculum.

The Collaborative Masters in African Agriculture Economics (CMAAE) pilot involves two institutions, Moi University of Kenya and Haramaya University, building commodity focused case studies that will be used in the program in Agricultural and Applied Economics. Cases will focus on the Economic Role of Prices and Approaches to the Study of Agricultural Market Organization for Coffee in Ethiopia and Maize in Kenya. Students will be involved in creating the cases and community-wide partners will be involved as subjects in the cases and also in the review of the materials. This collaborative approach to the creation of case studies can be a potentially cost-effective way of producing cases that can be used by all CMAAE institutions, as well as other universities.

**The AgShare Eco-system.** Running parallel to the university-based pilots is the development of tools and processes to make agriculture OER and the work of AgShare more visible in the global knowledge pool. OER Africa is working with Creative Commons, MSU, and a range of other technical partners to begin constructing an online 'eco-system' for AgShare, designed to enable flow of content through existing systems and increase the discoverability of that content online by people searching for agricultural content. The AgShare 'eco-system' will include support for preparing published OER for distribution as well as pushing content to other distribution channels and platforms (see http://www.oerafrica.org/agricultureoer). For example, AgShare output will be easily findable in Google and other search engines, and indexed in Agris. We are also using the Food and Agriculture Organization AGROVOC thesaurus, a multilingual, structured and controlled vocabulary designed to cover the terminology of all subject fields in agriculture, forestry, fisheries, food and related domains (e.g. environment).

A key component in the AgShare eco-system is DiscoverEd, an open source search tool developed by Creative Commons. Creative Commons is improving DiscoverEd as part of the AgShare project to better support the alignment of OER production with demand, and to support easy and accurate discovery of OER relevant to the project. To support the alignment of OER production, DiscoverEd is being improved to provide demand metrics, which will enable the direction of resource-development to those areas where demand (from MSc agriculture faculty, for example) is greatest. Support for additional metadata fields and metadata provenance tracking has already been completed as part of the project. Features improved and added to DiscoverEd for AgShare will become part of the core distribution, simplifying the customization process in the future for other communities of interest.

Results of the AgShare ecosystem and the DiscoverEd component are critical for enabling time-saving discovery of relevant open educational resources in African agriculture and to make visible African contributions to the global knowledge pool.

Unfortunately for AgShare, Creative Commons has redirected its resources to support development of external grant funded projects and no further development of DiscoverEd is scheduled. However, the work on metadata tagging by OER Africa and the application of Creative Commons licensing to all resources have provided a successful repository for the resources. Several initiatives appear to be taking root and are intended to enhance work started by DiscoverEd in the agriculture OER search and discoverability domains.

## 2.3 Project Three: AFRICA LEAD Database and OER Resources

The purpose of the program is to support the capacity building program of the US Government's Feed the Future (FTF) Initiative, which aligns US Government assistance with Africa-owned agriculture development plans that are, in turn, aligned with the African Union's Comprehensive Africa Agriculture Development Program (CAADP). CAADP seeks to ensure that governments throughout the continent apply 10% of their public budgets to agriculture in an effort to achieve the 6% annual growth in agricultural productivity that is needed to achieve sustainable food security throughout the continent. AFRICA LEAD provides leadership training, capacity assessments, logistical support for training and innovative short courses and internships/twinning arrangements prioritized in consultation with missions and partner countries and institutions, and a database of training offerings and open education resources on the continent that can be matched to the leadership training and capacity building needs.

The application was implemented using open source frameworks and industry best practices by Venturit, Inc. A private test environment of the application will be hosted in order to review progress with the stakeholders. Open source technologies will be used to develop the Africa Lead Search Indexing Tool. The following technologies have been employed to build the tool: WordPress- a content management framework developed using PHP programming language; MySQL- a database server which is used as the persistence storage of the WordPress content; Nutch framework- mainly used for scraping resource sites; the DiscoverEd Nutch plugin- used to tag creative commons content while scraping the resource sites and; Solr engine powers the course and resource search, providing powerful full-text search, hit highlighting, and faceted search.

The search functionalities are used within the database but the indexing tools for OER, beyond links to Food Safety Knowledge Network, have not yet been used by the project staff. The CAADP organizational offices and staff are also beginning to review the use of the tool to support its work. Efforts to grow the use of built-in capacities of the tool are now being included in several new grants.

### 2.4 Project Four: Venturit Media Platform- A Platform Based on Learning Theory Rather Than Just the Power of Technology

The youngest concurrent project at MSUglobal is one that focuses on veterinary medicine and alumni networks. While not strictly agriculture, the investment and planning will continue to push development of how we tag, find and make information discoverable. Big Ideas at MSU, a project created by MSUAA, the Evening College, and MSUglobal, is an application of the Venturit Inc. media platform that is designed to support activities at different stages of idea development (e.g., creation, incubation, dissemination). The platform enables MSU faculty, staff, students, and alumni the opportunity to share ideas and engage with others about ideas. The platform enables users to discover ideas, develop a deeper understanding of ideas, engage with the idea creators, and support the development of ideas. This project is designed to engage the MSU community, along with a global audience, change the way people think about how ideas evolve over time, and be used as a tool that supports teaching and learning.

Within Michigan State University's College of Veterinary Medicine (CVM) three separate subprojects are using the media platform. The goal of the USDA-funded project is to build a global online community of food systems for veterinary students community related to the study of food systems for veterinary students. This application enables learners to share experiences, resources and opportunities. The platform enables community members to share images, videos and documents, tag each media asset with descriptive information, mark content as private or public, have discussions related to topics and search the media library using a variety of filters. The next project supports Radiology Teaching and Learning and is designed to be used as a component of web-based radiology professional development web-based training events and courses. The third project is designed to be a showcase College of Veterinary Medicine Media Knowledge Assets. This application is designed to showcase CVM/Radiology-focused media assets, with an emphasis on video and audio lectures. The application enables users, including students, alumni and professionals enrolled in credit and non-credit courses and programs, to easily find and use the knowledge assets that CVM/Radiology has to offer and locate professional development opportunities.

**The Venturit Media Platform.** The Venturit platform holds an array of media (e.g., video, photos, audio, text), enables learners to easily search interconnected content, features social tools (e.g., discussions and sharing features), integrates with multiple web applications, is widgetized so content experts can embed the platform in other platforms, and has a content management driven design that enables non-programmers to manage site content. The media platform is designed to enable content experts to organize bits of information, located both online and offline, and organize the bits in ways that are both meaningful and cognitively manageable for learners and do it in such a way that learners work seamlessly with different types of media.

The media platform has three layers: interface; business; and storage. The media platform interface and business layers were built using Ruby on Rails. The asset management was done using Kaltura's Community Edition server. Learning theory is built into the business layer of the application. The business layer is owned and copyrighted by Venturit, Inc.

***MSUglobal's Search Indexing Tool.*** The media platform has an API. This enables the platform to integrate a tool such as MSUglobal's curated search indexing tool. MSUglobal's search indexing tool enables users to easily search on multimedia content that is curated by domain experts and aligned to a set of core competencies. Users are able to retrieve a variety of indexed items, examine content from multiple perspectives, and view text surrounding searched topics in order to place the topics in context. Admins are able to harvest content from other sites and manage aggregated content on their own.

The curated search indexing tool integrates the following technologies: Wordpress; Apache Nutch framework; DiscoverEd Nutch plugin; and the Solr enterprise search platform. The user interface was built on the WordPress 3.1+ CMS framework (written in PHP). A custom plugin which integrates Apache Nutch (written in Java) and Apache Solr (written in Java) frameworks was integrated into the framework. Through the admin interface users of the system are able to: Add a resource URL to fetch and assign tags and competencies to the resource; Run the indexer on demand. The indexer executes the Nutch crawler to crawl the URLs. The Nutch crawler crawls the URLs posts relevant content from URL web pages to Wordpress using the XMLRPC interface. As the content is scraped the DiscoverEd plugin filters and adds cc licensing information to the scraped content. We have created a custom Nutch plugin to post the content to Wordpress using XMLRPC. A Wordpress custom plugin imports the respective URL content by listening to XMLRPC posts and assigns the tags and competencies associated with the resource url. The Wordpress custom plugin indexes the resource content in the Solr search engine and provides an interface that enables users to do text searches, hit highlighting, and faceted search.

The platform integrates multiple open source technologies in order to take advantage of the strengths of different technologies.

**Based on Learning Theory.** The Venturit media platform is based on Cognitive Flexibility Theory (CFT), a constructivist learning theory that "emphasizes an adaptively creative response to new situations, rather than the more mechanical following of routinized schemes that already exist in memory" [6]. The theory is designed to support learning in domains where general principles do not account for enough of the variability in the way knowledge has to be applied. Instead, learners must have experience with a large number of examples to see the different ways that conceptual knowledge is combined and applied in real world contexts. New media now enables us to provide learners with opportunities to work in an online learning environment where content is organized and managed by content experts and explore content in both structured and nonlinear fashion. Cognitive Flexibility Hypermedia systems (CFHs) have been created for medicine, biology, civics, history, literature, military strategy, and teacher preparation in the literacy and physics domains.

Principal tenets of CFT and its applications to CFH learning environments include

*Manageable Chunks of Knowledge*. CFT argues that bite-size chunks of information that have the basic features of complexity but are smaller and easier to work with are easier to compare and contrast.

*Conceptual Variability and Context Awareness*. Concepts are defined through their use, so it is essential that learners see multiple examples of how concepts are applied in real world situations. If learners only examine one example of how a concept has been applied then this might lead to early closure, which "interferes with the development of complex understanding and the ability to apply knowledge to a wide variety of real-world contexts" (Spiro et al., 2003). Learners must have experience working with a large number of cases to see the different ways that conceptual knowledge is combined and applied in real world contexts.

*Multiple Perspectives*. The theory argues that in order to understand the complexity of events learners should examine events from different perspectives. Work in a learning environment where knowledge is interconnected might enable learners to follow different routes through content. This means that learners might return to the same event, but come from a different direction, bringing a different set of perspectives.

*Structured and Less Structured Learning Experiences*. CFT advocates that its learning environments offer learners opportunities to explore content in different ways. This means that learners should have structured and less structured opportunities to explore content.

More structured approaches to learning sometimes leave students with memorized knowledge that is neither deep nor easily applicable to new situations. Unstructured approaches, on the other hand, can leave the acquisition of some important aspects of knowledge to chance and may leave learners confused and discouraged. CFT attempts to bridge these two approaches by arguing for the need of both structured and unstructured learning experiences. CFH systems feature guided explorations in the form of tours that are designed to help learners understand specific paths through domain knowledge and opportunities for learners to take explore the landscape in an unguided manner thus enabling learners to examine more cases and form new connections.

**Key Platform Features**

- *Post Creation.* Content experts are able to create posts that feature different types of media uploaded from a computer or content grabbed from other sites, such as YouTube and Flickr, and create links between posts and related resources. Metadata related to embedded from online sources, such as YouTube, are brought in when available (i.e., title, author, description, date published, closed captioning, creative commons license, and source). This is more efficient for the admin because she does not need to reenter this data. In order to make post content more cognitively manageable admins are able to mark video in and out points. This means that admins are able to create multiple posts from a longer video.
- *Multiple Viewing Modes.* Learners are able to view content in four different viewing modes: 1-window; 2-window; 4-window; and Search Results. The 2 and 4-window modes are designed to support comparing and contrasting and creative thinking, because they enable users to place cases side by side.
- *Search Indexing Tool.* A Search Indexing Tool, developed by MSUglobal, has been integrated into the platform. This tool enables users to easily search on external resources that have been selected by domain experts and aligned to core competencies.
- *Widgetized.* Content experts are able to embed the platform, along with selected topics, tags, and competencies into other platforms (e.g., Blackboard, Drupal, Facebook, Wordpress, Moodle, Plone).
- *Social.* Learners are able to email and download content, share content on Facebook and Twitter, and have discussions about specific content.
- *Site Analytics Dashboards.* Content experts are able to develop a deeper understanding of how people are using the site by viewing data related to site usage in a user friendly dashboard.

## 3 Results

To date only the FSKN portal has shareable results based on use of the documents and implementation is pilot programs. The launch of both AgShare and AFRICA LEAD occurred within the last three months. AgShare has embedded in the grant funding monitoring and evaluation. AFRICA LEAD has a performance monitoring plan which includes both inputs and outcomes. The Venturit media platform will be measured on usage, appeal, and cost effectiveness.

The findings of the Pilot phase included results of training, use and re-use of OER, refinement of the assessments and testing the FSKN platform. The FSKN pilot activities in India, Egypt and China all demonstrated the effectiveness of the training programs to significantly improve participant knowledge regardless of previous educational attainment, previous training, or organization type. Assessment scores improved between the pre-and post-assessments by 6-16% across all pilots. Groups with lower performance on the pre-test tended to show the greatest improvement. Analysis of the results from the testing is continuing in order to learn more about how to improve the training, refine the learning materials and identify key determinants of participant performance. Participants also completed satisfaction surveys at the end of the online pilot experience. There was strong overall support for the value and effectiveness of the online training. Participants rated the PowerPoint and PDF files as most effective followed by an Mpeg4 file for iPods which included audio and video. The majority of participants ranked the quality of learning resources as excellent or good.

As part of the pilots, the original pre- and post-assessment instruments continued to be used in online environments. A total of 264 assessment questions were developed to evaluate knowledge against the 89 competencies in Basic Level for Food Manufacture. These 264 items were randomly allotted to two assessment forms, which then were randomly administered to participants in a manner such that each participant was exposed to all 264 items during the course of a single workshop. The online tools mimicked a similar assessment approach for the eLearning pilot participants. For face to face workshops, the questions were assigned by subject matter experts. Only in China, due to restrictions of time, the question set was refined to 140 items (70 per instrument) aligned with the all competencies. In all cases, statistical analyses demonstrated that the two assessment forms were equivalent with regard to participant performance.

## 4 Discussion

While all projects have occurred in the last two years and MSUglobal has had the ability to embed principle guidelines of using open education resources, using open source technologies, implementing meta data as needed for the projects, and with its external partner writing software improvements for indexing, search and discoverability, there is still is a need for a coherent plan. The coherence is needed for integration into rich web of agriculture and veterinary repositories. While federated search and scraping is becoming more commonplace, there is still a huge need for more discoverability within this community. In addition, the use of AGROVOC the agriculture vocabulary tool from the Food and Agriculture Organization provides an extensive base of words but it does not organize either in expert taxonomies, agreed upon competencies nor folksonomies.

While none of those intentionally limit the scope without more comprehensive organizing frameworks, agriculture repositories will continue to grow in a more ad hoc way. This should not be misunderstood as a need for a single all purpose repository, that simply is not realistic. Instead, it would be more beneficial to agree to an architecture and preferred metadata schemes/standards that would allow for more successful federated search.

## References

1. Spiro, R.J., Collins, B.P., Ramchandran, A.R.: Modes of Openness and Flexibility in Cognitive Flexibility Hypertext Learning Environments. In: Khan, B. (ed.) Flexible Learning in an Information Society. Idea Group, Hershey (2006)
2. Geith, C., Vignare, K., Bourquin, L., Thiagarajan, D.: Designing Corporate Training in Developing Economies Using Open Educational Resources. Journal of Asynchronous Learning Networks 14(3) (2010)

3. Geith, C., Butcher, N., Vignare, K., Yergler, N.: AgShare: Building Community and Content with Multiple Partners. In: Proceedings OpenEd 2010: 7th Annual Open Education Conference, Barcelona, November 2-4 (2010)
4. Geith, C., Vignare, K.: Access to Education with Online Learning and Open Educational Resources: Can They Close the GAP? Journal of Asynchronous Learning Networks 12(1) (2008)
5. Manso-Callejo, M., Wachowicz, M., Bernabé-Poveda, M.: The Design of an Automated Workflow for Metadata Generation. In: Sánchez-Alonso, S., Athanasiadis, I.N. (eds.) MTSR 2010. CCIS, vol. 108, pp. 275–287. Springer, Heidelberg (2010)
6. Spiro, R., Collins, B.P., Thota, J., Feltovich, P.: Cognitive Flexibility Theory: Hypermedia for Complex Learning, Adaptive Knowledge Application, and Experience Acceleration. Educational Technology 43(5), 5–12 (2003)
7. Ebner, H., Manouselis, N., Palmer, M., Enoksson, F., Palavitsinis, N., Kastrantas, K., Naeve, A.: Learning Object Annotation for Agricultural Learning Repositories. In: Ninth IEEE International Conference Advanced Learning Technologies, ICALT 2009, pp. 438–442 (2009)
8. Stracke, C.M.: The Benefits and Future of Standards: Metadata and Beyond. In: Sánchez-Alonso, S., Athanasiadis, I.N. (eds.) MTSR 2010. CCIS, vol. 108, pp. 354–361. Springer, Heidelberg (2010)

# Developing a Network of Cultural Heritage Objects Repositories for Educational Purposes

Kerstin Schmidt, Hans-Christian Schmitz, and Martin Wolpers

Fraunhofer Institute for Applied Information Technology FIT, Schloss Birlinghoven, 53754 Sankt Augustin, Germany
{kerstin.schmidt,hans-christian.schmitz, martin.wolpers}@fit.fraunhofer.de

**Abstract.** Natural history museums collect and provide access to digital representations of artifacts from their vast collections. The representations are used, among others, to facilitate their use in educational settings. For example, school teachers use them to prepare class visits to the museums. In the context of the European project Natural Europe, these repositories are to be bridged in order to enhance the usage experience in learning scenarios. Based on a thorough requirements engineering process, respective architectures are defined and services provided. In this paper, we discuss the outcomes of the requirements engineering process. While teachers appreciate the provision of fairly complex learning paths through museums, they also need direct access to the individual learning resources aka artifact representations. Finally, the technical architecture of the Natural Europe system is presented as a result of the requirements engineering process.

**Keywords:** Natural history, requirements engineering, learning repository, learning path, content harvesting, metadata.

## 1 Introduction

Throughout the past decades, natural history museums (NHM) have undergone fundamental changes in means of storing information on the objects in their repositories, those are cultural heritage objects (CHO). With the development of information technology and data management systems, the museums applied new technologies for the creation and the maintenance of digital data collections. While digital storage facilitates the administration and accessibility of the data collections compared to the previous analog methods, they also imply great challenges to the architecture of the respective storage systems concerning their interoperability. Not only is there a need for those systems to cooperate NHM-internally, but also for the cooperation on higher, national and cross-national levels as demanded by modern formal and informal educational methods. The Natural Europe project aims to overcome these barriers by providing a coordinated solution on a European level, which will connect the different NHMs' repositories and provide educational services on top of these federated repositories to enhance teaching and learning activities.

E. García-Barriocanal et al. (Eds.): MTSR 2011, CCIS 240, pp. 337–348, 2011.

The following exemplary use case describes a scenario, where Natural Europe facilitates the utilization of cultural heritage contents for educational purposes: according to the curriculum of his school, a teacher wants to explain the concept of evolution to his pupils. He plans a visit to the natural history museum of Lisbon to vividly explain the idea of evolution and its implications. The teacher develops one or even several learning paths for the museum visit. For preparing the visit, he collects various materials on the topic, including information on the Archaeopteryx, a dinosaur found in Bavaria and considered to be the evolutionary link between dinosaurs and birds. The material on the Archaeopteryx is provided by the museum in Eichstätt and made accessible via Natural Europe. In a post-visit phase the pupils develop presentations reflecting the museum visit. To this end, they use the Natural Europe repositories as a source for their research.

To implement the Natural Europe objectives, several NHMs, pedagogues, educational technologists, metadata experts and user groups work on European-wide integration and standardization processes, bringing together high quality cultural heritage contents of European NHMs and from digital libraries like Europeana [1], in order to enhance natural history and environmental education. The technical realization of the project does not start from scratch, but rather uses valuable experiences from related European projects. An enhanced version of the tool for authoring metadata on multimedia files that has been developed in the DELOS project [2] is used in Natural Europe for authoring digital cultural heritage contents, and the tools for the design of learning activities – like the aforementioned pathways – are based on similar tools used in the LOGOS project [3] and the e.KnowNet Network of Knowledge [4]. For the search and navigation components the interface developed in the MACE project [5] for architectural contents is adapted to Natural Europe's cultural heritage and learning contents.

After a description of the project's context and its main objectives in section 2, section 3 focuses on the results of the technical and educational requirements elicitation processes, shaping the overall Natural Europe system architecture, which is be presented in section 4. Section 5 concludes the paper and addresses the next steps.

## 2 The Natural Europe Project

The Natural Europe project, fully titled "Natural History & Environmental Cultural Heritage in European Digital Libraries for Education" [6], is supported by the European Commission under the Information Communication Technologies Policy Support Programme[1]. It focuses on two main topics, namely the federation of the participating natural history museums' digital data collections on the repository level, and the design and implementation of educational services and tools consuming the contents of these repositories.

### 2.1 Federated Cultural Heritage Repositories

There are several difficulties to overcome in order to build up a repository available for higher level, Europe-wide services. These technological aspects result from the

[1] http://ec.europa.eu/cip/ict-psp/index_en.htm

way the respective NHMs' data collections have been built up and have evolved over time. The process of creating the data collections – by digitization of analog contents or by direct creation of digital contents – has been implemented by each institution autonomously and according to its individual requirements, resulting in a large variety of different data collections of a different technological nature. This hinders the integration of digital data collections into commonly accessible repositories and thus vast amounts of contents remain unexploited. To overcome these restrictions, Natural Europe aims at federating the diverse data collections into one comprehensive repository, providing unified access to the contents originating from various sources. This enables the interaction with other CHO federations, like the Europeana digital library. Europeana collects digital contents originating from museums, libraries, archives and other audio-visual collections from all over Europe. The Natural Europe federation will serve as an aggregator for Europeana and vice versa, since both federations will consume contents of and upload contents to each other. Currently, six natural history museums are members of the Natural Europe consortium: the Arctic Centre in Rovaniemi, Finland [7], the Hungarian National History Museum in Budapest, Hungary [8], the Jura Museum in Eichstätt, Germany [9], the National Museum of Natural History in Lisbon, Portugal [10], the Natural History Museum of Crete, Greece [11] and the Estonian Museum of Natural History in Tallinn, Estonia[12].

## 2.2 Educational Services

Apart from preserving cultural heritage, one of the main interests of the NHMs is education. Analogous to the technical difficulties described before, learners and educators are facing similar problems. Usually the NHMs are connected to educational institutions and offer different educational services on a local level. Due to the lack of interoperable systems, they do not exploit the huge amount of information residing in other institutions.

Based on the federated repository developed in the context of Natural Europe, the project will develop educational tools, allowing the search, retrieval and creation of educational objects related with the contents of the cultural heritage repositories. The major objective from this educational point of view is the concept of learning paths, also called educational pathways. In this context, the term learning path refers to a sequence of learning activities. These are combined by educators in a certain order, ensuring that the learner increases his knowledge progressively by following the path. The educator is free to choose any structure of the learning path and its contents. While the term cultural heritage object refers to the digital objects in the NHM collections, the learning modules contained in a path as well as the path itself are called learning object (LO). These learning paths reflect the various usage scenarios of the cultural heritage contents to be implemented:

1. Educators want to enrich their curricula with structured or unstructured learning activities containing digital scientific material from the NHMs. Natural Europe will facilitate the access to LO repositories and also provide the tools for creating such learning paths.
2. Learners require access to rich scientific material to support their learning and research activities, and they need access to LO repositories (or parts of it) to follow the paths implemented by their educators.

3. Visitors want to enhance their visit by following structured activities through the NHM, enhance their experiences through pre- and post-visit activities.

In the following section we will describe the requirements for the project, reflecting the educators' concrete needs and serving as input for the design process of the system architecture.

# 3 Requirements

The technical and educational constraints have been set by classical methods of requirements engineering, such as workshops, interviews and questionnaires. After basic steps like setting the scope of the project and the identification of stakeholders, detailed analyses of the present situation in the NHMs nowadays have been made. Building on these, the explicit requirements for the features of future NHM portals have been elicited. In these processes representatives from the NHMs were involved, i.e. museum curators and collection experts from the particular departments in the NHMs for the technical part, as well as NHM-internal and external pedagogues of various educational levels for the educational part.

## 3.1 Content Requirements

The technical requirements mainly deal with the identification of the nature of existing collections in the NHMs participating in the project and the constraints these impose on the layout and implementation of the Natural Europe system.

It turned out that only few similarities could be found in the characteristics of the NHM collections, meaning that Natural Europe has to face the task of finding means of unifying the contents of the different collection for the use in a common repository.

Starting with data storage, many distinct methods have to be considered. In the relevant collections file-based storage systems can be found, like Microsoft Excel or plaintext XML files, as well as database management systems, like Microsoft Access, MySQL, Filemaker, and other third party content management systems. Apart from that, the data structures are also of different nature – while for many collections self-defined metadata schemas not conforming to any standards are used, others are compliant to common standards, e.g. Darwin Core[2], Dublin Core[3] or ABCD[4], and some do not even have the option of providing collection metadata in XML format. The accessibility aspect also implies great difficulties on the Natural Europe project since the collections are far from being completely web accessible. Some of them are, but most of them hosted on NHM internal servers, reside on personal computers or are stored on DVDs.

---

[2] http://www.tdwg.org/activities/darwincore/

[3] http://dublincore.org/

[4] http://www.tdwg.org/activities/abcd/

### 3.2 Technical Requirements

Looking at the curators' requirements for the Natural Europe repositories, it can be said that they need means for uploading content to the federated repository at least in a semi-automatic way: due to the amounts of CHOs it would be impossible to enter the data manually into the repository and due to the different metadata formats on the NHM side the data transfer cannot be done automatically, since the data have to be transformed into a standardized Natural Europe format first. This standard must have general metadata fields for capturing information on the CHOs as required by the interoperability with Europeana, and possibly more fields to capture additional information coming from the nature of the CHO's collection. For the maintenance of the data, including the creation of digital CHOs in the federated repository as well as the modification or extensions at a later time, the curators require an authentication and authorization system to protect their data from unauthorized access, and also a rights management system for the CHOs. Generally, the Natural Europe Intellectual Property Rights (IPR) schema, which is to be developed, should be applied to the contents, but there are other cases in which the access could be restricted to certain user groups. Lastly, they asked for Web2.0 features for the discussion on the CHOs, e.g. commenting on the content, notifying the author of errors or changes, Wikipedia discussions and Amazon reviews have been mentioned as examples for that.

### 3.3 Pedagogical Requirements

Previously the targeted outcomes of Natural Europe have been summarized as educational tools allowing search, retrieval and creation of learning objects. The educational requirements are far more complex than that and will be described in the following paragraphs.

**Educational Metadata.** Firstly, the learning objects require general metadata, just as the CHOs described in the previous subsection do. Second, regarding the requirements for the actual LOs, one of the most required features is the adaptability to the local curricula. Since the latter vary not only between the types of educational institutions but also in the different European countries, classifications are needed, clearly stating for which purpose and target audience they are suited. These classifications include educational characteristics like the type of the educational activity, target audience, educational level, school type, topics/context, duration of the activity and difficulty.

These classifications are necessary for the usability of the LOs because only by having these educational metadata the user perform reasonable actions on the datasets, like browsing for a specific pathway e.g. an elementary school course in biology on vertebrates of the Stone Age, or a university lecture on geology etc.

**Learning Objects.** Through the various procedures of the requirements engineering it turned out that there is a general need for learning paths of different educational levels and for different target groups. Having a repository containing learning paths (LPs) enriched with content from NHMs all over Europe will facilitate the educators' tasks of creating attractive structured learning activities including pre- and post-visit activities. This will help the educators to raise the learners' interest in the respective topic, it will increase the reusability and the dissemination of the educational pathways, which would otherwise might only exist on paper in somebody's desk, and additionally, it will help educators who are not that familiar with the adaption of certain pedagogical models to apply the pathways in their teaching activities. Private visitors of the NHMs that are interested in the pathways may also use them to plan their visits of a certain museum.

Apart from the learning paths, there is also a need for less structured educational activities. In this case, 'less structured' means that educators require a repository of educational contents to not only provide pre-structured activities but also smaller learning modules, which are not bound to an explicit educational model. This allows them to implement complex learning paths as well as smaller educational activities. Finally, the need for unstructured learning activities may even be drilled down to the usage of the CHOs in learning activities because in many scenarios educators might only want to use information on just one single object to their own learning activities.

**Platform Features.** The requirements elicited so far can be summed up by the following basic features the Natural Europe platform should offer. Further requirements resulting from the discussions on these features during the requirements elicitation process will be listed as well.

- The creation of new learning paths: this rather general task requires the platform to offer features like a user management system for authentication and authorization issues, options to choose between various learning path templates to be adapted by the learning paths, a repository for the storage and access of learning paths and most importantly the connection to CHO portals for objects to be integrated in the pathways. Apart from the template selection and adaption tasks, these features also apply for the revising and editing of existing learning paths.
- The usage of existing learning paths: to enhance the usability of the repository, the Natural Europe platform requires features like searching for learning objects, the navigation through the federated repositories, access and digital rights management. Similar to the technical requirements for the CHO repository, educators also would like to have the features of commenting and contacting the author to discuss and enhance their respective contents.

After one year of the project, the system architecture, which is described in the next section, has been designed according to the given requirements. While for editorial work on CHOs and LOs there are already prototype tools being brought to use, the specification of external content integration is currently being finalized.

## 4 The Natural Europe Architecture

From a technical point of view, the requirements for the Natural Europe architecture, whose main components are shown in figure 3, can be grouped into the following three categories, representing the main target outcomes of the project:

- A cultural heritage object repository: this repository provides means for the integration and the maintenance of the digital data collections coming from various

NHMs into one standardized Natural Europe repository. Also, it ensures the interoperability of the project repository with the Europeana digital library.

- An educational object repository: this repository provides means for hosting and maintaining the learning paths as well as the learning path templates, coming from the NHMs and the users of the Natural Europe portal. Beyond that it provides interfaces for exposing the educational contents to other learning object repositories.
- Graphical user interfaces (software and hardware) providing access to the platform services and allowing the application of the learning paths virtually or inside the NHMs.

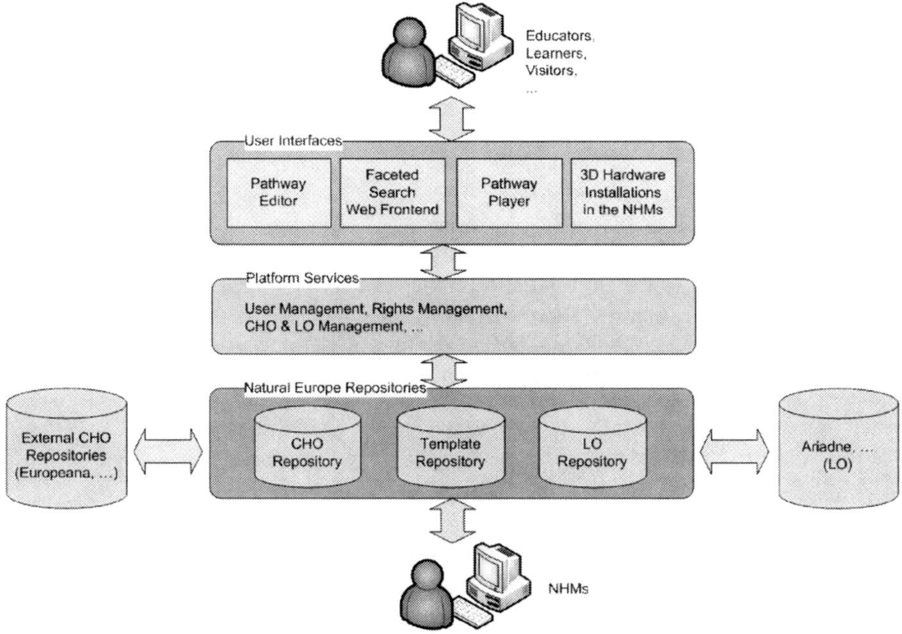


**Fig. 1.** Overview of the Natural Europe Architecture

## 4.1 Cultural Heritage Portal

The Natural Europe cultural heritage portal serves the purposes of a digital collection repository. While the NHMs are the main content producers for the repository, the project's learning objects portal, the Europeana digital library and other LO portals are the main content consumers of this portal.

**Metadata Standards.** To ensure the interoperability of the necessary software tools and components, common standards for the data formats and for the interfaces connecting the specific components have to be agreed upon.

After the evaluation of eligible standards, it was decided to use the Europeana metadata standard as the storage format cultural heritage object metadata in Natural Europe, since the exchange of data to and from Europeana, like harvesting, searching and retrieving, is a crucial requirement of the project. Thus the project supports the current Europeana metadata standard Europeana Semantic Elements (ESE) [13]. The ESE standard is a metadata standard based on Dublin Core [14], containing generic data fields for information on library, culture, museum and other archive objects.

The Europeana Data Model (EDM) [15] is a new approach of the Europeana federation which will support the full richness of the providers' metadata. It does not map the various standards to one central metadata schema, but it follows an open semantic web approach, by which the various metadata standards can be accommodated in an interoperable way. Natural Europe currently supports the ESE standard, but it will adapt EDM as soon as it is released by Europeana.

**Portal Connectivity.** To support the standardized metadata exchange between the Natural Europe repository and not only Europeana but also other CHO repositories, the project will make use of the Open Archives Initiative Protocol for Metadata Harvesting (OAI-PMH). The OAI-PMH protocol [16] is used to exchange Dublin Core-based metadata of records stored in a digital archive between other archives.

**Portal Services.** The services of the CHO portal can be grouped together into repository, harvesting and administration services. The repository services provides basic features like storage and maintenance of the CHOs and indexing the stored data for search services. Furthermore it controls the access to the repository contents which will be protected depending on the users' rights and on the digital rights for the CHOs. It is also responsible for the necessary functions to produce and consume CHO information transferred to and from other repositories or services. The harvesting service manages and performs the actual harvesting processes, i.e. is the data import from participating institutions like the NHMs in the Natural Europe project. Finally the administration service implements features like the management of users and access rights and the administrative control of the other services.

**Data Flow.** The lifecycle of a cultural heritage object in the CHO portal of a NHM can be illustrated by the following steps of the data flow from the participating NHMs to the common portal. First the NHMs will have to prepare the contents of their collections, which includes a) the creation of digital representations of the objects in case they are only present in an analog shape and b) the publishing of the contents, meaning that the objects have to be web-accessible for consumers like the users of the Natural Europe portal. Each participating NHM selects contents from its various collections and contributes the chosen CHOs to the project. For the use in the Natural Europe portal and others like Europeana the CHOs have to be supplied with persistent unique identifiers, ensuring that the original digital CHO can always be accessed via a URL that is not changing. Before the CHOs can be uploaded into the NHM's Natural Europe repository, their metadata has to be transformed into the common metadata schema which was determined to be the ESE/EDM schema. To do so, for each metadata schema which is used by the NHM a mapping has to be defined which is then used during the harvesting process to transform the original metadata into the ESE format – the EDM format will be used as soon as it's released by Europeana. Once the CHOs are imported to the repository, the NHM curators can administrate them, including modification or deletion as well as the enrichment with additional metadata. The step of mapping the metadata, importing the CHOs to the local NHM repository and the further administration and enrichment of the CHOs will be performed through a web-based authoring tool, which is already done by the museums. Finally the collected metadata in the NHM repositories are harvested into the common Natural Europe CHO repository which will provide the connection to and exploitation by other services and portals.

An overview of the components involved in the CHO portal, including the CHO data sources and the interaction with external portals, is shown in figure 2.

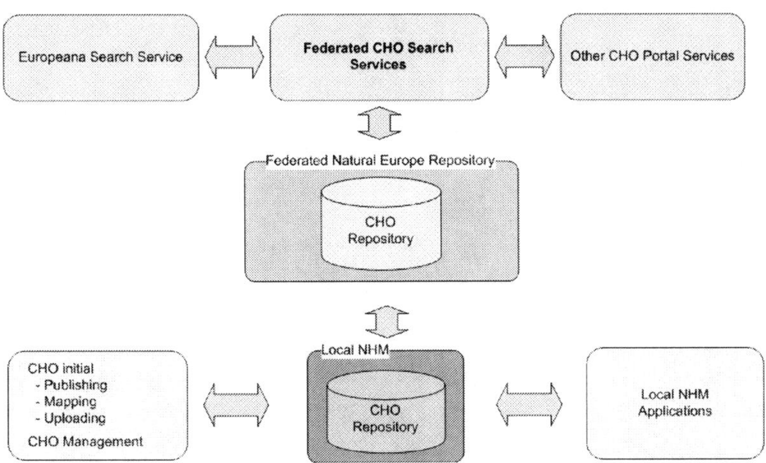


**Fig. 2.** The Natural Europe CHO Repository

## 4.2 Learning Object Portal

The second main target of the Natural Europe project is the creation of a learning object repository, hosting diverse learning activities which are enriched with contents from the internal cultural heritage object repository as well as external sources.

In contrast to the cultural heritage object repository, where the NHMs serve as main content producers, the learning object repository content will be produced by NHM educators, external educators (schools, universities, lifelong learning institutions etc), and other interested individuals. The content consumers on the other hand will be learners (pupils, students etc.) and also the educators (reusing existing contents) and other learning object federations.

**Metadata Standards.** In the context of the Natural Europe learning objects repository the Learning Objects Metadata (LOM) standard [17] will be used to describe the learning paths as well as the learning path templates. LOM is an internationally accepted standard published by IEEE for the description of learning objects, which can be any entity, digital or non-digital, that is used, re-used or referenced during technology-supported learning. This standard specifies a minimal set of fields necessary for the management of those learning objects, and at the same time provides a wide set of extensible data fields to adapt the standard to the needs of the usage scenario. The following categories of data elements are described by the LOM standard:

- General – general information on the LO
- Lifecycle – history and versioning of the LO
- Meta-Metadata – information on the LO's metadata
- Technical – technical requirements of the LO
- Educational – educational and pedagogical information on the LO
- Rights – intellectual property rights of the LO
- Relation – relation to other LOs
- Annotation – comments on educational use of the LO
- Classification – description of the LO according to a particular classification system

**Portal Connectivity.** Taking over the decision from the cultural heritage objects portal, the learning objects portal also uses OAI-PMH as the protocol for harvesting metadata to and from external learning repositories.

**Portal Components and Tools.** The repository layer of the LO portal consists of two components. On the one hand this is the learning path template repository, holding structured path templates which represent particular pedagogical models, as well as open templates, which do not force the use of a certain structure but allow the user to freely design the components of their learning activities. On the other hand there is the learning path repository, storing the actual implemented learning paths.

On top of these repositories search services provide the features of searching for and retrieval of cultural heritage object contents, and secondly the searching for and retrieval of LOs like the structured and open learning paths and their templates. In both cases project-internal sources as well as external portals' contents can be discovered, like Europeana for the CHOs and the OSR portal for the LOs, and the internal portals' contents are also provided for consumption by these external portals.

Finally there will be a rich set of tools and user interfaces exploiting the portal's features, like a tool for editing the CHOs' metadata, which at the same time integrates certain sets of vocabularies according to the respective CHO's collection background, to ensure the use of common classifications for the metadata. Apart from that there will be web-based tools for the implementation of LP templates and the actual LPs, mainly to be used by museum and school educators and also incorporating the discovery of the CHO and LO contents to be used during the implementation. Finally web frontends provide the actual application of the pathways to the end users, like following the contained learning activities or using the pathway for navigation through the actual museum. A faceted search interface similar to the one developed in the MACE portal will allow the convenient search and navigation through the portals' underlying CHO and LO contents. Apart from that it is also intended that these tools and features will be used by interactive installations on 3D interfaces inside the NHMs which allow 'minority-report-like' to navigate through and interact with the contents. Figure 3 shows the basic layers of the LO repository with the respective components contained, their interaction and the target users.

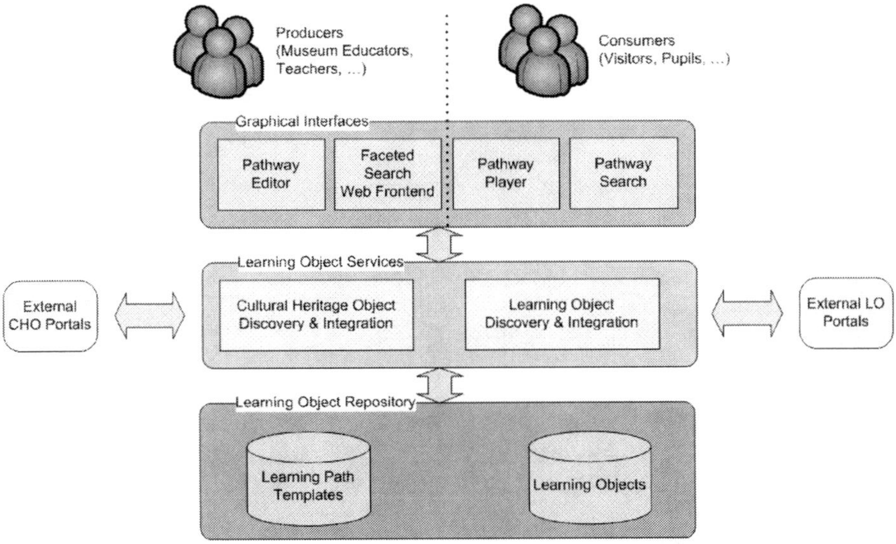

**Fig. 3.** The Natural Europe LO Repository

## 5 Conclusion

In this paper we presented the Natural Europe project and its targeted goal of exploiting the digital collection of cultural heritage items in natural history museums and connecting them to federated repositories for educational purposes. To achieve this, the architectural design is based on a continuous requirements engineering process whose outcomes, as we have shown, define the features of the Natural Europe system architecture, connecting a cultural heritage portal and a learning objects portal. Those are to be connected with other existing portals, CHO as well as LO, which also influenced the system architecture. The Natural Europe architectural design has by now reached a first completed status, the basic components of the CHO portal already being applied by the participating NHMs and the LO portal components being introduced and brought to use to the stakeholders. The next step will be the implementation of the interaction of the particular CHO and LO components with each other, followed by bi-directional content exchange with external portals. These steps will be accompanied by regular validations and performance tests, whose results affect the further implementation of the architecture and ensure the required functionality of the system.

## References

1. Europeana: think culture, `http://www.europeana.eu/`
2. DELOS Network of Excellence on Digital Libraries, `http://www.delos.info/`
3. LOGOS: Knowledge-on-Demand for Ubiquitous Learning, `http://www.logosproject.com/`

4. e.KnowNet, http://www.e-knownet.eu/
5. MACE – Metadata for Architectural Contents in Europe, http://portal.mace-project.eu/
6. Natural Europe, http://www.natural-europe.eu/
7. Arctic Centre – University of Lapland, http://www.arcticcentre.org/
8. Hungarian Natural History Museum, http://www.nhmus.hu/
9. Jura Museum Eichstätt, http://jura-museum.de/
10. National Museum of Natural History – University of Lisbon, http://www.mnhn.ul.pt/
11. Natural History Museum of Crete – University of Crete, http://www.nhmc.uoc.gr/
12. Estonian Museum of Natural History, http://www.loodusmuuseum.ee/
13. Europeana Semantic Elements Specification, Version 3.4, http://www.version1.europeana.eu/c/document_library/get_file?uuid=77376831-67cf-4cff-a7a2-7718388eec1d&groupId=10128
14. Dublin Core Metadata Initiative: Dublin Core Metadata Element Set Version 1.1, http://dublincore.org/documents/dces/
15. Definition of the Europeana Data Model Elements, Version 5.2.2, http://www.version1.europeana.eu/c/document_library/get_file?uuid=7ef62859-4846-4697-9285-6bde99d976f5&groupId=10605
16. The Open Archives Initiative Protocol for Metadata Harvesting, Version 2.0, http://www.openarchives.org/OAI/openarchivesprotocol.html
17. IEEE 1484.12.1-2002 Draft Standard for Learning Object Metadata, http://ltsc.ieee.org/wg12/files/LOM_1484_12_1_v1_Final_Draft.pdf

# A Learner Model for Learning Object Based Personalized Learning Environments

Galip Kaya and Arif Altun

Hacettepe University, Computer Education and Instructional Technology, Ankara, Turkey
{galipk,altunar}@hacettepe.edu.tr

**Abstract.** A personalized learning experience is possible with the use of learner models in intelligent tutoring systems, adaptive educational hypermedia systems and semantic web based learning environments. Although there are standards to overcome complexities or lack of fulfilling user needs (such as, IEEE or IMS), these models are either too generic or too complex to handle. In addition, there are various alternative learner or user models in the literature. Yet, these models do not address how to model learning object (LO) based instructional systems. Therefore, in this paper, an ontology based learner model is proposed for e-learning systems which use instructional learning objects.

**Keywords:** Learner model, learning object, ontology, personalization.

## 1 Introduction

Personalized learning removes time, location and other constraints in teaching and tailors teaching for each learner's constantly changing needs and skills [1]. In another definition, personalization is described as adapting learning experience to different learners due to the analysis of knowledge, skills and learning preferences of individuals [2]. Concept of personalized learning has changed tradition of "one design for all" of the traditional learning environments. In learning environments, instructional design has evolved from "one instructional design for many learners" to "one design for one learner" or "many designs for one learner". In traditional learning environments, materials are prepared for an average learner. But in personalized learning environments, materials can be adapted due to academic records, psychological attributes, skills, learning environment preferences of the learner. Content knowledge to be learned can be made more complicated or simpler according to the needs and the demands of the learner [3].

Personalized learning experience is possible for the learners in modern e-learning systems like intelligent tutoring systems, adaptive educational hypermedia systems, adaptive educational systems, semantic web based education systems. The underlying technology of these systems is the use of a learner model, a learner profile or a user model where the information of learner goals, preferences or needs is kept in learner data [4]. A learner model consists of meta-knowledge which includes the instructional decisions about a learner. So a learner model can be defined as an abstract image of the learner in the system [5].

E. García-Barriocanal et al. (Eds.): MTSR 2011, CCIS 240, pp. 349–355, 2011.

Although there are standards suggested by consortiums like IEEE (Personal and Private Information, PAPI) or IMS (Learner Information Package, LIP) for learner models, complexities in application of the model is articulated by researchers due to the details in standards or different user needs that cannot be fulfilled by these standards; hence, researchers have proposed new learner or user models for educational or other adaptive systems [6], [7], [8], [9], [10], [11].

The use of LOs for educational purposes brings various advantages, including reusability of the content knowledge, and preparing educationally sound content and delivery. It is expected that a learning environment is to handle the standards with existing LOs. Having considered the constraints that standards bring, an applicable learner model with LO-based systems could not be found in the reviewed literature. For this reason, in this paper, an ontology based learner model is proposed for e-learning systems that use learning objects.

## 2 Literature Review

In personalized learning environment literature, personalization is well separated in types by Martinez [3] according to its complexity parameter; personalization is divided into five groups:

1. Name based personalization: System addresses user by his/her name when user logs into system by username and password.
2. Self-described personalization: System takes user's preferences, attributes and past experiences by tools like questionnaires, pre-tests and registration forms.
3. Segmented personalization: Learners are grouped by common attributes (class, department, degree etc.) and demographic information. In this method, teaching is applied to whole group.
4. Cognitive personalization: In this method, content and teaching is delivered according to cognitive process, strategy, skill and preferences of learners. System will adapt content by user's working memory capacity, user's preference of text or image based representation etc.
5. Whole-person personalization: This method is a combination of cognitive based personalization and psychological resources that affect learning and performance. In this method, the system inferences over user model in learning process and constantly updates user model. So, user can be represented in all aspects.

In user model literature, models generally have similar constructs as described in [3]. Most of the researchers suggest personalization should be in whole-person personalization. On the other hand, segmented personalization is defined as stereotyping and this approach is either advocated or criticized in the literature. It is criticized because of stereotyping of users is an error-prone process.

In literature, there are different suggestions to keep which data will be kept and how it is kept:

Devedzic [2] suggests that IMS LIP and IEEE PAPI standards are a good starting point but they may not be sufficient for all requirements of the system. Researcher suggests to keep the following information in user model:

1. Objective data: It is provided by learner. e.g. learning history, learning preferences
2. Subjective data: This is related with cognitive personalization attributes. This data is kept by system and frequently updated.
3. Learner performance, and
4. Learning history

Devedzic's [2] research is an example for using ontologies in learner models. In this research, an ontology based personalized learning environment named TANGRAM is exemplified. Although researcher suggests that by using TANGRAM, it is possible to provide a better personalization via breaking learning objects into pieces to be used via ontologies. This method is limited with only MS Word and Powerpoint based learning objects. This limitation will cause problems in general validity and usability of system.

Razmerita et al. [6] developed OntobUM, which is ISM LIP based user modeling system for knowledge management systems. Similar to other user modeling studies, the system uses both users' self-reported data and information that emerges in user-system interaction. Another ontology based user model for knowledge management systems is proposed in [12]. In this model, data about user is classified due to related domain and defined as domain-dependent (data special to related domain) and domain-independent (data special to user). The proposed models may be a basis for a learner model, but there are many deficiencies like modeling learning goals, recording learner progress etc. when using these models in adaptive learning environments.

Vogiatzis et al. [13] proposed a user model for adaptive hypermedia systems. Researchers listed data to be stored as: Demographic data about user, learner goals, learner preferences and system usage data. Researchers also suggested that user should be placed in a stereotype like beginner, intermediate or advanced. As stereotyping is not a flexible technique and all attributes of a stereotype is not valid for all users in that stereotype, it is questionable whether this model is an adequate solution.

Hend and Maia [10] have developed an AHAM-based (Adaptive Hypermedia Application Model) adaptive tutoring system. They have used IMS LIP and IEEE PAPI standards. In their research, the learner model has been classified in four parts:

1. Machine based: keeps data of delivery format, security, access attributes etc.
2. Learner based: Learner selects courses or curriculum to define general purpose of learning. Sub-goals, demographic information, stereotype-data such as cognitive profile, learning preference etc. are also selected by learner. Role of teacher is to update these data
3. System based: History of interaction, teaching portfolio, proficiencies are kept
4. Teacher based: Teacher determines sub goals.

In Hend and Maia's [10] research, teachers have working load. It is obvious that tutors cannot be available all the time and teacher cannot guess learners' cognitive style or other cognitive attributes properly. In adaptive systems, learners should access tutoring systems without support of teachers. But in this system, teacher has a role in every step of learning.

There are several upper ontologies to be able to use in learner models [8] [14]. Theses ontologies may be a basis for learning model ontologies since they have general classes and relationships. Yet, they are inadequate to be used as modeling the learning process in order to meet the fundamental educational requirements.

Paneva [15] proposed an ontology based on a user model that uses IMS LIP and IEEE PAPI standards for digital libraries. Paneva [15] supports in the model that two-way information shall be kept in the system. General data like personal information, goals, interests, presentation method etc. and learner choices about multimedia digital library. Research has a good point of view by relating the model with multiple intelligence framework. Yet, it has weaknesses as it uses stereotyping or putting forward the importance of model in computer context instead of cognitive convenience of the model to users.

In literature, overlay model is another model mentioned frequently [5], [16], [17], [18], [19], [20]. In overlay model, user information is stored as a subset of domain model. For all sub-domains, user knowledge about this sub domain is expressed as quantitatively or qualitatively.

Overlay model is criticized as it is too simple [19]. It is mentioned that user model cannot always be defined as a certain subset of an expert model. From this point of view, bug model is proposed. Basis of this model is that a user may have both right and wrong information. The purpose of the model is not only to point out the buggy information gathered from the user, but also to adapt the user by identifying wrong information that he/she has. One step further of bug model is the genetic model. In this model, user knowledge is identified from simple to complex and from special to general. Although these two models have stronger capabilities than overlay model, it is reported to be harder to apply them to tutoring systems [17].

To sum up, in practical terms, there are no systems that genetic model is used in. Use of bug models are only limited with simple problem solving related intelligent systems. On the other hand, overlay model is very popular in web based adaptive education environments and adaptive hypermedia systems. Shortcomings of the models in literature for learning object and semantic web based tutoring systems are considered and to overcome deficiencies, a learning model is needed.

## 3 A Learner Model for Learning Object Based Personalized Learning Environments

First problem encountered in learning models is how to gather data about learners. It is mandatory to get personal information from learners by direct input. In addition, the system should guess preliminary knowledge for each user to adapt the learning environment for them. At this stage, pretests may be a solution; but, taking pre-tests as a first step might decrease learners' motivation [7]. This situation is called as "cold start problem" and to overcome, the system should gather knowledge about learner's academic history without disturbing him/her.

In our learner model, cold start problem will be solved by curriculum ontologies. By use of these ontologies, system will infer learner's current knowledge by taking the current grade of learner. Model will use both curriculum and domain ontologies to place a new learner in a proper learning location.

The proposed model is constructed by answering three fundamental questions: What will be modeled about learners, how will it be modeled and how the sustainability of the model would be maintained?

### 3.1 What Will Be Modeled, How Will It Be Modeled?

The information which will be kept about user embedded in the model is listed as follows:

- Demographic information: learner related demographic information will be kept in this part.
- Current learner status: learner's current knowledge about the domain will be kept in overlay model by use of ontologies. Interaction of domain ontologies and user model, current status of learner will be inferred.
- Expectations: The expectations (or learning goals), which learners are expected to acquire, will be provided by the curriculum ontology. In this ontology, the expectations will be kept hierarchically and be related to each other by ontological rules. Expectations and LO relationship will be provided by inferring and reasoning over curriculum. Domain ontologies are to provide the course and subject list to the learner.
- Individual attributes: Individual related data such as cognitive attributes, learning styles, presentation types etc. will be kept in this area. Cognitive attributes are frequently cited in the literature [2], [3], [10], [17], [19]. Yet it is not clearly stated how to keep cognitive attributes or how to correlate the cognitive attribute and the domain. In our model, cognitive attributes will be modeled by means of CogSkillNet cognitive skills ontology which is developed by Askar and Altun [21]. A new learning profile ontology will be developed and domain ontology, cognitive skills ontology and learning profiles ontology will work together to provide proper learning content to learners according to their cognitive skills and learning profiles.
- Performance: Completed courses, resolved tests, progress status, achieved gains and other results which emerge in the interaction of system and learner will be kept in this area.
- Context Attributes: Technical data such as connection speed, operating system of learner etc. will be kept in this area according to the learner's preferences.

### 3.2 How the Sustainability of the Model Would Be Maintained?

When users completed their study on LOs, they will be provided feedback via the system, such as the difficulty of the course, presentation strategy, relevance of the LO with learner or with subject etc. Learner model will be updated by these feedbacks and current status and progress of the learner will be updated in the system. Initial UML diagram of the user model interacting ontology classes is shown in Figure 1.

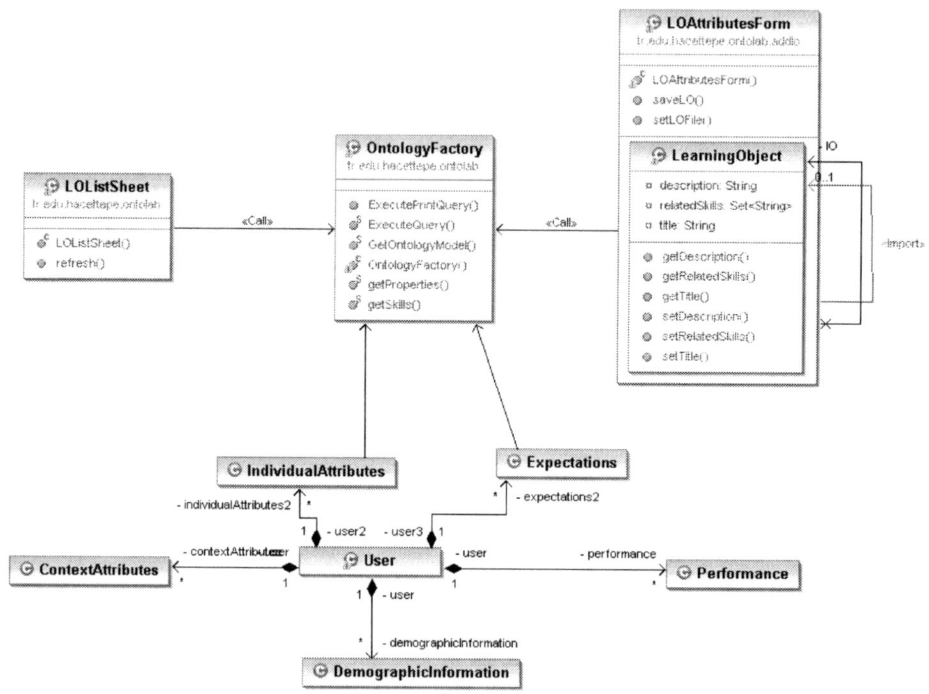


**Fig. 1.** UML diagram of user model

## 4 Conclusion

In this paper, a learner model is proposed to be applied in personalized e-learning environments. It is suggested that this model should be ontology based to provide reasoning and inference functionalities.

Currently, we have developed the CogSkillNet ontology. Also CogSkillNet based LO navigation software is being implemented. This navigation system will provide ontological support to search, navigate and find LOs, and will suggest related LOs related to queries of learners. When the model is implemented it will be verified and tested in the LO navigation system.

Another component of the model, learner ontology and supportive curriculum and learning profile ontologies are to be developed as future work

**Acknowledgments.** This research is supported by TUBITAK SOBAG, with the grant number of 110K602.

## References

1. Sampson, D., Karagiannidis, C., Kinshuk: Personalized learning: Educational, technological and standardization perspective. Interactive Educational Multimedia 4, 24–39 (2002)
2. Devedzic, V.: Semantic Web and Education. Springer-Verlag New York, Inc., Heidelberg (2006)

3. Martinez, M.: Designing Learning Objects to Mass Customize and Personalize Learning. In: Wiley, D.A. (ed.) Instructional Use of Learning Objects: Online Version (2000) (retrieved November 19, 2010)
4. Martins, A.C., Faria, L., et al.: User Modeling in Adaptive Hypermedia Educational Systems. Educational Technology & Society 11(1), 194–207 (2008)
5. Holt, P., Dubs, S., et al.: The state of student modelling. In: Student Modelling: The Key to Individualized Knowledge-Based Instruction, pp. 3–35. Springer, Heidelberg (1994)
6. Razmerita, L., Angehrn, A., et al.: Ontology-Based User Modeling for Knowledge Management Systems. In: Brusilovsky, P., Corbett, A., de Rosis, F. (eds.) UM 2003. LNCS, vol. 2702, pp. 148–148. Springer, Heidelberg (2003)
7. Denaux, R., Dimitrova, D., et al.: Interactive Ontology-based user modeling for personalized learning content management. In: Proc. of AH 2004 Semantic Web for E-Learning Workshop (2004)
8. Heckmann, D., Schwartz, T., et al.: Gumo – The General User Model Ontology, pp. 428–432 (2005)
9. Andrejko, A., Barla, M., et al.: Ontology-based User Modeling for Web-based Information Systems. In: Wojtkowski, W., Wojtkowski, W.G., Zupancic, J., Magyar, G., Knapp, G. (eds.) Advances in Information Systems Development, pp. 457–468. Springer, US (2007)
10. Hend, M., Maia, W.: Semantic Learning Model And Extended Student Model: Towards An Aham-Based Adaptive System (2008)
11. Martins, A.C., Faria, L., et al.: User Modeling in Adaptive Hypermedia Educational Systems. Educational Technology & Society 11(1), 194–207 (2008)
12. Andrejko, A., Barla, M., et al.: Ontology-based User Modeling for Web-based Information Systems. In: Wojtkowski, W., Wojtkowski, W.G., Zupancic, J., Magyar, G., Knapp, G. (eds.) Advances in Information Systems Development, pp. 457–468. Springer, US (2007)
13. Vogiatzis, D., Tzanavari, A., et al.: The Learner's Mirror. In: Ninth European Conference on Pattern Languages of Programs, EuroPloP (2004)
14. Yudelson, M., Gavrilova, T., Brusilovsky, P.: Towards user modeling meta-ontology. In: Ardissono, L., Brna, P., Mitrović, A. (eds.) UM 2005. LNCS (LNAI), vol. 3538, pp. 448–452. Springer, Heidelberg (2005)
15. Paneva, D.: Use of Ontology-based Student Model in Semantic-oriented Access to the Knowledge in Digital Libraries. In: Proceedings of the Fourth HUBUSKA Open Workshop Semantic Web and Knowledge Technologies Applications, Varna, Bulgaria (2006)
16. Brusilovsky, P., Cooper, D.W.: Domain, task, and user models for an adaptive hypermedia performance support system. In: Proceedings of the 7th International Conference on Intelligent User Interfaces, pp. 23–30. ACM, San Francisco (2002)
17. Brusilovsky, P., Millán, E.: User Models for Adaptive Hypermedia and Adaptive Educational Systems. In: Brusilovsky, P., Kobsa, A., Nejdl, W. (eds.) Adaptive Web 2007. LNCS, vol. 4321, pp. 3–53. Springer, Heidelberg (2007)
18. Antonio, A.d., Ramirez, J., et al.: A proposal for student modelling based on ontologies. In: Proceeding of the 2008 Conference on Information Modelling and Knowledge Bases XIX, pp. 298–305. IOS Press, Amsterdam (2008)
19. Brut, M.: Ontology-Based Modeling and Recommendation Techniques for Adaptive Hypermedia Systems. University "Alexandru Ioan Cuza" of Iaşi Faculty of Computer Science, Iaşi (2009)
20. Nguyen, L., Do, P.: Combination of Bayesian Network and Overlay Model in User Modeling. In: Allen, G., Nabrzyski, J., Seidel, E., van Albada, G.D., Dongarra, J., Sloot, P.M.A. (eds.) ICCS 2009. LNCS, vol. 5545, pp. 5–14. Springer, Heidelberg (2009)
21. Askar, P., Altun, A.: CogSkillnet: An Ontology-Based Representation of Cognitive Skills. Educational Technology & Society 12(2), 240–253 (2009)

# User-Oriented Content Retrieval Using Image Segmentation Techniques

Pythagoras Karampiperis

National Center of Scientific Research "Demokritos", Athens, Greece
pythk@ieee.org

**Abstract.** The need for applying advanced social information retrieval techniques for personalizing web-based information discovery has been identified as a key challenge. Until now, significant R&D effort has been devoted aiming towards applying collaborative filtering techniques for educational content retrieval. However, limited attention has been given to the use of educational metadata as a mean to enhance social filtering techniques via educationally informed filtering decisions. In this paper we propose the use of an add-on filtering service on existing social filtering systems/applications so as to create a data post-filtering mechanism that makes use of intelligence stored in TEL metadata. The proposed methodology starts with the generation of a matrix that represents the educational characteristics of the resources suggested by typical social filtering techniques and applies post-filtering using the educational "footprint" of the resources already used by the targeted end-user.

## 1 Introduction

The high rate of evolution of Web 2.0 applications implies that on the one hand, increasingly complex and dynamic web-based learning infrastructures need to be managed more efficiently, and on the other hand, new type of learning services and mechanisms need to be developed and provided. To meet the current needs, such services should satisfy a diverse range of requirements, as for example, personalization based on social filtering [1]. In this context, the need for applying advanced social information retrieval techniques for personalizing web-based information discovery and retrieval has been identified as a key challenge. This has become more critical in the case of Technology Enhanced Learning applications, since on the Web a vast variety of digital learning resources exist that have the potential to facilitate teaching and learning tasks. Until now, significant R&D effort has been devoted aiming towards applying collaborative filtering techniques for educational content retrieval [2]. These techniques are using usage log files over a set of educational resources to provide personalized recommendations by comparing the profile of the learner in hand with similar persons/groups recorded in the historical log data [3]. However, limited attention has been given to the use of educational metadata as a mean to enhance social filtering techniques via educationally informed filtering decisions.

In this paper we propose the use of an add-on filtering service on existing social filtering systems/applications so as to create a data post-filtering mechanism that

E. García-Barriocanal et al. (Eds.): MTSR 2011, CCIS 240, pp. 356–362, 2011.

makes use of intelligence stored in TEL metadata. The main driver of this work was inspired by the idea of using visualization information for accessing Learning Object Repositories [4]. Our goal was to investigate how image segmentation techniques could be applied in order to enhance the social filtering process of educational content. More precisely, the proposed methodology starts with the generation of a matrix that represents (in visual form) the educational characteristics of the resources suggested by typical social filtering techniques and applies post-filtering using the educational "footprint" of the resources already used by the targeted end-user. For the generation of the resource filter we utilize image segmentation techniques, taking into account the spatial coherence of the created visual representation. We treat the filtering problem as an inference problem, assuming that each pixel in the educational "footprint" (visualization) has a hidden binary label associated with it which specifies if it is appropriate for the targeted learner or not. In order to solve the inference problem, we use a variation of the EM algorithm which incorporates the spatial constraints with just a small computational overhead. Moreover, a potential drawback when applying social filtering techniques is that the models used are not fully transparent to the end user, thus, affecting the end-users' trust on the provided recommendations [5]. Since the generated filter by the proposed approach is represented visually, end-users can directly observe the core of the educational filtering process and make modifications/updates if desired.

The paper is structured as follows: In section 2, we discuss how educational metadata could be used in order to generate the educational "footprint" (visualization) of a set of educational resources. Finally, we demonstrate the application of the proposed visualization and filtering process on an easy-to-understand real life scenario.

## 2 Social Filtering via Educational Metadata Visualizations

Social filtering is a method for making automatic recommendations (filtering) about the preferences of a user by collecting preference information from many users. The underlying assumption of social filtering is that the users with similar preferences in the past tend to have similar preferences in the future. Our proposed method is an add-on filtering service on existing content-based filtering systems/applications, utilizing intelligence stored in TEL metadata. The main idea of the proposed approach is to post-filter the recommendations provided by typical passive social filtering techniques using the educational "footprint" of the resources already used by the targeted end-user. To achieve this, we create a matrix that represents (in visual form) the educational characteristics of the resources already recorded in the historical log files. Based on this matrix, we generate another matrix that represents the educational preferences of the targeted user. The latter matrix acts as an educational post-filter on the resources suggested by a typical social filtering system. This post-filtering is made by comparing the generated filter with the educational "footprint" of the resources suggested by a passive social filtering technique. Next paragraphs present how educational metadata are used to create the educational "footprint" of a single resource, as well as, of a set of resources. It is clear that this method is used for creating both the educational representation of the resources already used by the targeted user (which is the input for the filtering generation process), and the educational representation of the resources suggested by a passive social filtering technique (which is the input for the post-filtering process).

## 2.1 Creating the Educational Footprint of a Learning Resource

In order to generate the educational footprint (representation) of an educational resource we use the corresponding metadata record, a subset of the IEEE Learning Object Metadata (LOM) standard elements. The metadata elements used were selected in such a way that each element uses a specific state vocabulary, as illustrated in Table 1.

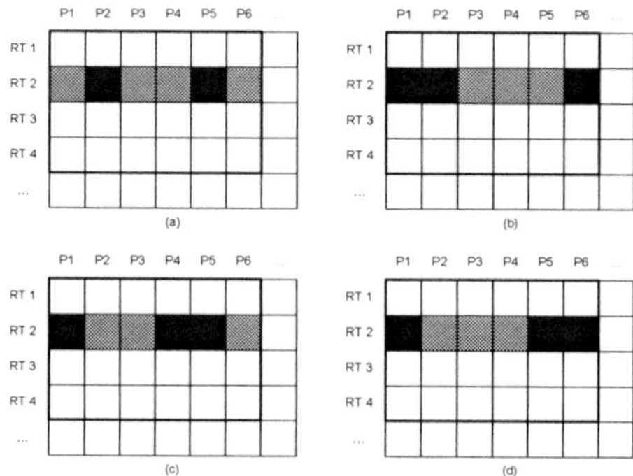


**Fig. 1.** Examples of representing the educational footprint of individual learning resources with Learning Resource Type (RT) equal to "simulation"

The educational footprint of a learning resource is a 15x8 pixels image where the first dimension (lines) stands for the states of the Learning Resource Type attribute and the second dimension (columns) stands for the rest eight attributes used.

Fig.1, presents examples of the produced representations for different cases of educational content, with the same learning resource type. For presentation simplicity, we have used resources that use only two values (states) per each metadata attribute (represented with gray and black colors accordingly).

Each pixel is colored according to the value of the corresponding attribute of the second dimension. The color coding used for each metadata attribute $j$ is defined by the formula:

$$Color^{j}_{RED} = Color^{j}_{GREEN} = Color^{j}_{BLUE} = \left(1 - \frac{k^{j}}{N}\right) \times 255,$$

where N stands for the number of vocabulary states of metadata attribute $j$, and $k^{j}$ is the state code of attribute $j$ for a given educational resource.

**Table 1.** Educational Resource Description Model and Color Coding used

| Metadata Element Used | Vocabulary State | State Code | Color Code (R-G-B)=(X-X-X) | Color |
|---|---|---|---|---|
| Interactivity Type | active | 1 | X=(2/3)*255 | |
| | expositive | 2 | X=(1/3)*255 | |
| | mixed | 3 | X=0 | |
| Interactivity Level | very low | 1 | X=(4/5)*255 | |
| | low | 2 | X=(3/5)*255 | |
| | medium | 3 | X=(2/5)*255 | |
| | high | 4 | X=(1/5)*255 | |
| | very high | 5 | X=0 | |
| Semantic Density | Same Vocabulary and Color Coding with "Interactivity Level" | | | |
| Typical Age Range | K12 | 1 | Custom Vocabulary (not defined in IEEE LOM). In our simulations we used the same Color Coding with "Interactivity Type" | |
| | 13-18 | 2 | | |
| | Adults | 3 | | |
| Difficulty | Same Vocabulary and Color Coding with "Interactivity Level" | | | |
| Intended End User Role | teacher | 1 | X=(3/4)*255 | |
| | author | 2 | X=(2/4)*255 | |
| | learner | 3 | X=(1/4)*255 | |
| | manager | 4 | X=0 | |
| Context | school | 1 | Same Color Coding with "Intended End User Role" | |
| | higher education | 2 | | |
| | training | 3 | | |
| | other | 4 | | |
| Typical Learning Time | Custom Vocabulary (not defined in IEEE LOM). In our simulations we used the same Vocabulary and Color Coding with "Interactivity Level" | | | |
| Learning Resource Type | exercise | 1 | This metadata element was used as the second dimension for the creation of the resource visual matrix. Thus, no color coding was used for this metadata element since each line (or set of lines) in the visual matrix represents directly the value of the "Learning Resource Type" | |
| | simulation | 2 | | |
| | questionnaire | 3 | | |
| | diagram | 4 | | |
| | figure | 5 | | |
| | graph | 6 | | |
| | ... | ... | | |

## 2.2 Creating the Educational Footprint of a Set of Learning Resources

In order to generate the representation of a set of learning resources, we start from the representation of the first learning resource in the set and extend the resolution of the generated image for each $n \times n$ resources, with $n \geq 2, n \in N^*$ per learning resource type. So the size of the generated representation for a set can be: $(15k) \times (8k)$pixels, where $k \in N^*$. As a result the generated visualizations can be (15 x 8), (30 x 16), (45 x 24), ... pixels. Fig.2, presents the aggregated representation of the resources demonstrated in previous section (Fig.1).

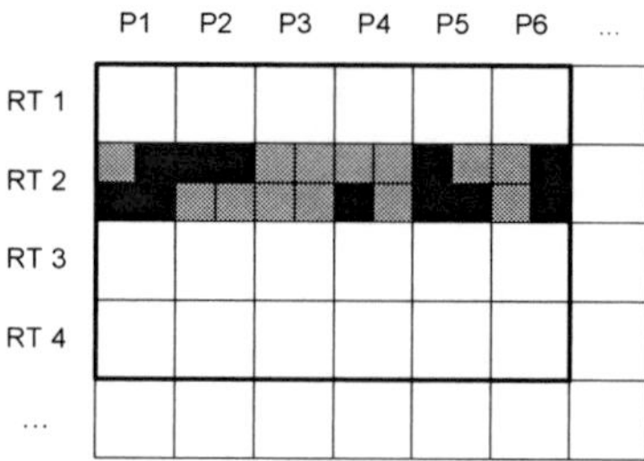


**Fig. 2.** Example of aggregated representation of a set of learning resources

Next section presents the methodology for generating the educational post-filter (that is, a matrix which represents the educational preferences of the targeted user) for the resources suggested by a typical passive social filtering system.

# 3 Demonstration

In order to make a preliminary evaluation of the effectiveness the proposed approach we used 10 Learning Object sets consisting of 135 learning object metadata records,

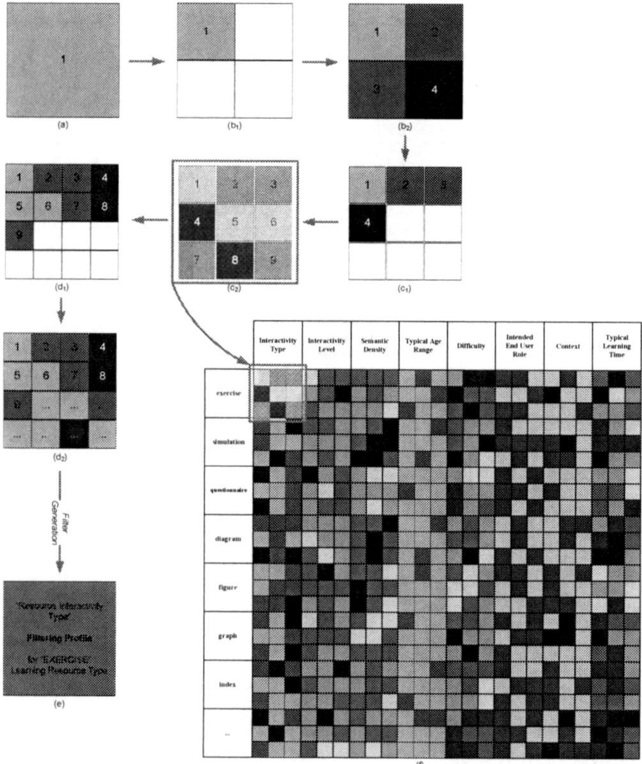


**Fig. 3.** Generating the educational footprint of a set of learning resources

that is, 9 Learning Objects per Learning Resource Type (simulating 10 different end-user's historical log files) and a set of 20 learning object metadata records (simulating recommendations from a passive social filtering system), with normal distribution over the value space of each metadata element. The goal of the evaluation was to test the ability of filtering out learning resources with educational footprint that does not match the educational preferences of a given end-user. From this preliminary evaluation, we have evidence that such an add-on service has the potential to enhance social filtering techniques via educationally informed filtering decisions.

Fig.3 presents an example of how the educational footprint for a set of 9 Learning Objects per Learning Resource Type is generated, depicting the step-by-step result of this process for the case of "Interactivity Type" metadata attribute. As we can observe, this is an incremental process starting with the representation of the educational footprint of the first learning object in the set (Fig.3a), continues with the representation for the first 2x2 learning objects (Fig.3b), then with the representation of the first 3x3 learning objects (Fig.3c), and so on for larger sets of learning objects (Fig.3d).

This representation is used as an input for generating the resource filter for the educational post-filtering of the resources suggested by a typical social filtering system/application. An example of such a filter is presented in Fig.4.

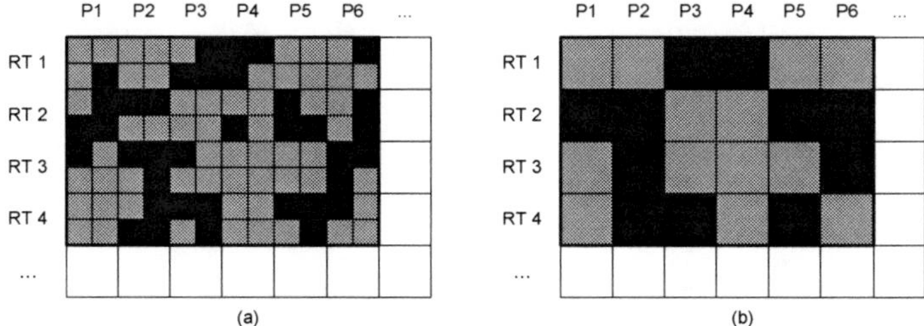


**Fig. 4.** (a) Example of representing a set of 16 learning objects – 4 per each learning resource type, (b) result of the proposed algorithm acting as a post-filter for future recommendations

# 4 Conclusions

In this paper we propose the use of an add-on filtering service on existing social filtering systems/applications so as to create a data post-filtering mechanism that makes use of intelligence stored in TEL metadata. The proposed methodology starts with the generation of a matrix that represents the educational characteristics of the resources suggested by typical social filtering techniques and applies post-filtering using the educational "footprint" of the resources already used by the targeted end-user. We treat the filtering problem as an inference problem, assuming that each pixel in the educational content visualization has a hidden binary label associated with it which specifies if it is appropriate for the targeted learner or not.

## References

[1] Ahn, J.W., Farzan, R., Brusilovsky, P.: Social Search in the Context of Social Navigation. Journal of the Korean Society for Information Management 23(2), 147–165 (2006)

[2] Recker, M.M., Walker, A., Wiley, D.A.: Collaboratively filtering learning objects. In: Wiley, D.A. (ed.) The Instructional Use of Learning Objects: Online Version (2000)

[3] Mobasher, B., Dai, H., Nakagawa, M., Luo, T.: Discovery and evaluation of aggregate usage profiles for web personalization. Data Mining and Knowledge Discovery 6, 61–82 (2002)

[4] Klerkx, J., Duval, E., Meire, M.: Using Information Visualization for Accessing Learning Object Repositories. In: Proc. of the 8th IEEE International Conference on Information Visualisation, pp. 465–470 (2004)

[5] Pierrakos, D., Paliouras, G., Papatheodorou, C., Spyropoulos, C.D.: Web usage mining as a tool for personalization: A survey. User Modeling and User-Adapted Interaction 13, 311–372 (2003)

# Semantic Annotation for Indexing Archaeological Context: A Prototype Development and Evaluation

Andreas Vlachidis and Douglas Tudhope

Hypermedia Research Unit, Faculty of Advanced Technology, University of Glamorgan, Pontypridd, CF37 1DL, Wales, UK
{avlachid,dstudhope}@glam.ac.uk

**Abstract.** The paper discusses the process of developing Semantic Annotations, a form of metadata for assigning conceptual entities to textual instances, in this case archaeological grey literature. The use of Information Extraction (IE), a Natural Language Processing (NLP) technique is central to the annotation process. The paper explores the use of Ontology Oriented Information Extraction (OOIE) methods for the definition of rich semantic-aware indices of archaeology documents. The annotation process follows a rule-based information extraction approach using GATE. In particular the report discusses a prototype development that adopts the core ontology, CIDOC CRM, together with an English Heritage archaeological extension, to inform and direct the information extraction effort. The prototype evaluation, supports the assumptions made, about the capability of the method to construct rich indices of grey literature documents empowered by Semantic Annotations.

**Keywords:** Natural Language Processing, Ontology Based Information Extraction, Semantic Annotation, GATE, Digital Archaeology.

## 1 Introduction

The complexity of human language results in a challenging environment for computations to provide solutions for the whole range of language related processes. On the other hand, advances in language engineering and computer technology of the past decades have made it possible for modern computer systems to perform natural language engineering tasks that previously were impossible to build and execute.

Information Extraction (IE) is a Natural Language Processing (NLP) technique which analyses a textual input and produces structured textual output of information capable for further manipulation. Such data manipulation can be directed towards automatic database population, machine translation, term indexing analysis and text summary generation. [1][2][3][4].

The fundamentally different role of IE does not compete with Information Retrieval (IR); on the contrary the potential combination of the two technologies promises the creation of new powerful tools in text processing. In particular, IR could benefit from the construction of sensitive indices of extracted information closer related to the "actual meaning" of a given text [1].

E. García-Barriocanal et al. (Eds.): MTSR 2011, CCIS 240, pp. 363–374, 2011.

### 1.1 The Aims of the IE Prototype Development

The paper discusses the details of a pilot study, which prototyped both the development and evaluation methods of an Information Extraction system aimed at the delivery of semantic annotation metadata. The main objective of the pilot study was to explore and evaluate the capability of the CIDOC CRM [5] and CRM-EH [6] ontologies for modelling and resolving free text information from grey literature documents into semantic annotation. In order to accomplish the above objective, the pilot study investigated the capacity of the GATE [7] framework to accommodate the task of information extraction of grey literature documents of archaeological reports with respect to the above ontologies. Earlier results of the prototype development effort have confirmed the feasibility of GATE and JAPE grammars to support an initial IE task targeted at archaeological grey literature reports [8].

The prototype development explored the flexibility of GATE for modification and adaptation to a chosen semantic annotation task. The adaptation was concerned with the ability of JAPE grammar rules to target CIDOC CRM and CRM-EH concepts, as well as the capability of GATE gazetteers to accommodate resources like thesauri and glossaries which contain terms that enjoy unique terminological references. In addition, the study applied a pilot evaluation method for assessing the overall performance of prototype information extraction system. The evaluation method was informed by literature and followed established evaluation measurements for assessing the performance of semantic annotation systems. The delivered semantic annotations contribute to the STAR project which aims to achieve semantic interoperability over diverse archaeological resources [9]

### 1.2 The Platform of the IE Prototype Pipeline

The prototype pipeline was developed in the GATE (General Architecture for Text Engineering) [10] environment, utilising hand crafted JAPE rules and exploiting domain vocabulary that was made available as gazetteer listings.

GATE is described as an infrastructure for processing human language, which provides the architecture and the framework environment for developing and deploying natural language software components [7]. Offering a rich graphical user interface, it provides easy access to language, processing and visual resources that help scientists and developers to produce natural language processing applications.

JAPE (Java Annotation Pattern Engine) [11] is a finite state transducer, which uses regular expressions for handling pattern-matching rules. Such expressions are at the core of every rule-based IE system aimed at recognising textual snippets that conform to particular patterns while the rules enable a cascading mechanism of matching conditions that is usually referred as the IE pipeline. JAPE grammars are constituted from two parts; the LHS (Left Hand Side) which handles the regular expressions and the RHS (Right Hand Side) which manipulates the results of the matching conditions and defines the semantic annotation outcome.

### 1.3 The Role of Semantic Annotation

Semantic Annotation is the process of tying ontological definitions to natural text by providing class information to textual instances [12]. Described as a mediator platform between concepts and their worded representations, Semantic Annotation as metadata can automate the identification of concepts and their relationships in documents. It is proposed that a mechanism responsible for connecting natural language and formal conceptual structures could enable new information access methods and enhance existing ones.

Semantic annotation enriches documents, enabling access on the basis of a conceptual structure. This aids information retrieval from heterogeneous data enabling users to search across resources for entities and relations instead of words. Semantic Annotation has the potential to bridge the gap between natural language text and formal knowledge expressed in ontologies, as evident from a number of IE projects [13].

### 1.4 The CIDOC CRM – EH Ontology

Ontologies are conceptual structures that formally describe a given domain by defining classes and sub-classes of interest and by imposing rules and relationships among them to determine a formal structure of 'things' [14][15]. The size and the scope, defines whether an ontology is called light-weight, core or upper level but all ontologies model a particular reality.

Ontological concepts can enrich information retrieval tasks by facilitating rich, semantic information seeking activities, both during query formulation and during retrieval. Inferences across diverse sources are supported by ontological structures, which are capable of mediating retrieval from heterogeneous data resources [16]. In addition, ontologies can be incorporated both in rule-based and machine-learning information extraction tools for supporting their semantic annotation operation. Usually such information extraction systems are described as ontology based (OBIE) or ontology oriented (OOIE), depending on the level of ontology engagement [17].

The CIDOC CRM is ISO standard ('ISO 21127:2006) core ontology for cultural heritage information aimed at enabling information exchange between heterogeneous resources by providing the required semantic definitions and clarifications. The CRM is the result of 10 years effort by the CIDOC Documentation Standards Working Group [18]. It is a comprehensive semantic framework designed to promote shared understanding of cultural heritage information.

The extended CRM model CRM-EH, is developed by the English Heritage EH, an organisation that has a major role in the dissemination of standards in cultural heritage domain both at national and international level. The extension resulted from the need to provide a common ground of shared meanings for what has been described as '*an archipelago of diverse, specialised and rather isolated and independent information systems and databases*' of the current archaeological systems [19]. The extended ontology comprises 125 extension sub-classes and 4 extension sub-properties. Based on the archaeological notion of context, modelled as place, the CRM-EH describes entities and relationships relating to a series of archaeological events, such as stratigraphic relationships and phasing information, finds recording and environmental sampling.

### 1.5 Semantic Technologies for Archaeological Resources (STAR) Project

The Semantic Technologies for Archaeological Resources (STAR) project aims to develop new methods for linking digital archive databases, vocabularies and associated unpublished on-line documents, often referred to as 'Grey Literature'. The project supports the efforts of English Heritage (EH) in trying to integrate the data from various archaeological projects and their associated activities, and seeks to exploit the potential of semantic technologies and natural language processing techniques, for enabling complex and semantically defined queries over archaeological digital resources [9].

To achieve semantic interoperability over diverse information resources and to support complex and semantically defined queries, the STAR project has adopted the English Heritage extension of the CIDOC Conceptual Reference Model (CRM-EH). The adoption of CRM-EH ontology by the project is necessary for expressing the semantics and the complexities of the relationships between data elements [20].

The project developed a CRM-EH based search demonstrator which cross searches over disparate datasets (Raunds Roman, Raunds Prehistoric, Museum of London, Silchester Roman and Stanwick sampling ) and a subset of archaeological reports of the OASIS grey literature corpus. Also the project delivered a set of web services for accessing the SKOS[1] terminological references and relationships of the domain thesauri and glossaries which are employed by the project.

### 1.7 OASIS Grey Literature Reports

The term grey literature, it is used by librarians and research scholars to describe a range of documents and source materials that cannot be found through the conventional means of publication. Preprints, meeting reports, technical reports, working papers, white papers are just a few examples of grey literature documents which are not always published by conventional means. The expansion of the Web and the advent of sophisticated workstations increased the possibilities for disseminating information on a large scale. Thus the need for solutions targeted at accessing information with the volume of available grey literature documents is becoming more and more apparent [21].

A considerable volume of grey literature documents falls within the scope of the STAR project, constituting a valued resource for enabling access to diverse archaeological resources. Grey literature documents hold information relative to archaeological datasets that have been produced during archaeological excavations and quite frequently summarise sampling data and excavation activities that occurred during and after major archaeological fieldwork. Integration of grey literature in STAR is intended for enabling cross-searching capabilities between datasets and grey literature documents, with respect to the semantics defined by the adopted CRM-EH ontology.

The collection of grey literature documents (corpus) that concerns the prototype development originates from the Online AccesS to the Index of archaeological

[1] Simple Knowledge Organization System (SKOS) is a standard language built upon RDF(S)/XML W3C technologies for the formal representation of knowledge organization systems, such as thesauri.

investigations (OASIS) project [22]. The OASIS project is a joint effort of UK archaeology research groups, institutions, and organisations, coordinated by the Archaeology Data Service (ADS) [23], University of York, aiming to provide an online index to archaeological grey literature documents.

## 2 The Prototype Development of IE Pipelines

Two separate information extraction pipelines were developed to address particular objectives of the information extraction task. Both contribute to the main aim of the provision of semantic indexing with respect to the CRM-EH ontology.

The first pipeline (pre-process) is intended to reveal commonly occurring section titles of the grey literature documents and to extract the summary sections of grey literature documents. Summaries were identified from archaeology experts as important document sections containing rich information worth targeting by the main ontology oriented information extraction phase.

The use of the ontologies was examined and explored by the second, information extraction phase, which aimed at identifying pieces of information from grey literature documents, which could be associated with CRM and CRM-EH ontological entities. In particular, the pipeline explored the potential of the ontology to inform the construction process of the JAPE rules and its capacity to assign ontological definitions to the delivered semantic annotation metadata.

### 2.1 Pre-processing Corpus Collection

The pre-processing phase (Figure 1) employed domain neutral information extraction techniques for the identification of specific document sections, which were then used as input at later stages of the prototype IE pipeline. During project discussions, archaeology experts suggested the exclusion of Headings and Table of Contents (TOC) from the semantic annotation process. It was made clear that such sections enjoy limited ontological commitment to CIDOC CRM, since they do not make use of terms in a rich discussion setting but instead use terms in isolation. Detecting headings also serves the purpose of revealing different document sections, such as summary sections, which contain rich discussion worth revealing. Heading annotations are used by the pipeline as input to detect the beginning and the end of each summary section.

The identification of the heading spans is based on a collection of eight different pattern-matching rules. Two rules annotate heading areas that commence with a numerical prefix followed by a capitalised or upper initial word, which might be followed by more words not necessarily in capital or upper initial case, such as "*3.1 Prehistoric phase*". Another set of rules targets single worded headings that have upper initial or capitalised case and do not commence with numerical prefixes, such as "*Introduction*". Also a specific set of rules targets headings that are followed by a sequence of dots and a number, which are very frequently found in table of contents.

The identification of TOC is based on a simple pattern that joins four or more previously identified Heading annotations together. Similarly, identification of the summary sections is also based on a simple JAPE grammar, which matches sections wrapped between two heading annotations. The first heading annotation contains any of the words; "summary", "abstract" or "overview" independently of their case and the second heading annotation is simply the next occurrence of a heading in the document

### 2.1.1 JAPE Rules of the Pre-Processing Pipeline

As an example, consider the rule for matching headings beginning with a numeral. The rule matches phrases which commence with numbers like 1, 1., 1.1, 1.1.1. etc. followed by a non lowercase word Token, which is then followed from any number of Tokens including sequence of Dots (previously identified) until the end of line (EL) token.

```
({BL, Token.kind==number, Token.length <= 2}
(({Token.string == "."})?
({SpaceToken.kind == space})?
({Token.kind == number, Token.length <= 2})?)*
({SpaceToken.kind == space})+
({Token.orth !="lowercase", Token.kind == word})
({Token.kind==word}|{Token.kind==number}|
{Token.kind==punctuation}|{SpaceToken.kind==space}|
{Dots})*{EL})
```

The following rule matches document summary sections which commence with a heading annotation of the type *Summary* (matched by a previous rule) and ends with the next available heading annotation of the document. It is possible to identify large chunks of text by configuring the rule to process only the heading annotations not Tokens or other annotation types, simplifying this way the grammar of the pattern.

```
{Heading.type=="Summary"}
{Heading}
```

## 2.2 Domain Oriented Information Extraction Pipeline

The domain-oriented pipeline (Figure 1) extracts specific archaeological information utilising the available English Heritage terminology resources and the domain ontologies, CIDOC CRM and CRM-EH. The choice of ontological entities targeted by the process is based on project discussions with English Heritage and specifically with project archaeological collaborator, Keith May, and the study of available use case scenarios. It was decided that the prototype system should focus on the extraction of the following concepts:

a) E19 Physical Object described as "*items having physical boundaries that separate them completely in an objective way from other objects*"

b) E49 Time Appellation described as "*appellation of all forms of names or codes, such as historical periods, and dates, which are characteristically used to refer to a specific temporal extend that has a beginning an end and a duration*"

c) E53.Place with emphasis on EHE0007.Context described as "*Spatial elements that constitute an individual archaeological unit of excavation including both primitive contexts and larger groupings of contexts*"

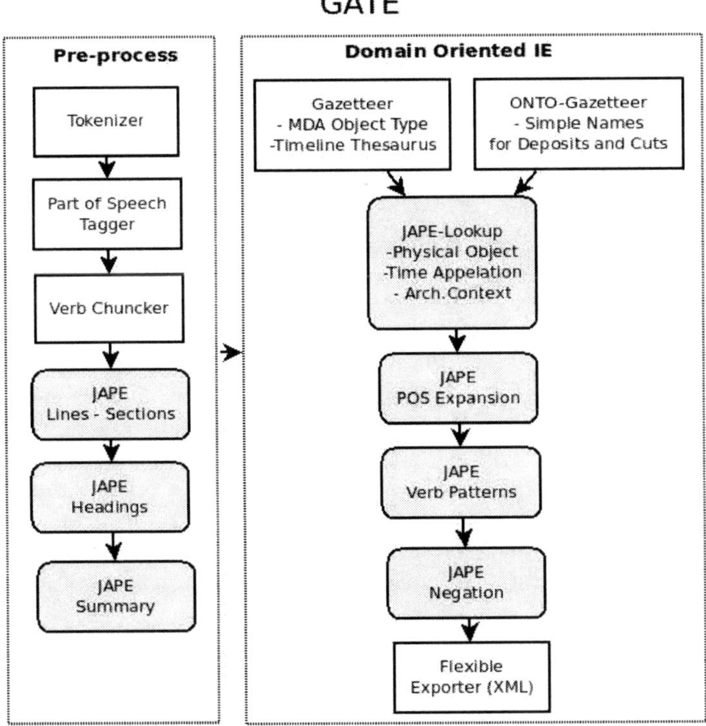


**Fig. 1.** The Information Extraction pipeline developed in GATE. Two separate exercises are shown here the Pre-processing and the Domain Oriented IE. Bespoke JAPE rules shown in Grey boxes, white boxes are used for GATE modules

### 2.2.1 Terminology Resources for Information Extraction

A range of different terminology resources such as thesauri and glossaries were made available to the STAR project by English Heritage, which were transformed to SKOS XML files (Skosified) in a previous exercise [24]. In addition, the "Skosified" resources were transformed into GATE gazetteer listings using XSLT transformation templates. The pilot system utilizes three particular resources; the Archaeological Object Type thesaurus, the Time-line thesaurus and the glossary Simple Names for Deposits and Cuts [25]. In addition, the pipeline exploits an RDF version of the CRM-EH ontology which is imported into GATE using the OWLIM ontology repository plug-in of GATE.

All gazetteer resources were assigned a *Major Type* and a *Minor Type* property, which were accessed by JAPE grammars for the definition of Lookup rules. The *Major Type* property was used for declaring the conceptual definition of each resource. Hence, all terms originating from the Archaeological Object Type thesaurus were given a *Major Type* "Physical Object", the terms of the Time-line thesaurus were given a *Major Type* "Time Appellation" while the glossary populated the OWLIM repository with instances of the CRM-EH class EHE0007 Context. Each individual term was also assigned a *Minor Type* property corresponding to its *skos:concept* terminological reference, maintaining in this way a link between GATE gazetteer and SKOS structure.

### 2.2.2 JAPE Rules of the Domain Oriented Information Extraction Pipeline

The prototype pipeline implemented fifteen different JAPE grammars for identifying the three main ontological concepts (Physical Object, Archaeological Context, and Time Appellation). The grammars exploited the *Major Type* gazetteer property for assigning the corresponding ontological reference to the matches, with the exception of the matches of Archaeological Context which instead of Major Type property used the CRM-EH class property. This is because the Lookup mechanism for Archaeological Context experimented with the use of the OWLIM ontology plug-in.

Additional rules were used for extending the initial Lookup annotations to include meaningful moderators. In the case of Time Appellation, two gazetteer listings were used for expanding over prefix terms (Earlier, Later, etc.) and suffix terms (Period, Century, etc.)

The following rule matches three different cases of Time Appellation expansion. a) Expansion towards prefix and suffix i.e. “Early Roman Period”, b) Expansion only towards prefix i.e. “Early Mediaeval” and c) Expansion only towards suffix i.e. “Prehistoric period”

```
({Lookup.minorType==Date_Prefix}{TimeAppellation}
{Lookup.minorType==Date_Post})|
({Lookup.minorType==Date_Prefix}{TimeAppellation}|
{TimeAppellation}{Lookup.minorType == Date_Post})
```

Moreover, JAPE patterns were employed to identify pairs of Lookups, such as Time Appellation and Physical Object i.e. “Roman Pottery” or Time Appellation and Archaeological Context, i.e. “Mediaeval Deposit”. This last approach was elaborated further by the definition of JAPE patterns that matched linguistic evidence of combinations between Lookups and verb phrases in the form of <Lookup><verb><Lookup>. Such patterns were aimed at matching combinations between Time Appellation and Physical Object as for example “...coins dating to Roman period...”, Time Appellation and Archaeological Context as for example “...pits are of prehistoric date...”, and Archaeological Context and Physical Object as for example “…pits containing pottery...”. This above pattern-matching approach was aimed at supporting the assumption that text phrases carry information which describes relations between CRM-EH entities and that linguistic evidence in form of pattern matching rules can be employed to extract such textual instances.

The following rule matches phrases that connect Lookup annotation via verb phrases i.e. “pits containing pottery”

```
{Context}({Token.kind==word}|{Token.category==","})*
{VG}({Token.kind==word}|{Token.category==","})*
{PhysicalObject}
```

## 3 Evaluation of Semantic Annotations

The effectiveness of Information Extraction systems is measured in *Recall* and *Precision* rates. The measurement units originate from the IR domain but they have been redefined during the Machine Understanding Conference MUC to reflect matching and mismatching of the information extraction process [26].

The evaluation task aimed at measuring the performance of the prototype information extraction mechanism with regards to the concepts of Time Appellation, Physical Object and Archaeological Context. The task had a largely formative and less summative character, aiming not just to evaluate the performance of the prototype system but also to suggest the necessary development improvements that have to be taken on board by a full scale Information Extraction exercise.

For the purposes of the evaluation, a manually annotated versions of the intended IE results was created and made available to the GATE Corpus Benchmarking Utility. Four versions of individual manual annotation sets were produced by four project members including the developer and an archaeology expert.

Since the major aim of the pilot evaluation was to inform a later larger scale IE the evaluation exercise did not conclude in a single definitive "gold standard" [27] version, which was not considered necessary for the purposes of the pilot study. Instead, the evaluation task used all four manual annotation versions in order to get a pluralistic view for the system performance, informed by the differences of manual annotation between individual sets.

The summary sections which participated in the evaluation task were extracted during the pre-processing phase. The selected extracts originated from five archaeological Evaluation reports and five archaeological Excavation reports, which were identified by the archaeology expert as sections carrying rich and relevant information to the aims of the prototype evaluation.

### 3.2 Evaluation Results

A closer examination of the overall system's performance (Table 1) revealed encouraging results regarding Precision, Recall and their weighted average F-measure. When including AV's version of manual annotation to the overall score the system's performance score improved even further. This was to be expected since AV was involved in the development of the system and so was more aware of the capabilities of the extraction mechanism regarding coverage of gazetteer resources and pattern matching rules. On the other hand, the system delivered some positive results against KM, who is an archaeology expert involved in the definition of the CRM-EH ontology and so his judgment is considered to be more definite and closer to the ontological definition than the rest of annotators.

Examining the performance of the system against manual annotators reveals that there is some basic agreement between annotators about the system performance. Excluding AV, the system delivered an average fMeasure score of 56%, marking the system's ability to target specific concepts with some success.

**Table 1.** System's performance against the gold standard Annotations

| | AV | CB | DT | KM |
|---|---|---|---|---|
| **Precision** | 0.85 | 0.68 | 0.73 | 0.69 |
| **Recall** | 0.85 | 0.69 | 0.61 | 0.71 |
| **F-Measure :** | 0.76 | 0.56 | 0.56 | 0.56 |

**Table 2.** System's performance for three ontological entities showing differentiation between annotators

| | Entity | Correct | Missing | Precision | Recall | F-Measure |
|---|---|---|---|---|---|---|
| | E49 | 51 | 5 | 0.99 | 0.90 | 0.94 |
| **AV** | E19 | 14 | 4 | 0.53 | 0.75 | 0.62 |
| | EHE007 | 56 | 2 | 1.00 | 0.96 | 0.98 |
| | E49 | 46 | 20 | 0.93 | 0.68 | 0.79 |
| **CB** | E19 | 13 | 20 | 0.50 | 0.40 | 0.44 |
| | EHE007 | 35 | 66 | 0.62 | 0.35 | 0.44 |
| | E49 | 44 | 16 | 0.90 | 0.71 | 0.80 |
| **DT** | E19 | 9 | 19 | 0.41 | 0.35 | 0.37 |
| | EHE007 | 35 | 78 | 0.66 | 0.31 | 0.42 |
| | E49 | 45 | 35 | 0.93 | 0.55 | 0.70 |
| **KM** | E19 | 10 | 17 | 0.42 | 0.39 | 0.40 |
| | EHE007 | 31 | 96 | 0.60 | 0.25 | 0.36 |

The system's performance on individual ontological entities reflects the differentiation between individual annotation sets. The system performed well against E49 Time Appellation entities delivering high *Precision* varying from 90% to 99% and *Recall* rates varying from 55% to 90% (Table 2). Precision was also good for EHE0007 Archaeological Context entities (60% to 66%). However, Recall for the same entity type was low, varying from 25% to 35% due to the limited coverage of the ontology list of instances for this particular entity. The coverage of the gazetteer listing for the E19 Physical Object entity was also problematic affecting the overall performance of the system for this particular type. The system for Physical Object annotations, delivered a precision score around 50%, indicative of the volume of false positive matches identified.

## 4 Conclusions

The prototype development has reached its aim of implementing a prototype IE system, capable of extracting concepts with respect to a given domain ontology and generating rich semantic annotations of grey literature documents. The initial evaluation results are encouraging and have revealed the capacity of the method for identifying rich textual instances that correlate to a set of ontological entities and properties. Extraction of ontological phrases that combine more than one ontological entity also looks promising. Such phrases carry the potential of extracting CRM-based Event type entities. In addition, the results of the pilot study suggest that utilization of the hierarchical relationships of the available thesauri is required by a full-scale system. A sophisticated exploitation of thesaurus relationships could benefit the IE outcome by enabling a selective use of the terminology resources that does not harm recall by using too little or precision by using too much of the available vocabulary. Further elaboration of the method is required for revealing further the capabilities of NLP techniques to provide rich semantic indices at an operational level.

**Acknowledgements.** The STAR project is funded by the UK Arts and Humanities Research Council (AHRC). Thanks are due to Phil Carlisle & Keith May (English Heritage), Ceri Binding (University of Glamorgan), Renato Souza (Universidade Federal de Minas Gerais, Brazil).

## References

1. Cowie, J., Lehnert, W.: Information extraction. Communications ACM 39(1), 80–91 (1996)
2. Lewis, D., Jones, K.: Natural language processing for information retrieval. Commun. ACM 39(1), 92–101 (1996)
3. Moens, M.F.: Information Extraction Algorithms and Prospects in a Retrieval Context. Springer, New York (2006)
4. Gaizauskas, R., Wilks, Y.: Information extraction: beyond document retrieval. Journal of Documentation 54(1), 70–105 (1998)
5. CIDOC-CRM, `http://www.cidoc-crm.org/`
6. CRM-EH, `http://hypermedia.research.glam.ac.uk/resources/crm/`
7. Cunningham, H., Maynard, D., Bontcheva, K., Tablan, V.: GATE: A Framework and Graphical Development Environment for Robust NLP Tools and Applications. In: Proceedings of the 40th Anniversary Meeting of the Association for Computational Linguistics ACL 2002 (2002)
8. Vlachidis, A., Binding, C., May, K., Tudhope, D.: Excavating Grey Literature: a case study on the rich indexing of archaeological documents via Natural Language Processing techniques and Knowledge Based resources. ASLIB Proceedings Journal 62(4&5), 466–475 (2010)
9. Tudhope, D., Binding, C., May, K.: Semantic interoperability issues from a case study in archaeology. In: Kollias, S., Cousins, J. (eds.) Semantic Interoperability in the European Digital Library, Proceedings of the First International Workshop SIEDL 2008, Associated with 5th European Semantic Web Conference, Tenerife, pp. 88–99 (2008)
10. General Architecture for Text Engineering GATE, `http://gate.ac.uk/`
11. Cunningham, H., Maynard, D., Tablan, V.: JAPE a Java Annotation Patterns Engine, 2nd edn. Technical report CS–00–10, University of Sheffield, Department of Computer Science (2000)

12. Bontcheva, K., Cunningham, H., Kiryakov, A., Tablan, V.: Semantic Annotation and Human Language Technology. In: Semantic Web Technology: Trends and Research in Ontology Based Systems, John Wiley and Sons, Sussex (2006)
13. Uren, V., Cimiano, P., Iria, J., Handschuh, S., Vargas-Vera, M., Motta, E., Ciravegna, F.: Semantic annotation for knowledge management: Requirements and a survey of the state of the art. Web Semantics: Science, Services and Agents on the World Wide Web 4(1), 14–28 (2006)
14. Guarino, N.: Formal Ontology and Information Systems. In: Guarino, N. (ed.) Formal Ontology in Information Systems, pp. 3–15. IOS Press, Amsterdam (1998)
15. Wilks, Y.: The Semantic Web as the apotheosis of annotation, but what are its semantics? Intelligent Systems 23(3), 41–49 (2008)
16. Kiryakov, A., Popov, B., Terziev, I., Manov, D., Ognyanoff, D.: Semantic annotation, indexing, and retrieval. Web Semantics: Science, Services and Agents on the World Wide Web 2(1), 49–79 (2004)
17. Bontcheva, K., Duke, T., Glover, N., Kings, I.: Semantic Information Access. In: Semantic Web Semantic Web Technology: Trends and Research in Ontology Based Systems. John Wiley and Sons, Sussex (2006)
18. Crofts, N., Doerr, M., Gill, T., Stead, S., Stiff, M.: Definition of the CIDOC Conceptual Reference Model, http://cidoc.ics.forth.gr/docs/cidoc_crm_version_5.0.1_Mar09.pdf
19. Cripps, P., Greenhalgh, A., Fellows, D., May, K., Robinson, D.E.: Ontological Modelling of the work of the Centre for Archaeology. CRM – EH model diagram (2004), http://cidoc.ics.forth.gr/docs/AppendixA_DiagramV9.pdf
20. STAR project, http://hypermedia.research.glam.ac.uk/kos/star/
21. Debachere, M.C.: Problems in Obtaining Grey Literature. IFLA Journal 21(2), 94 (1995)
22. Online AccesS to the Index of archaeological investigations OASIS, http://oasis.ac.uk/
23. Archaeology Data Service ADS, http://archaeologydataservice.ac.uk
24. Binding, C., Tudhope, D., May, K.: Semantic Interoperability in Archaeological Datasets: Data Mapping and Extraction via the CIDOC CRM. In: Christensen-Dalsgaard, B., Castelli, D., Ammitzbøll Jurik, B., Lippincott, J. (eds.) ECDL 2008. LNCS, vol. 5173, pp. 280–290. Springer, Heidelberg (2008)
25. EH National Monuments Records Thesauri, http://thesaurus.english-heritage.org.uk/
26. Grishman, R., Sundheim, B.: Message Understanding Conference-6; a brief history, pp. 466–471. Association for Computational Linguistics, New Jersey (1996)
27. Maynard, D., Peters, W., Li, Y.: Metrics for Evaluation of Ontology-based Information Extraction. In: Procceding of WWW 2006 Workshop on Evaluation of Ontologies for the Web (2006)

# Practice-Based Ontologies: A New Approach to Address the Challenges of Ontology and Knowledge Representation in History and Archaeology

Emad Khazraee and Michael Khoo

Drexel University, College of Information Science and Technology, 3141 Chestnut St., Philadelphia, PA, 19104, USA
{Emad,khoo}@drexel.edu

**Abstract.** Data production in history and archaeology far outpaces data processing. In order to apply computers to this problem, historical data must be converted to machine-readable forms. This process is easy for domains of knowledge that have explicit terminology, but history and archaeology lack these characteristics. This study therefore proposes a phenomenological approach to requirements gathering for knowledge representation and ontology systems for historians and archaeologists. The approach utilizes qualitative and ethnographic research methods to gather data on practitioners' reasoning and knowledge practices. The design requirements for ontology design can be extracted from the 'thick description' produced by this process, and used to build 'practice-based ontologies.' This paper presents the theoretical framework and early outcomes of ethnographic research with archaeologists in practice at the University of Pennsylvania.

**Keywords:** Ontology, knowledge representation, practice-based ontology, ethnographic method, archaeology.

## 1 Introduction

Historical and archaeological data are very diverse. They are dispersed in different institutions, and across different countries. The pace of data production in this field is far higher than the amount of data being processed. This situation leaves large amounts of data to be processed, and many research questions unsolved. Increasing computational power can help to solve more complicated questions, but a precondition to benefit from computers is that data must be converted and formalized into machine-readable formats.

Knowledge engineering is "the application of logic and ontology to the task of building computable models" that "analyzes knowledge about some subject and transform it to a computable form for some purpose" [1]. It includes the study of a domain and the mechanics and dynamics of knowledge in that domain, to ascertain whether it is possible to achieve computable forms for that purpose. Davis et al. [18] summarized following five principles for knowledge representation. A knowledge representation: is a surrogate, is a set of ontological commitments, is a fragment

E. García-Barriocanal et al. (Eds.): MTSR 2011, CCIS 240, pp. 375–386, 2011.

theory of intelligent reasoning, is a medium for efficient computation, and is a medium of human expression. These principles are easier to achieve in domains of knowledge that have explicit terminology, well-formed definitions, generic forms of analyses, and techniques for verification. However, history as a domain to some extent lacks all the above characteristics [2-3]. Uncertainty, contradictory data, and different analytical perspectives are intrinsic characteristics of historical investigation. These make the historical reasoning process complex: the problem with history is that what we know is vague. Domains such as history and archaeology have therefore not been addressed very well in knowledge representation.

This paper introduces new directions for the study and design of knowledge representation systems and ontologies for history and archaeology. First, some existing systems, approaches and challenges for history and archaeology knowledge will be discussed. We then describe an opportunity for addressing some of these challenges through the ethnographic study of the communities of practice amongst archaeologists. We call this approach 'practice-based ontology design.' The findings of a pilot empirical study will be presented.

## 2 Ontologies for History and Archaeology

This paper considers more recent approaches to archaeology known as post-processual and interpretative archaeology. Processual archaeology was inspired by "hypothetic-deductive positivism derived from Hemple (deducing statements from general theories and test them against observable data)" (Renfrew & Bahn, 2005, p. 208). In contrast, post-processual archaeology emphasizes the interpretation instead of explanation. In this paradigm the interpretation process is done considering hermeneutic cycle, a back and forth process between interpretation and data, between whole and part. Fitting is used instead of hypothesis testing in this paradigm. The main goal of this paper is to understand the requirements for ontologies to support the generation of cohesive interpretive narratives from archaeological data processing.

A number of ontologies already exist to facilitate information sharing and exchange, including the CIDOC Conceptual Reference Model (CRM) for cultural heritage, developed by the International Council of Museums (ICOM) and adopted as ISO standard in 2006 [5-6]. However, these ontologies are not enough to accelerate historical data processing. A more ambitious vision for ontologies in history and archaeology includes knowledge acquisition, reasoning, hypothesis generation and verification, and narrative storytelling. It is therefore necessary to study how to employ formal ontologies for these goals. First however we need to understand the characteristics of ontologies for history and archaeology, and how these characteristics differ from those of scientific domains.

### 2.1 Complexities and Obstacles of Knowledge Representation in History and Archaeology

The nature of historical investigation is different from that of scientific investigation. The works of historians is more like the work of detectives, findings and assembling clues from the available data and proposing a possible reconstruction of the past [5].

Any different selection set of information pieces may entail a different reconstructed model. In many cases, the historical objects do not contain the facts that historian needs, but it is the story and contexts which formed around the object which give meaning to it. Different interpretation of the information pieces may lead to different models which can be at the same time valid.

Archaeologists and historians deal with enormous amounts of data [3]. Studying, comparing and evaluating of this data manually is time consuming and expensive, and requires high levels of expertise and human resources. Scholars often specialize on specific civilizations and/or geographical areas, and so it is difficult to achieve mastery across different periods and regions. Analysis of large amounts of data for patterns and hypothesis verification can be a long process, that can be exacerbated by poor communication and data sharing. Historical fields suffer from lack of consensus on terminology and classification, and this can impede research.

A major consideration in the use of knowledge engineering methods in history is the difference between history and science. Every scientific field has three features in common: generic forms of analyses, and shared language for communication and techniques of verification. White argues that history is in a proto-scientific situation [2], and does not have a system of stipulated meaning for lexical elements and a system of syntactical rules of inference [3]. There are different levels of observation, from empires to villages to individuals, and utilizing different approaches and considering different cultural contexts can lead to alternative views; and different cultural perspectives may result in different interpretations [9]. Methods of reasoning based on deduction are not directly applicable to historical interpretation. This difference has direct impacts on knowledge representation and ontology design processes in history. While technical terminologies are requirements in a scientific field, Doerr [5] notices the proliferation of specialized terminologies in archaeology and history and the lack of consensus in this regard. The problem here is the fluid, dynamically changing nature of knowledge, which Sowa refers to as "knowledge soup" [1, 7]. Overgeneralizations, abnormal conditions, incomplete definitions, conflicting defaults, and unanticipated applications are innate complexities of knowledge soup [7]. These complexities are omnipresent in historians' works, where interpretation plays an important role, both in the selection of the required information and in the synthesis of a narrative to represent the history. Finally, historical studies are often incomplete. New discoveries can reveal more information and more pieces of the puzzle, and disprove former propositions. This ever-changing nature of historical data is closer to non-monotonic logics than the monotonic ones. This point should be considered in design and implementation of any ontology intended to deal with the flow of information in historical data.

Doerr [8] indicates that due to the quality and quantity of data in historical knowledge and the way it is acquired, the main complexities in the design of conceptual models lies in epistemological arguments rather than engineering models. It is therefore necessary to clarify our epistemological assumptions (what is usually taken for granted) for ontologies for history and archaeology. One question to ask is whether the formalization of knowledge in the form of a series of well-formed propositions is the only way of representing historical knowledge. Abduction, induction and analogy are the other forms of reasoning are also possible [7].

## 2.2 Historiography and Archaeology

A main concern of the philosophy of history is "what is involved in our knowing, representing, and explaining history?" [12]. Different trends in the philosophy of history and historiography have looked at different ways of knowing, such as the learning of facts about the circumstances in the past; providing a narrative of our understanding of the sequence of events by putting together actors, events and actions; and aiming to achieve causal explanations for historical events under general laws. This study adopts an approach in historiography that defines history as "a continuous process of interaction between the historian and his facts, an unending dialogue between the present and the past" [14]. This perspective towards history falls into the realm of hermeneutic approach to history and has consequences regarding the objectivity of historical narrative: the "archaeologist mediates between past and present, neither simply describing data nor simply translating it into our terms" [15].

A central figure here is Hayden White, whose approach is presented *Metahistory* [2]. According to White any historical work consist of five levels of conceptualization: chronicle, story, mode of employment, mode of argument and mode of ideological implication. Chronicle and story refer to primitive elements and represent the process of selection and arrangement of data from unprocessed historical accounts. In the chronicle, the data are arranged in temporal order and in the story, they receive further arrangements by characterizations of events as inaugural, termination and transitional motifs. In these two steps the unprocessed historical data are converted to a more comprehensible format for an audience. The mode of employment is an attempt to provide an explanation for the history by selecting a kind of story that can explain the sequence of the events. (White uses the four different mode of employment indicated by Northrop Frye in his Anatomy of Criticism, as Romance, Tragedy, Comedy, and Satire.) In this step, the work of historian is still to a create a narrative operation of what happened.

At the next level, historian looks for an explanation by a formal account and tries to explicate "the point of it all" through a formal discursive argument. In this step, historian is involved in an investigative operation. He tries to explain the events of history by construction of a nomological-deductive argument, in the same way natural events are explained by identification of a universal causal law. In this step the argument moves from "what happened" to "why it happened as it did". The last level of conceptualization in historical work is explanation by ideological implication. It seeks implications that can be drawn from the study of past events to the understanding of the present ones, either to change the world or maintain its current state.

The levels of conceptualization presented by White [2] are similar to the Sowa's proposed model of reasoning [7], involving collecting enough records to form a story (starting with induction), finding a mode of explanation in form of a narrative (continuing with abduction), and finally seeking principles which can explain the situation and can be extended to explain similar ones (achieving a theory for deduction). The last level is an attempt to apply the findings to change or maintain the existing state of the world, which can be compared with the action in the Sowa's model. This approach to the study of the ontologies for history and archaeology can open a new direction for further research in this field.

## 3 Empirical Study of Archaeologists in the Practice

To investigate these ideas further, the authors are applying qualitative research methods to collected empirical data on the practices of archaeologists and historians. This section of paper reports preliminary findings of a pilot study in this regard.

### 3.1 Theoretical Framework

**Knowledge Representation and Phenomenology.**To study the ontological aspect of a domain of knowledge, one has to investigate the objects of human activity and inquiry within that domain. In history and archaeology, this includes cultural objects. One of the complexities of cultural objects is that they do not just consist of physical parts (as considered by many ontologies). According to Ingarden's phenomenological approach [19], cultural objects consist of two aspects; a physical objective aspect, and a mind dependent aspect. Ingarden argues against identifying cultural objects with merely physical things that constitute them [20], and a phenomenological approach can help to distinguish between the manner the world represented in scientific descriptions and the way we experience it as human beings. The existing study can therefore be categorized as a phenomenological approach to knowledge representation in history and archaeology. Some of the questions being considered are the extent to which we can describe a domain of knowledge as the way humans perceive it; what are the considerations and constraints in generating such descriptions; and whether the outcome can be utilized by machine to address our problems more efficiently.

The research follows a qualitative approach. According to Smith and Klagges [9] qualitative research is relatively rare in ontology and knowledge representation work. Some important exceptions include Diana Forsythe [4, 21, 22], who challenged some common assumptions in AI and expert system community. Her background as an anthropologist helped her to study knowledge engineering labs by ethnographic method [23], and her work revealed some important outsider assumptions and their effects on the design of knowledge base systems. Recently, qualitative and ethnographic research methods have been used in the field of ontology design. Pattuelli [24] used a user-centered approach in design and evaluation of an ontology for cultural heritage resources. Kuziemsky, Downing, Black, and Lau [25] used grounded theory for designing an ontology in health care for palliative care sever pain management. Furthermore, grounded theory and qualitative methods have been used for ontology design and knowledge representation in the earth science project GEON [26-27]. This shows that qualitative paradigm started to be adopted in the ontology design community.

Ethnographic study of historians and archaeological practice can therefore provide a better insight into different aspects of these domains. As the process of reasoning in history has significant tacit and implicit dimensions, qualitative research can be a good way to provide in depth description of this process. This research can focus on how historians collect their data; how they organize their data; how they discriminate among data pieces to select a specific set of them; and finally how they interpret them in a broader context.

**Ethnographic Method.** As has been described, this study is focused on trends in historical research which try to understand the meaning of historical events, rather than defining causal explanations. This approach is reflected in the phenomenological component of the theoretical framework. An important concept here is that of *verstehen* which was introduced by Dilthey in an important effort to liberate humanities from the methodology of natural sciences (*verstehen* means understanding and relates to our understanding of others [28-29]). An ethnographic approach is an appropriate method to achieving such a deep understanding, through describing and interpreting social practices and the system of meaning in a different culture (an archaeological community of practice can be perceived as a cultural milieu here).

The aim of collecting ethnographic data is to study and show how historians reconstruct possible pasts by selecting and discriminating amongst heterogeneous archaeological information pieces. Hodder [15], one of the pioneers of post-processual archaeology, indicates that our knowledge about the process of archaeological reasoning is limited and this issue has not been addressed well: "the problem is it is extremely difficult to document archaeological reasoning process" [15]. He concludes that "by looking at the practices of archaeologists, by looking at what they (or at least some of them) do rather than what they think they ought to do, it has been possible to characterize some components of the reasoning process involved" [15]. This statement by one of the thought leaders of archaeology justifies the application of ethnographic method as an efficient method to understand how archaeologists reason. Since archaeological practice - the process of selecting, merging and synthesizing a possible world from data - is mostly an internal process and somehow a black box, a qualitative and ethnographic approach can be helpful here.

### 3.2 Research Questions

With the main challenges of the knowledge representation and ontology design in the mind, we are approaching the problem through an empirical study of archaeologists' work practices. The main research questions of the study focus on the process of constructing narratives based on the analysis of recorded data form archaeological sites: how do archaeologists organize their information? How do they discriminate among the information pieces? How do they connect the selected information pieces together to generate a narrative? and, What are their considerations (explicit and implicit) during these processes? One aim of the research is to provide a "thick description" of how archaeologists reason and 'do' archaeology.

### 3.3 Pilot Study, Research Site and Data Collection Methods

Scholars affiliated with the Department of History of Art at the University of Pennsylvania are being studied for this research. Availability of this research site in one of the most important archaeological institutions of the world, in addition to the teams of archaeologist and historians actively working on collections and excavations, is a unique opportunity to pursue this research. Different specialists actively participate in the projects, including a zooarchaeologist, an archaeobotanist, an art historian, an epigrapher, and a philologist. This diverse group of participants reveals the diverse nature of data and different existing perspectives in the domain.

This study utilizes different qualitative and ethnographic methods for data collection such as observation (direct observation and participant observation), think-aloud protocols, existing documents (field notebooks), semi-structured interviews. Ethnographic observation will be used to observe historians, archaeologists, researchers and staff in 'real life' conditions, in order to understand their work practices, especially with regard to the collecting, organizing and processing of historical data. Ethnographic observation is a "systematic observation" as described by Weick as "sustained, explicit, methodical observing" to understand "social situations" in their context, and participant observation is the "craft of experiencing and recording events in social settings" [28].

Think-aloud protocols are research methods used to understand the subjects' cognitive processes based on their verbal reports [30]. They can be used during processes such as describing metadata for digitized objects or during categorizing pottery shreds. Think-aloud process can be used to review a discovery process, too. Hodder [15] emphasizes that archaeological reports are usually "cleaned-up" and do not contain iterations, guesses and debates. In this way, researchers can review the process of generating a fit model to the data. Through this method, it would be possible to reveal the real process of archaeological reasoning and achieve an in-depth description of this process. Another major resource is reviewing archaeologists' fieldwork documents (day-books and field-books). As Hodder [15] indicated archaeological work is a data-led process, therefore, it is important to understand what did excavators noticed in the field and why they made some decisions. It is important to see how gradual discovery process, in the end, transforms into a narrative in a form or a report.

A significant data collection instrument is the semi-structured interview, which includes a number of foreshadowing questions. As emphasized by Baumeister and Newman in [28], the goal of interviews in this case, is to achieve context, action and intentionality of the interviewee's experience through a narrative. Following early orientation, leading actors in the research site were selected as interviewees, and snowball sampling was used to select further interviewees [28]. In this stage of research ten face-to-face interviews were carried out with eight participants. The Interviews lasted an average of one hour. Interviewees were asked about their practice, how they defined raw data, how they collected their raw data, how they processed their data to make sense out of it. Interviewees were asked to respond in the context of at least one real project.

### 3.4 Preliminary Findings of Pilot Study

In this section we present some emergent themes from the first rounds of data analysis. These are illustrated with anonymous excerpts from the interview data.

**Heterogeneous datasets.** Archaeological practice today includes many disciplines, including genetics, geography, hydrology, art history and philology. Many different perspectives and multiple scales of analysis are involved in this practice. The objects of study are spread in a wide range from DNA, micro-organisms, animal remains (fauna), flora, and geomorphology data to human expressions in form of works of art. The scale of analysis differs from molecule to landscapes. Therefore, such a diverse set of material can hardly being accommodated in a single static structure. Moreover, data is stored and represented in a variety of forms such as quantitative, qualitative, visual, structured and open discursive forms. Some examples provided by interviewees are as follows:

> "Archaeology, I'm sure everybody has said this to you, archaeology cannot be done by a single person and it is by its very nature a collaborative enterprise. No one person can do all of the jobs that need to be done in any archaeological project."

> "Since taking over the directorship of the project, I am working on all the material. I have to decide who's going to do what - [...] works with the seeds, the bones goes to [...], the pottery is being worked on by [...], the architecture by two people [...], and I and [...] and [...] will do stratigraphy. Then I will be responsible for the glyptic ... and seal impressions ... and there are inscriptions those have all been already read by [...]."

**Emerging understandings.** During observation sessions and interviews with archaeologists one of the interesting themes is how they make sense of the raw data. This process is often a tacit and emergent process involve heuristics. Over time and involvement with the data, archaeologists start to see patterns that can be transformed into narratives. This ability is partly built on the experience and professional training.

> "I did one-to-one size drawing ... I draw each one (bones), sometimes I try to get a little more systematic, I make little cross section [drawings], ... I measure as many of the complete dimensions as it had, [numbers] and I talk about where the traces are, and if there are traces of manufacture then I talked about what seems to have happened. Was it weathered, was it broken during use, is there a trace of something sticky on it or something that would give me some feeling that what it was. Is it an ornament, is there a picture?"

**Forcing data.** Hodder [15] and Hamilton [31] have emphasized a tension between codification and interpretation in practice. In the pilot study, the introduction of new information technologies can present new categories that can shape interpretations:

> "I wanted to record those individual, ... I was taking long pages of notes that, if you look closely, cannot be linked directly to these other observations (in the old forms)... those have become unlinked by being forced to these [new] forms. ... These forms (old forms) were never designed to be able to capture that kind of individual difference."

> "Making the transfer between these kinds of two or three dimensional arrays (recording forms) to a more of an inventory of information where each bone can have an infinite number of pieces of information, gathered about it, to make it more flexible, took a long time, and I cannot still do things with this data set (recorded in old forms) that I can do with my current data set (recorded in new way) ... in my new strategy, even there I find myself again entering things twice in order to accommodate the new insights that I have. So, after this I decided, no more forms! It's just too confining, it does not let me gather enough information about individuals."

**Building models.** An interesting observation is the dynamic formation of categories through work processes, and the evolving nature of ontological models. These models form in a circular process, starting with general emerging themes and then being refined through time. This process is very similar to open coding and axial coding process in qualitative research [32], and evolves over time.

> "[At the beginning] frankly, I did not know what I was doing, I just describe everything I can see [laughing] and after the first couple of years people said: "Oh! Your work … is so interesting and important." I just thought that just shows that you know how to do it and you do it in a different way."

In some cases, due to the dynamic nature of archaeological work, categorical description does not work well for further analysis and comparison because the unique nature of a data object needs interpretive descriptions instead of an assigned keyword. While mst data entry forms and simple classifications are good for general objects, particulars can provide clues to interesting differences and discoveries.

**Writing stories.** Narrative is a central component in archaeology, as in history. Knowledge representation systems and ontologies should assist archaeologists to become better storytellers. The measurement of the success of a system then would be to what extent the parts (the different data types) can be integrated into the whole (the story). Contemplating the role of narrative and storytelling will be a new direction for further study in this regard.

> "You take all of those pieces of tiny bits of data, the evidence, and plug it into the story ... one of the things about archaeology is that you cannot ever think that you are right or you cannot think that the story has been fully told ... every time there is more material, then you either incorporate it in your story, or you scrap your story and start all over again with a new story that fits better."

An important issue to mention here is that the collective nature of archaeological work has direct consequences on the outcome. Archaeology as a practice cannot be done in isolation because the multi-aspect nature of archaeological work needs involvement of different specialties; therefore, it is a collective practice and each team member brings his/her own perspective to the team. Accordingly, there is a synergy in collaborative team work that usually creates a story different from each single expert story while accommodating their insights.

> "Getting us (different specialists) together, and having the others as witnesses for what we have said, clearly, has been important to us, and getting people to harmonize the impression that we all take away from the same data. We tried; we did write a paper with five different authors. ... I gave them the data and then they wrote the paper and I saw it. [I said:] "It is not what it says to me!" Just one step outside my narrow perception of material they saw something completely different. They were willing totally to overcome the weaknesses of the data and take the general patterns ... I was looking on this thing projected on the wall, I thought: is that what I did, is that what I said, is that what I think? I thought, well, you know, it is the best fit with the rest of the data, it must be true. But it was a very provocative moment for me and trying to achieve this goal of integration."

### 3.5 Discussion: Practice-Based Ontologies

Early emerging themes in the analysis have important implications for knowledge representation and ontology design in history and archaeology. Far from dealing with tidy facts and concepts that can be represented in databases without problem, these archaeologists are working in multi-disciplinary teams, and with heterogeneous datasets. They spend considerable time working with their own and others' data, in order to see emergent patterns, from which they begin to construct narratives. These findings reflect the claims of post-processual archaeology, which does not assume a universal type for archaeological sites that can be applied to all situations, but which views each site as a unique setting.

To share data and to work with large data sets, archaeologists must use general codification schemes. This can cause a tension between their knowledge and the coding scheme. Therefore, we need to find better ways to preserve the diversity and richness of data and efficiently deal with complexity, but over large scales and multiple sites. One possible strategy that can address the tension between existing codification systems and the complex practices of archaeologists is to recognize two types of categorization and classification: categorization for organization, and categorization for understanding. The former is mainly used for information organization and retrieval. Data sharing and integration also can be seen as further functions. In contrast, categorization for understanding is an inductive approach to codification. Archaeologists categorize things in different schemes to make sense of data and find a fitting scheme that can accommodate their data. Accordingly, this categorization has a dynamic nature and evolves through time and improves by development of our understanding. The main goal of this process is to make sense out of data, to let us understand the phenomena.

This distinction between categorization for organization and categorization for understanding is an important issue to be considered in ontology design process. The latter may not help us to efficiently retrieve information or link data but it can assist archaeologists in the reasoning process. Therefore, as well as ready-to-use ontologies for archaeologist, we should also provide frameworks to create their own ontologies in the process of their work based of the requirements of their practice and perspective. We call these ontologies *practice-based ontologies*, and will continue to refine their definition and investigate their requirements in future research.

## 4 Conclusion

Data production in history and archaeology far outpaces data processing. To apply computers to this problem, historical data must be converted to machine-readable forms. This ethnographic study has illustrated some of the tensions between machine-readable forms and the complex dimensions of archaeologists' practical knowledge. We have proposed a qualitative phenomenological approach to understanding the knowledge practices of archaeologists, based on a thick description of archaeologists in practice at the University of Pennsylvania. This his has led to the identification of preliminary design requirements for systems to support practice-based ontologies for archaeologists.

**Acknowledgement.** We are grateful to Professor Holly Pittman at the University of Pennsylvania who kindly provided access to the research site to the authors; and to the anonymous reviewers' insightful comments, which greatly contributed to enhancing the quality of this article.

## References


1. Sowa, J.F.: Knowledge representation: logical, philosophical, and computational foundations. Brooks/Cole, Pacific Grove (2000)
2. White, H.: Metahistory: the historical imagination in nineteenth-century Europe. Johns Hopkins University Press, Baltimore (1973)
3. White, H.: Interpretation in History. New Literary History 4, 281–314 (1973)
4. Forsythe, D.E.: Studying Those Who Study Us: An Anthropologist in the World of Artificial Intelligence. Stanford University Press, Stanford (2001)
5. Doerr, M.: Ontologies for Cultural Heritage. In: Handbook on Ontologies, International Handbooks on Information Systems, pp. 463–486. Springer, Heidelberg (2009)
6. Doerr, M., Hunter, J., Lagoze, C.: Towards a core ontology for information integration. Journal of Digital Information 4, 169 (2003)
7. Sowa, J.F.: The challenge of knowledge soup. In: Ramadas, J., Chunawala, S. (eds.) Research Trends in Science, Technology and Mathematics Education, p. 55. Homi Bhabha Centre, Mumbai (2006)
8. Doerr, M.: The CIDOC Conceptual Reference Module - An ontological approach to semantic interoperability of metadata. AI Magazine 24, 75–92 (2003)
9. Smith, B.: Klagges: Philosophy and Biomedical Information Systems. In: Munn, K., Smith, B. (eds.) Applied Ontology: an Introduction, pp. 21–38. Ontos Verlag, Piscataway (2008)
10. Pasin, M., Motta, E.: Ontological requirements for annotation and navigation of philosophical resources. Synthese, 1–33 (2009)
11. Pasin, M., Motta, E., Zdrahal, Z.: Capturing knowledge about philosophy. In: 4th International Conference on Knowledge Capture, pp. 47–54. ACM, Whistler (2007)
12. Little, D.: Philosophy of History (2007), `http://plato.stanford.edu/entries/history/`
13. Hempel, C.G.: The Function of General Laws in History. The Journal of Philosophy 39, 35–48 (1942)
14. Carr, E.: What is history? Penguin, London (1990)
15. Hodder, I.: The Archaeological Process: An Introduction. Blackwell, Oxford (1999)
16. Legg, C.: Ontologies on the Semantic Web. Ann. Rev. Info. Sci. Tech. 41, 407–451 (2008)
17. Chomsky, N.: Language and Thought. Moyer Bell (1995)
18. Davis, R., Shrobe, H., Szolovits, P.: What is a knowledge representation? AI Magazine 14, 17 (1993)
19. Ingarden, R.: Ontology of the Work of Art: The Musical Work, The Picture, The Architectural Work, The Film. Ohio University Press, Athens (1989)
20. Thomasson, A.L.: Ingarden and the ontology of cultural objects. In: Existence, Culture, and Persons: The Ontology of Roman Ingarden. Ontos, Frankfurt (2005)
21. Forsythe, D.E.: Engineering Knowledge: The Construction of Knowledge in Artificial Intelligence. Social Studies of Science 23, 445–477 (1993)
22. Forsythe, D.E., Buchanan, B.G.: Knowledge acquisition for expert systems: some pitfalls and suggestions. IEEE Transactions on Systems, Man and Cybernetics 19, 435–442 (1989)
23. Forsythe, D.E.: It's Just a Matter of Common Sense: Ethnography as Invisible Work. Computer Supported Cooperative Work (CSCW) 8, 127–145 (1999)
24. Pattuelli, M.C.: Modeling a domain ontology for cultural heritage resources: A user-centered approach. J. Am. Soc. Inf. Sci. n/a (2010)

25. Kuziemsky, C.E., Downing, G.M., Black, F.M., Lau, F.: A grounded theory guided approach to palliative care systems design. International Journal of Medical Informatics 76, S141–S148 (2007)
26. Ribes, D., Bowker, G.C.: A Learning Trajectory for Ontology Building
27. Ribes, D., Bowker, G.C.: Between meaning and machine: Learning to represent the knowledge of communities. Information and Organization 19, 199–217 (2009)
28. Lindlof, T.R., Taylor, B.C.: Qualitative Communication Research Methods. Sage Publications, Thousand Oaks (2002)
29. Ramberg, B., Gjesdal, K.: Hermeneutics (2005), http://plato.stanford.edu/entries/hermeneutics/
30. Wildemuth, B.M.: Applications of Social Research Methods to Questions in Information and Library Science. Libraries Unlimited, Westport (2009)
31. Hamilton, C.: Faultlines: the construction of archaeological knowledge at Catalhoyouk. Presented at the Theoretical Archaeology Group Conference, Liverpool (1996)
32. Strauss, A.L., Corbin, J.M.: Basics of Qualitative Research: Grounded Theory Procedures and Techniques. Sage Publications, Newbury Park (1990)

# A Semantic Mapping of VRA Core 4.0 to the CIDOC Conceptual Reference Model*

Panorea Gaitanou and Manolis Gergatsoulis

Database & Information Systems Group (DBIS),
Laboratory on Digital Libraries and Electronic Publishing,
Dept. of Archives and Library Science, Ionian University, Corfu, Greece
{rgaitanou, manolis}@ionio.gr

**Abstract.** In the recent decades, we have witnessed a tremendous proliferation of metadata schemas, particularly in the cultural heritage domain. As a consequence, there is a growing need to integrate these schemas to solve a number of problems, mainly concerning interoperability issues and loss of implicit knowledge. In this paper, we present a semantic mapping of VRA Core 4.0, a cultural heritage metadata schema describing visual resources, to CIDOC CRM. The mapping rules are expressed formally using a path-based mapping language called *Mapping Description Language (MDL)*. This work is based on a semantic integration scenario, where CIDOC CRM acts as a mediation schema.

**Keywords:** Cultural Information, Metadata Interoperability, Ontology-based Integration, Mapping Language, VRA Core 4.0, CIDOC CRM.

## 1 Introduction

Cultural heritage institutions (otherwise called "memory institutions") use various metadata schemas for the documentation of cultural collections, to facilitate access and retrieval to cultural information via the web. The complexity of the cultural information imposes the development of several different metadata standards (such as DCMI, VRA Core 4.0, EAD, Spectrum etc.), which exhibit significant diversity. This heterogeneity often results in data exchange failure, as the end user cannot access an integrated information system and retrieve the desired information. In order to address all the aforementioned issues and achieve a unified and standard-independent access to the relative information, it is necessary to integrate all these schemas. One of the most important and continuously evolving methods implemented in the interoperability domain is the ontology-based integration [7]. Ontologies can play a leading role in this area, as they provide the means for defining common vocabularies, representing the

* This research has been co-financed by the European Union (European Social Fund ESF) and Greek national funds through the Operational Program "Education and Lifelong Learning" of the National Strategic Reference Framework (NSRF) Research Funding Program: Heracleitus II. Investing in knowledge society through the European Social Fund.

E. García-Barriocanal et al. (Eds.): MTSR 2011, CCIS 240, pp. 387–399, 2011.

domain knowledge, and facilitating knowledge sharing and reuse among heterogeneous and distributed application systems. As a consequence, ontologies are considered to be important building block for integration architectures [8].

The basic component of an information integration system is the mapping of the various metadata schemas to a schema or a core ontology, acting as a mediation schema, so that (meta)data integration is successfully accomplished. In the integration scenario proposed by our research group [8], the CIDOC CRM ontology is used as a mediation schema, to which different metadata schemas (such as MODS, DC, MARC, EAD etc.) are mapped.

In this paper, we present the mapping of the VRA Core 4.0 schema to the CIDOC CRM ontology. Our mapping methodology is based on a path-oriented approach and is formally defined through a Mapping Description Language (MDL), which defines semantic mapping rules from the source schema to the target schema. Each element of the source schema (with its subelements and attributes) is represented as a VRA path (expressed in XPath form) and is then semantically mapped to an equivalent path of classes and properties of CIDOC CRM. It is important to note that the mapping procedure focuses on the restricted version of VRA Core 4.0, which imposes controlled vocabularies and type lists. As a consequence, each attribute assigned to an element of the metadata schema may lead to the generation of different semantic paths on the ontology, depending on the values of that attribute, and produces a plethora of conceptual expressions corresponding to the same element.

## 2 Mapping VRA Core 4.0 to CIDOC CRM

### 2.1 Brief Description of VRA Core 4.0

VRA Core 4.0 [9] is a metadata schema for the cultural heritage domain, initially developed by the Visual Resources Association's Data Standards Committee. Currently, it is hosted by the Network Development and MARC Standards Office of the Library of Congress (LC) [5] in partnership with the Visual Resources Association. VRA Core 4.0 provides guidance on describing works of visual culture, collections, and images that document them. It allows for three broad groups of entities, which are works (element `work`), images (element `image`), and collections (element `collection`). A work may represent a painting, sculpture or other artistic product. An image is a visual representation of a work that may appear in a wide range of formats including digital image formats such as JPEG, GIF and TIFF or physical photographs, slides, etc. Finally, a collection represents a group of works or images. VRA Core 4.0 contains a set of top level elements (`agent`, `culturalContext`, `date`, `description`, `inscription`, `location`, `material`, `measurements`, `relation`, `rights`, `source`, `stateEdition`, `stylePeriod`, `subject`, `technique`, `textref`, `title` and `worktype`) and several optional global attributes (`dataDate`, `extent`, `href`, `pref`, `refid`, `rules`, `source`, `vocab`, `xml:lang`), which are applied additionally to any element or subelement, when necessary. Two XML Schema versions have been proposed for VRA Core 4.0. An *unrestricted version*, which specifies the basic structure of the schema and imposes no restrictions on the values of its elements, sub-elements, or attributes, and a *restricted version*, which extends the unrestricted one by imposing controlled type lists and date formats.

*Example 1.* In this example we present a fragment of a simplified VRA document, describing a textual manuscript of the 18th century, taken from
`http://www.vraweb.org/projects/vracore4/example017.html`.

```
<?xml version="1.0" encoding="UTF-8" ?>
<vra>
 <work id="w_4" source="Core 4 Sample Database (VCat)" refid="4">
  <agentSet>
   <agent>
    <name vocab="ULAN" refid="500017255"
                       type="personal">Jefferson, Thomas</name>
    <dates type="life">
     <earliestDate>1743</earliestDate>
     <latestDate>1826</latestDate>
    </dates>
    <culture>American</culture>
    <role>author</role>
   </agent>
  </agentSet>
  <measurementsSet>
   <measurements type="height" unit="cm">75.56</measurements>
   <measurements type="width" unit="cm">62.23</measurements>
  </measurementsSet>
  <stylePeriodSet>
   <stylePeriod vocab="LCSAF"
                refid="85041401">Eighteenth century</stylePeriod>
  </stylePeriodSet>
  <techniqueSet>
   <technique vocab="AAT"
              refid="300053162">calligraphy(process)</technique>
   <technique vocab="AAT" refid="300054698">writing(process)</technique>
  </techniqueSet>
  <titleSet>
   <title type="popular" xml:lang="en">Declaration of Independence</title>
  </titleSet>
  <worktypeSet>
   <worktype>manuscript (document genre)</worktype>
  </worktypeSet>
 </work>
</vra>
```

### 2.2 The CIDOC CRM Ontology

*CIDOC Conceptual Reference Model* [3] emerged from the CIDOC Documentation Standards Group in 1999, as a formal extensible ontology which provides a conceptual representation of cultural heritage domain promoting semantic interoperability and integration. It is an object-oriented model comprised of a class hierarchy of 86 named classes interlinked by 137 named properties. CIDOC CRM defines the complex interrelationships between objects, actors, events, places, and other concepts used in the cultural heritage domain [2]. A *class* (also called *entity*), identified by a number preceded by the letter "E" (e.g. E39 Actor), groups items (called *class instances*) that share common characteristics. A class may be the *domain* or the *range* of *properties*, which are binary relations between classes. Properties are identified by numbers preceded by the letter "P" (e.g. P2 has type (is type of)). A property can be interpreted in both directions (active and passive voice), with two distinct but related interpretations. A *subclass* is a class that specializes another class (its *superclass*). A class may have one or more immediate superclasses. When a class $A$ is a subclass of a class $B$ then all instances of $A$ are also instances of $B$. A subclass inherits the properties of its superclasses without exception (*strict inheritance*) in addition to having none, one or more properties of its own. A *subproperty* is a property that specializes another property. A sample of CIDOC CRM properties is shown in Fig. 1.

| **Property Id & Name** | **Entity - Domain** | **Entity - Range** |
|---|---|---|
| P1 is identified by (identifies) | E1 CRM Entity | E41 Appelation |
| P2 has type (is type of) | E1 CRM Entity | E55 Type |
| P4 has time-span (is time-span of) | E2 Temporal Entity | E52 Time-Span |
| P14 carried out by (performed) | E7 Activity | E39 Actor |
| P58 has section definition (defines section) | E18 Physical Thing | E46 Section Definition |
| P108 has produced (was produced by) | E12 Production | E24 Physical Man-Made Thing |

**Fig. 1.** A sample of CIDOC CRM properties

## 2.3 Mapping Methodology

The proposed mapping methodology is based on a path-oriented approach. A mapping from a source schema to a target schema transforms each instance of the source schema into a valid instance of the target schema. Hence, we interpret the metadata paths to semantically equivalent CIDOC CRM paths. As we are interested in metadata schemas, which are based on XML, the paths in the source schemas that are based on XPath [10], in fact extend the XPath *location paths* with *variables* and stars (meaning *data transfer*). To formally express the mappings we have developed a *Mapping Description Language* (MDL). The syntax of its sentences, called *MDL mapping rules* is given bellow in EBNF:

(R1) R ::= Left '−−' Right
(R2) Left ::= $\mathrm{A}_{Path}$ | $\mathrm{V}_{Path}$
(R3) $\mathrm{A}_{Path}$ ::= $\epsilon$ | '/' $\mathrm{R}_{Path}$
(R4) $\mathrm{R}_{Path}$ ::= L | L '*' | L '{' $\mathrm{V}_l$ '}' | L '*' '{' $\mathrm{V}_l$ '}'

(R5) $V_{Path}$ ::= '\$' $V_l$ '/' $R_{Path}$ | '\$' $V_l$ '{' $V_l$ '}'
(R6) Right ::= $E_t$ | $E_e$ '→' O | '\$' $V_c$ '→' O | '\$' $V_p$ '→' $P_p$ '→' $E_{t55}$
(R7) O ::= $P_e$ '→' $E_t$ | $P_e$ '→' $E_e$ '→' O
(R8) $E_e$ ::= E | E '{' $V_c$ '}'
(R9) $E_t$ ::= E | E '{' $V_c$ '}' | E '{=' String '}'
(R10) $E_{t55}$ ::= E55 | E55 '{' $V_c$ '}' | E55 '{=' String '}'
(R11) $P_e$ ::= P | P '{' $V_p$ '}'

The terminals used in these rules have the following semantics:

- L: represents an XPath *location path.*
- $V_l$: represents the *location variables*, which are used to declare the "branches" of the XML trees (XPath paths).
- $V_c$: represents the *class variables.* They are used to declare that a class can be the starting point of one or more CIDOC CRM paths.
- $V_p$: represents the *property variables.* They declare that a property is the starting point of a new CIDOC CRM path, which consists of a property of a property linking the property with an instance of the E55 Type class.
- E: represents the identifier of the class.
- E55: represents the identifier of the class E55 Type.
- P: represents the identifier of the property.
- $P_p$: represents the identifier of the property of a property.
- *String*: represents a string.

## 3 Mapping VRA Core 4.0 to CIDOC CRM

As VRA Core 4.0 is XML-based, we use XPath to locate its elements/attributes. A *VRA path* is a sequence of VRA elements and subelements, starting from the root element `vra` separated by the slash symbol (/). For instance, the path `/vra/work/titleSet/title` denotes the title of a work. A *CRM path* is defined as a chain in the form entity-property-entity, such that the entities associated with a property correspond to the property's domain and range. VRA Core 4.0 defines three basic top elements: `work`, `collection` and `image`. A work is defined as a physical entity that exists, existed in the past, or may exist in the future. It might be an artistic creation, such as a painting or a sculpture, a performance, a building or other construction, etc. Therefore, we associate each `work` element with an instance of the class[1] `E24 Physical Man-Made Thing`, which comprises all persistent physical items that are purposely created by human activity.

In the following subsections, we present the mapping of VRA Core 4.0 to CIDOC CRM and particularly we focus on the mapping of the `work`'s subelements `agent` and `inscription` that present significant diversity and complexity.

[1] For brevity reasons, when we say that an element/attribute is mapped to a class we mean that the value of the element/attribute is mapped to an instance of that class.

### 3.1 Mapping the Agent Element and Its Subelements

The agent element denotes a person, group or corporate body that has contributed to the production or creation of the work being described. It contains the following five subelements:

- name: specifies the names and appellations, assigned to an individual, group or corporate body. A type attribute is assigned to this subelement, with possible values "personal", "corporate", "family", or "other".
- culture: refers to the nationality or culture of the person, group, or corporate body that participated to the work being described.
- dates: contains the subelements earliestDate and latestDate, which refer to the dates associated with the agent. A type attribute is also assigned to this subelement (with possible values "activity", "life" and "other").
- role: denotes the specific role of the individual, group or corporate.
- attribution: relates a characteristic or a specific attribute to the agent.

*Mapping the agent element:* Examining the semantics of CIDOC CRM, we see that the appropriate class to map agent is E39 Actor. Instances of E39 Actor corresponding to specific agents must be related to the instance of E24 Physical Man-Made Thing representing the work, to denote that an agent "contributed to the production or creation of the work being described". However, as CIDOC CRM is event-centric, it has no properties to directly relate instances of these two classes. Instead, these instances can be related indirectly, through an event (instance of the class E12 Production) during which the object was created. In this way, the specific work (i.e. the instance of E24 Physical Man-Made Thing) is related through the property P108B was produced by to this event. This event is then related to the instances of E39 Actor (representing agents), through the property P14 carried out by producing CIDOC CRM paths of the form:

E24 Physical Man-Made Thing → P108B was produced by →
E12 Production → P14 carried out by → E39 Actor

which semantically correspond to VRA paths of the form:

/vra/work/agentSet/agent

Note that, in case there are more than one agents (i.e. more than one agent subelements of the element agentSet), different subpaths of the form:

→ P14 carried out by → E39 Actor

will be rooted to the (same) instance of E12 Production to relate it with the different agents (instances of E39 Actor) that took part in this production event.

*Mapping the name subelement of agent:* The name subelement is mapped to the class E82 Actor Appellation and is linked to the corresponding class E39 Actor through the property P131 is identified by. Thus the VRA path:

/vra/work/agentSet/agent/name

is mapped to the semantically equivalent CIDOC CRM path:

E24 Physical Man-Made Thing → P108B was produced by →
E12 Production → P14 carried out by → E39 Actor →
P131 is identified by → E82 Actor Appellation

*Mapping the* `type` *attribute of the* `name` *subelement:* This `type` attribute is quite remarkable as it determines if an agent is a person (when `type` = `"personal"`), a corporate or an organization (when `type` = `"corporate"`), a family (when `type` = `"family"`), or none of the above (when `type` = `"other"`). To map the attribute `type` in CIDOC CRM, we have investigated two different approaches:

**First approach:** An instance of the class `E55 Type` having the same value as the `type` attribute, is linked to the corresponding instance of `E39 Actor` through the property `P2 has type` giving the following CIDOC CRM path[2]:
`E24 Physical Man-Made Thing` → `P108B was produced by` →
`E12 Production` → `P14 carried out by` → `E39 Actor` [→ `P2 has type` → `E55 Type`] → `P131 is identified by` → `E82 Actor Appellation`
which semantically correspond to the VRA paths:
`/vra/work/agentSet/agent/name[@type]`

**Second approach:** The value of `type` is used to refine the mapping of the corresponding agent by replacing the class `E39 Actor` with an appropriate subclass of this class. More specifically, if `type` = `"personal"`, the agent is mapped to the class `E21 Person` and the CIDOC CRM path becomes:
`E24 Physical Man-Made Thing` → `P108B was produced by` →
`E12 Production` → `P14 carried out by` → `E21 Person` →
`P131 is identified by` → `E82 Actor Appellation`
which semantically corresponds to the VRA path:
`/vra/work/agentSet/agent/name[@type="personal"]`
Similarly, if the value of `type` is `"corporate"` (resp. `"family"`), the agent is an instance of the class `E40 Legal Body` (resp. `E74 Group`).

Notice that the attribute `type` of `name` actually refers to the `agent` element to which we believe that should have been assigned and not to `name`.

*Mapping the* `role` *subelement of* `agent`*:* This mapping is expressed using the subproperty `P14.1 in the role of`, which links the property `P14 carried out by` to an instance of the class `E55 Type` giving the following CIDOC CRM path:
`E24 Physical Man-Made Thing` → `P108B was produced by` →
`E12 Production` → `P14 carried out by` [→ `P14.1 in the role of` → `E55 Type`] → `E39 Actor`
which semantically corresponds to the VRA path:
`/vra/work/agentSet/agent/role`

*Mapping the* `culture` *subelement of* `agent`*:* This element can be modelled as a membership of the agent to a group expressed as an instance of the class `E74 Group`. `E39 Actor` is related to `E74 Group` through the property `P107B is current or former member of`, giving a CIDOC CRM path of the form:
`E24 Physical Man-Made Thing` → `P108B was produced by` →
`E12 Production` → `P14 carried out by` → `E39 Actor` →
`P107B is current or former member of` → `E74 Group`

[2] The notation [...] denotes that a new branch is rooted on the `E39 Actor` class node.

which semantically corresponds to the VRA path:
`/vra/work/agentSet/agent/culture`

*Mapping the* **`dates`** *subelement of* **`agent`**: The `dates` is one of the most complex subelements to map for three reasons: a) it contains a `type` attribute, with possible values `"life"`, `"activity"`, and `"other"`. Thus, it can define either the dates that span the known activity of an individual, group or corporate body, or the birth and death dates of a person (or even none of the above, when `type` = `"other"`), b) it is strongly related to the `name` subelement, and especially, to the value of the `type` attribute of `name`. For instance, if the value of the `type` attribute of the `name` is `"corporate"`, then the value of the `type` attribute of the `dates` subelement is either `"activity"` or `"other"`, denoting eg. the foundation dates of a corporate body, c) it contains two additional subelements, the `earliestDate` and the `latestDate`, which also define different semantic mappings. Here, we present the mapping of `dates` (and its subelements) when the `type` attribute of `name` has the value `"personal"`, while the `type` attribute of the `dates` element has the value `"life"`. Then, `earliestDate` represents the birth date of an agent, while the `latestDate` represents the date of his/her death. In this case, `earliestDate` corresponds to an instance of the class `E67 Birth` (denoting the birth event of a person), and is related to an instance of `E21 Person` through the property `P98B was born`, while `latestDate` corresponds to an instance of `E69 Death` (denoting the death event of a person), and is related to an instance of `E21 Person`, through the property `P100B died in`. In both cases, an instance of the class `E52 Time-Span` is linked to the instance of `E67 Birth` (resp. `E69 Death`), through the property `P4 has time-span`. Finally, to denote the specific date of the birth/death event, an instance of the class `E50 Date` is linked to an instance of `E52 Time-Span` through the property `P78 is identified by`. Based on this analysis, we get the following CIDOC CRM path:

`E24 Physical Man-Made Thing` → `P108B was produced by` →
`E12 Production` → `P14 carried out by` → `E21 Person` →
`P98B was born` → `E76 Birth` → `P4 has time-span` →
`E52 Time-Span` → `P78 is identified by` → `E50 Date`

which semantically corresponds to:
`/vra/work/agentSet/agent/name[@type="personal"]`
`/dates[@type="life"]/earliestDate`

and the following CIDOC CRM path:
`E24 Physical Man-Made Thing` → `P108B was produced by` →
`E12 Production` → `P14 carried out by` → `E21 Person` →
`P100B died in` → `E69 Death` → `P4 has time-span` →
`E52 Time-Span` → `P78 is identified by` → `E50 Date`

which semantically corresponds to the VRA path:
`/vra/work/agentSet/agent/name[@type="personal"]`
`/dates[@type="life"]/latestDate`

The mapping presented in this section is depicted in Fig. 2, when the value of the `type` attribute of `name` is `"personal"` and the second approach is applied. In this figure, the upper part of each box indicates the VRA path mapped to the CIDOC CRM class shown in the lower part. Boxes are linked with arrows that represent CIDOC CRM properties appearing as labels to these arrows. This mapping can be formally expressed in MDL as shown in Table 1. In this

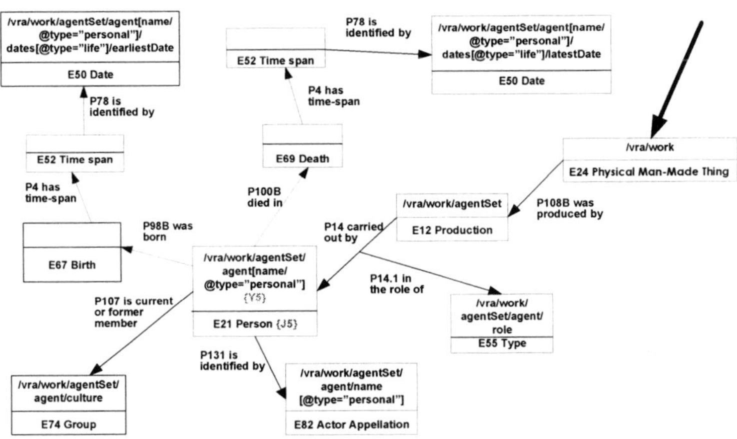


**Fig. 2.** The mapping of the value `personal` of the `type` attribute of the element `agent`

table, Rule R1 states that `/vra/work` is mapped to the class `E24`. R2 states that `agentSet` corresponds to the class `E12`, which is related to `E24` through the relation `P108B`. Rules R3, R4, R5 describe the three different versions of the `agent` element, according to the three possible values of the `type` attribute of `name`, which correspond to the three different subclasses (`E21`, `E40`, `E74`), respectively. We should note that variables Y5, Y10 and Y15 on the left part of the rules, and variables J5, J10 and J15 on the right part, denote branching points as it is also shown in Fig. 2 and in Table 1.

### 3.2 Mapping the `Inscription` Element and Its Subelements

The `inscription` element denotes all marks or written words added to an object at the time of its production, or in its subsequent history and includes dedications, signatures, dates, texts, stamps from the printers or publishers etc. It contains the following subelements:

- `author`: specifies the person or corporate body that created the inscription.
- `position`: the location of the text or symbol of the object being described.
- `text`: refers to the written text or the sign language of `inscription`. A `type` attribute is assigned to this subelement, with possible values: `"mark"`, `"caption"`, `"signature"`, `"date"`, `"text"`, `"translation"`, and `"other"`.

**Table 1.** Mapping the VRA element `agent` to the CIDOC CRM using MDL

| RuleNo | VRA paths | CIDOC CRM paths |
|---|---|---|
| R1: | /vra/work{X1} | E24{C1} |
| R2: | $X1/agentSet{Y1} | $C1→P108B→E12{J1} |
| R3: | $Y1/agent[name/@type="personal"]{Y5} | $J1→P14{*S2*} →E21{J5} |
| R4: | $Y1/agent[name/@type="corporate"]{Y10} | $J1→P14{*S3*} →E40{J10} |
| R5: | $Y1/agent[name/@type="family"]{Y15} | $J1→P14{*S4*} →E74{J15} |
| R6: | $Y5\|$Y10\|$Y15/name* | $J5\|$J10\|$J15→P131→E82 |
| R7: | $Y5\|$Y10\|$Y15/culture* | $J5\|$J10\|$J15→P107→E74 |
| R8: | $Y5\|$Y10\|$Y15/role* | $S2\|$S3\|$S4→P14.1→E55 |
| R9: | $Y5/dates[@type="life"]/earliestDate* | $J5→P98→E67→P4→E52→ P78→E50 |
| R10: | $Y5/dates[@type="life"]/latestDate* | $J5→P100B→E69→P4→E52→ P78→E50 |

*Mapping the* `incription` *element:* The `inscription` element can be mapped to the class `E37 Mark`. The class `E24 Physical Man-made Thing` is related to `E37 Mark` through the property `P128 carries`, giving the CIDOC CRM path:

`E24 Physical Man-Made Thing` → `P128 carries` → `E37 Mark`

which semantically corresponds to the VRA path:

`/vra/work/incriptionSet/inscription`

*Mapping the* `author` *subelement of* `inscription`*:* To map `author` the class `E39 Actor` is employed in conjunction with `E65 Creation` denoting a creation event during which the inscription was created. `E37 Mark` is related to `E65 Creation` through the property `P94B was created by`. `E65 Creation` is related to `E39 Actor` through the property `P14 carried out by`. Finally, the value of `author`, is associated to `E82 Actor Appellation` which is related to `E39 Actor` through `P131 is identified by` giving the CIDOC CRM path:

`E24 Physical Man-Made Thing` → `P128 carries` → `E37 Mark` →
`P94B was created by` → `E65 Creation` → `P14 carried out by` →
`E39 Actor` → `P131 is identified by` → `E82 Actor Appellation`

which semantically corresponds to the VRA path:

`/vra/work/incriptionSet/inscription/author`

*Mapping the* `position` *subelement of* `inscription`*:* The `position` subelement, which refers to the location of the text or symbol of the object being described, is mapped to an instance of the class `E46 Section Definition`, which is linked to the instance of the class `E37 Mark` through the property `P58 has section definition`. The CIDOC CRM path created has the form:

`E24 Physical Man-Made Thing` → `P128 carries` → `E37 Mark` →
`P58 has section definition` → `E46 Section Definition`

while the equivalent VRA path has the form:

`/vra/work/incriptionSet/inscription/position`

*Mapping the `text` subelement of `inscription`:* The `text` subelement is mapped to the class `E33 Linguistic Object`. The class `E37 Mark`, to which the inscription element has been mapped, is related to the class `E33 Linguistic Object` through the property `P138 represents`, giving the CIDOC CRM path:

`E24 Physical Man-Made Thing → P128 carries → E37 Mark →`
`P138 represents → E33 Linguistic Object`

while the equivalent VRA path has the form:

`/vra/work/incriptionSet/inscription/text`

To map the `type` attribute of `text`, `E55 Type` is employed. The instance of `E55 Type` is given the value of the `text` attribute. `E33 Linguistic Object` is related to `E55 Type` through `P2 has type`. The CIDOC CRM path created:

`E24 Physical Man-Made Thing → P128 carries → E37 Mark →`
`P138 represents → E33 Linguistic Object → P2 has type → E55 Type`

corresponds to the equivalent VRA path:

`/vra/work/incriptionSet/inscription/text/@type`

When the global attribute `xml:lang` is assigned to `text` element to specify its language, this is mapped to the class `E56 Language`, which is related to `E33 Linguistic Object` through the property `P72 has language`, giving the path:

`E24 Physical Man-Made Thing → P128 carries → E37 Mark →`
`P138 represents → E33 Linguistic Object →`
`P72 has language → E56 Language`

while the equivalent VRA path has the form:

`/vra/work/incriptionSet/inscription/text/@xml:lang`

This mapping of `inscription`, expressed in MDL, is shown in Table 2.

**Table 2.** Mapping the VRA element `inscription` to CIDOC CRM through MDL

| RuleNo | VRA paths | CIDOC CRM paths |
|---|---|---|
| R1: | /vra/work{X1} | E24{C1} |
| R11: | $X1/inscriptionSet/incription{Z1} | $C1→P128→E37{D1} |
| R12: | $Z1/author* | $D1→P94B→E65→P14→E39→ P131→E82 |
| R13: | $Z1/text*{Z3} | $D1→P138→E33{D3} |
| R14: | $Z3/@type* | $D3→P2→E55 |
| R15: | $Z3/@xml:lang* | $D3→P72→E56 |
| R16: | $Z1/position* | $D1→P58→E46 |

## 4 Conclusions and Related Work

Mapping VRA Core 4.0 to CIDOC CRM proved to be a difficult and time-consuming activity requiring a deep and conceptual work. CIDOC CRM provides very rich structuring mechanisms and an abstract but fine-grained conceptualization for events, objects, agents, things, etc. This generated a large number of conceptual expressions that should be studied carefully to select the semantically closest one to map a metadata element. The influence of value of the `type` attribute assigned to several elements makes the mapping even more complex. Due

**Table 3.** Mapping the other VRA elements to the CIDOC CRM expressed in MDL

| RuleNo | VRA paths | CIDOC CRM paths |
|---|---|---|
| R1: | /vra/work{X1} | E24{C1} |
| R17: | $X1/materialSet/material*{T1} | $C1→P45→E57{A1} |
| R18: | $T1/@type* | $A1→P2→E55 |
| R19: | $X1/measurementsSet/measurements*{W1} | $C1→P43→E54{B1}→P90→E60 |
| R20: | $W1/@type* | $B1→P2→E55 |
| R21: | $W1/@unit* | $B1→P91→E58 |
| R22: | $X1/titleSet/title*{Q1} | $C1→P102{S1} →E35 |
| R23: | $Q1/@type* | $S1→P102.1→E55 |
| R24: | $X1/rightsSet/rights{F1} | $C1→P104→E30{G1} |
| R25: | $F1/@type* | $G1→P2→E55 |
| R26: | $F1/rightsHolder* | $G1→P75B→E39 |
| R27: | $F1/notes* | $G1→P3→E62 |
| R28: | $F1/text* | $G1→P1→E75 |
| R29: | $X1/techniqueSet/technique* | $C1→P108B→E12→P32→E55 |

to lack of space, we give without analysis in Table 3 the mappings of the VRA Core 4.0 elements `title`, `material`, `measurements`, `rights` and `technique`.

Several papers deal with ontology-based integration. In [1] a mechanism for integrating cultural information is proposed which uses a rule-based mapping language similar to ours to map XML fragments to the concepts and roles of a domain specific ontology. In [4] XML data are transformed to a global ontology through mapping rules based in OWL. The integration of the CIDOC CRM ontology in the core model of the BRICKS project is presented in [6].

## References

1. Amann, B., Beeri, C., Fundulaki, I., Scholl, M.: Ontology-Based Integration of XML Web Resources. In: Horrocks, I., Hendler, J. (eds.) ISWC 2002. LNCS, vol. 2342, pp. 117–131. Springer, Heidelberg (2002)
2. Kakali, C., Lourdi, I., Stasinopoulou, T., Bountouri, L., Papatheodorou, C., Doerr, M., Gergatsoulis, M.: Integrating Dublin Core Metadata for Cultural Heritage Collections Using Ontologies. In: DC 2007, pp. 128–139 (2007)
3. ICOM/CIDOC CRM Special interest Group. Definition of the CIDOC Conceptual Reference Model, Version 5.0.2 (January 2010), `http://www.cidoc-crm.org`
4. Lehti, P., Fankhauser, P.: XML Data Integration with OWL: Experiences and Challenges. In: SAINT 2004, pp. 160–170. IEEE, Los Alamitos (2004)
5. Library of Congress (LC). VRA Core: a Data Standard for the Description of Works of Visual Culture (2011), `http://www.loc.gov/standards/vracore/`
6. Meghini, C., Risse, T.: BRICKS: A Digital Library Management System for Cultural Heritage. ERCIM News (61) (April 2005)
7. Noy, N.F.: Semantic Integration: a Survey of Ontology-Based Approaches. SIGMOD Record 33 (2004)

8. Stasinopoulou, T., Bountouri, L., Lourdi, I., Papatheodorou, C., Doerr, M., Gergatsoulis, M.: Ontology-Based Metadata Integration in the Cultural Heritage Domain. In: Goh, D.H.-L., Cao, T.H., Sølvberg, I.T., Rasmussen, E. (eds.) ICADL 2007. LNCS, vol. 4822, pp. 165–175. Springer, Heidelberg (2007)
9. VRA Core 4.0 Element Description and Tagging Examples (2007), http://www.loc.gov/standards/vracore/schemas.html
10. World Wide Web Consortium (W3C). XML Path Language (XPath) 2.0 (2007), http://www.w3.org/TR/xpath20/

# Bringing Environmental Culture Content into the Europeana.eu Portal: The Natural Europe Digital Libraries Federation Infrastructure

Konstantinos Makris, Giannis Skevakis, Varvara Kalokyri, Nektarios Gioldasis, Fotis G. Kazasis, and Stavros Christodoulakis

TUC-MUSIC, Lab. Of Distributed Multimedia Information Systems, Technical University of Crete, University Campus, 73100, Chania, Greece
{makris,skevakis,vkalokyri,nektarios, fotis,stavros}@ced.tuc.gr

**Abstract.** The aim of the Natural Europe project [1] is to improve the availability and relevance of environmental culture content for education and life-long learning use, in a multilingual and multicultural context. Cultural heritage content related with natural history, natural sciences, and nature/ environment preservation, is collected from six Natural History Museums (NHMs) around Europe into a federation of European Natural History Digital Libraries that is directly connected with Europeana.eu. We present here the Natural History Digital Libraries Federation infrastructure along with the appropriate tools and services that (a) allow the participating NHMs to uniformly describe and semantically annotate their content according to international standards and specifications, (b) interconnect their digital libraries, and (c) expose metadata records for Natural History cultural heritage objects to Europeana.eu.

**Keywords:** Europeana, cultural federation, digital curation, preservation metadata.

## 1 Introduction

In an era where natural history and environmental education inadequacy in formal and informal contexts is becoming an increasingly challenging issue, harvesting the potential of European digital libraries appears as a very attractive option. Natural History Museums (NHMs for short) can play an important role to the reconstruction of our environmental culture since they are "key tools" of social and cultural development and can promote the harmony between humanity and nature. NHMs are generally well prepared to meet the above challenge. Most of them are quite innovative institutions dealing at the same time with conservation of collections, scientific research, and cultural and educational activities. Consequently, they have the possibility to conceive their own special ways of mediation and they have the opportunity to deal with environment as a natural part of our general culture.

However, an impressive abundance of high quality digital content that is available in NHMs around Europe remains largely unexploited due to a number of barriers, such as: the lack of interconnection and interoperability between the management

E. García-Barriocanal et al. (Eds.): MTSR 2011, CCIS 240, pp. 400–411, 2011.

systems of NHMs, the lack of centralised access through a European point of reference like Europeana, as well as the inadequacy of current content organization and the metadata used.

The Natural Europe project offers a coordinated solution at European level that aims to overcome the aforementioned barriers in order to amplify the exploitation of natural history heritage into formal and informal learning processes. Among the main objectives of the Natural Europe project is the development of a Natural History Digital Libraries Infrastructure which offers appropriate tools and services that (a) allow the participating NHMs to uniformly describe and semantically annotate their content according to international standards and specifications, (b) interconnect their digital libraries, (c) expose their metadata records for Natural History (NH for short) Cultural Heritage Objects (CHO for short) to Europeana.eu.

The rest of this paper is organised as follows. Sections 2, 3, and 4 present the infrastructure developed to support the aggregation of Environmental Heritage resources (spread in European Natural History Museums) and their further dissemination to external organizations. Section 5 presents an indicative information flow in our case study. Section 6 concludes and presents the work that is planned to take place in the next future.

## 2 The NH Digital Libraries Federation Infrastructure

The NH Digital Libraries Federation Infrastructure aims to aggregate natural history content metadata from NHMs and Europeana.eu in order to exploit them for learning purposes. Moreover, it aims to expose the aggregated metadata records to Europeana.eu. To achieve these objectives a set of tools and services have been developed and deployed in order to ensure that possibly diverse legacy natural history metadata will be unified, semantically enriched, aggregated, preserved, and finally harvested by Europeana.eu. The federation's stakeholders and their basic requirements are:

1. **The Natural History Museums.** The participating NHMs are the main content providers whose metadata are aggregated into the federation. With respect to NHMs the federation architecture ensures that at least the following requirements are met: a) *Legacy Metadata Migration.* Any existing metadata is properly migrated (mapped, imported, and handled) to the environment; b) *Metadata Unification.* All the contributed metadata records are encoded following the ESE [2] and EDM [3] schemas; c) *CHO Publishing.* Each contributed CHO or a representation needs first to be published and web accessible (in its original quality or not). d) *Semantic Multimedia Annotation.* NHMs have the ability to classify and/or annotate/link their contributed CHOs with terms from common (widely-accepted) controlled vocabularies (with concepts, places, persons, etc.) and/or taxonomies; e) *Standardized Metadata Dissemination.* The metadata records of the contributed CHOs are exposed through standard protocols, schemas, and formats.
2. **The Europeana.eu Cultural Heritage Portal.** The NH metadata that the Natural Europe aggregates will be made available through the Europeana.eu Portal. On the other side, the federation also harvests from it NH related metadata that is already available there. Currently the Europeana.eu metadata dissemination process is

based on web service and widget technologies that allow to search and display the Europeana collections in external applications[1]. For the metadata submission process, the following requirements have to be met: a) *Conformance to ESE/EDM*. Currently, the Europeana's metadata submission process requires that providers should encode their metadata records according to the Europeana Semantic Elements specification (ESE for short), while in the near future it will be able to accept also metadata that conform to the Europeana Data Model (EDM for short) specification; b) *OAI-PMH Compliant Metadata Dissemination*. In order to submit the aggregated metadata to Europeana.eu, the federation offers appropriate OAI-PMH [6] service interfaces; c) *CHO Persistent Identification*. Each contributed CHO is persistently identified and addressed so that local (to museum) identification changes do not affect Europeana's resolving process.

3. **Third-Party Federations.** Integration with third-party federations (e.g. BHL [4], Organic.Edunet [5], etc.) imposes specific requirements. For that the NH Digital Libraries Federation infrastructure offers an OAI-PMH [6] based metadata dissemination service.

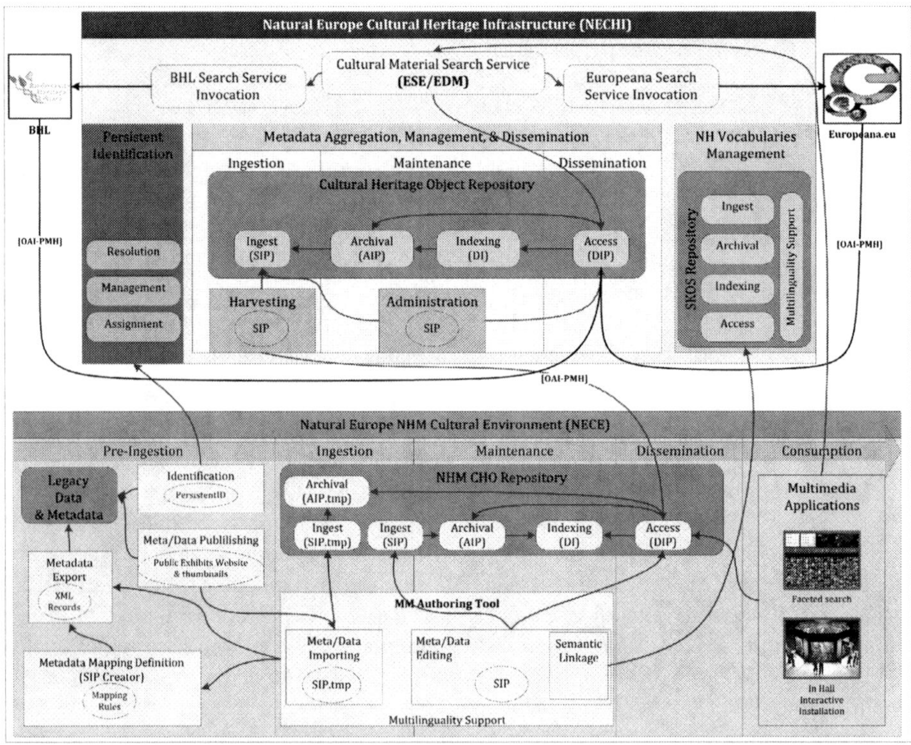


**Fig. 1.** The Natural History Digital Libraries Federation Architecture

[1] Europeana also aims to provide an OAI-PMH target service that will enable external applications to harvest the Europeana collections.

Fig. 1 illustrates the Natural History Digital Libraries Federation Architecture consisting of two nodes: a) The Cultural Heritage Infrastructure (the Federal) Node where the metadata aggregation, management, and dissemination is taking place. Moreover, this infrastructure provides NHMs with appropriate services for semantic annotation and persistent identification of their CHOs; and b) The NHM Cultural Environment (the Federated) Node that provides appropriate CHO curation and management tools and services, while also defines specific practices and processes in order to achieve the requirements discussed above.

## 3 The Natural Europe Cultural Heritage Infrastructure

The Cultural Heritage Infrastructure (NECHI for short) accommodates software components that can be categorized under the following three high level functions:

1. Metadata Aggregation, Management, and Dissemination of CHOs contributed by the participating NHMs.
2. Shared Knowledge Management: Management of shared vocabularies as well as brokerage of other vocabulary or taxonomy accessing services.
3. Persistent Identification and Resolution: Services that allow local identification modifications to remain transparent.

### 3.1 Metadata Aggregation, Management and Dissemination

Metadata Aggregation, Management, and Dissemination is the core theme of the NECHI. This functionality actually refers to a complete information preservation life-cycle as defined in the OAIS Reference Model [7], where the information producers are the participating NHMs and information consumers are the Europeana.eu Portal, and other third-party federations like the BHL and Organic.Edunet Portals.

In the NECHI the OAIS reference model is materialised by the Cultural Heritage Object Repository (CHO Repository for short) that accommodates appropriate software components that span the Ingestion, Maintenance, and Dissemination phases for the CHO metadata contributed by NHMs. In each phase, appropriate information packages are manipulated by the repository.

**Information Packaging.** A clear definition of information is central to the ability of an OAIS to preserve it and is one of the first steps in designing such a system. Information is manipulated by an OAIS in the form of Information Packages. In Natural Europe an information package may accommodate both content information (CHOs) as well as descriptive information (CHO metadata and administrative information).

Every submission of information to an OAIS by a Producer, and every dissemination of information to a Consumer, occurs as one or more discrete transmissions of Information Packages:

- The Submission Information Package (SIP for short) refers to the information package that is submitted to the system by a Producer in the Ingestion Phase. A SIP for the CHO Repository is created as a result of a metadata harvesting process or as a result of a metadata modification process from the CHO

Metadata Administration Interface. In the first version of the CHO repository, SIPs conform to the ESE Specification Version 3.4. However in future releases, SIPs for this repository will be based on the EDM specification.

- The Archival Information Package (AIP for short) refers to the information packages that are archived by the CHO Repository in the Maintenance Phase. Within the CHO Repository one or more SIPs are transformed into one or more AIPs for storage according to its internal storage specification (e.g. collection-based, record based, aggregation-based, etc.). Each AIP has a complete set of administration information for the associated descriptive information. AIPs will be also indexed on various descriptive information attributes in order to be efficiently accessed and disseminated.
- The Dissemination Information Package (DIP for short) refers to the information package that the CHO Repository provides in response to a request for accessing all or a part of an AIP from a Consumer in the dissemination phase. A DIP may include part of an AIP (e.g. as a result of a search request), a complete AIP (as a result of an access request), or collections of AIPs (e.g. as a result of a harvesting request), and it may or may not contain complete descriptive information.

**Software Modules & Services.** As illustrated in Fig. 1, the Metadata Aggregation, Maintenance and Dissemination functionality is implemented by a) the CHO Repository, b) the Harvesting Module, and c) the Administration Module.

The CHO Repository is responsible for the maintenance (storage, management) and dissemination of the aggregated CHO Metadata. It accommodates modules for ingestion, archival, indexing, and accessing of CHO Metadata.

The Harvesting module is responsible for the scheduled harvesting of CHO metadata from the participating NHMs. For that, this module acts as an OAI-PMH Harvester application that is issuing specific harvest requests to NHMs' metadata repositories following a predefined schedule.

The Administration module allows administrators to manage the CHO Repository AIPs and their administration information. It also allows them to manage user accounts and privileges, as well as to schedule harvests from specific NHMs. This module allows also content providers to inspect (preview) and manage (accept, reject) harvested metadata collections.

The Cultural Material Search Service exploits functionality provided by the Access module (and backed up by the Indexing module) in order to search for CHO descriptions (in the form of ESE/EDM) aggregated by the participating NHMs and stored in the CHO Repository. Moreover, it is routing search queries to other federations or portals like Europeana.eu and BHL, in order to allow the exploitation of Natural History related content (the metadata of which are stored in such federations) for other purposes. The results of this service are returned as DIPs in the form of ESE/EDM descriptions or references to them.

### 3.2 NH Shared Knowledge Management and Dissemination

In the context of Natural Europe the participating NHMs enrich the description of their contributed CHOs through a semantic annotation process. In order for this semantic annotation process to be effective in terms of exploitation (i.e. others can understand the annotations), a community-wide accepted taxonomy [8] has been adopted. Mappings between the classifications that the NHMs use and the standard taxonomy have been defined.

To support this, NECHI accommodates a specific sub-system for managing these taxonomies and/or classification schemes (vocabularies) as well as allowing museum curators to search and locate specific entries into them. Moreover, this sub-system may also act as a proxy for accessing vocabularies from other NH related communities like BHL, uBio [9], etc. if such a requirement is considered important for the relevant stakeholders (e.g. NHMs). In the future when EDM will be widely adopted, the Shared Knowledge Management and Dissemination sub-system will offer access to shared vocabularies and/or authority files officially adopted by the Europeana Semantic Layer (ESL for short) [10] in order to allow a fully-fledged EDM usage.

Natural Europe shared vocabularies have been represented as RDF resources based on the SKOS model [11]. Thus, this sub-system offers functionality for publishing (ingest), storing (archival), indexing, and searching SKOS vocabularies.

The adoption of the SKOS model for the representation of the shared vocabularies is motivated by the fact that SKOS has been also adopted for the development of the ESL.

### 3.3 Persistent Identification and Resolution

Persistent Identification of material is a well identified requirement in many application domains and becomes more vital as we move to the web of data. Europeana.eu has already identified [12][13] this requirement and has put persistent identification of aggregated content in top priority.

Persistent Identification is very important as it enables digital objects to have a stable access point over the Internet. Persistent identifiers (PIDs for short) are resolved by appropriate resolution services that resolve them to the current valid address of a digital object.

Following that, Natural Europe has taken care so that the aggregated CHOs are persistently identified. For that, NECHI accommodates a specific sub-system that allows museums to define PIDs for their contributed CHOs (if they don't already do so) using the Purl system [14]. This sub-system provides a specific identification namespace for each participating NHM, under which the NHM persistently identifies its contributed CHOs. Moreover, this sub-system provides a resolution service which is be able to resolve a PID to the registered CHO address.

This approach is aligned with the specification of the Europeana Resolution and Discovery Services (ERDS for short) [15] which keeps track of namespaces of persistent identifiers and of the institutions that are responsible for those namespaces. With the help of this information, the ERDS can forward resolving requests to the right local resolver and present the returned link.

## 4 The NHM Cultural Environment

The NHM Cultural Environment (NECE for short) node provides: a) the specification of the Metadata Management Life-cycle process along with the tools and services needed in order to ensure that existing collections and their descriptions will be migrated, annotated, preserved, and properly disseminated (mainly to NECHI), and b) the advanced Cultural Heritage Multimedia Consumption Applications that enable NHMs to present (acting as Information Consumers) Natural History content to their visitors (physical or virtual ones); content not only from their collections, but also from other collections reached through the NECHI.

### 4.1 The NHM Metadata Management Life-Cycle Process

In Natural Europe the participating NHMs contribute (metadata about) a large number of Natural History related CHOs which are aggregated by the project and exploited for educational purposes. In order to ensure that a) any existing metadata will be exploitable in this context and b) the new knowledge will be built on the same basis for all participating NHMs, a metadata unification task is required. Moreover, these unified metadata need to be preserved in a dedicated metadata repository able to support the enrichment activities over them. For supporting these requirements, NECE provides the NHM CHO Repository which is able to handle metadata in the form needed by Natural Europe (ESE/EDM). This repository along with the MM Authoring Tool[2] which has been implemented on top of it, are the cornerstone components of the NECE at each NHM side as far as the Metadata management and dissemination is concerned. CHO metadata is ingested to, preserved at, and disseminated from, this repository.

The complete life-cycle process that the NECE defines for NHM (contributed) metadata management comprises four phases: a) pre-ingestion phase, b) ingestion phase, c) maintenance phase, and d) dissemination phase.

Except for the preparatory pre-ingestion phase, the other three phases (ingestion, maintenance, dissemination) are standard phases defined for an Open Archival Information System. In these phases each NHM uses as a front-end Application (the MultiMedia (MM) Authoring Tool) for the semantic annotation of its CHOs, and as a back-end infrastructure the NHM CHO Repository that offers standard OAIS functionality for the enriched metadata. In the sub-sections below we discuss the various phases of the metadata management life-cycle process along with the tools, repositories, and services of the NECE that NHMs will be using in each phase.

**Pre-Ingestion Phase.** In the pre-ingestion phase, each NHM ensures that it's existing CHO collections and their metadata descriptions will be appropriately migrated into Natural Europe. This includes:

1. Meta/Data Publishing of contributed CHOs: Unless already did so, each NHM can currently use the provided MM Authoring Tool in order to publish each contributed CHO to the web, so that it will be accessible in one click from end users (users of Europeana.eu, Natural Europe, BHL, etc.). The tool automatically publishes the CHO and creates a thumbnail for the contributed CHO.
2. Identification of contributed CHOs: Unless already did so, each NHM will be able in the upcoming release of the MM Authoring Tool to persistently identify the CHOs that will contribute to the project using appropriate services provided by the Persistent Identification sub-system of NECHI. As mentioned, each NHM is provided with a specific identification namespace under which it will permanently identify its contributed (at least) CHOs.

---

[2] The running version of this tool is available at `http://147.27.41.103:8080/mmat`

3. Metadata Unification: Metadata unification (as discussed above), is implemented during the pre-ingestion phase in two steps (following the Europeana.eu content provision practices); firstly, the legacy NHM metadata for the contributed CHOs is provided as record based XML documents and secondly, NHM metadata experts, along with Natural Europe technical experts map these legacy metadata to the ESE specification. Metadata Mapping Definition is done using the Europeana.eu SIP Creator application [16]. SIP Creator is a web-launched GUI tool that allows any record based XML schema (inspected from XML data) to be mapped to ESE schema. The output of this process is a set of mapping rules that are used in the ingestion phase from the MM Authoring Tool for transforming the legacy metadata into ESE format and importing them into NHM CHO Repository.

**Ingestion Phase.** The ingestion phase refers to both importing existing metadata (after applying the mapping rules) to the Natural Europe environment and enriching metadata through a semantic annotation process.

1. Regarding the importing of existing metadata, the MM Authoring Tool provides the functionality to load existing collection metadata, load the appropriate mapping rules, apply these rules to the loaded metadata producing a Submission Information Package (for the underlying repository) that conforms to ESE specification. This

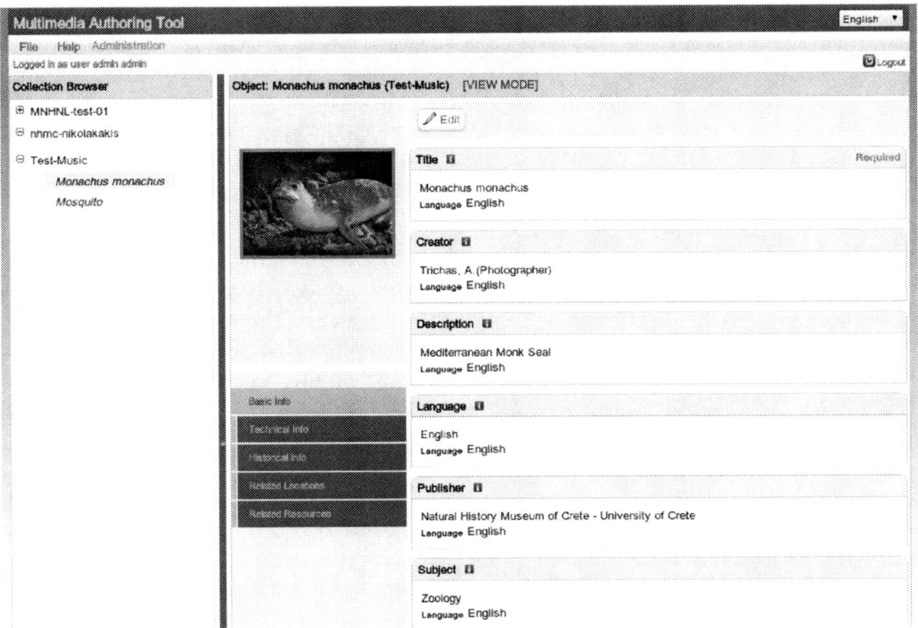


**Fig. 2.** The MultiMedia Authoring Tool in use

SIP, is stored into the NHM CHO Repository with a pending status (i.e. as a temporary AIP). Museum curators have then the ability to inspect the imported collection metadata, modify, reject, or accept it. Once accepted, the imported collection is permanently stored and properly indexed into the NHM CHO Repository as a normal AIP. In this importing phase, the MM Authoring tool allows museum curators to define the identification scheme for the imported records, taking into account any existing identification policy. An indicative screen-shot of this tool is illustrated in Fig. 2.

2. As far as ingestion through the normal metadata curation/annotation activity is concerned, the MM Authoring tool allows museum curators to maintain (create/view/modify/enrich) CHO metadata. This activity is facilitated by access and concurrency control mechanisms provided to ensure security, integrity, and consistency of contents.

**Maintenance Phase.** The maintenance phase refers to storage and management of CHO metadata. At each museum, a complete metadata management solution is provided consisting of: a) the NHM CHO Repository as a back-end infrastructure, and b) the MM Authoring Tool as a front-end Metadata Authoring Tool. The NHM CHO Repository is implementing the OAIS Reference Model with all its discrete entities. For the Maintenance phase, the Archival, and Indexing modules provide the required functionality. As mentioned, the stored AIPs are currently based on the ESE specification, while in future releases they will be based on the EDM specification.

On the other hand, maintenance covers also the day-to-day metadata curation activities of a museum. For that, the MM Authoring tool provides museum curators with a intuitive Graphical User Interface that allows them to easily a) view, edit, delete and create CHO metadata; b) create and manage CHO collections; and c) semantically annotate (interlink) CHOs using shared knowledge systems (SKOS vocabularies and taxonomies, or other RDF resources). The MM Authoring tool currently supporting the ESE specification (i.e. it is producing and manipulating Information Packages conforming to ESE); however, in future releases, it will be able to work with EDM allowing NHMs to exploit the full potential of semantic web and interlink their CHOs with other resources on the web of data (e.g. link them with ESL resources, other published CHOs, authority files for places, events, persons, etc.).

Finally, the tool is multilingual and has been provided to the NHM partners' languages. Moreover, it allows the definition of metadata in any language. Elaborating more on this, multilingual CHO descriptions are provided using the XML "Lang" attribute although Europeana.eu does not currently directly exploits it.

## 5 Indicative Flow of Information

As a pilot case we consider a digital representation of an exhibit of the Natural History Museum of Crete [17] along with its descriptive metadata. Below (Fig. 3) we provide its description as defined in the ESE specification that is currently supported by Europeana.eu as created by the MM Authoring Tool. The same description will be provided in EDM (Fig. 4), when fully supporting it, and interlinked with other resources of the semantic web.

| Node | Content |
|---|---|
| metadata | |
| xmlns:dc | http://purl.org/dc/elements/1.1/ |
| xmlns:europeana | http://www.europeana.eu |
| xmlns:dcterms | http://purl.org/dc/terms/ |
| record | |
| dc:coverage | Greece |
| dc:creator | Trichas,A. |
| dcterms:issued | 09/11/2010 |
| dc:description | Mediterranean monk seal (common name). NHMC exhibit. |
| dc:format | TIFF |
| dc:identifier | 35651 |
| dc:language | English |
| dc:publisher | Natural History Museum of Crete-University of Crete |
| dc:rights | NHMC |
| dc:source | Natural History Museum of Crete-University of Crete |
| dc:subject | Zoology |
| dc:subject | MAMMAL |
| dc:title | Monk seal Monachus monachus |
| dc:type | image |
| dcterms:alternative | Monachus monachus |
| dcterms:created | 12/10/2002 |
| dcterms:spatial | NHMC, Knossou Avenue, Irakleiou, Kriti |
| europeana:collectionName | NHMC data_set_1 |
| europeana:country | greece |
| europeana:dataProvider | Natural History Museum of Crete-University of Crete |
| europeana:isShownBy | http://www.nhmc.uoc.gr/zoology/35651.html |
| europeana:language | el |
| europeana:object | http://www.nhmc.uoc.gr/zoology/35651_tn.jpg |
| europeana:provider | Natural History Museum of Crete-University of Crete |
| europeana:type | IMAGE |
| europeana:uri | http://www.europeana.eu/resolve/record/nhmc_dataset_1/568A74590C376DEF30D8E9CD6C23671902245EA6 |

**Fig. 3.** Example Description of Natural History Museum Object (ESE)

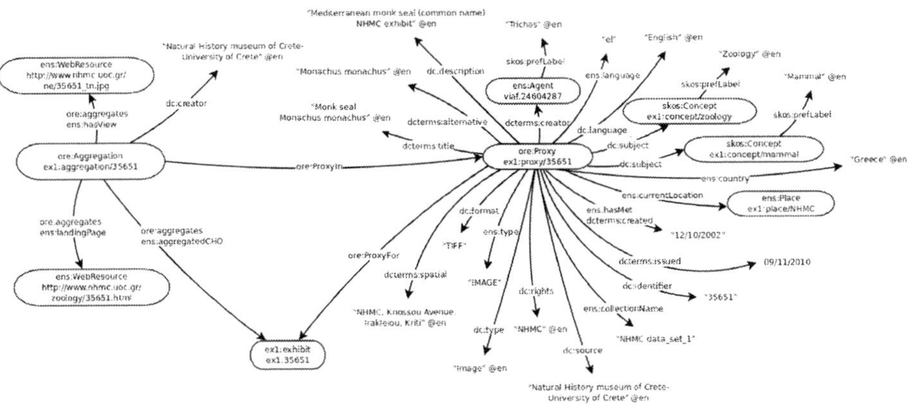


**Fig. 4.** Example Description of Natural History Museum Object (EDM)

# 6 Conclusions

In this paper we presented the infrastructure developed in the context of the Natural Europe project to support the aggregation of Environmental Culture resources (spread in European Natural History Museums), their exploitation for education, and their further dissemination to external organizations. The Natural Europe federation infrastructure consists of: a) The federated node that accommodates tools and services to support the complete metadata (and optionally data) life-cycle management process (ingestion, preservation, dissemination). Apart from the deployed tools and services on the federated nodes, the project has set up specific work-flow processes for the publication, migration (from existing systems), and semantic description, of contributed resources. b) The federal node accommodating services and applications to support the aggregation, exploitation, and further dissemination of resource metadata using standard OAI-PMH services. The federal node provides also services for shared knowledge management (SKOS vocabularies and authority control files), and persistent identification (PURL maintenance and resolving).

Currently the federated node has been provided for use to six European NHMs with a positive feedback regarding the usability and functionality of the tools and the provided functionality. It is expected that the federal node, which is currently under development, will begin to harvest the contributed CHOs within the next few months. According to the Natural Europe project's work-plan, the contributed CHOs will be exposed to Europeana.eu portal by the end of this year.

**Acknowledgements.** This work has been carried out in the scope of the Natural Europe Project (Grant Agreement 250579) funded by EU ICT Policy Support Programme.

## References

1. The Natural Europe Project, http://www.natural-europe.eu
2. Europeana Semantic Elements Specification V.3.4, http://version1.europeana.eu/c/document_library/get_file?uuid=77376831-67cf-4cff-a7a2-7718388eec1d&groupId=10128
3. Europeana Data Model Definition V.5.2, http://version1.europeana.eu/c/document_library/get_file?uuid=aff89c92-b6ff-4373-a279-fc47b9af3af2&groupId=10605
4. Biodiversity Heritage Library (BHL), http://www.biodiversityheritagelibrary.org/
5. Organic.Edunet Portal. Learning Material on Organic Agriculture in Europe, http://portal.organic-edunet.eu/
6. Open Archives Initiative – Protocol for Metadata Harvesting V.2, http://www.openarchives.org/OAI/openarchivesprotocol.html
7. ISO 14721:2003 Open Archival Information System (OAIS) Reference Model, http://www.iso.org/iso/iso_catalogue/catalogue_tc/catalogue_detail.htm?csnumber=24683
8. The Catalogue of Life, http://www.catalogueoflife.org
9. Universal Biological Indexer and Organizer (uBio), http://www.ubio.org
10. The Europeana Semantic Data Layer, http://www.europeanaconnect.eu/documents/D1.1.2_eConnect_Semantic_Layer.v1.0_UVA.pdf
11. W3C: Simple Knowledge Organisation System (SKOS), http://www.w3.org/TR/2009/REC-skos-reference-20090818/

12. Cousins, J.: Europeana: Access to Culture for All?, http://www.conference.bvoe.at/presentations/cousins.pdf
13. The Europeana Strategic Plan 2011-15: http://version1.europeana.eu/c/document_library/get_file?uuid=c4f19464-7504-44db-ac1e-3ddb78c922d7&groupId=10602
14. Persistent Uniform Resource Locators (PURL), http://www.purl.org
15. EuropeanaConnect: D5.4.1Europeana Resolution Service, http://www.europeanaconnect.eu/documents/D5.4.1_eConnect_EuropeanaResolution_Service_v1.0_DNB.pdf
16. Europeana Labs: SIP Creator, http://europeanalabs.eu/sip-creator/
17. The Natural History Museum of Crete (NHMC), http://www.nhmc.uoc.gr

# A New Architecture and Approach to Asset Representation for Europeana Aggregation: The CARARE Way

Christos Papatheodorou[1,2], Costis Dallas[1,3], Christian Ertmann-Christiansen[4], Kate Fernie[5], Dimitris Gavrilis[1], Maria Emilia Masci[6], Panos Constantopoulos[1,7], and Stavros Angelis[1]

[1] Digital Curation Unit – IMIS, Athena Research Centre, Athens, Greece
[2] Department of Archives and Library Sciences, Ionian University, Corfu, Greece
[3] Department of Communication, Media and Culture, Panteion University, Athens, Greece
[4] Heritage Agency of Denmark, Copenhagen, Denmark
[5] MDR Partners, UK
[6] Laboratorio LARTTE, Scuola Normale Superiore di Pisa, Italy
[7] Department of Informatics, Athens University of Economics and Business, Athens, Greece
{c.papatheodorou,c.dallas,d.gavrilis,p.constantopoulos, s.angelis}@dcu.gr, chrert@kulturarv.dk, kfernie@tiscali.co.uk, e.masci@sns.it

**Abstract.** This paper presents a new metadata aggregation approach based on a mediating repository that intends to ensure the integrity, authenticity and semantic enrichment of metadata provided to Europeana by heterogeneous collections. Primary metadata are mapped to CARARE schema, a schema suitable for describing archaeological and architectural heritage assets, digital resources, collections, as well as events associated with them. The paper specifies the proposed schema and discusses the overall architecture of the proposed approach.

**Keywords:** Cultural heritage metadata, Composite content, Metadata integration, Infrastructures for sharing content.

## 1 Introduction

Archaeological and architectural monuments and sites constitute an important part of tangible European cultural heritage. Aligned with the initiative to establish the Europeana digital library as a single point of access to Europe's cultural heritage, the CARARE Best Practice Network was established in 2009 in order to increase the quantity and quality of digital content for the archaeological and architectural heritage that is available to users of Europeana, while addressing some significant issues specific to archaeological and architectural heritage. An overview of the CARARE project, its workplan and the challenges it purports to meet is presented in [1].

The objective of CARARE is to integrate in Europeana an estimated 2 million digital resources, corresponding to approximately 1 million unique monuments, buildings, landscapes, heritage sites and artefacts. Content belongs to a broad and diverse set of heritage organizations across Europe, each adhering to different organizational

E. García-Barriocanal et al. (Eds.): MTSR 2011, CCIS 240, pp. 412–423, 2011.

principles, descriptive standards and management procedures with regard to their data. The cultural assets to be made available are very diverse, from prehistoric and Iron age archaeological survey results to complex Mediterranean archaeological sites, and historic buildings. The digital resources representing such assets are also heterogeneous, ranging from paintings and prints to photographs, archaeological and architectural plans, sections and drawings, and, increasingly, digital 3D models and renderings thereof [2]. Heritage assets are as a rule associated with geographic information, both in the form of geographic coordinates according to some grid standard, and in the form of named geographic entities such as historical locations and areas. In addition, some heritage assets – for instance, an archaeological site such as Pompeii – are characterized by a nested mereological structure, being composed of buildings, each of which is also composed by particular architectural elements; resources may, indeed, represent architectural and archaeological assets at quite different levels of complexity. In general digital resources are as a rule associated with a heritage asset but in some cases the monuments and buildings captured in historical images are no longer known.

Earlier initiatives to provide metadata and links to digital resources for aggregation in Europeana are based on a mapping architecture, by virtue of which original metadata records are translated – following a set of atomic rules for individual resources and data elements – into a common output schema such as the Europeana Semantic Elements (ESE) [3]. CARARE represents a significant departure from this architecture. It introduces the notion of an information broker – an intermediate repository acting as a mediator– intended to ensure the integrity, authenticity and content enrichment of metadata provided to Europeana by heterogeneous collections. This paper presents the rationale, the background, as well as the specific requirements in the field of schema definition, overall architecture and technical approach, as well as content enrichment, adopted by CARARE to meet the challenges presented above.

## 2 Background

Lightweight Information Describing Objects (LIDO) [4] is an output of the work of CIDOC Working Group Data Harvesting and Interchange and the result of a joint effort by the CDWA Lite - museumdat Working Group, the SPECTRUM community, the Documentation Committee of the German Museums Association and the ATHENA project. LIDO is intended to harvest information for museum objects, and as such it aims to aggregate information from several standards and schemes that have been developed in Europe and the rest of the world. These include the following schemes and standards: CDWA Lite, CIDOC CRM, museumdat and SPECTRUM. Its perspective is event-oriented, a view originating from the contribution of CIDOC CRM [5]. The schema provides seven areas in each record for an object. Each metadata record includes (a) a Metadata Record ID, (b) a Category, (c) the descriptive metadata and (d) administrative metadata of the record.

On the other hand, MIDAS Heritage [6] is a standard, which indicates the kind of information that should be recorded to support effective sharing and long-term preservation of the knowledge of the historic environment. MIDAS has a three-level structure: (a) Themes, which are the highest level information blocks of interest for the historic environment community, (b) Information Groups, which set the specific standard for what should be included in an entry covering a particular subject and (c) Units of Information, which are the basic items that make up an entry.

The main conclusions of a comparative study between LIDO and MIDAS schemas are that MIDAS is a more general standard covering several areas of Cultural Heritage Assets Management, while LIDO focuses mainly on the description of museum objects, providing a semantically rich carrier for information exchange. MIDAS organizes the Cultural Heritage Assets to the classes Monuments, Areas and Artefact-Ecofact and provides elements for the rich description of their geospatial features, without focusing on the digital representations. It provides a significant set of elements for cartography. The added value feature of MIDAS is that it provides analytical information on the interventions, works and studies referring to a Cultural Heritage Asset. Moreover the Theme Information Sources allows for the rich documentation of the objects and the interventions on them. LIDO provides elements for the detailed description of museum objects, as well as their digital representations. Since it focuses on museum objects, it does not cover significant aspects of the cultural heritage assets documentation, such as their geographical identification and the history of the interventions and restorations. In general, the two schemas have different orientations and serve different purposes and functions. MIDAS is more general and attempts to cover a wide range cultural heritage asset types as well as the activities performed on them, while LIDO provides a descriptive expressiveness for museum objects. Although both schemas share many common entities, these are interpreted differently according to the orientation and scope of each schema.

A deeper comparison between the two schemas is needed to reveal the elements needed to constitute the CARARE schema. This comparison should be based on CIDOC CRM ontology, due to its power to express clearly the semantics of the cultural heritage metadata schemas [2]. In particular the study and comparison of the schemas with the CIDOC CRM, is based on the POLIS DTD [7], a DTD derived directly by the CIDOC CRM. The POLIS DTD was produced as part of an EU funded Greek national research project to develop an interoperability framework for the cultural heritage. It provides a language to describe immovable monuments, a domain that covers adequately the domain of CARARE project. It focuses on both the objects and their digital representations, but it is a descriptive schema and therefore it is not so analytical in handling information concerning the management of the cultural heritage assets, such as the description of particular investigative, consultation, research, etc. activities. However due to its event-based character, it is powerful and flexible enough to handle the information for these activities as events.

Given that the two metadata schemas include a lot of elements and the comparison of their semantics requires a lot of effort, the existence of a DTD, which directly implements particular CIDOC CRM expressions covering almost completely the CARARE domain, facilitates significantly the process of the comparison that aims to the development of an appropriate schema. The added-value features of the approach we followed are: (a) homogeneity: the metadata schemas are compared and discussed under the prism of another metadata schema, (b) facility: there is no need to develop all the CIDOC CRM expressions and to correlate them with LIDO and MIDAS schemas. The mapping emerged the following conclusions: (i) The proposed structure of the CARARE schema should be an application profile based on MIDAS, which covers several activities for the Cultural Heritage Assets Management and ensures metadata provenance. Moreover MIDAS is almost completely aligned (mapped) to CIDOC CRM paths, while the elements that are not covered by the schema are considered of minor importance (ii) LIDO provides a powerful framework for the detailed description of the digital resources that correspond to heritage assets and the related activities and interventions on them.

## 3 CARARE Schema

The CARARE schema is designed to support the delivery of metadata to the CARARE aggregator and to Europeana. It is an application profile based on MIDAS Heritage, LIDO and the POLIS DTD and the rationale for this decision comes from the nature of the problem it is designed to solve.

CARARE is aggregating content for Europeana from a wide variety of organizations and digital sources with varied and particular histories, which yet draw on common practices that have been developed over the years and both have informed and have been informed by standards, in particular the CIDOC CRM and the CIDOC core data standard for archaeological sites and monuments. These sources include, e.g., administrative/scientific national registries of sites and monuments, archaeological museum collections, collections of 3D models describing any of these types of objects, as well as digital historical document collections such as the Visual Fortune of Pompeii archive described below. At the same time, all of these sources must be aggregated and delivered into the common format, which Europeana now uses to describe its content, the Europeana Data Model (EDM) [8], a target which is developing and changing over time.

Implementing the CARARE schema has followed an iterative approach with an initial adaptation to account for the particular focus of Europeana on digital resources, and the implementation of the schema followed an iterative approach grounded in specific real cases as detailed in the next section.

The CARARE schema's focus is on the detailed description of monuments, digital resources and related events in which the monument has been involved. The Schema follows the structure of MIDAS with the inclusion of elements from the LIDO Resource Set and ESE (Europeana Semantic Elements) to cover the information needed for the digital resources being made accessible to the CARARE and Europeana service environments. Conceptually the top-level themes in the CARARE schema are:

- Heritage asset Identification – the descriptive information and metadata about the monument, historic building, archaeological landscape area, shipwreck, artifact, ecofact etc.
- Digital resource – these are digital representations and sources of information (images, texts, videos, audio, 3D models) about the heritage asset being provided to the service environment (e.g. Europeana). They are often digital representations of monuments or of parts of monuments.
- Activity – these are events that the heritage asset has taken part in, such as Creation, Field investigation, Research and analysis Historical events, etc.
- Collection – this is a collection level description of the data being provided to the service environment.

An XSD has been developed to support the harvesting of metadata from the CARARE content providers to the CARARE aggregator for supply to Europeana. In the CARARE XSD each of these themes are global wrappers which are contained within a CARARE wrapper. An object in the CARARE schema consists of the Heritage Asset Identification (HA) wrapped together with the related Digital Resources (DR), Activities (A) and Collection information (C) (Figure 1). The cardinality of themes and elements has been specified to enable the harvesting of the real data actually present in CARARE content providers' datasets and to meet Europeana's requirements. Some CARARE content providers have collections of historical images where the heritage asset which was photographed is unknown. These collections are interesting for Europeana and for the study of archaeology and archictecture in a region. Since Digital Resources are required for Europeana, the XSD specifies that it is mandatory to include at least one Digital Resource in a CARARE record, while it is highly recommended that each CARARE record should include a single Heritage Asset.

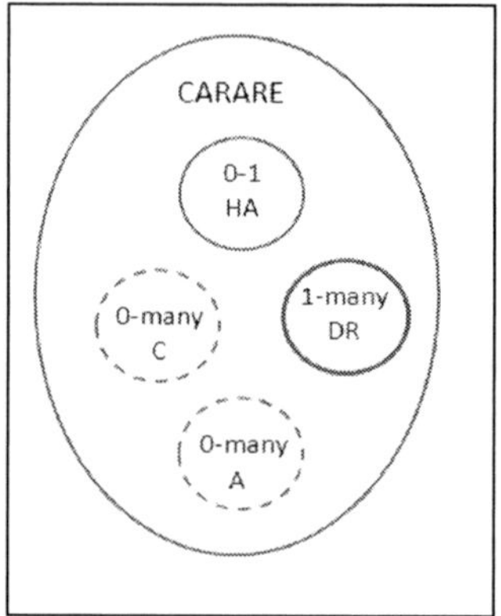


**Fig. 1.** Representation of a CARARE object containing zero or one heritage asset and its related one to many digital resources, zero to many collection descriptions and zero to many activities

For each of the 4 top-level themes the CARARE schema defines a rich set of descriptive information and also administrative metadata. A series of global information types are used globally across the schema to define elements including:

- Appellation – the name (or title) and identification number
- Description – a textual description of the features of the asset
- Spatial – place name data, geo-spatial coordinates and other spatial data.
- Temporal – dating or other information relating to the time span.
- Actors – data about people, organisations or groups and their roles.
- Address – postal address
- Rights – associated with the object or metadata
- Relations – between the top-level themes i.e. heritage asset, activity, digital resource and collection
- Record information – metadata about the record.

In addition to these global elements, the Heritage asset identification element wraps the following additional elements:

**Designations** - information about any designations for a monument or building which provide it with protection in law. Its sub-elements are Protection type, Grade (the level of protection), and related dates.

**Conditions** - information about the condition of a monument or building. Its sub-elements comprise a detailed assessment of the condition of a Heritage Asset, any treatment required and an estimation of the percentage of the monument affected, as well as the date when the assessment of condition was made.

**Characters** - it wraps the main descriptive metadata of the asset and comprises:

- Heritage asset type: classification of the monument, building, landscape feature, artefact or ecofact primarily with respect to its function or use, e.g. house.
- Materials: the basic materials of which a monument is composed, e.g. brick, stone, tile.
- Inscriptions: text inscribed on a monument or building, if any. The element may be repeated using the xml:lang attribute if the element value is available in alternate languages. A preferred/alternate attribute may be used to indicate which value is preferred. The type of inscription may be indicated using an attribute.
- Dimensions: Holds the measurement type (e.g. height, length, width, depth), shape of the heritage asset (e.g. oval) as well as the measurement units (e.g. metres, centimetres), scale and of course the corresponding values.
- Repository location: It identifies the institution with custody of the artifact and possibly the current location.

In addition to the global elements the information about a digital resource includes:

- Format – the file format of the resource, e.g. MIME type.
- Subject – the subject or topic of the resource.
- Publication statement – the name of the publisher, place and date of publication
- Type – The nature or genre of the resource.
- Link– the URL of the resource. A reference to the digital object on the content provider's web site in the best available resolution/quality (i.e. a link to the resource as a text, image, sound, or video file, not to the webpage that contains it). The data given here will allow the automatic generation of a thumbnail by Europeana for its functionality.
- Object – A URL to a thumbnail.
- IsShownAt – A URL to the digital object on the content provider's website in its full information context (i.e. a link to the webpage that contains the digital object and contextual information).
- Resource metadata location – pointer to other information about the resource making the resource available.
- Rights – the rights associated with the digital object itself (copyright, access rights, reproduction rights).

An outline of the CARARE schema together with its implementation in an XSD was published on the project website, and was implemented for testing by project partners in the CARARE metadata mapping and ingestion tool. Following testing some adaptations were made to the implementation of the schema in the XSD, including for example definition of the mandatory and repeating elements, definition of the XML type of the elements (text, date, integer etc).

## 4 Testing and Improvements

The definition of the CARARE schema is the result of an interactive process among metadata experts and content providers. In fact, a collaborative testing phase based on the mappings from the original databases provided by the CARARE Partners and the CARARE Application Profile, led to its progressive modification and improvement. Such methodology ensures a strong applicability of the CARARE schema to the domain chosen for the aggregator, mainly centered on archaeological and architectural sites heritage.

The main mapping test was carried out on contents provided by the Scuola Normale Superiore of Pisa (hereafter SNS), which participates in the project both as content provider and as metadata expert, because it will provide metadata describing 2D images and 3D models related to archaeological monuments, thus offering a complete example. Metadata provided by SNS will describe resources from the following two inter-related collections:

- The Visual Fortune of Pompeii: a digital archive of ancient visual documents representing the monuments of Pompeii [9];
- The 3D model of Pompeii Civil Forum: a multi-resolution survey documenting the present status of the buildings in the Forum of Pompeii, carried out in 2009 by the Politechnique of Milan [10].

In the first case, contents of the Visual Fortune of Pompeii database have been directly mapped to the CARARE schema. In the second case, 3D models are shown through 3D PDF documents and further enriched with explicative texts, then metadata describing both the 3D PDF documents and the survey activity are created ad hoc.

The Visual Fortune of Pompeii database is articulated as follows:

- Iconographic sources: entity describing 2D images (ancient prints, drawings and photographs) representing the monuments of Pompeii;
- Bibliographic sources: entity describing digitized ancient books from which are taken the Iconographic sources;
- Archival sources: entity describing digitized manuscripts and inedited documents from which are taken the Iconographic sources;
- Places: authority file describing each of the monuments of Pompeii;
- Persons: authority file describing physical and moral persons related to the various types of sources with different roles (e.g. invention, creation, publication);
- Subjects: thesaurus of subjects represented in the Iconographic sources.

The above listed entities and authority files are related to each other with various bi-directional relations. In particular, Iconographic sources (2D images) and Places (monuments) are linked with a many-to-many relation, as one monument can be represented in many 2D images, and one 2D image can represent many monuments. In addition, as previously pointed out, metadata will also be created to describe: 3D PDF documents including 3D models and related texts on a given monument; and activities of survey and creation of 3D models representing the monuments.

A modeling issue encountered during the initial testing was how to shape a CARARE record that is which elements to include, how to manage elements cardinality and their reciprocal relations. The initial CARARE schema outline allowed many-to-many relation among monuments (Heritage Assets) and 2D/3D digital resources. This allowed different possible approaches: (a) to include within a CARARE record one monument and the related digital resources; (b) to include only one digital resource and the related monuments; (c) to wrap within one record all of the monuments and all of the related digital resources.

Approaches (a) and (b) cause repetition of information: in the first case information on digital resources can be duplicated in many records, while in the second case there is a duplication of information on monuments. Approach (c) solves the duplication problems, but the risk is to include an entire collection within a single record. After an accurate evaluation, as the CARARE repository is focused on heritage assets and monuments, approach (a) was chosen. This choice resulted in the refining of the CARARE schema and the architecture of the system as follows: duplications will be managed within the repository at the Dissemination Information Package (DIP) of OAIS model level, when packaging contents for Europeana. This also resulted in the decision to include mandatory unique identifiers within each of the CARARE sub-elements, and the rule to represent exclusively direct relations, assuming that inverse relations are always inferred, in order to better manage bi-directional relations. The resulting high-level mapping is resumed in table 1.

**Table 1.** High-level mapping of CARARE to SNS metadata

| **Cardinality** | **SNS contents** | **CARARE elements** |
|---|---|---|
| 1 | 4. Places (monuments) | heritageAsset |
| 0-to-N | 2. Bibliographic sources<br>3.Archival sources | heritageAsset/<br>references |
| 0-to-N | 1. Iconographic sources<br>7. 3D PDF documents | digitalResource |
| 0-to-N | 8. Survey and 3D model creation | Activity |

Physical and moral persons (n. 5) are mapped within various elements in the sub-element <actors>, using different <roles>. The testing also resulted in the addition of attributes to elements in the XSD. For instance, the availability of SNS contents and of other contents in more than one language led to the addition of the "xml:lang" attribute to be added to all text elements to allow alternate language labels to be identified; the main language of the metadata is specified within the record information sub-element <language>. In this way title, description, keywords and other elements are multi-lingual. Another example is the choice to allow for repeatable <title> sub-elements, with an attribute to indicate which of the titles is "preferred". This is due to the fact that different titles and translated titles are often used for cultural digital resources, and monuments, as it came out from the mapping tests.

In conclusion, the testing phase of the CARARE schema was carried out on real examples from content providers, brought to add modifications and improvements to the schema and to draw up practical guidelines for its usage, leading to a stable and consistent version of the XSD, which effectively responds both to the objectives of the project and to the real situations of content providers.

## 5 CARARE Architecture and the MoRe System

Typical Europeana aggregation projects involve the use of a metadata mapping system and an OAI-PMH export system to map native metadata schemas to Europeana Semantic Elements (ESE) schema and recently the Europeana Data Model (EDM) and then expose it through OAI-PMH for harvesting. Despite the fact that this approach is tested and is robust, due to its simple architecture it lacks the added value services that can be provided in this kind of projects. The CARARE architecture (Figure 2) introduces a central repository where all metadata are stored and enriched before being mapped to EDM and provided to Europeana. The concentration of all metadata objects into one system creates added value by allowing the content providers to enrich their objects using semantic relations, measure the quality of their metadata, etc. The use of a central system also facilitates the preservation of the providers' metadata since it holds not only the transformed metadata but also the native.

MoRe, the CARARE repository expands the MOPSEUS [11] functionalities, which is a fedora-commons based repository, fully compliant to OAIS. The repository allows the ingestion of metadata not only through traditional OAI-PMH based methods but also using submission packages that contain complex datastreams and information in order to preserve as much information as possible and also to allow for the implementation of added value services. All information that exists in the repository takes the form of three kinds of packages: submission, archival and dissemination as defined by the OAIS model. The various services use this information and create their own packages, which are then ingested into the repository by creating new versions.

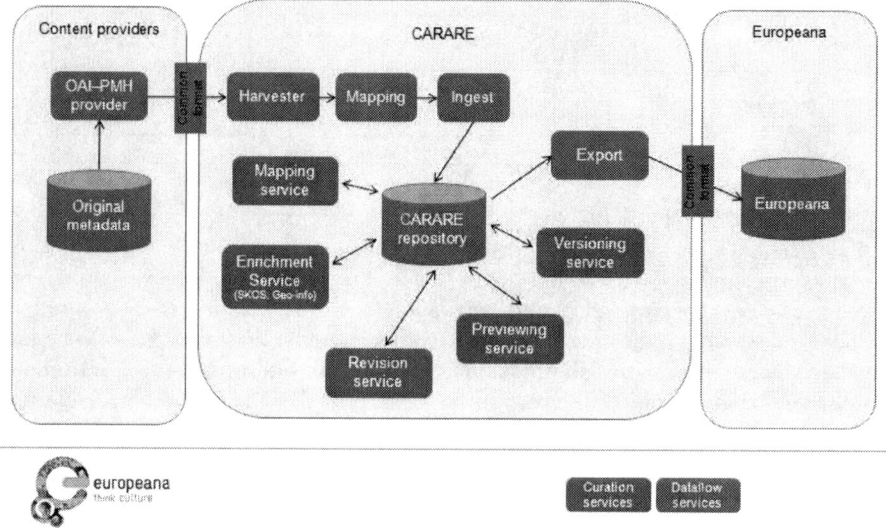


**Fig. 2.** The CARARE system architecture

MoRe provides a series of services for CARARE: geo-information normalization, semantic enrichment, metadata completeness monitoring, etc. The main functionality of MoRe: semantic enrichment, is a process where objects are enriched using semantic relations derived from EDM. The user is provided with the means to quickly locate relevant objects and add semantic relations between them. During the enrichment process, the enriched metadata datastreams are created in order to preserve the whole information that exists inside MoRe. Other services provide the users with the means to quickly locate geographical and temporal information that exists in their objects, and see how this information will be parsed by Europeana services, normalize it if necessary etc. The constant monitoring and reporting of metadata completeness ensures the quality of metadata delivered to Europeana by performing checks that go beyond the traditional mandatory field checks.

# 6 Discussion

CARARE schema builds on a long history of schema development amongst the archaeology/architecture heritage community. It is influenced by early work on the CIDOC core data standard for archaeological sites, by the CIDOC CRM, by work on LIDO and the Europeana Data model. It has information resources about monuments and buildings, and their digital representations at its centre.

Place, or the location of a monument or building, is one of the most frequent starting points for enquiries about the archaeological and architectural heritage and is important for both research and preservation of the historic environment. Monument inventories normally include the details of both named locations and geospatial coordinates to enable sites to be located in the real world. The availability of geospatial data is allowing heritage datasets to be included in geo-portals used in environmental planning and monuments and buildings to be viewed alongside the details of other protected resources. The CARARE metadata schema provides a rich set of spatial data elements covering named places (such as Stonehenge or the temple of Dion) and geospatial coordinates (including points, bounding boxes, height and other aspects). The widely available place name gazetteers typically stop at the level of the names of cities, villages, administrative areas, and the more famous buildings and monuments. This means that CARARE has the potential to develop linked data for place-names and to contribute to the development of geo-parsers used by Europeana to enrich records which include place names in free text but lack spatial coordinates.

Realizing this potential involves addressing an issue posed by the native systems used by the CARARE content providers which is specific to geographic data that is the use of national map reference systems. Integration of the data into an international service such as Europeana requires the conversion of geographic data from the national coordinate systems (such as the OS national grid system in use in the UK) to a world geodetic system (such as WGS84). Converting the coordinates in this way enables them to be plotted on world maps and provides a consistent base for information retrieval. CARARE will offer a service to its content providers to convert their coordinates to WGS84 after data has been ingested to the CARARE repository and before it is provided to Europeana.

The mapping of CARARE schema to Europeana Data Model (EDM) [8] was a complex process since the former is an XML Schema, while the latter is a conceptual model expressed by OWL and RDF Schema. Europeana has recently released an XML Schema for the EDM, for data providers to use when uploading metadata based on it. Our conceptual mapping approach adopted the following rules:

- The CARARE elements are mapped either directly to EDM Classes and ESE elements, or to particular EDM paths. An EDM path is defined as a sequence of Class /subclass /ESE Element -> EDM Property -> Class /subclass /ESE Element.
- We consider that a monument is an instance of the EDM class Physical Thing. A monument is identified by a set of particular characteristics that refer to its nature, location, related events, etc. Thus these characteristics are grouped under the class Physical Thing.
- The information about a Heritage Asset or a digital resource is an instance of the class Europeana Object. This information could be either textual metadata (such as title, etc.), thumbnails and other digital objects representing the monument.

The XML Schema for the EDM which was recently released by Europeana provides the framework for its implementation in the Europeana interface in autumn 2011. The exclusion of the classes ‘Physical Thing’ and ‘Activity’ from this implementation has implications for CARARE. Our conceptual mapping of Heritage Asset Identification information and of Digital resources to the class Europeana Object (or Provided Cultural Heritage Object) provides the framework for data supply to Europeana. While the maintenance of the full metatada (including data about Activities and other elements which cannot be mapped to EDM) on the CARARE MoRe repository will facilitate future developments.

## 7 Conclusions

In this paper we presented the innovative approach introduced by the CARARE Best Practice Network to the aggregation of metadata for architectural and archaeological assets, and related digital resources, in Europeana. The CARARE approach is based on the introduction of MoRe (Monument Repository), an intermediate, OAIS-compliant, metadata aggregator acting as a mediator between heterogeneous primary collections and the Europeana digital library. Primary resources are mapped to CARARE schema, a common metadata schema suitable for the representation of metadata describing heritage assets, digital resources, collections, and events associated with them, thus encompassing adequately the scope of information typically found in relevant primary collections. The MoRe repository manages ingested metadata (including those related to 3D digital resources) and related geographic information in an OAIS-compliant way, providing for the semantic enrichment, long-term preservation and curation [12] of metadata, and for their flexible delivery to emerging new standards within the Europeana ecosystem, such as EDM.

## References

1. Hansen, H.J., Fernie, K.: CARARE: Connecting Archaeology and Architecture to Europeana. In: Ioannides, M., Fellner, D., Georgopoulos, A., Hadjimitsis, D.G. (eds.) EuroMed 2010. LNCS, vol. 6436, pp. 450–462. Springer, Heidelberg (2010), `http://www.springerlink.com/content/978-3-642-16872-7`

2. Cruz, I.F., Xiao, H.: The role of ontologies in data integration. Engineering Intelligent Systems 13(4), 245–252 (2005)
3. Europeana v.01: Europeana Semantic Elements Specifications, ver. 3.2.2 (2010), http://www.version1.europeana.eu/c/document_library/get_file?uuid=c56f82a4-8191-42fa-9379-4d5ff8c4ff75&groupId=10602
4. Data Harvesting and Interchange Working Group: LIDO - Lightweight Information Describing Objects (2009), http://www.cidoc.icom/museum/WG_Data_Harvesting(en)(E1).xml
5. ICOM/CIDOC Documentation Standards Group, CIDOC CRM Special Interest Group: Definition of the CIDOC Conceptual Reference Model (2011), http://cidoc.ics.forth.gr/definition_cidoc.html
6. English Heritage: MIDAS Heritage: the UK historic environment data standard. English Heritage, 2nd edn. (2007), http://www.english-heritage.org.uk/
7. Constantopoulos, P., Bekiari, C., Doerr, M.: POLIS DTD (2005) (in Greek), http://www.ics.forth.gr/CULTUREstandards/paradotea/paradotea_final/K7_odigos_akiniton_perigrafi_2006.pdf
8. Europeana v.01: Definition of the Europeana Data Model elements ver. 5.2.2 (2011), http://version1.europeana.eu/web/europeana-project/technicaldocuments/
9. La Fortuna visiva di Pompei, http://pompei.sns.it
10. Guidi, G., Remondino, F., Russo, M., Rizzi, A., Voltolini, F., Menna, F., Fassi, F., Ercoli, S., Masci, M.E., Benedetti, B.: A multi-resolution methodology for archaeological survey: the Pompeii Forum. In: Proc. VSMM 2008 – Conference on Virtual Systems and MultiMedia dedicated to Digital Heritage. Project papers, pp. 51–59 (2008)
11. Gavrilis, D., Papatheodorou, C., Constantopoulos, P., Angelis, S.: Mopseus – A Digital Library Management System Focused on Preservation. In: Lalmas, M., Jose, J., Rauber, A., Sebastiani, F., Frommholz, I. (eds.) ECDL 2010. LNCS, vol. 6273, pp. 445–448. Springer, Heidelberg (2010)
12. Angelis, S., Benardou, A., Constantopoulos, P., Dallas, C., Gavrilis, D.: A curation-aware repository supporting information-intensive scholarly research. In: 6th Intl. Digital Curation Conference: Participation & Practice: Growing the Curation Community Through the Data Decade (2010)

# Harvesting Metadata for Immovable Monuments from "Local Inventories" and Mapping Them to Standards

Michail Agathos and Sarantos Kapidakis

Laboratory on Digital Libraries and Electronic Publishing,
Department of Archives and Library Science, Ionian University,
Ioanni Theotoki 72, 49100, Corfu
{agathos,sarantos}@ionio.gr

**Abstract.** This work focuses on a harvesting of various metadata elements that exists in informally and heterogeneous schemas for the documentation of immovable monuments used from many Greek public services, working in the field of the built heritage. A series of crosswalks between the metadata collected and schemas of official standards that used widely to describe architectural works, allowed us to identify which official metadata schema, would fit better in these records, in order to reduce syntactic and semantic heterogeneity and in parallel to reveal gaps and shortcomings of these standards in the description of immovable monuments.

**Keywords:** Cultural Heritage, Historic Buildings, Metadata Standards, Immovable Monuments, Switching (Pivot) Schema, Semantic Interoperability.

## 1 Introduction

The role of national inventories in the management of the cultural heritage has long been recognised. They are indispensable, for purposes of identification, protection, interpretation, and physical preservation of movable objects, historic buildings, archaeological sites, and cultural landscapes [1]. A recent survey (Agathos, Kapidakis 2011)[1], involving a large number of distributive "local inventories for immovable monuments in Greece (mostly of the Greek Ministry of Culture and Tourism), indicate that records for these inventories compiled with heterogeneous approaches, using diverse schemas and description methods and in most of the cases with a complete absence of official metadata standards. Majority of the participants, in the survey mentioned above, prepare and use their own schema in order to record, inventory or identify immovable monuments located within the jurisdiction of the Organization, making thus a "local" inventory for "local" use in most of the cases.

In order to exam the metadata elements used in these records we asked from many Greek public services (43), responsible for the built cultural heritage in Greece, to sent us a completed example of their inventory record. Finally 31 different forms[2] were collected, representing 31 different schemas and therefore different "inventory

[1] Available at: http://hdl.handle.net/10760/15847

[2] All the Participants keep in store a total of 900,000 records.

E. García-Barriocanal et al. (Eds.): MTSR 2011, CCIS 240, pp. 424–432, 2011.

systems" [2]. These forms used for the recording of immovable monuments, especially historic buildings such as houses, churches, mansions, industrial buildings etc.

## 2 Constructing the Switching (Pivot) Schema

All the elements from the above schemas clustered into 15 information categories (groups), according to their semantic meaning: Titles, Location, Functional Type - Use, Names and Roles, Dating, Conservation and Treatment History, Physical Condition, Physical Appearance, Building Parts: Materials & Techniques, Measurements, Protection-Legal Status, General Notes, Related References, Illustrative Material: Images-Plans-Sketches and Administrative Information about the record. Categories[3] are inspired from the sections of Core Data Index to Historic Buildings and Monuments of the Architectural Heritage[4] and Categories of CDWA, but for many collected elements a new clustering required, creating thus new categories (e.g. Physical Appearance). With this method a new element set derived (a superstructure), consisting of the above 15 categories and 135 elements[5].

In order to answer the question which official and widely used metadata standard for the description of material of our culture (especially for the build environment), would cover semantically this new schema, demonstrating in parallel deficits of these standards, a crosswalk practice was adopted and a switching mechanism was created.

Currently, crosswalks are by far the most commonly used method to enable interoperability between and among metadata schemas [3]. In this work, because the number of schemas involved was big, the new schema described above, is used as the switching mechanism among each of following individual schemas (targets): MIDAS Heritage, CDWA, CDWA Lite, VRA Core 4.0 and CDI.

The switching (pivot) schema is available at: `http://dlib.ionio.gr/standards/immovable_crosswalks.htm`.

Each of the individual metadata schemas (targets) is mapped to the collected elements only (source). Additionally, attempts were made to map or create crosswalks between equivalent metadata elements (or closely equivalent), as our primary goal was to demonstrate weakness and deficits of the Standards in the description of immovable heritage. The absolute crosswalking[6] was necessary for another reason: as the source schema has a richer structure than many of the target schemas, there was a risk, converting from a rich structure to a simpler structure, many of the target schemas to lose or distort their meaning. The reality is that crosswalks constructed based on real data conversion (as in our case) might be very different from those based on metadata specifications (Zeng and Xiao, 2006).

[3] Categories are slightly expanded (e.g. Building Parts: Materials & Techniques instead of Materials/Techniques (CDWA).

[4] For brevity's sake will be referred as CDI.

[5] The new schema includes all the elements from the 35 collected schemas described above, regardless their repetition frequency or their descriptive value and presented without any adaptation.

[6] The absolute crosswalking approach requires exact mapping between the involved elements of a source schema and a target schema.

## 3 Reviewing the Standards

Since the 1960s, the Council of Europe has worked to protect and enhance the architectural and archaeological heritage, through the exchange of ideas and through developing guidelines and standards. Among their efforts is the design of CDI (1992). The standard defines the core information for documenting historic buildings, archaeological sites and monuments.

The basic aim of the CDI, is to make it possible to classify individual buildings and sites into 9 information groups (sections) [4]. These 9 sections are supported by sub-sections and a set of 45 data fields, some of which are mandatory. The CDI is designed to enable the compiler to make cross-references to the more detailed information about a building, including written descriptions and photographs; associated archaeological and environmental information; details of fixtures, fittings, and machinery installed within individual buildings.

MIDAS Heritage (1998) is a data standard for information about the historic environment which was developed for use in the UK and Ireland and is maintained by the Forum on Information Standards in Heritage [5]. It states what information should be recorded to support effective sharing and long-term preservation of the knowledge of the historic environment. It consists of 9 Themes: the broadest level areas of interest, 16 Information Groups, these set the specific standard for what should be included in an entry covering a particular subject and 138 Units of Information the basic 'facts' or items that make up an entry. 'Monument' information group in MIDAS Heritage usage, among built, buried and underwater heritage of all dates and types, includes buildings (both ruined and in use).

In the late 1990s the Getty Institute and the Art Information Task Force (AITF) developed CDWA (Categories of the Description of Works of Art). The standard created as a need for the art documentation and museum communities for a data structure standard specifically designed for describing unique works of art, architecture, and material culture. CDWA is an extensive set of metadata elements (includes 532 categories and subcategories) and guidelines, which can describe the content of art databases by articulating a conceptual framework for describing and accessing information about works of art, architecture, other material culture, groups and collections of works, and related images.

What was still missing was an "AACR for art objects" [6], and CCO (Cataloguing Cultural Objects) was the response to this need. CCO designed specifically to deal with unique items of art, architecture, and material culture. Is a data content standard specifically for unique museum and special collections-type objects and built works, and a technical format or data interchange standard for expressing and exchanging metadata records about those kinds of works.

The CDWA Lite schema (2006) which corresponds to CCO, is a response to later needs. It is a distillation of the very ample, exhaustive set of elements and sub-elements of CDWA. The purpose of this schema is to describe a format for core records for works of art and material culture, based on the data elements and guidelines contained in the CDWA and CCO. Like VRA Core, CDWA Lite offers an XML format in which to store metadata about works of visual culture in accordance with CCO. CDWA Lite XML schema has a total of twenty-two top-level elements. It is

OAI-harvestable, relatively simple, and much more appropriate for expressing metadata records for art and material culture.

The VRA Core 4.0 XML (2007) is a descriptive metadata standard for the description of culture works (paintings, sculptures, photographs, buildings etc) as well as the images that document them. It consists of nineteen elements and twenty-three subelements.

## 4 Exploring the Missing Metadata Elements from the Standards

Switching schema, allowed us a hierarchical rating according to informative coverage provided: Specifically MIDAS Heritage Standard and CDWA is able to cover much of the collected elements. Comparing these two standards, MIDAS Heritage come short of functionality, as is a set of closely integrated data standards, rather than one single stand alone standard. User communities, who want to design any particular information system or dataset based on MIDAS, have to develop first a shared compliance profile assisting them to develop a standard that meets their needs. VRA Core 4.0 and CDWA Lite, provide almost the same level and method of description for these records, while CDI as will see from the results bellow, must be evolved and changed, trying to keep pace with rapid technology changes in the online environment and staying aware of more global standards initiatives.

From the switching schema came out useful conclusions for the most complex information categories of the source schema:

### 4.1 Titles – Building Names

Titles are an extremely important access point for the resources and are frequently used in brief record displays to assist end users in deciding whether to investigate a resource further. Specifically historic buildings frequently carry many names over the course of their lives [7], for instance, a building may have a name at the time it is being built, another name when completed, a local name and then several names as it moves through successions of ownership. It' s obvious that Building names, which acts as the title of the record, needs to declare what type of title is recorded, as a way to distinguish between the various types of titles including a "type" refinement. As we can see from the switching schema, all the target schemas, except CDI, which has not the potential to declare the type of a building name, could describe adequately all these types of titles, identifying what type of title is recorded.

### 4.2 Location

The "Location" category of the source schema includes elements that describe where the building is currently located. Elements in this category for commonly known non-administrative units, such as "locality" or "exact location", provide a more precise location for the building. These elements are valuable for a building record, especially when the building is located in sparsely populated or poorly mapped areas. In their majority elements of the same category (in other target schemas) in our study, lack of elements that provide space for a free-text explanation and directions of the location of the site or monument (except from MIDAS that provides "Directions" unit of information).

Information's about the location of a building in a record should not allow for uncertainty and ambiguity and must be as detailed and precise as possible. In its simplest form, record must identify address details and reconciling the building with a National Grid map base so that a grid reference can be obtained. As shortcoming spatial coordinates elements of the source schema ("x, y coordinate") have no equivalent in CDWA, CDWA Lite and VRA Core. Equally, important elements of the source schema as "Block Number" and "Cadastral Register Number", which act as official and obligatory identifiers for each building, have no equivalents in these standards (semantic equivalent only to "Cadastral Reference Value" unit of information of MIDAS and "Cadastral Reference - Land Unit" data field of CDI).

Moreover a building record must provide a satisfactory description of the site, in which a building is located: Within this site there may be smaller areas, that have been shaped by physical, administrative or property boundaries, which have distinct and different characters and needs also a description [8] (e.g. a building located in an historic town centre, that in turn is divided into fairly small areas or neighbourhoods). In the source schema recorded elements such as "Area Characteristics" or "surrounding" with no equivalents in most of the target schemas. The inclusion in a building record of such elements could provide a deeper understanding of the historical development and the existence of a building in its wider landscape.

### 4.3 Functional Type-Use

This category in the source schema accommodates elements describing the type of the building and its category, its current and past uses as its occupancy and ownership status. All the targets schemas were able to describe the type (broad functional category) and the category of a building (precise building type defined by function), but its current, past or proposed use, could not be described semantic (especially in CDWA Lite and VRA Core), as there are no exact equivalent elements. An important omission which revealed from the mapping, is that there are no equivalent or comparable metadata elements to describe the Occupancy[7] Status of a building (e.g. if the building is vacant, part occupied, occupied or unknown) or the Ownership Status, which acts as a broad category for ownership (e.g. private, government, municipal, etc). Only MIDAS provides "Occupancy" information unit that is semantic equivalent to the "Occupancy Status" element that recorded in the source schema.

### 4.4 Conservation - Treatment History

The Category "Conservation-Treatment History" of the source schema accommodate elements that records the type of conservation works applied to a monument, narrative descriptions for related procedures applied to that ("Conservation comments" element), as proposals by local authorities related to the conservation of the building ("Approvals", "Restoration Proposals" elements). Regretfully all these elements have no equivalents in CDI, VRA Core and CDWA Lite. Only MIDAS provides semantic

[7] The data fields occupancy and ownership are proposed from the expanded version of CDI.

equivalent units of information (e.g. "Management Proposal Work Proposed" equivalent of "Restoration Proposal" of the source scheme). CDWA provides only a narrative description subcategory that concerns procedures or actions that a building has undergone for repair or conserve, but this is far from semantic for many elements of the source schema.

## 4.5 Physical Condition

This information category provides an assessment of the overall physical condition of the monument as sources of damage for the various parts of a building. CDWA provides "Condition-Examination Description" subcategory for a prose description of the overall physical condition, characteristics, and completeness of a work, equivalent to "Condition Statement" unit of MIDAS which can be used for a detailed assessment of the condition and any treatment required. MIDAS is semantically more precise from all the others in this category, as it has the potential to record from the overall current condition ("Condition Statement" unit) to more specific physical, biological or other source of damage ("Agent of Damage" unit of information, is semantic equivalent to many elements of the source schema in the corresponding category). CDI from the other, provides the repeated data field "General Condition" in order to distinguish between the integrity of the building (demolished, ruined, remodelled, restored) and its state (good, fair, poor, or bad), while CDWA Lite and VRA Core does not provide any element for the description of the condition of the monument.

## 4.6 Physical Appearance

The collected elements in this category deal with description of salient aspects of the physical appearance of a building and its decoration, including morphological elements, the existence of inscriptions and its artistic value. From the mapping revealed that only CDWA provides a semantic equivalent subcategory for the elements of the source schema for a detailed description of the physical appearance of an architectural work ("Physical Appearance" subcategory). In all other target schemas many of the collected elements fall within a type of "description" element, but this is far from semantic.

## 4.7 Building Parts: Materials and Techniques

Many of the recorded elements in this category of the source schema accommodate information both for the substances or materials used in the creation of various parts, as well as the techniques, processes, or methods incorporated in their fabrication. Buildings can be made up of many different parts, each composed of different materials, and made using different techniques. A record must include information's about the materials that was used for the different constituent parts of the building. In MIDAS, CDWA, CDWA Lite, VRA Core the various structural parts of a building described with a similar way (with an *extent* refinement for CDWA and VRA Core, with the "Extent Materials Techniques" sub-element in CDWA Lite and with a "component" unit for MIDAS). This is a shortcoming of the standards, as for the complexity of the various parts may be required more sophisticated and semantic elements. CDI, on the other hand, recommends only two data fields to record the various parts of a building: "Main Materials and Structural Techniques" sub - section, for the main walling material, and "Covering Materials" element to record the main roofing material.

### 4.8 Measurements

The Measurements category contains information about the dimensions and size of the building. Measurements may be recorded according to different criteria, depending upon the type and the part of the work being measured. In VRA CORE and CDWA Lite schema, measurement information recorded with analogous way, qualifying their "Measurements" element and "Measurement Set" subelement respectively, with an "extent" attribute in order to explain the part of the work being measured. The same happens with CDWA. In Midas the various parts of the building that measured could be described adding the Information Unit "component" of the standard. CDI has not equivalent elements for such information.

### 4.9 Protection – Legal Status

This information category of the source schema includes elements for statements on whether the building is protected, giving in parallel the relevant legislation with which the building is protected, and if so, the type of protection, the grade and the date at which it was granted. Moreover accommodate data about the government body whish is responsible for the building as specific elements for the type and delimitation of a buffer zone that serves to provide an additional layer of protection to the monument. The crosswalk for this category showed as that only MIDAS provides semantic equivalent information units for this category.

Specifically Information Units as: "Statutory Name", "Statutory Description", "Protection Type", "Grade", "Protection Start/ End Date" could cover many of the data described above. Moreover MIDAS, it is the only from the target schemas that is able to accommodate information about the government body ("Authorisation Required" unit of MIDAS equivalent to "Protection Body" element of the source schema) which is responsible for the building, as the radius or width of a Buffer Zone around the monument ("Buffer Zone Width" equivalent to "Zone Delimitation" element of the source schema). "Protection/Legal Status" section of CDI, is quite limited to record information's such as if the building is listed on a statutory list and the Grade of Protection (equivalent to "Grade Protection" of the source schema) to show their relative architectural or historic interest, and the date at which this protection was granted.

Description for the legal status and protection of a building in CDWA is limited to "Legal Status" subcategory that allows for general statements as "public property" "scheduled property" "registered property". As a shortcoming there are no equivalent elements in CDWA Lite and VRA Core for the elements of the source schema in this category.

### 4.10 General Notes - Description

In a "note" or "description" type of element, in many metadata standards, it is possible to include very specific details for a work that cannot be described in other elements, including comments, description, or interpretation that gives additional information for the content and context of that work, or qualify information in a number of other particular elements (especially "note").

In a crosswalk "description" and "notes" types of elements should be clearly distinguished because the semantic mapping will be far from accurate. Specifically CDI provides "Historical Summary" data field, (a description type element), that allows for a brief textual summary of the historical development of the building. Furthermore in the expanded version of CDI[8] the historical summary of the building's development can be separated from the descriptive summary. Although in CDWA "Descriptive Note Text: subcategory is equivalent to "Descriptive Note Wrapper" in CDWA Lite, the last can accommodate additional description for physical characteristics for a work, overlapping this way with other elements. "Description" element in VRA gives additional information not recorded in other categories. In MIDAS, "Description" unit of information, acts only as "note" element (like optional subelement "notes" in VRA Core) expand upon the indexing terms used in other units of information using in parallel "Description Type" unit to specialise the nature of the description, by level of detail or intended use.

It is obvious that building records require additional elements-specific[9] free-text notes to explain or qualify information as well as distinctly types of "description" elements, as a way to avoid overlapping of other elements.

### 4.11 Illustrative Documentation: Photographs – Maps - Drawings

Photographs, like drawings, sketches and plans amplify and illuminate an historic building record as is generally the most efficient way of presenting the appearance of a building providing a better understanding for this. Many of the records collected, comes with a surrogate (digital or print) of a photograph, sketch or an extract of map. VRA Core 4 is the only schema from the reviewed above that is uniquely able to capture descriptive information about works and images using the same set of elements and to describe both the building and its image and to indicate relationships between the two. CDWA Lite provides "resourceWrap" element to capture information about the images or other resources that serve as visual surrogates of the work, equivalent CDWA use "Image References" subcategory to identify the images in which the work is depicted. Prototypes of these surrogates, that are considered works in their own right such as drawings, prints, paintings, or photographs considered art, could be described via the "Related Works Wrapper" element of CDWA Lite which is equivalent to "Relation" element of VRA Core 4.0 and equivalent to "Related Work Label/Identification" of CDWA.

MIDAS and CDI could provide illustrative material for a building, only with references to external information held in databases, documentation centres, and elsewhere, enabling the compilers to conceptualise the route from microcosm to macrocosm and allowing the users of the information to make the same connections.

---

[8] Technical Co-operation and Consultancy Programme of the Council of Europe suggests a slightly expanded version of the *CDI*, with additional recommendations for the sections Physical Condition and Notes, as and a new section called Illustrations.

[9] Like MODS, which provides a *type* attribute to identify what type of note is recorded, making these types available to the community.

In reality it is practical for many memory institutions that keeping such records, to record in the same schema, information about the original object and its related image or digital surrogate thus creating a sort of hybrid work- image or work-digital surrogate record, allowing end users to negotiate between the work and a digital surrogate [9].

## 5 Future Work

It is recognized that records of immovable monuments is less examined than records of other material of our culture and must be researched from the aspect of metadata schemas. In many cases, especially with complex objects or hierarchically structured as buildings, a combination of schemas working together may be the best solution, as there is no "one-size-fits-all" metadata schema or controlled vocabulary or data content standard. Specifically a new harmonized profile will reduce the "description injustice" for immovable monuments, introducing new unified concepts and values. In order to achieve this, our study will continue exploring the descriptive needs of immovable monuments and the coverage provided by other standards and schemas, with CIDOC-CRM to play the role of "semantic mediator" between all these standards and new descriptive concepts that will be revealed.

## References

1. Council of Europe: Guidance on inventory and documentation of the cultural heritage. Council of Europe, Strasbourg (2009)
2. Sykes, M.H.: Manual on Systems of Inventorying Immovable Cultural Property. Unipub., Lanham (1984)
3. Chan, L.M., Zeng, M.L.: Metadata interoperability and standardization: A study of methodology Part I+II. D-Lib Magazine 12, 6 (2006)
4. Thomas, R., Bold, J. (eds.): Documenting the Cultural Heritage. J. Paul Getty Trust, Los Angeles (1998), `http://archives.icom/museum/object-id/heritage/index.html`
5. English Heritage: MIDAS a Manual and Data Standard for Monument Inventories, 3rd edn. English Heritage, London (2003)
6. Baca, M.: CCO and CDWA Lite: Complementary Data Content and Data Format Standards for Art and Material Culture Information. VRA Bulletin 34(1), 69–75 (2007)
7. TOBuilt, A Database in Toronto - Data values and standards, Canada, `http://www.tobuilt.ca/DataStandards.html`
8. Jones, D. (ed.): Understanding Place: Historic Area Assessments: Principles and Practice. English Heritage Publishing, London (2010)
9. Gill, T., Gilliland, J., Whalen, M., Woodley, M., Baca, M. (eds.): Introduction to Metadata, 3d Version. Getty Publications, Los Angeles (2008)

# Agriculture-Related Concepts in Non-agricultural and General Thesauri

Tomaz Bartol

University of Ljubljana,
Department of Agronomy, Biotechnical Faculty,
Jamnikarjeva 101, 1001 Ljubljana, Slovenia
tomaz.bartol@bf.uni-lj.si

**Abstract.** We identified general indexing concepts, based on descriptors (preferred-terms) which contain terms 'agriculture' and 'agricultural', in several non-agricultural and general information systems/databases, and respective thesauri (controlled vocabularies), covering fields of civil, mechanical, chemical engineering, physics, psychology, medicine, biomedicine, education, business, economics, finance, library science, sociology, social and political sciences and related disciplines (for example, CSA-Illumina ERIC (Education Resources Information Center), LISA (Library and Information Science Abstracts), Sociological Abstracts (Sociological Indexing Terms), Ebsco Academic Search Complete (Subject Terms), Medline (MeSH), Political Science Complete, Ei Engineering Village Compendex and Inspec, PsycINFO (Thesaurus of Psychological Index Terms), ABI/Inform Global (ProQuest Thesaurus). We compared characteristics, strengths and limitations of thesauri in each respective system with regard to searching (word-, phrase-indexing, truncation (wildcard), stemming, autostemming, hierarchical structure (tree-structures) and relations among preferred-, non-preferred terms, Narrower, Broader and Related Terms. We assessed database coverage of general agriculture-related topics based on retrieval with these terms.

**Keywords:** Agriculture, thesauri, controlled vocabularies, indexing, subject headings, preferred terms, databases.

## 1 Introduction

The scientific field of agriculture has been for almost four decades served by three major general agricultural information systems, Agris (FAO), Agricola (NAL-National Agricultural Library of the USDA), and CAB Abstracts (CABI), which offer quality indexing, based on thesauri (subject headings). There also exists a smaller database FSTA (Food Science and Technology Abstracts) specialized in food- and nutrition. In addition, important information on food, nutrition, medicinal plants and animal health can be found in Medline (or associated PubMed), compiled by the National Library of Medicine (NLM). In the more recent period, the three agricultural databases have been collecting an increasing amount of information that is unique to each system. CAB Abstracts is considered as the most comprehensive database with a very complex thesaurus (CAB Thesaurus). Its access, however, is limited only to

E. García-Barriocanal et al. (Eds.): MTSR 2011, CCIS 240, pp. 433–444, 2011.

subscribing institutions. Agricola (NAL Thesaurus) is free and has specialized in coverage of documents compiled by the NAL, offering access to many full-text documents. Access to Agris is also free. It compiles and indexes many non-conventional full-text publications which may not be indexed by other databases. Its multilingual Agrovoc thesaurus has become an important system for organization of knowledge in open archive networks [1]. Some of these agricultural vocabularies are being remodeled as ontologies [2], [3]. Besides standard bibliographic information systems, agricultural information is increasingly organized in other types of databases, for example learning depositories for educational purposes [4].

The thesauri and the above databases have frequently been assessed in articles with regard to the coverage of specific aspects, for example animal welfare [5] or food and nutrition [6]. But agricultural sciences are a very broad area, and can also involve economical aspects (agricultural economics), social aspects (rural sociology), mechanical engineering (agricultural machinery), education (extension services), and more. Agriculture-related information is thus expected to be scattered across many different disciplines, and respective information services.

A study that briefly mentioned agriculture in contexts other than agriculture, on the example of Humanities Index, Academic Search Elite (EBSCO) and Periodicals Research II (Micromedia ProQuest), found that although there were only minor variations in some broader headings, there is frequently little consistency between the terms that occur in the titles or abstracts, and the subject headings [7]. An extensive bibliography on animal health compiled 900 citations from the three general agricultural databases, as well as EBSCO Academic Search Complete, LISA, PsychInfo, databases [8], which were also used in our study. Author-address or affiliation was assessed on an example of Web of Science, Biosis Previews, CAB Abstracts, Chemical Abstracts, Compendex/Inspec, Francis, Medline, Pascal, and Sociological Abstracts databases [9]. CAB thesaurus was addressed along with INSPEC and LISA thesauri with regard to comparison between database subject headings (descriptors) and author keywords [10]. The coverage of LISA, Medline, Compendex, and other databases was investigated, although with regard to issues other than agriculture [11]. Some biomedicine-specific issues have been investigated on the example of Inspec thesaurus [12]. The Sociological Indexing Terms have been compared with biomedical MeSH, and Thesaurus of Psychological Index Terms [13]. The accessibility of a particular database and utility of a thesaurus, through different hosts, was addressed on the example of Inspec, available through Ebsco, Engineering Village, and the Institute of Information Science (ISI) [14]. Records retrieved from the same database can sometimes vary, depending on the interface, what was shown on the examples of Dialog DataStar; EBSCOhost and OVID, so it is advisable to use truncation in simple searches [15]. However, in a user-study on Compendex (Ei Thesaurus) only one student paid attention to truncation [16]. But users' comprehension of thesauri can be limited what was shown on the example of the ProQuest Controlled Vocabulary [17].

Only a few authors have compared agricultural and non-agricultural databases and thesauri, especially those outside the scope of life-sciences. We thus wish to investigate also some other bibliographic systems, for example in the fields of social sciences, civil engineering, physics, etc. which also possess hierarchical thesauri, in order to assess the structure of major subject headings, on the example of terms *agriculture* and *agricultural*. We expect these terms to be also used in these non-agricultural information systems. The context needs to be investigated. The structure may be quite different in each respective system. Retrieval based on these terms may reflect some specifics which we also wish to assess. It is probable that end-users are not aware of such particularities. Our research suggests some techniques for enhanced information retrieval, encouraging end-users to also use other information systems, not necessarily specialized in agriculture, for better knowledge discovery.

# 2 Materials and Methods

## 2.1 Outline of Information Systems

It is getting increasingly difficult to obtain up-to-date information on the total numbers of records in databases and information systems because the records include a growing number of non-scholarly publications. Official fact-sheets sometimes omit the information on numbers of records and prefer to offer only the information on coverage of journal titles and other publications. Many of the respective information systems specialize in providing access to full-text journals, including newspapers, general magazines and wire feeds. For example, in the database ABI/Inform Global (ProQuest) more than 2.200.000 records are derived from the 'Wall Street Journal' alone (with backfile coverage to 1984).

In Table 1 we present original names for subject headings as are called in respective databases along with search syntax. on the example of truncated term *agricultur**. Field-name prefixes are case sensitive in all databases hosted by Ebsco. In the ensuing short presentation of databases we provide some concise information on databases and respective thesauri. Some databases are compiled by professional associations, agencies etc. (for example, ERIC, PsycINFO, Medline), and some are a product of providers which frequently host also other databases (for example, Ebsco, Proquest).

**Table 1.** Original names for subject headings in respective database, and search syntax

| *Database* | *Heading Name* | *Search Syntax* |
|---|---|---|
| C-Eric | Descriptors | de=agricultur* |
| C-Lisa | Descriptors | de=agricultur* |
| C-SocAb | Descriptors | de=agricultur* |
| Eb-ASC | subject terms | SU agricultur* |
| Eb-Medl | word in subject heading | MW agricultur* |
| Eb-PSC | subject terms | SU agricultur* |
| Ei-Comp | Ei controlled term | agricultur* wn cv |
| Ei-Insp | Inspec controlled term | agricultur* wn cv |
| O-PsyInf | Heading Word | agricultural.hw |
| P-ABInf | Subject headings | SU(agriculture) |

Some databases can be accessible through several different hosts. The abbreviations, preceding the host names, will be, for the purpose of conciseness, used throughout the text, and in the tables and figures. We will used the expressions 'preferred terms' (descriptors) and 'non-preferred terms' (non-descriptors, entry terms) to indicate the type of a heading.

## Hosts, Databases and Thesauri under Study

***C-Eric*** - *CSA Illumina ERIC - ERIC Thesaurus*
Education, schools and teaching. Journals (1085 titles) + many non-journal resources (books, educational reports); 1,4 million records (330,000 full-text) (1966-2011).
*Thesaurus:* 6,000 preferred, 4,500 non-preferred terms

***C-Lisa*** - *CSA Illumina -LISA - LISA Thesaurus*
Library and information science. Mostly journals (440 titles); 350,000 records (1969-2011). *Thesaurus*: more than 6000 total terms (exact data are not available).

***C-SocAb*** - *CSA Illumina - Sociol. Abstracts - Sociological Indexing Terms (SIT)*
Sociology, social and behavioral science. Journals (1,800 titles), books, book chapters, dissertations, proceedings; 1 million records (1952-2011).
*Thesaurus:* (SIT):4088 preferred (1456 top terms:), 2739 non-preferred terms

***Eb-ASC*** - *Ebsco-Academic Search Complete - Subject Terms (ST)*
General scientific database - emphasis on life sciences, also social sciences and humanities. Journals (13,200 titles, incl. 8,600 full-text), proceedings, books, reports (1887-2011) / *Thesaurus* (ST): 196,000 preferred and 204,000 non-preferred terms

***Eb-Medl*** - *Ebsco - Medline - MeSH (Medical Subject Headings )*
All fields of biomedicine (incl. veterinary science). Mostly journals (4,800 titles), 18 million records (1946 to 2011).
*MeSH Thesaurus*: 26,142 preferred, 177,000 non-preferred (entry) terms.

***Eb-PSC*** - *Ebsco - Political Science Complete - Political Science Thesaurus*
Political sciences, humanitarian issues, law and legislation. Journals (2,900 titles, including 530 full text), proceedings, books (340 full-text reference books), (1895-2011). / *Thesaurus*: 7,366 preferred, 10,175 non-preferred terms

***Ei-Comp*** - *Ei (Engineering Village) - Compendex - Ei Thesaurus*
Chemical engineering, civil eng., electrical eng., mechanical eng., mining eng., etc. Journals and trade magazines (5,600 titles), proceeding; 12 million records (1970-2011) + 1.7 million abstract records (1884-1969).
*Thesaurus*: 10,200 preferred, 9,420 non-preferred terms

***Ei-Insp*** - *Ei (Engineering Village) - Inspec - Inspec thesaurus*
Physics, electrical engineering & electronics, computers & control, information technology, mechanical & production engineering, etc. Journals (4,000 titles), proceedings, books, reports and dissertations; 11 million records (1969-2011).
*Thesaurus*: 9,573 preferred, 8,826 non-preferred terms

***O-PsyInf*** - *OVID - PsycINFO - Thesaurus of Psychological Index Terms*
Psychology and related disciplines. Journals (2,460 titles); books and book chapters (11% of database); 3 million records (1880-2011, including records dated as far back as 17th Century) / *Thesaurus*: 5,613 preferred, 2,609 non preferred terms

***P-ABInf*** - *ProQuest - ABI/Inform Global - ProQuest Thesaurus*
Business, economics, management, marketing , etc.. Documents: newspapers, trade journals, scholarly journals, magazines (3,500 titles, incl. 2,500 full text), wire feeds, reports, dissertations & theses, proceedings (1923-2011).
*Thesaurus*:11,000 preferred, 5,600 non-preferred terms

## 2.2 Characteristics of Thesauri

In the following overview some more important characteristics of thesauri are identified, prior to the subsequent analysis. Thesauri which are available through different platforms exhibit some common features in each platform. The most important distinction can be attributed to the differences in the so called *phrase-* and *word-indexing*. In the case of phrase-indexed search mode, the term *agricultural economics* will for example not be retrieved, neither with the term *agricultural* nor *economics*. In the ensuing analysis we always employed the utility of 'word-indexing' in order to harmonize comparison of search results. Word-indexing comes about under different names, such as 'Rotated Index', 'Term Contains' or 'Permuted index'. In some thesauri such a permuted index will show preferred-terms only, but in other thesauri it will display also non-preferred terms. These differences are explained. These principles may differ among platforms what may yield quite different search results, especially if ignored by end-users which are frequently not aware of such differences. Not all thesauri offer an automated possibility to 'explode' a term in retrieval. The acronyms UF, BT, NT, RT will be used throughout the text for the concepts of Used For, Broader Term, Narrower Term, Related Term.

*CSA Illumina thesauri*

1. *Alphabetical List* - list of terms with no indication of relationships; all preferred and non-preferred terms (which begin with particular letters) will be listed.
2. *Hierarchy* - list of preferred terms with relationships. Complete terms (subject headings) need to be used. Term *agriculture* will retrieve only the heading *agriculture. Alternative agriculture* will not be retrieved. *Agricultural* will not retrieve anything. A complete heading must be entered, such as *agricultural economics.*
3. *Rotated Index* - list of all *preferred* and *non-preferred* terms that contain a particular word, but not if occurring in BTs or NTs of the term. RTs are also shown bellow a heading term. Both *alternative* and *agriculture* will retrieve *alternative agriculture.* The symbol [+] after a term suggests that this contains further narrower terms.
   *Explode* - this function includes all narrower terms in database searches.

*Ebsco thesauri*
Ebsco uses a complex system of descriptors which can come about as major headings accompanied with specific qualifiers or subheadings. Dozens of agriculture-related individual descriptors exist, such as *agriculture & energy* (BT: *power resources*), *agriculture-Social Aspects* (BT: *social change*), which are not linked to *agriculture* neither with a BT nor a RT.

1. *Term Begins With* - alphabetic list of preferred and non-preferred terms.
2. *Term Contains* - all preferred and non-preferred terms will be listed, including those terms where *agriculture* appears in annotation/scope note, BTs, UFs (But not if it appears in RTs or NTs of the heading term).

3. *Relevancy Ranked* - similar function to 'Term Contains', but will also retrieve some other terms which are related to the term *agriculture* (e.g. *agricultural*).

*Explode* function includes all narrower terms in database searches. Ebsco MeSH has some additional utilities: symbol [+] before a term implies further NTs. It is also possible to limit searches to a heading as a major topic.

*Ei - Engineering Village thesauri*

1. *Search* - a list of preferred terms, also those where *agriculture* appears in BTs, NTs, RTs and UFs (UFs are italicized).
2. *Exact Term* - only exact phrase, e.g. *agricultural products*, but not *agricultural.*
3. *Browse* -alphabetical list of all preferred and non-preferred terms.

In order to explode-search, the narrower terms (1. level) need to be appended by manual selection of terms in the thesaurus.

*OVID thesauri*

1. *Thesaurus* - alphabetic list of all preferred and non-preferred terms. *Workers* will not return *agricultural workers* but will return '*personnel* Used For *Workers'*.
2. *Permuted index* - list (word-indexed) of all preferred and non-preferred terms.
3. *Scope Note* - phrase indexed - will return only those exact heading or non-de which have a scope note. Workers will retrieve preferred term *personnel*, but not *agricultural workers*.

*Explode* - this function includes all narrower terms in database searches.

*ABI/Inform Global (ProQuest) thesauri*

ProQuest products use ProQuest Thesaurus; other databases-specific thesauri are available through ProQuest platform, such as MeSH.

1. *Contains word(s)* - permuted list of preferred and non-preferred terms, where "agriculture" is a part of a descriptor.
2. *Begins with* - alphabetic list of preferred and non-preferred terms.

In order to explode-search, the narrower terms (1. level) need to be appended by manual selection of terms in the thesaurus.

## 3 Results and Discussion

### 3.1 Major Subject Heading *Agriculture* - Tree Structures

In Figure 1, and the following Table 2, we present the number of headings and simplified tree structures for the principal heading (preferred term) *agriculture*. There is no such heading in O-PsyInf. Also, in Eb-PSC there is no independent heading *agriculture*-only. However, in this thesaurus there are 8 multi-word headings containing *agriculture*, many supplied with qualifiers.

**Table 2.** Tree Structures for the heading *Agriculture* and number of terms in respective thesauri

| |
|---|
| ***C-Eric*** |
| UF (11): agcl safety, agcl science, agcl sciences, agcl supplies, agcl trends, agriscience, farm supplies, feed industry, feed stores, fertilizers, livestock feed stores BT1 (1): technology<br>NT1 (4): agronomy, animal husbandry, gardening, horticulture<br>RT (19): agribusiness, agcl colleges, agcl education, agcl engineering … natural resources, ornithology, rural sociology, seasonal employment, veterinary medicine |
| ***C-Lisa*** |
| UF (1): farming / BT1 (1): food industry<br>NT1 (6): agcl economics, agcl engineering, alternative agriculture, animal husbandry, horticulture [1], tropical agriculture / RT (1): farmers |
| ***C-SocAb*** |
| UF (2): agronomy/agronomists, farming / BT (0)<br>NT1 (2): animal husbandry, part time farming<br>RT (24): agrarian societies, agrarian structures , agribusiness, agcl development , agcl economics … industry , land use, plants (botanical), rural areas, soil conservation |
| ***Eb-ASC*** |
| UF (2): farming, husbandry / BT1 (2): industrial arts, life sciences<br>NT1 (106): acclimatization (plants), aerial photography in agriculture, aeronautics in agriculture, agcl ability, agcl chemistry … volunteer workers in agriculture, water in agriculture, wetland agriculture, women in agriculture, zinc in agriculture<br>RT (24): agrarian societies, agcl colleges, agcl education, agcl exhibitions, agcl extension work … land use, rural, physiocrats, rural industries, 'sociology, rural' |
| ***Eb-Medl*** |
| UF (2): agcl development, agcl workers / BT1 (1): 'technology, industry, and agriculture'<br>NT1 (9):agcl irrigation, animal husbandry, aquaculture [1], beekeeping, dairying, gardening, hydroponics, organic agriculture, weed control / RT (0) |
| ***Ei-Comp*** |
| UF (2): agcl applications, limestone--agcl applications / BT1 (1): industry<br>NT1 (8): agcl products [11], agronomy, crops, cultivation, farms [3], forestry [3], harvesting, irrigation [2]<br>RT (14): agcl chemicals, agcl engineering, agcl machinery, agcl runoff, animals … nitrogen fertilizers, orchards, rural areas, soil conservation, veterinary medicine |
| ***Ei-Insp*** |
| UF (0) / BT1 (1): farming / NT1 (1): irrigation<br>RT (19): agcl engineering , agcl machinery , agcl pollution , agcl products , agcl safety agrochemicals … natural resources, organic farming, pest control, soil, vegetation mapping |
| ***P-ABInf*** |
| UF (0) / BT (0)<br>NT (17): agribusiness , agcl banking, agcl biotechnology , agcl checkoffs , agcl economics … planting , selective breeding , sustainable agr , tillage , urban farming<br>RT (21):agcl commodities, agcl education, agcl engineering, agcl lending, agcl management … irrigation, organic farming, pastures, plantations, soil fertility |

The heading *agriculture* is arranged in quite different tree structures in different thesauri, and has considerably different BTs, NTs, RTs, and UFs. C-Eric has as many as 11 UFs. Some thesauri have no UFs (Ei-Insp, P-ABInf). BTs also differ among thesauri, ranging from *technology* to *food industry* and *industrial arts*. The greatest difference is exhibited in NTs. Ei-Insp has only one NT, Eb-ASC has as many as 106 NTs (in Figure 1 we limited the X-axis because no other heading exceeded 24 terms). Eb-Medl has no RTs, but Eb-ASC and C-SocAb both have 24 RTs. Interestingly, among the 24 RTs, these two thesauri share only 2 RTs: *agrarian societies* and *agricultural technology*. Table 2 presents only the first and the last five terms if there are more than 10 different headings in a thesaurus. The terms *agriculture-* (agr) and *agricultural* (agcl) are abbreviated. There is at least one similar term - *farming* - which could also have been examined, but it was omitted at this point, also because other, non-agricultural concepts can be associated with it, for example *wind farming*.

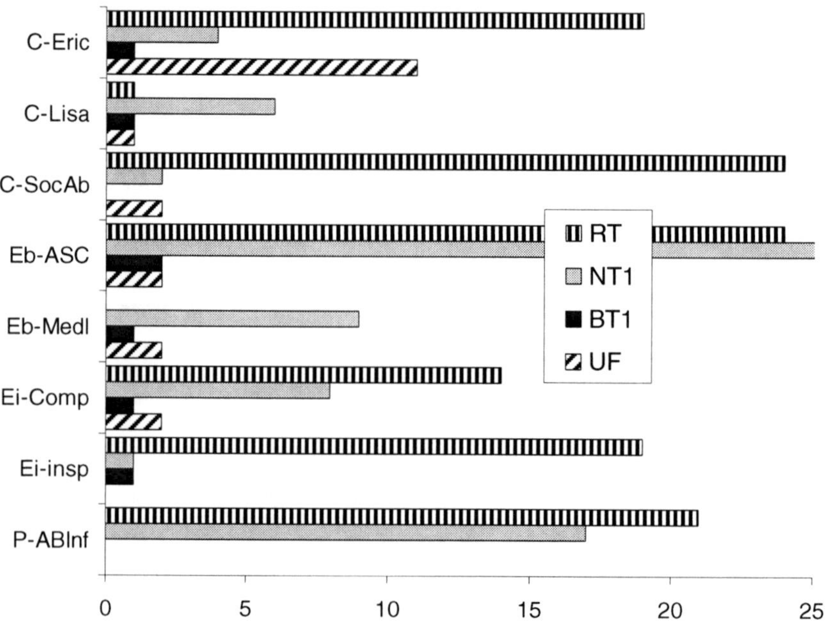


**Fig. 1.** Number of non-descriptors (UF), broader terms (BT), narrower terms (NT), related terms (BT) in the respective database thesauri for the subject heading *Agriculture*

## 3.2 Subject Headings Containing *Agriculture* and *Agricultural*

In the following list (Table 3) we present all headings (preferred-terms) which contain terms *agriculture* (agr) and *agricultural* (agcl) in respective thesauri. Only the first and the last five headings are listed if there are more than 10 different headings. Thesauri vary quite significantly also with regard to the purpose of these two terms. In C-Eric, C-SocAb, Ei-Comp, Ei-Insp there is only one term containing *agriculture*, but in Eb-ASC there are as many as 121.

**Table 3.** Subject headings containing *Agriculture* and *Agricultural*

| |
|---|
| ***C-Eric*** |
| agr (1): agr<br>agcl (8): agcl colleges, agcl education, agcl engineering, agcl laborers, agcl machinery, agcl occupations, agcl production, agcl skills |
| ***C-Lisa*** |
| agr (6): agr, agr libraries, alternative agr, tropical agr, un food and agr organization, united states department of agr<br>agcl (4): agcl economics, agcl engineering, national agcl library, national agcl library |
| ***C-SocAb*** |
| agr (1): agr<br>agcl (14): agcl collectives, agcl development, agcl economics, agcl enterprises, agcl mechanization, agcl policy, agcl production, agcl technology, agcl workers |
| ***Eb-ASC*** |
| agr (121 occ): agr, agr & state, sustainable agr, agr & politics, agr & civilization … sulfur in agr, trace elements in agr, volunteer workers in agr, water in agr, zinc in agr<br>agcl (213): african american agcl laborers, agcl ability, agcl administration, agcl administrators … 'water-supply, agcl', women agcl economists, women agcl engineers, women agcl laborers, women agcl students |
| ***Eb-Medl*** |
| agr (4): agr, organic agr, 'technology, industry, and agr', united states department of agr<br>agcl (4): agcl irrigation, agcl workers' diseases, 'chemistry, agcl', 'crops, agcl' |
| ***Eb-PSC*** |
| agr (8): agr & state, agr & politics, agr & energy, collectivization of agr, communism & agr, national socialism & agr, agr & state-environmental aspects, agr-international cooperation<br>agcl (22): colonial agcl laws, agcl subsidies, migrant agcl laborers, agcl extension work-gov. policy … agcl price supports-law & legisl., agcl prices-gov. policy, agcl prices-gov. policy-simulation games, agcl resources-law & legisl., agcl subsidies-law & legisl. |
| ***Ei-Comp*** |
| agr (1): agr<br>agcl (7): agcl chemicals, agcl engineering, agcl implements, agcl machinery, agcl products, agcl runoff, agcl wastes |
| ***Ei-Insp*** |
| agr (1): agr<br>agcl (5): agcl engineer., agcl machinery, agcl pollut., agcl products, agcl safety |
| ***O-PsyInf*** |
| agr (0)<br>agcl (2): agcl workers, agcl extension workers |
| ***P-ABInf*** |
| agr (4): agr, agr teachers, community supported agr, sustainable agr<br>agcl (16): agcl banking, agcl biotechnology, agcl checkoffs, agcl chemicals, agcl commodities … agcl management, agcl policy, agcl pollution, agcl production, agcl subsidies |

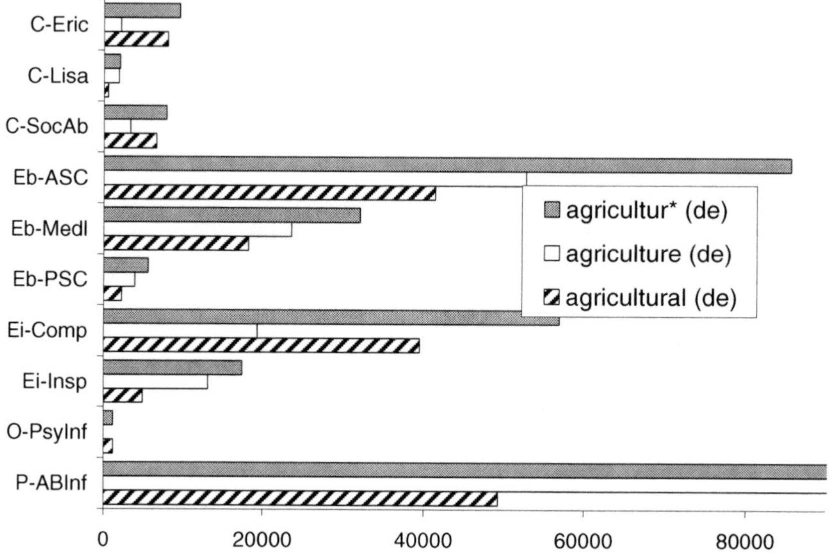


**Fig. 2.** Occurrence of the term *'agriculture'*, *'agricultural'* and *'agricultur*'* (truncated) in the subject heading fields in the respective databases

No O-PsyInf headings contain *agriculture*. Only two contain *agricultural*. But there are as many 213 *agricultural* terms in Eb-ASC. Some headings are deeply specialized, such as *women agricultural engineers* or *women agricultural laborers*. This is reflected in a very high number (196,000) preferred terms in Eb-ASC thesaurus. Eb-PSC also contains very specialized headings, such as *'agricultural extension work-government policy'*. Figure 2 presents records retrievable with *agriculture* and *agricultural*, occurring anywhere in the heading. We employed advanced-search syntax (shown in Table 1). Additionally, we compared both *agriculture/agricultural* with retrieval based on the truncated stem *agricultur**. P-ABInf needs explanation. Almost three times more records get retrieved with *agriculture* than *agricultural*. It is not possible to retrieve records based on strict subject-headings. Search for *agriculture* retrieves not only headings but also all records classified with the *Classification Code 8400: Agriculture industry* (more than 100.000 records). *Agriculture* thus retrieves 142.000 records, and *agricultural* 49.000 in P-ABInf. In Eb-ASC, Eb-Medl, Ei-Insp, and P-ABInf more records are retrieved with headings containing *agriculture*, whereas in C-Eric, C-SocAb, Ei-Comp and O-PsyInf more records are retrieved with *agricultural*. These two concepts are very similar so truncation could be employed. *Agriculture* could quite easily be substituted with *agricultural* in many headings, such as *agriculture libraries*. Truncation will not diminish precision, and recall will be better. Ei databases and thesauri (Inspec and Compendex) need further comment. Search employs default automatic stemming (autostemming), also when subject headings are used as a search criterion, so there is no difference in retrieval with either *agriculture* or *agricultural*. Exact numbers of documents, indexed with either term, can only be ascertained by switching off autostemming.

## 4 Conclusions

Our research addressed some non-agricultural information systems and databases, and respective thesauri, especially those covering topics of business, economy, social sciences, engineering etc., which may offer some important information with regard to, for example, agricultural buildings and machinery, agricultural economics, rural sociology, agricultural extension services, etc. These systems may frequently be overlooked by end-users seeking agricultural information. Because of the many databases included in the study we investigated only the most general terms *agriculture* and *agricultural* which, however, have restricted connotations, thus facilitating a uniform comparison among thesauri. These terms are contained in most non-agricultural thesauri.

There exist quite significant differences among databases in retrieval of documents based on thesaurus-derived subject headings, not only among different databases which are field-specific anyway, but also in the same database available through different hosts, what was also noticed by other authors [15]. The usage varies both in number of terms as well as the structure, and employment of respective non-preferred, broader, narrower or related terms. Some thesauri include as many as 11 non-preferred terms, some other offer none. The number of related terms may range between none to 24, and the number of narrower terms between 1 (one) to as many as 106. Some multiple-word subject headings can be very complex and precise, and can include as many as six words. But multi-word phrases are prone to variation [18]. In many such phrases, both terms, *agriculture* or *agricultural*, could be used so it is preferable to employ truncated (*agricult**) searching, based on the utility of 'Term Contains' or 'Permuted'. Phrase-indexing principles will in some systems reduce results considerably because a term will not be retrieved if it is a part of a phrase. On the other hand, the default stemming will in some systems return many more results than expected. It is thus important to carefully choose the most appropriate heading as well as the most appropriate technique of retrieval. End-users need to be aware of these differences among thesauri, databases and providers, if they wish to optimize retrieval and interpret search results correctly.

## References


1. Subirats, I., Onyancha, I., Salokhe, G., Keizer, J.: Towards an architecture for open archive networks in agricultural sciences and technology. Online Information Review 32(4), 478–487 (2008)
2. Sicilia, M.A.: Linking learning technology with agricultural knowledge organization systems. In: 4th International Conference on Information and Communication Technologies in Bio and Earth Sciences (HAICTA 2008) AgroLt Workshop (2008)
3. Dagobert, S., Lauser, B., Liang, A., Fisseha, F., Keizer, J., Katz, S.: Reengineering Thesauri for New Applications: the AGROVOC Example. Journal of Digital Information 4(4) (2004)
4. Manouselis, N., Najjar, J., Kastrantas, K., Salokhe, G., Stracke, C.M., Duval, E.: Metadata interoperability in agricultural learning repositories: An analysis. Computers and Electronics in Agriculture 70(2), 302–320 (2010)

5. Grune, B., Fallon, M., Howard, C., Hudson, V., Kulpa-Eddy, J., Larson, J., Leary, S., Roi, A., van der Valk, J., Wood, M., Dörendahl, A., Köhler-Hahn, D., Box, R., Spielmann, H.: Report and recommendations of the international workshop Retrieval approaches for information on alternative methods to animal experiments. Altex 21(3), 115–127 (2004)
6. Bartol, T.: Assessment of food and nutrition related descriptors in agricultural and biomedical thesauri. In: Sartori, F., Sicilia, M.Á., Manouselis, N. (eds.) MTSR 2009. CCIS, vol. 46, pp. 294–305. Springer, Heidelberg (2009)
7. DeLong, L.: Subscribing to databases: how important is depth and quality of indexing? Acquisitions Librarian 19(37/38), 99–106 (2006)
8. Croft, V.F.: Animal Health Libraries, Librarians, and Librarianship: A Bibliography (2009), http://hdl.handle.net/2376/1469
9. Bartol, T., Hocevar, M.: The capital cities in the ten new European Union countries in selected bibliographic databases. Scientometrics 65(2), 173–187 (2005)
10. Gil-Leiva, I., Alonso-Arroyo, A.: Keywords given by authors of scientific articles in database descriptors. Journal of the American Society for Information Science and Technology 58(8), 1175–1187 (2007)
11. Jacso, P.: Content Evaluation of Databases. Annual Review of Information Science and Technology 32, 231–267 (1997)
12. Morris, T.: Visualizing the structure of medical informatics using term co-occurrence analysis: II. INSPEC perspective. In: Proceedings of the 64th ASIST Annual Meeting, vol. 38, pp. 489–497 (2001)
13. Carolyn, G., Weaver, C.G.: Gerontology and geriatrics: a multidisciplinary approach to indexing. In: Hornyak, B. (ed.) Indexing Specialties: Psychology, pp. 41–48. Information Today & American Society of Indexers, Medford (2002)
14. Salisbury, L., Gupta, U.: A Comparative Review of INSPEC on EBSCOHost, Engineering Village (EV2), and Institute for Scientific Information (ISI). Charleston Advisor 6(1), 5–11 (2004)
15. Younger, P., Boddy, K.: When is a search not a search? A comparison of searching the amed complementary health database via EBSCOhost, OVID and DIALOG. Health Information & Libraries Journal 26(2), 126–135 (2009)
16. Anghelescu, H.G.B., Yuan, X., Zhang, X.: Domain knowledge, search behaviour, and search effectiveness of engineering and science students: an exploratory study. Information Research 10(2) (2005)
17. Greenberg, J.: User comprehension and searching with information retrieval thesauri. Cataloging and Classification Quarterly 37(3/4), 103–120 (2004)
18. Savary, A., Jacquemin, C.: Reducing Information Variation in Text. In: Renals, S., Grefenstette, G. (eds.) Text- and Speech-Triggered Information Access. LNCS (LNAI), vol. 2705, pp. 145–181. Springer, Heidelberg (2003)

# Intellectual Property Rights in Environmental and Natural History Collections: A Preliminary Discussion

Effie Tsiflidou[1,2], Alexios Dimitropoulos[1,2], Zoi A. Makrodimitri[1,3], and Nikos Palavitsinis[1,2]

[1] Greek Research & Technology Network (GRNET), Greece
[2] Agro-Know Technologies, Greece
[3] National Centre for Scientific Research "Demokritos", Institute of Physical Chemistry, Molecular Thermodynamics and Modelling of Materials Laboratory, Greece
{effie.tsaki,alexios.dimitropoulos}@gmail.com, zmakrod@chem.demokritos.gr, palavitsinis@grnet.gr

**Abstract.** In an era where digitization and networking technologies offer a new way for environmental and natural history institutions, like natural history museums and science centers, to promote their services to the general public, new challenges rise on the field on intellectual property rights. Natural Europe project aim is to make natural history knowledge accessible on an open access basis to a wide spectrum of end-users, through Europeana portal. The value the project delivers is not merely in making works available online, but in the open access terms under which the works are available. Therefore understanding Intellectual Property considerations is fundamental in achieving this goal. The aim of this paper is to enable a further discussion on issues concerning the intellectual property rights for environmental and natural history collections.

**Keywords:** Intellectual Property Rights, cultural collections, natural history.

## 1 Introduction

The last few years the way that intellectual property rights (IPR) operate has been completely changed due to the technological development and more particular the spread of online activities. IPR comprise patents, trademarks, designs and geographical indications, as well as copyright (authors' rights) and rights related to copyright (for performers, producers and broadcasters) [1]. Today's digital technologies allow perfect, inexpensive and unlimited copying and dissemination of content – legal or otherwise. On one hand stands the need for adequate protection of creators and their work, while on the other hand free distribution of work boosts the development of the Information Society as a whole - as they say "knowledge stands on the shoulders of giants". IPR rules therefore need to be adapted to our digital times. A balance must be found between the interests of right holders and users, in order to protect the originality and enable legal or licensed re-use by others.

In general, copyright grants exclusive but limited rights to the creator of an original work to copy, reproduce, perform and distribute it. From its inception, however,

E. García-Barriocanal et al. (Eds.): MTSR 2011, CCIS 240, pp. 445–452, 2011.

copyright law has been as much about the promotion and circulation of knowledge and good ideas as it is about the protection and rewarding of creators [2]. Limitations and exemptions to creators' copyright protection are as important to society as the protection itself.

Significant users and creators of copyright content are cultural institutions, like libraries, museums and archives. Cultural institutions provide a site where the interests and activities of many people converge, including entities with very different powers in relation to the circulation of cultural works; authors, artists and other creators; commercial publishers; researchers; educators; and the general public. As well as their public interest missions of preservation, access, research and education, cultural institutions may also have quasi-commercial arms involved in publishing and product development [3].

According to the National Initiative for a Networked Cultural Heritage (NINCH) working group on best practices, cultural institutions are primarily interested in two of the many issues that surround copyright: how they can legally digitize material in which they may not hold the copyright and how they can ensure that no one else can use the materials they have digitized without their approval [2].

Europeana[1], Europe's digital library, museum and archive, is the focus of a number of IPR issues, as the portal provides access to three sets of assets: the Open Source code base, the authority-controlled metadata and the digitized content [4]. The main rights concerns are around the digitization of public domain content and orphan works. Compared to the aforementioned literature aim of this paper is to focus on the IPR issues that arise for environmental and natural history digital collections.

First chapter stands as an introduction on IPR issues in relation to cultural digital collections and related literature. In the second chapter Natural Europe project is presented. In the third chapter similar projects managing natural history, environmental or biodiversity content are examined, in relation to IPR issues and the policy that they follow. In chapter four, after presenting the types of natural history content provided for Natural Europe, it is analyzed the methodology followed for the selection of IPR scheme in the context of the project. In the same chapter follows a discussion on the IPR and the metadata, in first place how IPR are described in the metadata schema used by Natural Europe and then under what licenses the metadata generated within the project are classified. In the last chapter

## 2 Background

Natural history Museums (MNHs) form a rich source of knowledge about the natural history and the biodiversity of a particular region, thus they provide content which can be characterized as cultural heritage content. Natural Europe project[2] aims to bring together digital collections from NHMs around Europe and connect them with Europeana portal. Additionally, on the intentions of the project is to study the educational methods and deploy the necessary software tools that will

[1] http://europeana.eu/portal/

[2] http://www.natural-europe.eu/

allow museum educators to design innovative online pathways through the digital collections of NHMs. To this end, the federation of the Natural Europe digital libraries aims to facilitate: storage, search and retrieval of digital content that is related to Natural History.

In Natural Europe project participate five (5) NHMs and a Science Center (SC) from six different European countries (Hungary, Greece, Portugal, Estonia, Germany and Finland). Concerning the digital content provided by NHMs in the context of Natural Europe project, this estimated on 16.000 resources of natural history content.

**Table 1.** NHMs participating and providing content in the context of Natural Europe Project

| Country | Name of Institution |
|---|---|
| Hungary | Hungarian Natural History Museum (HNHM) |
| Greece | Natural History Museum of Crete (NHMC) |
| Portugal | Natural History Museum of Lisbon (MNHNL) |
| Estonia | Estonian Natural History Museum (TNHM) |
| Germany | Jura-Museum Eichstatt (JME) |
| Finland | Arctic Center (AC) |

The next chapter is an attempt to study similar projects managing natural history, environmental or biodiversity content, in relation to IPR issues and the policy that they follow.

## 3 IPR in Cultural Collections

This chapter describes the way in which relevant initiatives work with IPR issues. In order to understand particularities inherent in cultural heritage objects we studied the policy followed by other similar projects. For the purpose of this paper three different but relevant projects are presented, the BHL- Europe project, the Sterna initiative and the OpenUp! project.

BHL-Europe project aims to make available Europe's biodiversity information to everyone by improving the interoperability of European biodiversity digital libraries. BHL-Europe content is available under the Creative Commons[3] licensing.

Birds and all kinds of information related to birds are at the core of the STERNA digital library. It brings together and presents all sorts of multimedia resources on birds, bird species and their habitats in one information space. STERNA IPR issues regarding a source are the sole responsibility of the partner who provides that source. The following picture presents an example of a copyright statement on a STERNA library object.

[3] http://creativecommons.org/

Opening Up the Natural History Heritage for Europeana is another initiative that provides natural history multimedia content with specific biological and multilingual functionality to a wide European cultural audience through EUROPEANA. For the IPR framework OpenUp! is adapting the IPR agreements currently used by its partners when releasing their multi-media content.

| Project name | URL | Type of license |
|---|---|---|
| BHL-Europe | http://www.bhl-europe.eu/ | Creative Commons |
| Sterna library | http://www.sterna-net.eu/ | Partners IPR agreements |
| OpenUp! | http://open-up.eu/ | Partners IPR agreements |

## 4 IPR in Natural Europe

### 4.1 Types of Natural History Content Provided for Natural Europe

The main aim of the Natural Europe project is to make natural history knowledge accessible on an open access basis to a wide spectrum of end-users, through Europeana portal. The value the project delivers is not merely in making works available online, but in the open access terms under which the works are available. Therefore understanding Intellectual Property considerations is fundamental in achieving this objective.

The digital content that will be available online through the project mainly consists of text, like scientific descriptions of animals or plants; images depicting natural history objects in low or high resolution; videos and animations of natural phenomena and animals; sounds of animals; educational pathways that will be created by museum educators or teachers. Moreover, copyright issues concern also the metadata provided for the description of the digital content and for the educational pathways.

### 4.2 Selected IPR for Natural Europe Digital Content

In order to select the appropriate IPR schemes for the Natural Europe project we have examined existing IPR schemes that gain wide acceptance, like Creative Commons and Europeana Classification.

Additionally, a discussion with Natural Europe museum partners has taken place during the 2$^{nd}$ project meeting focused on copyright issues related to the digital content of NHMs. The discussion was based on three parameters/ questions

- Is commercial use allowed?
- Are modifications allowed?
- Is the content public/ free for use?

Museum curators and educators mentioned that licensing is much depended on the types of the content resolution available and on the indented use. Thus, for example, for content that is used for educational purposes predominates the idea of free use. Fig. 1 conceptualizes the discussion with museum curators and educators on IPR during the 2$^{nd}$ meeting project.

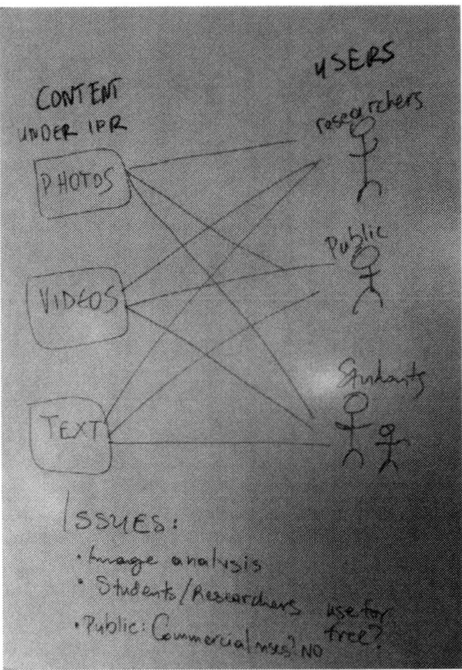


**Fig. 1.** Drawing the concept of IPR issues for NHMs during the 2nd project meeting in Athens

Concerning the requirements for Natural Europe, museum partners mentioned that they would like to have the possibility of applying different IPR schemas to their items. Most of the provided content will be under the Creative Commons license “Attribution Non-commercial” (BY-NC), but occasionally might appear some restrictions. Rather rarely some content will be free of any restrictions. Additionally, they would like the target users of the Natural Europe services to be able to add suitable data or metadata to the existing objects, but only in a way that the original data can't be changed.

Eliciting the IPR requirements related to the digital collections from the project meeting an IPR Scheme is under development that will be comprised by the following schemes:

- Creative Commons
  - Public Mark
- Europeana Classification
- Copyright Statement (free text)

Examining all the different IPR schemes available as well as others used by similar initiatives, we have concluded that although a single type of IPR scheme may be applied to describe all cultural content in Natural Europe, it definitely has to be distinguished amongst the different types of material being licensed. In Natural Europe licenses are provided for two different categories of content. One is the cultural heritage objects provided by NHMs collections and the other is the educational material i.e. the educational pathways. Table 2 depicts all the types of material available in Natural Europe, along with the proposed copyrights for each type.

**Table 2.** CC Licenses for Natural Europe

| | | CC Licenses | | | | | |
|---|---|---|---|---|---|---|---|
| | | **BY** | **BY-NC** | **BY-ND** | **BY-NC-ND** | **BY-SA** | **BY-NC-SA** |
| **Content Type** | Text | | | | | | X |
| | Images (law resolution) | | | | | | X |
| | Images (high resolution) | | X | | | | |
| | Videos | | | | | | X |
| | Sounds | | | | | | X |
| | Educational Pathways | | | | | | X |
| **Metadata** | Metadata for CHO | | | X | | | |
| | Metadata for Pathways | | | | | | X |

## 4.3 IPR and Metadata

### 4.3.1 IPR in Metadata

For the description of cultural heritage objects that will be populated in the context of Natural Europe project has been developed an application profile based on the Europeana Semantic Elements specification [5]. More specifically, for the indication of the usage and the access rights to the digital object has been used the elements Europeana:rights and dc:rights. The dc:rights element (Table 3) is used as a free text, where the Europeana:rights element (Table 4) applies a controlled vocabulary [6].

**Table 3.** Dc:rights element from ESE specification version 3.4

| Element name: rights | | | | | |
|---|---|---|---|---|---|
| Namespace | Dc | | | | |
| URI | http://purl.org/dc/elements/1.1/rights | | | | |
| Label | Rights | | | | |
| Definition | Information about rights held in and over the resource. | | | | |
| Europeana note | Information about intellectual Property Rights, access rights or license arrangements for the digital object (digitized or born digital). | | | | |
| Obligation & Occurrence Europeana search and display features | Simple search | Facet | Timeline | Advance search | Full search result display |
| | X | | | | X (Rights line) |
| Example | <dc:rights>Copyright © British Library Board, BBC and contributors</dc:rights> | | | | |

### 4.3.2 IPR for Metadata

Metadata in Natural Europe concern (a) the metadata generated from museum curators annotating Cultural Heritage Objects (CHO) and (b) the metadata generated from educators creating educational pathways. All the metadata produced by Natural Europe are characterized under the Creative Commons licensing and more specifically under the Non-Commercial –Non derivatives for CHO metadata and Non-Commercial –Share Alike (NC-SA) for Educational metadata (as shown in Table 2).

**Table 4.** Dc:rights element from ESE specification version 3.4

<table>
<tr><td colspan="6">Element name: rights</td></tr>
<tr><td>Namespace</td><td colspan="5">Europeana</td></tr>
<tr><td>URI</td><td colspan="5">http://www.europeana.eu/schemas/ese/rights</td></tr>
<tr><td>Label</td><td colspan="5">Europeana Rights</td></tr>
<tr><td>Definition</td><td colspan="5">Information about copyright of the digital object as specified by isShownBy and isShownAt</td></tr>
<tr><td>Europeana note</td><td colspan="5">The value is a URL constructed according to the specifications in the "Rights Guidelines" document on the Technical Requirements page at:<br>http://version1.europeana.eu/web/guest/technical-requirements/<br>The URLs are constructed by adding a code indicating the copyright status of an object to the domain name where that status is defined.<br>For users of Europeana.eu this copyright information also applies to the preview specified in europeana:object.<br>The rights statement will be represented as a badge on the object page and as a text string in the metadata display.<br>The provision of this element has an obligation level of "recommended" in this version of ESE. It will be changed to "Mandatory" in a later version.</td></tr>
<tr><td rowspan="2">Obligation & Occurrence Europeana search and display features</td><td>Simple search</td><td>Facet</td><td>Timeline</td><td>Advance search</td><td>Full search result display</td></tr>
<tr><td></td><td></td><td></td><td></td><td>X</td></tr>
<tr><td>Example</td><td colspan="5"><europeana:rights>http://creativecommons.org/licenses/by/3.0/ </europeana:rights></td></tr>
</table>

## 5 Conclusions and Directions for Future Research

This paper presented the process of defining IPR for environmental and natural history digital content. In order to select the appropriate IPR schemes for the Natural Europe project existing IPR schemes that gain wide acceptance, like Creative Commons and Europeana Classification, have been extensively examined. Furthermore, the discussion with museum curators has been considered very valuable for understanding in depth the needs of the natural history museums rights holders and users. According to this, what is notable is the fact that museum curators find important to preserve the copyrights in the institution, but use the Share Alike option from Creative Commons, for educational purposes. In general, is very common in cultural institutions to offer flexibility in copyrights for educational purposes. In addition, a further discussion should be encouraged concerning the Public Mark option provided by Europeana, facilitating access to cultural heritage among general public.

The present study was based on a preliminary survey of existing IPR schemas and it combined user generated requirements to define an IPR scheme for a specific natural history collection. An interesting extension of this paper could examine the actual use of IPR at collection level by other projects / museums that offer their content online.

**Acknowledgments.** The work presented in this paper has been funded with support by the European Commission, and more specifically the project "Natural Europe: Natural History & Environmental Cultural Heritage in European Digital Libraries for Education" of the ICT PSP Programme. All authors have been supported with funding by GRNET during the implementation of this work, in the context of the above mentioned EU project.

## References

1. European Commission: Commission sets out "blueprint" for Intellectual Property Rights to boost creativity and innovation, IP/11/630 (2011)
2. The NINCH Guide to Good Practice in the Digital Representation and Management of Cultural Heritage Materials
3. Hudson, E., Kenyon, A.: Communication in the Digital Environment: An empirical study into copyright law and digitisation practices in public museums, galleries and libraries, Intellectual Property Research Institute of Australia, Working Paper No. 15/05 (2005) ISSN 1477-1795
4. Purday, J.: Intellectual property issues and Europeana, Europe's digital library, museum and archive. Legal Information Management 10, 174–180 (2010)
5. The Europeana Semantic Elements Specification, version 3.4 (March 31, 2011), `http://www.version1.europeana.eu/c/document_library/get_file?uuid=77376831-67cf-4cff-a7a2-7718388eec1d&groupId=10128`
6. The Guidelines for the europeana:rights metadata element, version 4.0 (November 30, 2010), `http://version1.europeana.eu/web/guest/technical-requirements/`

# Developing a Metadata Application Profile for Sharing Agricultural Scientific and Scholarly Research Resources

Nikos Diamantopoulos[1,2], Cleo Sgouropoulou[1], Kostas Kastrantas[3], and Nikos Manouselis[3,4]

[1] Department of Informatics, Technological Educational Institute of Athens, Agiou Spyridonos, 12210, Egaleo, Greece
diamanto87@gmail.com, csgouro@teiath.gr
[2] Greek Research & Technology Network, 56 Messogeion Av., 11527, Athens, Greece
[3] Agro-Know Technologies, 17 Grammou St., 15235 Athens, Greece
kkastrad@agroknow.gr
[4] Information Engineering Research Unit, Computer Science Department, University of Alcalá, Ctra. de Barcelona km 33.6, 28871, Alcalá de Henares (Madrid), Spain
nikosm@ieee.org

**Abstract.** This article describes the development of the VOA$^3$R Metadata Application Profile to facilitate the description, dissemination and reuse of research results in the fields of Agriculture and Aquaculture within a federation of open access repositories. VOA$^3$R is a research project aspiring to deploy an advanced, community-focused integrated platform for the retrieval of relevant open content and data that supports explicit models of the scholarly lifecycle and the practical tasks targeted by applied research. The article details the phased implementation of the core VOA$^3$R application profile relating to the description of scholarly resources. The development process has been based upon the methodology and components of the Singapore Framework for Dublin Core Application Profiles: functional requirements, domain model, description set profile, usage guidelines and data format. Maximum interoperability, reuse of existing mature metadata standards, long-term quality control and extensibility for addressing further community needs, constitute the main benefits of this approach.

**Keywords:** Metadata, Dublin Core Application Profile, requirements, domain model, agricultural research resources.

## 1 Introduction

The advent of the ICT revolution is undoubtedly transforming the way research is conducted. Scientific publications, data and other types of research results are increasingly being stored in institutional and consortial repositories, the widespread growth of which is further intensified by opportunities for discovering, sharing, exchanging and reuse of scholarly information across research communities.

Novel search engines as Google Scholar, databases as PubMed, citation systems as CiteULike and indexes as DOAJ are access enablers for large masses of open

E. García-Barriocanal et al. (Eds.): MTSR 2011, CCIS 240, pp. 453–466, 2011.

scholarly content, but they still require significant effort for research stakeholders to retrieve relevant information for targeted needs and cover only some of the processes and tasks that are inherent to the nature of research work, such as highly specific literature analysis and contrast, preprint publishing, informal feedback, different forms of peer evaluation and post-archival assessment.

The Virtual Open Access Agriculture & Aquaculture Repository (VOA$^3$R) project aspires to improve the propagation of European agriculture and aquaculture research results by developing and providing services that integrate existing open access repositories and scholarly publication management systems into a single point of access to scholarly research, by means of a federation approach.

The VOA$^3$R platform aims at deploying an advanced, community-focused integrated platform for the retrieval of relevant open content and data that includes explicit models of the scholarly methods and procedures used and of the practical tasks targeted by applied research. The VOA$^3$R platform aims to facilitate researchers to formulate their information needs in terms of elements of the scientific methods established in their field, combined with topical descriptions. The community approach will foment the enhancement of information seeking with extended evaluation elements (as for example, ratings, public reviews, social tagging and links to supporting or conflicting reports) that complement and go beyond the traditional, anonymous peer review process.

Metadata figures prominently in the efforts towards the development of such advanced services. The re-use of mature metadata standards and the specification of domain-specific metadata profiles for integrating existing agriculture and aquaculture repositories is one of the key aspects of VOA$^3$R. Devising and adopting a standardized and consistent metadata approach will ensure interoperability among diverse research resource stores and sources, thus enabling content providers to federate with the platform, and allow the implementation of seamless, transparent and user-friendly research information access services.

This article details the phased development of the core VOA$^3$R Application Profile (AP) for the description, dissemination and reuse of scholarly resources within a federation of open access repositories. The development has been based upon the methodology and building components defined by the Singapore Framework (SF) for Dublin Core Application Profiles (DCAP) [1]. Maximum interoperability, reuse of existing mature metadata standards, long-term quality control of the resulting schemas and extensibility for addressing further community needs, constitute the main benefits of this approach.

More specifically, the article starts with a background discussion on fundamental metadata application profiling issues and the rationale behind the decision to define the VOA$^3$R AP as a DCAP. In continuation, it introduces the actual AP building process within VOA$^3$R, integrating and further refining the SF generic guidelines. The elicitation of the metadata design principles and functional requirements as well as the specification of the VOA$^3$R metadata domain model are of particular focus for the article. Finally, a common metadata element set manifesting the underlying model in DC is presented. The element set constitutes the interoperability instrument for participating repositories to realize the project's content population and integration plans, also contributing to the first evaluation of the VOA$^3$R AP. The article concludes with a discussion on the challenges for future enhancements to the core VOA$^3$R AP.

## 2 Application Profiles and the VOA$^3$R Metadata Approach

Research activity and processes delineate a complex landscape within which stakeholders need to interoperate requiring support for a wide range of rich, resource-based functions and services. Repositories participating in the VOA$^3$R federation are online loci for collecting, preserving, and disseminating (in digital form) the research intellectual output of institutions or communities, including a broad range of resources: scholarly publications such as research journal articles, before (pre-prints) and after (post-prints) undergoing peer review, scientific datasets, digital versions of theses and dissertations, research project reports as well as other digital assets generated by normal scientific and academic life.

For repositories, metadata is a valuable asset that needs to be shared with external systems. Internally, a repository or other metadata-rich system may only consider its own requirements. However, when it comes to the efficient and reliable sharing and exchange of cross-repository content, well-designed, domain specific and agreed upon metadata specifications become a crucial consideration [2].

Efforts to develop such metadata specifications should start off by reviewing existing, mature standards, studying their underlying models, and by investigating their interoperability features and support of reuse. Well-established metadata standards for cross-domain and domain-specific information resource description have already been around for quite some time. Dublin Core (DC) Metadata [3] and IEEE Learning Object Metadata (IEEE LOM) [4], among others, are examples of such standards, providing semantic support for a broad range of purposes and business models.

Owing to the diverse needs of particular communities and applications, metadata standards generally support a notion of customisation through Application Profiles. Application profiling of metadata specifications in its simplest form supports the process of selection of a set of metadata elements from an element vocabulary, possibly extending the base element vocabulary as defined in the specification using locally defined elements, and choosing a set of useful value vocabularies for use with these elements [5]. However, since "*the days of one-size-fits-all standards are over... Domains are now overlapping and becoming liquid*" [6], the term AP encompasses broader forms of customisation, involving the "*assemblage of metadata elements selected from one or more metadata and combined in a compound schema. The purpose of an AP is to adapt or combine existing schemas into a package that is tailored to the functional requirements of a particular application, while retaining interoperability with the original base schemas.*" [7]. Enabling such customisation of metadata standards is one of the ultimate goals of metadata harmonisation.

Nevertheless, this mixing-and-matching of elements from several metadata specifications with different underlying abstract models leads to semantic confusion, producing significant barriers for the harmonisation that APs have been designed to enable. As shown in [8], the different abstract models of the DC and IEEE LOM metadata standards (entity-relationship or graph based model grounded in model-theoretical semantics vs. abstract hierarchical model with no formal semantics) produce problematic mixing scenarios, driving one to the conclusion that the notion of 'reusing elements' between metadata standards and formats using incompatible abstract models is fundamentally flawed.

So, what is the right course to be followed? As a first step towards the development of domain-specific, harmonized metadata specifications, it is essential to decide upon the nature of the abstract model within the framework of which the metadata will be defined and to pursue opportunities for leveraging existing stable specifications with compatible underlying models that can be combined to build sustainable APs. Implementers can then be safe in the knowledge that the assembled whole can be interpreted correctly by independently designed applications [1].

The reuse of existing, mature metadata standards, aligned with the Semantic Web principles and tailored to the specific requirements of the agriculture and aquaculture research communities has been one of the key constituents of VOA$^3$R's metadata approach. On this basis and taking into consideration the issues and best practices in the field, the VOA$^3$R metadata development team pursued the creation of a Dublin Core Application Profile (DCAP) complying with the Singapore Framework (SF). Conforming to the Dublin Core Abstract Model (DCAM) [9], VOA$^3$R aims to produce a sustainable and adaptable metadata architecture based on the resource-property-value model, moving from the resource-legacy approach representing an information package, to focusing on the component parts of a resource description. This approach will result in flexible and extensible metadata that integrate with a semantic web of linked data. Building on the abstract model, the VOA$^3$R AP will articulate what is intended and can be expected from data, thus promoting the sharing and linking of data within and between research communities.

The SF for DCAPs is providing a formalisation for building and documenting APs, including guidance for metadata creators and clear specifications for metadata developers. As described in the framework, a DCAP is a packet of documentation containing the following parts [1]:

- *Functional requirements*, describing the functions that the AP is designed to support, as well as functions that are out of scope. Ideally, functional requirements address the needs of metadata creators, resource users and application developers, so that resulting applications fully support the needs of the community [10].
- A *Domain Model* which defines the basic entities metadata and the relationships among them, using a formal or informal modelling framework. The domain model is the basic blueprint for the construction of an AP [10], [11].
- A *Description Set Profile*, providing a method for specifying structural constraints on the descriptions and statements held in a metadata record.
- *Usage guidelines*, describing how to apply the AP, how the used properties are intended to be used in the application context, etc.
- *Encoding Syntax guidelines*, defining AP-specific syntaxes, if any.

The ensuing sections of the paper describe the SF-based VOA$^3$R AP development process and present activities and outcomes related to the requirements elicitation and domain modelling tasks.

## 3 The VOA3R AP Development Process

Building APs is a complex and demanding task requiring a well-defined process which, apart from the technical development issues, puts emphasis on the adopting communities, namely the creators and consumers of metadata, the application developers, the profile authors, etc. [2]. On this basis, the VOA$^3$R metadata architecture team has established and orchestrated a four-phase building process (Table 1) for the development of a full-fledged AP specification which defines concrete tasks and expected outputs, adhering to and further refining the SF mandates. Due to its generic characteristics, the proposed process can be applied for the development of any DCAP.

Fig. 1 illustrates, by means of the Business Process Modelling Notation (BPMN), the ordered sequence of the process tasks and the key actors' contribution to the AP lifecycle evolution.

**Table 1.** Phases and tasks of the VOA$^3$R AP development process

| Phase 1. Domain analysis and requirements specification |
|---|
| • Set up working groups to involve technical and domain metadata experts and engage the stakeholder communities to ensure maximum completeness in the identification of requirements and foment community uptake<br>• Organize activities on review of related work on existing related mature metadata standards and schemes and their use cases. Analyse current practice within stakeholder communities<br>• Orchestrate requirements elicitation activities for identifying what kinds of functionality the application profile is intended to support<br>• Develop the functional requirements specification |
| **Phase 2. Metadata Design** |
| • Devise a domain model by means of a UML/entity-relationship diagram<br>• Elaborate the detailed description of the metadata properties of the application profile (Description Set Profile) corresponding to the model.<br>• Foster activities for the development of required community-specific vocabularies<br>• Produce initial set of usage guidelines<br>• Organize review and consensus activities |
| **Phase 3. Testing and validation** |
| • Conduct self and invited testing<br>• Collect examples and develop a set of best practices guidelines for each of the properties in the application profile<br>• Organize expert and community review of whole specification (AP, Vocabularies, Best Practices)<br>• Arrange for compatibility testing by the DC Architecture Working Group<br>• Release the final DCAP |
| **Phase 4. Community Support** |
| • Develop compliance guidelines<br>• Produce metadata mappings/crosswalks to allow interoperation<br>• Maintain and revise application profile according to evaluation activity results |

It should be noted that this process mainly follows a top-down approach, in accordance with the inherent DCAP development methodology. However, specific tasks within the process build upon a bottom-up approach, in the sense that stakeholders' existing and applied metadata practices need to be analysed and taken into consideration. Furthermore, in many cases provisional metadata solutions need to be devised in order to support interoperability and harmonization related experiments. Although this may seem like a diverging activity, it often proves most valuable for revealing important issues regarding the conceived usefulness, appropriateness and comprehensiveness of the designed AP. As pinpointed in [12], "*the two approaches are not necessarily at odds and could be instead viewed as two renderings of the same intellectual work along a continuum*".

In accordance with the proposed process, the next sections focus on the domain analysis and functional requirements elicitation as well as on the specification of the VOA$^3$R domain model. A core element set taking into consideration both the content providers' metadata practice and the principles of the underlying model is also presented.

# 4 Domain Analysis and Functional Requirements Specification

## 4.1 The VOA$^3$R AP Stakeholders

Domain analysis and requirements specification is a broad community task involving metadata experts, managers of services, domain experts, application developers, and potential end-users of the services. The formulation of functional requirements should, thus, start off with comprehensive identification of the stakeholder communities and the activities that should be supported by the resulting AP.

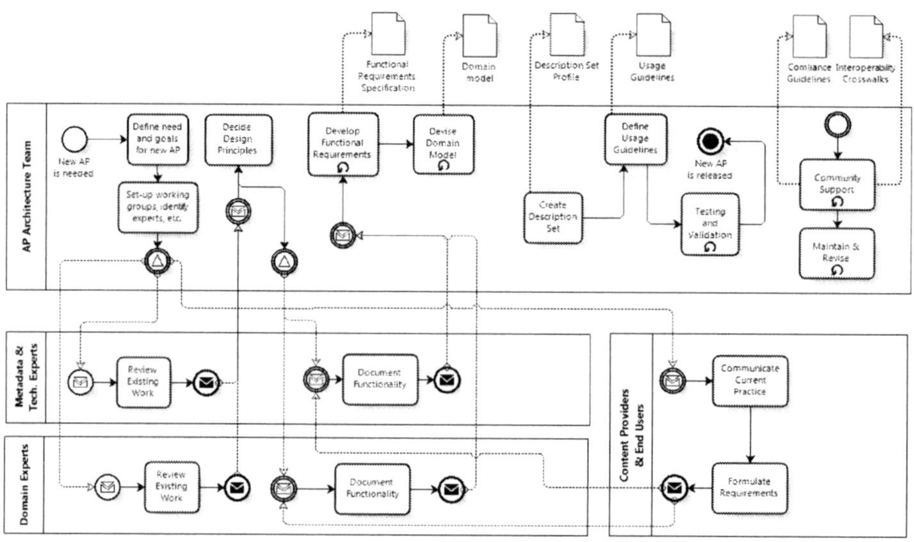


**Fig. 1.** BPMN model of the VOA$^3$R AP development process

As already described, the VOA$^3$R platform aspires to build content-based communities of interested peers, in which end users interact with content of the federation of repositories that is accessible through VOA$^3$R, either as producers or consumers of content.

*Producers* of content are mainly researchers and academics, but nonetheless they could also be students (acting as junior researchers) or even practitioners and industry representatives. The VOA$^3$R platform will support producers to identify existing content (aggregated through the associated repositories), describe and tag the content with appropriate metadata, allow for content-based collaboration in terms of feedback to readers and, last but not least, statistics and information on the readability, rating and overall impact of their publication (using various metrics). *Consumers* of content (i.e. readers) being practitioners, students, people from the industry, policy makers, end-consumers, etc. will need to locate material and then study it. They will also want to know-who (except to knowing how) and know-when (time-based searching).

The main innovation of VOA$^3$R is that it shall provide a "community and social-oriented" platform, in which all users will be able to exchange ideas, collaborate and engage in content-based discussion. In that sense, all members of the VOA$^3$R portal will be motivated to associate in networks of contacts, invite peers, provide details on their personal, research and work profile and, last but not least, 'cultivate' both their real and virtual (online) reputation, by means of attracting interest to their research, ideas, etc. and, also, participating in the online community, promoting the use of the portal by more people, enriching metadata and discussing with interested parties.

The aforementioned rich end user services may only be achieved by means of a metadata infrastructure designed to enable the efficient aggregation of content provided by participating repositories and sources. Thus, metadata creators and developers, domain experts, metadata curators and repository managers constitute the main technical stakeholder group of the VOA$^3$R AP.

## 4.2 Analysis of VOA$^3$R Providers' Metadata Practice

Review and exploitation of relevant work and current practice are important constituents of the overall effort towards well-defined requirements. Within VOA$^3$R, the application profile development evolves in parallel with actual implementation tasks, one of the most prominent being the population and integration of content (scientific and scholarly research resources) by the different providers. In a repository federation as VOA$^3$R, this introduces a key challenge in revealing and addressing the metadata harmonization and quality assurance issues for the participating repositories.

To this end, we initiated a task of analysing the providers' metadata in order to devise a balanced, common set of metadata elements, as an interoperability instrument for participating repositories to carry out the project's content population and integration plans, also contributing to the formulation of the VOA$^3$R AP requirements. Special attention was given in the reuse of existing formats in order to keep the metadata creation effort manageable and to safeguard member institutions' investment.

The resulting core element set mainly targets bibliographic metadata that are already captured in the providers' repositories. Metadata that can support the research-related advanced services of the VOA³R platform (rich semantics) implementing aspects related to the research methods, reviewing processes, social interaction, metadata quality, systematic taxonomies as well as metadata that can help monitoring and managing collections and their records within the federation are expected to result after the overall modelling effort and will be reflected in future extended formats.

To this end, the team first examined various content providers' metadata schemas and employed the content analysis methodology to identify relevant elements. For each schema, the following questions were asked:

- Which metadata schema is used internally?
- How is the schema defined?
- What are the recommended, mandatory, and optional elements?
- Are any controlled lists, authority lists or thesauruses being used and what are their corresponding vocabularies?
- Is there a language indication in the stored strings?
- Is there a policy regarding the creation and indexing of metadata?
- How many records are stored and what percentage of them is related to a full-text document?
- What type of documents and what file types are available?

The content analysis resulted in a proposed core VOA³R Element created by taking elements from the following namespaces: Dublin Core Metadata Element Set (DCMES) [3], IEEE LOM Metadata Set [4] and Agricultural Metadata Element Set (AgMES) [13].

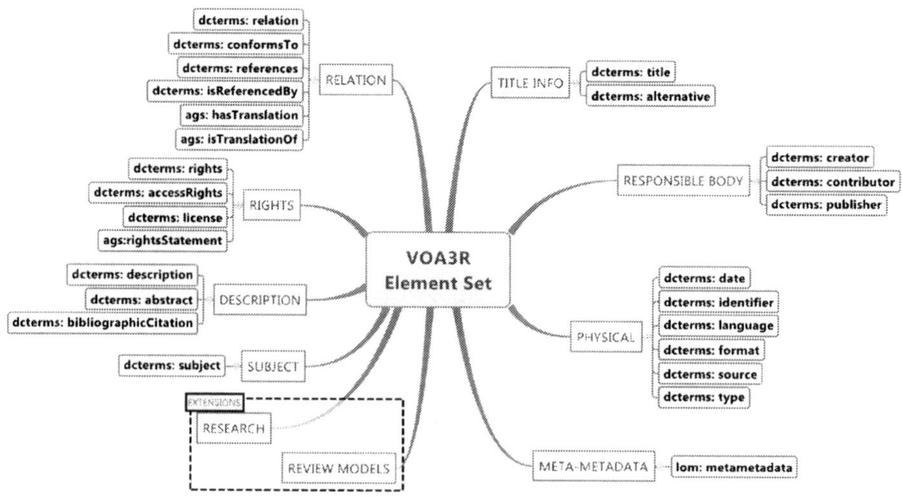


**Fig. 2.** The VOA³R Element Set

Most of the participating repositories are exposing simple DC records that can be consumed by means of the OAI-PMH harvesting protocol. An initial harvesting experiment realized within VOA³R demonstrated poor quality and inconsistency within these simple metadata descriptions as barriers to harmonization and to the provision of richer services, conclusion that further enhances reports from other similar initiatives implementing cross-search related services for resources in specific domains.

Poor quality and inconsistent metadata are a real problem for aggregator services, for a variety of reasons. Within the work on VOA$^3$R, these were analyzed to help inform the requirements specification process. From the analysis of the providers' schemas, a great deal of semantic and syntactic inconsistency was discovered. Semantic fuzziness is concentrated mostly around the elements of `<dc:contributor>`, `<dc:date>` and `<dc:alternative>`. Creators and contributors are often mixed and there is a level of ambiguity in the exact semantics of a date. Moreover `<dc:alternative>` in some cases is used to store translations while in others more complex forms of title information such as subtitles and transliterations. Identification proved to be a particular issue, with inconsistent practices employed in the use of `<dc:identifier>` to capture identifiers for full-text resources, metadata records and other related resources in a way that cannot be easily disambiguated. Additionally, simple Dublin Core does not allow for the specification of Syntax Encoding Schemes, so where a particular identifier scheme has been used, for example, URI, DOI, ISBN, these cannot be easily identified. Other syntactic issues that arise are: use of multiple `<dc:title>` elements and the inability to specify the language of the element contents, that is, for translated titles; where name elements such as `<dc:creator>` and `<dc:publisher>` are used it is not possible to indicate whether a normalized form has been used or whether the name is of a person or organization; and where `<dc:subject>` is used there is no means of indicating whether controlled terms from a specific vocabulary have been used.

In addition, to get round the limitations of simple DC to respond to more refined descriptive needs of the hosting scientific/scholarly resources, some repositories try to accommodate more information than necessary into the metadata fields. This varying use of metadata leads to semantic inconsistencies making implementation of harmonized end user resource discovery services across multiple repositories an impossible task.

### 4.3 Functional Requirements Specification

The requirements specification task within VOA$^3$R has been based on a two level approach. First, a set of high-level use cases were formulated (Table 2). These use cases have been extremely useful in shaping the VOA$^3$R metadata landscape, in terms of identifying the constituent entities regarding and relating to agricultural research resources.

Subsequently, a list of detailed functional requirements was elaborated to capture and document the broad range of sophisticated functionalities envisaged: e.g. association of resources with research activity, reflection of peer-reviewing processes in metadata, representation of complex resources, versioning identification, navigation through records, etc. The elicitation activities were mainly carried out within focused stakeholder workshops and consultation meetings; however, significant experience was also accumulated by the providers' content analysis experiments and harmonization activities.

**Table 2.** High-level use cases for the VOA$^3$R functionality

| | Use case description |
|---|---|
| UC#1 | Facilitate the description, sharing and re-use of agricultural research work within open access repositories |
| UC#2 | Support the discovery of agricultural research resources |
| UC#3 | Support the identification and classification of resources using existing ontologies and taxonomies |
| UC#4 | Enable the semantic interconnection of resources by supporting the exposure of metadata in linked data formats |
| UC#5 | Combine bibliographic metadata with domain-specific data related to the agricultural research communities practice and activities |
| UC#6 | Serve as an interchange format between various repositories using different metadata standards and formats |
| UC#7 | Use for harvesting metadata from diverse source repositories through a quality controlled process into curated "refaratories" |
| UC#8 | Support added-value services for the agricultural research community |

Requirements have been documented by means of a formal template introduced by the development team of the DC Scholarly Works Application Profile (SWAP) [14]. A key aspect of this expression is the formulation of usage scenarios delineating the goals and boundaries for the intended applications and services as well as limitations in current practice. Table 3 illustrates a representative VOA$^3$R AP requirement.

**Table 3.** Example of a VOA$^3$R AP Functional Requirement

| Functional Requirement | Reflect publication/lifecycle status in metadata |
|---|---|
| Description | Metadata about the publication status of a resource is needed, covering any type of status that can be found within (or outside) a repository. |
| Usage Scenario | A manuscript is pre-archived in some special service and then submitted to a journal for publication. The author publishes the submitted version at his/her personal web site. The paper is accepted by the journal and published after revisions at the publisher's web site. The publisher's copyright policy allows the institution of the author to include a pre-print version of the published paper in its collection and offer open access to it. |
| AP Requirement | A publication status element is needed, together with a rich vocabulary to allow the representation of all possible statuses. |

The requirements elicitation and documentation process strengthened our commitment to moving towards the development of a full DCAP, adopting a more complex model which offers the opportunity to move away from single and flat metadata descriptions to more flexible, extensible, semantically rich and sustainable metadata. Following the DCAP's underlying entity-relationship approach the potential for describing multiple, domain-related entities and their properties, and relating these to each other can be explored.

## 5 Modelling the Metadata

A domain model is a conceptual model identifying the entities we want to describe, the relationships between them and the attributes necessary to effectively describe the entities. It acts as a communication tool and should be understandable by technical and non- technical audiences. The domain model has a key role to play in providing shared understanding between different stakeholders [10]. By taking the modular approach of a domain model it is much easier to base metadata design on requirements rather than on the constraints of a particular metadata solution. This approach also moves us away from the traditional flat metadata descriptions, often containing information about a number of entities in a single description [2]. At the modelling stage, this approach is free from dependence on a particular metadata approach or metadata vocabulary: metadata vocabulary decisions come later.

The VOA$^3$R AP domain modelling task started with the exploration of a number of relevant modelling approaches for bibliographic and scholarly resources as well as for research activity and product related information. Among the reviewed models, the following were considered as a basis for VOA$^3$R:

- The IFLA Functional Requirements for Bibliographic Records (FRBR) conceptual model, focusing on the representation of the bibliographic universe, using an entity-relationship model [15]. It defines a set of four primary (Group 1) entities: *Work*, *Expression*, *Manifestation*, and *Item*. In addition, there are two "agent" entities (*Corporate body* and *Organization*) linked through a relationship with Group 1 entities such as those responsible for the creation, production, dissemination and ownership (Group 2), and a set of "subject" entities can be defined as anything that can be the subject of a work (*Concept*, *Object*, *Event*, *Place*) (Group 3).

  For bibliographic catalogues, the power of FRBR lies in the ability to group items logically and to facilitate the discovery of all instances of a particular work in a single search, while being able to distinguish between the different expressions, manifestations, and items and to navigate easily to the most appropriate.
- The Scholarly Works Application Profile (SWAP) [16] is a DCAP providing a model for describing scholarly works. A *Scholarly Work* is defined as a distinct intellectual or artistic scholarly creation. The SWAP model is a simplification of FRBR using fewer entities, relationships and attributes since its identified requirements do not call for the same degree of the FRBR complexity. The SWAP model comprises of the following five entities: *Scholarly Work*, *Expression*, *Manifestation*, *Copy* and *Agent* [17].
- The Common European Research Information Format (CERIF) is a formal model for interoperable research information systems. CERIF [18] is organized around three core entities: *Person*, *Project* and *Organization Unit*, and three result-related entities: *Result Publication*, *Result Patent* and *Result Product*.

FRBR functions as a generic model (sometimes known as FRBRER) on the basis of which several variant resource-related DCAPs have been developed (e.g. SWAP). The need for these variations has emerged due to the differing requirements of the heterogeneous resource types treated within each profile. The VOA$^3$R model builds upon FRBR, SWAP and CERIF, introducing certain modifications relating to the particular nature and attributes of the agricultural research resources.

The VOA$^3$R AP domain model takes the form of a lightweight entity-relationship model expressed as a UML diagram (Fig. 3) conveying the following semantics:

1. The *VOA$^3$R Resource* constitutes the core concept (illustrated as an abstract entity) of the overall model. The *VOA$^3$R Resource* is instantiated through an FRBR-based group of entities: *Research Work*, *Expression*, *Manifestation* and *Item*.
   - A *Research Work* is a distinct intellectual, research-related creation. It can be abstract, having no material object associated with it, or expressed through concrete expressions.
   - An *Expression* is an information object related to the type of the research work (scientific publication, social publication, presentation, research dataset, etc.)
   - A *Manifestation* is an embodiment of an expression of a work. Manifestations may denote the various versions of an expression produced throughout its publication lifecycle, its alternative language versions, or the different constituents of an expression.
   - Finally, *Item* regards an exemplar of a manifestation, i.e. the appearance and characteristics of the physical form of the resource as well as its actual location.
2. *Agents* are individuals or organizations influencing the lifecycle of a resource: its creation, publication, dissemination, validation, etc.
3. *Context* is an abstract entity, providing a placeholder for the representation of further aspects of the research lifecycle. A *Research Study* is an entity related to the context within which research resources are produced and accommodates information regarding research methods, protocols, variables, techniques, etc.

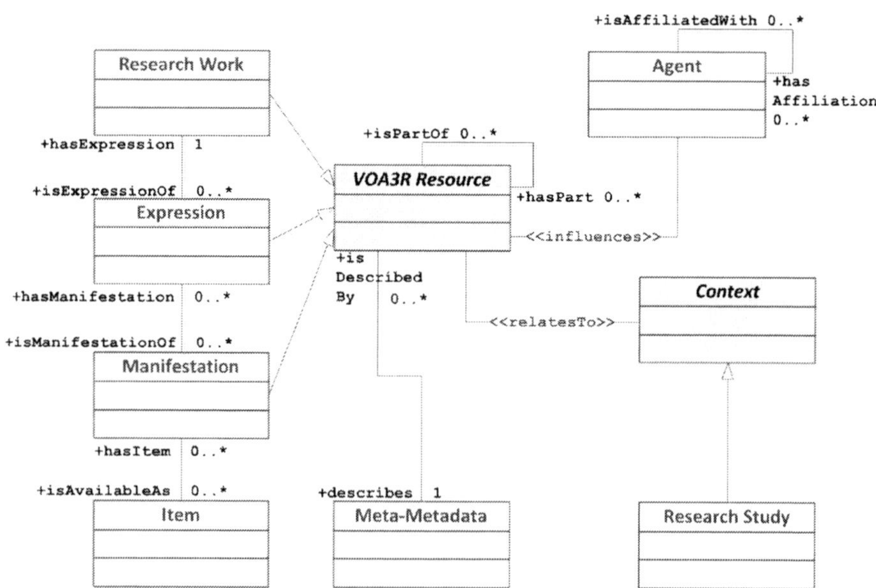


**Fig. 3.** The VOA$^3$R AP Domain Model

4. The *Meta-metadata* entity holds information necessary for collection management (e.g. creator, owner and date of creation of the metadata, annotation relevant to the metadata etc.). Some of this information could also be useful in helping users establish the basis for claims made in a metadata record [19].

It should be noted that for the sake of readability, the model figures general relationships between the *VOA$^3$R Resource - Agent* and *VOA$^3$R Resource - Context* entities. In particular implementations these relationships are expected to be instantiated in more specific ones, corresponding to the nature of 'impact' of (each of) the agent(s) to the resource.

## 6 Concluding Remarks

The VOA$^3$R project focuses on the dissemination of European agriculture and aquaculture research results by developing and providing innovative, semantically rich services to stakeholder research communities. For this purpose, existing open access repositories and scholarly publication management systems will be integrated into a single point of access to scholarly research resources.

Devising a metadata architecture and best practice for supporting this effort is a highly challenging task calling for approaches that respond to the dual objective of addressing the immediate implementation needs on the basis of current practice and taking forward the metadata research agenda by pursuing the long term goals of enhanced interoperability and semantic web alignment.

To this end, the VOA$^3$R metadata team has initiated the development of a DCAP for modelling and describing research resources and associated entities related to the project's research domain. This approach offers the opportunity to move away from single and flat metadata descriptions to more flexible, extensible, semantically rich and sustainable metadata.

In this article we have introduced the VOA$^3$R AP building process which adheres to and further refines the SF recommendations. Our work has so far resulted in the definition of comprehensive functional requirements and the design of a first, modular domain model drawing upon FRBR, SWAP and CERIF. In addition, a core element set, as the 'lowest common denominator' for basic interoperability, has been produced, enabling participating repositories to carry out the project's content population and integration plans.

Undoubtedly, a lot remains to be done. An enhancement of the domain model is already in line in order to address the advanced identified metadata requirements for capturing additional aspects of research work and the lifecycle of the research process from concept, to data collection, pre-publishing assessment, different review models, publishing through various open access channels and post-publishing review, rating and discussion. Future work also involves the compilation of a Description Set Profile for adding machine-readability to the VOA$^3$R AP, thus setting the ground for its validation and evaluation.

**Acknowledgments.** The work presented in this paper has been partially funded by the European Commission, in the context of the FP7 CIP-ICT-PSP.2009.2.4 "VOA$^3$R: Virtual Open Access Agriculture & Aquaculture Repository" project.

## References

1. Nilsson, M., Baker, T., Johnston, P.: The Singapore Framework for Dublin Core Application Profiles (2008), http://dublincore.org/documents/singapore-framework/
2. Allinson, J.: Describing Scholarly Works with Dublin Core: A Functional Approach. Library Trends 57, 221–243 (2009)
3. Dublin Core Metadata Initiative (DCMI): Dublin Core Metadata Element Set, http://dublincore.org/documents/dces/
4. IEEE Learning Technology Standards Committee (LTSC): IEEE Draft Standard for Learning Object Metadata (IEEE 1484.12.1) (2002)
5. Nilsson, M.: Harmonization of Metadata Standards. Deliverable of the PROLEARN IST-507310 European Project, http://ariadne.cs.kuleuven.be/lomi/images/5/52/D4.7-prolearn.pdf
6. Rust, G.: Thoughts from a different planet. Keynote Presentation at the FRBR Workshop in 21st Century Catalogues (2005), http://www.oclc.org/research/activities/past/orprojects/frbr/frbr-workshop/presentations/rust/
7. CEN Workshop on Learning Technologies (CEN WSLT): CWA 15555:2006 Guidelines and Support for Building Application Profiles in E-Learning (2006), ftp://ftp.cenorm.be/PUBLIC/CWAs/e-Europe/WS-LT/cwa15555-00-2006-Jun
8. Nilsson, M., Johnston, P., Naeve, A., Powell, A.: The future of learning object metadata interoperability. In: Harman, K., Koohang, A. (eds.) Learning Objects: Standards, Metadata, Repositories, and LCMS, pp. 255–313. Informing Science Press, California (2007)
9. Dublin Core Metadata Initiative (DCMI): Dublin Core Abstract Model, http://dublincore.org/documents/2007/04/02/abstract-model/
10. Coyle, K., Baker, T.: Guidelines for Dublin Core Application Profiles (DCAP), http://dublincore.org/documents/2008/11/03/profile-guidelines/
11. Zumer, M., Lei Zeng., M., Salaba, A.: FRBR: A Generalized Approach to Dublin Core Application Profiles. In: International Conference on Dublin Core and Metadata Applications, Pittsburg, pp. 21–30 (2010)
12. Greenberg, J., White, H.C., Carrier, S., Scherle, R.: A Metadata Best Practice for a Scientific Data Repository. Journal of Library Metadata 9, 194–212 (2009)
13. Agricultural Metadata Element Set (AgMES), http://www.fao.org/aims/agmes_intro.jsp
14. Eprints application profile functional requirements specification, http://www.ukoln.ac.uk/repositories/digirep/index/Functional_Requirements
15. IFLA Study Group on the Functional Requirements for Bibliographic Records (FRBR): Functional requirements for bibliographic records - final report. K.G. Saur, Munich (1998)
16. SWAP Working Group, JISC: Dublin Core Scholarly Works Application Profile, http://www.ukoln.ac.uk/repositories/digirep/index/Scholarly_Works_Application_Profile
17. Allison, J., Johnston, P., Powell, A.: A Dublin Core application profile for scholarly works, http://www.ariadne.ac.uk/issue50/allinson-et-al/
18. European Organisation for International Research Information (EuroCRIS): Common European Research Information Format (CERIF), http://www.eurocris.org/
19. Chaudhri, T.: Assessing FRBR in Dublin Core Application Profiles. Ariadne 58 (2009)

# Introducing a Content Integration Process for a Federation of Agricultural Institutional Repositories

Vassilios Protonotarios[1], Laura Gavrilut[1], Ioannis N. Athanasiadis[1], Ilias Hatzakis[1], and Miguel-Angel Sicilia[2]

[1] Greek Research & Technology Network (GRNET), Greece
[2] University of Alcala, Department of Computer Science, Spain
vprot@aua.gr, lgavrilut@yahoo.com, ioannis@athanasiadis.info, hatzakis@grnet.gr, msicilia@uah.es

**Abstract.** Aggregating metadata from various sources often raises practical issues, such as incompatibility between the different metadata application profiles (AP) used as well as quality aspects of the metadata used. In the case of repositories hosting agricultural-related content, the existence of various metadata AP with significant differences between them makes the effort of interconnecting these repositories a difficult task. This paper proposes a process for integrating different agricultural scientific content repositories and a workflow that should be followed in the context of populating the repository. Although the proposed solution refers to repositories with agricultural content, the same process can be followed in the case of repositories with different content. This process was proposed under the VOA3R project which is funded by the European Commission's ICT PSP Programme.

**Keywords:** content integration, aggregation, application profile, metadata.

## 1 Introduction

Despite the existence of a significant number of digital repositories hosting content related to agriculture, any effort towards interconnecting these repositories through metadata aggregators exhibits a high degree of complexity. This is due to the fact that the available resources have been annotated with metadata using application profiles with significant differences (e.g. Dublin Core, IEEE LOM, FAO AGRIS, custom etc.) therefore high incompatibility.

In the case of the VOA3R project (http://www.voa3r.eu), a number of repositories for agriculture and aquaculture will be aggregated and access to their resources will be provided through the VOA3R platform. However, the metadata used for the annotation of the resources in these repositories come from a wide variety of metadata AP, so there is a need for developing a technical solution for the harmonization of these differences and the usage of a common metadata AP, which will be used in all cases.

E. García-Barriocanal et al. (Eds.): MTSR 2011, CCIS 240, pp. 467–477, 2011.

As a result, the need for a process which will set up aggregators of content repositories, so that metadata is harmonized and quality-certified is raised. In order to facilitate the aggregation of resources coming from various sources, this paper proposes a process/workflow for the integration of institutional repositories, as well as a scenario illustrating its application.

## 2 Background

### 2.1 About VOA3R

VOA3R (Virtual Open Access Agriculture & Aquaculture Repository: Sharing Scientific and Scholarly Research related to Agriculture, Food, and Environment) is a 36-months CIP-ICT-PSP EU project which aims to improve access to European agriculture and aquaculture research results by using an innovative approach to sharing open access research results. One of the goals of the VOA3R project is to create a federation of scholarly content, by organizing and coordinating the population of the VOA3R repository with content coming from the participating repositories and communities. This content will be described with semantically rich and interoperable metadata. In order to achieve this, a workflow which should be followed by all the members of the VOA3R federation during the phase of content population in the VOA3R repository is developed. The scope of the process is to integrate new resources to the existing collections and populate the VOA3R federation with educational agricultural content.

In order to come up with a common and compatible framework for the annotation of resources aggregated from various repositories, thus minimizing any incompatibility issues, VOA3R aims to develop a metadata AP (the VOA3R AP) based on the feedback received from the content providers of the project. This profile will be used for the integration of content within the project and will remain available for future use by additional content providers which wish to make their collections available through the VOA3R federation and platform.

While the content providers of the VOA3R project with existing repositories need to map their metadata AP to the VOA3R AP, the content providers with new repositories can adopt the VOA3R AP and procedures for quality assurance from the very beginning. In the case of the content providers with existing repositories, an enrichment of the metadata description, in order to meet the criteria of the VOA3R metadata application profile might be necessary. As regards this enrichment, there are two options:

- To revise or to extend their internal metadata schema by adding content to their repository or
- To modify their existing metadata application profile, in order for it to be compliant with the VOA3R one.

The content provider will be free to use the option that suits best its collection's needs. It should be emphasized that both options are supported by the VOA3R project.

## 2.2 The VOA3R Collection

The VOA3R federation consists of content integrated from the following institutional repositories:

### 2.2.1 Epsilon - Swedish University of Agriculture Sciences (SLU), Sweden

SLU has national responsibility for research and education in veterinary medicine, forestry, and landscape planning. SLU will be provider of metadata to the VOA3R specifications as well as content provider through the Epsilon Open Archive, which contains a significant number of research documents in the form of theses (≈3700), papers and publications (≈500) and other documents (≈550).

### 2.2.2 OceanDocs - Hasselt University (UHASSELT), Belgium

Hasselt University will provide metadata and data from two different sources: OceanDocs (≈9000 resources) and the Center for Environmental Sciences (≈150 resources), as well as repository development (based on the OceanDocs experiences), and also at a more limited level in the development of tools.

### 2.2.3 Organic Eprints - International Center for Research in Organic Food Systems (ICROFS), Denmark

ICROFS is a "centre without walls", facilitating the multidisciplinary research in organic food and farming across institutes. It is hosted by the Faculty of Agriculture of the University of Aarhus (DJF). Its main purpose is to initiate, coordinate and monitor high quality organic research, and to stimulate international and transnational research. ICROFS hosts and administrates Organic Eprints (≈6900 resources), the open-access archive on organic research (www.orgprints.com), which will be used in the VOA3R project.

### 2.2.4 ProdINRA - French National Institute for Agricultural Research (INRA), France

INRA is the largest agricultural research institute in Europe. INRA has signed the Berlin Declaration on Open Access in July 2005. An institutional open archive called Prodlnra has been set up to make available the INRA research, which will contribute about 3800 publications (e.g. articles, reports, papers and books) to the VOA3R Repository. Since October 2006, ProdINRA is used by all researchers at INRA.

### 2.2.5 U-GOV - Consorzio Interuniversitario (CINECA), Italy

CINECA is a Consortium consisting of 36 Italian Universities. Its institutional mission is to support research conducted by the Italian scientific community through supercomputing and its applications. Cineca will participate on VOA3R as content and metadata provider through the U-GOV Repository, offering about 45000 scholarly resources to the VOA3R repository.

#### 2.2.6 ARI Repository - Agricultural Research Institute (ARI), Cyprus

The Agricultural Research Institute (ARI) was established in 1962 and is one of the Departments of the Ministry of Agriculture, Natural Resources and Environment. ARI undertakes applied and basic research within the wider domain of plant and animal production. Its mission is to provide high quality scientific research using methods that are financially, environmentally and socially sustainable. ARI repository includes full-text publications (in PDF format) from 1965-2007, with the metadata description in FAO AGRIS AP.

More information about the aforementioned institutional repositories and their corresponding contribution to the VOA3R network can be found in Table 1.

## 3 Generic Process

### 3.1 Aim of the Process

The VOA3R project has developed a methodology for the integration of content, in order to meet the specific needs of the VOA3R project. The methodology proposed by the VOA3R project consists of four phases, during which the VOA3R content repository will be populated. The Initial Phase will be a Testing Phase which may be considered as a kind of pre-phase, followed by other three phases. All these Phases include a series of generic steps that are proposed to be followed in all phases (the testing phase and the three phases) of the content population and are explained below:

| Phase | Date | Activities |
|---|---|---|
| Controlled testing phase | Jul – Aug '11 | Enrichment of test metadata records using Confolio |
| Phase 1 of Content Population | Sep – Dec '11 | Integrating the agriculture and aquaculture repositories using the common OAI-PMH specification |
| Phase 2 of Content Population | Jan– Aug '12 | Integration of repositories without OAI-PMH support and further contents from internal partners |
| Phase 3 of Content Population | Sep '12 – May '13 | Enriching the VOA3R portal by external partners and contents |

**Table 1.** VOA3R Content Providers and information on their collections

| **Partner name** | ***Swedish University of Agriculture Sciences (SLU)*** | ***International Centre for Research in Organic Food Systems (ICROFS)*** | ***Consorzio Interuniversitario (CINECA)*** | ***Hasselt University (UHASSELT)*** | ***French National Institute for Agricultural Research (INRA)*** | ***Agricultural Research Institute (ARI)*** |
|---|---|---|---|---|---|---|
| **Collection / Repository Name** | Epsilon Repository | Organic Eprints | U-GOV Repository | OceanDocs Repository | ProdINRA Repository | ARI Repository |
| **Resource type** | ~3700 Theses<br>~500 Papers and publications<br>~550 Reports/Others | ~10000 Papers<br>~1400 Organizational resources<br>~350 Others | ~45000 Papers, publications and other types<br>~1700 News items | ~9000 Papers and Reports | ~4300 Papers<br>~340 Books, book chapters & working papers<br>~ 240 Reports<br>~ 430 Other documents | ~315 Monographies |
| **Content language(s)** | Swedish and English | English, German and Danish | Italian, English | English, French, Spanish, Portuguese, Russian, Arabic | French & English | English |
| **Metadata language(s)** | There is a "main" language for the metadata record which is the language of the publication. | English | Italian, English | English | French & English | English, Greek |
| **Topics covered by the content** | Horticulture, Agricultural Science, Natural Resources, Forest Sciences | Research in Organic food and farming systems | Agriculture, rural development; agricultural economics, management, agribusiness, ICT, information systems, e-business, social economy & rural sociology | Oceanography, Water, Fisheries, Environment, Aquatic science | Agriculture, Environment, Food | Agriculture |

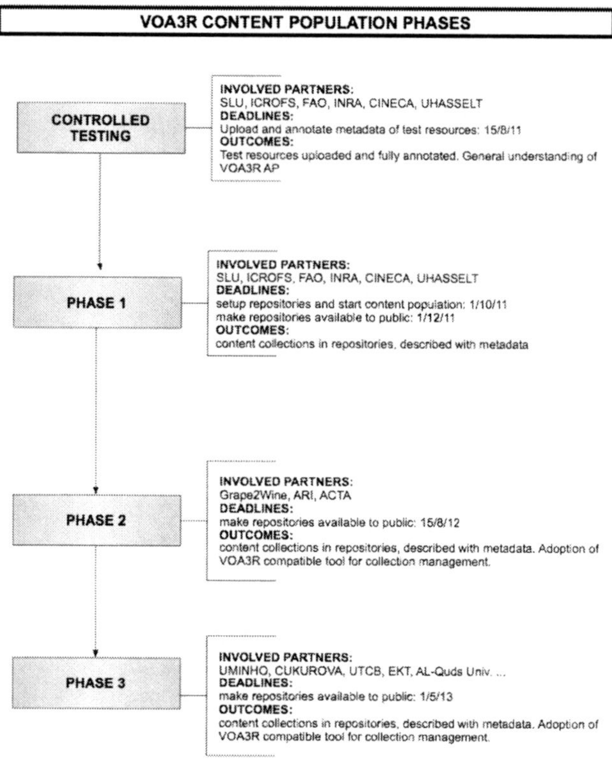


**Fig. 1.** The four phases of the VOA3R content integration procedure

## 3.2 Overview of the Process

The proposed four steps are the following:

**1. Uploading/Integration**

This step includes the uploading of the resource or the link pointing to this resource. According to our methodology, a pre-check of the content should be conducted before the uploading takes place, in order to assure that the following basic criteria are fulfilled, by the statement of the content provider:

- the resource does not contain any violent, pornographic or racist content;
- the resource is relevant to the repository it should be inserted (e.g. Agriculture & Aquaculture in the case of VOA3R project) and
- the Intellectual Property Rights do not prohibit the resource to be distributed.

**2. Enrichment**

The Enrichment step includes the annotatation and/or enrichment of the metadata elements, according to the VOA3R set of recommended metadata. The main language of the metadata should be English, while an additional translation of these metadata fields should take place.

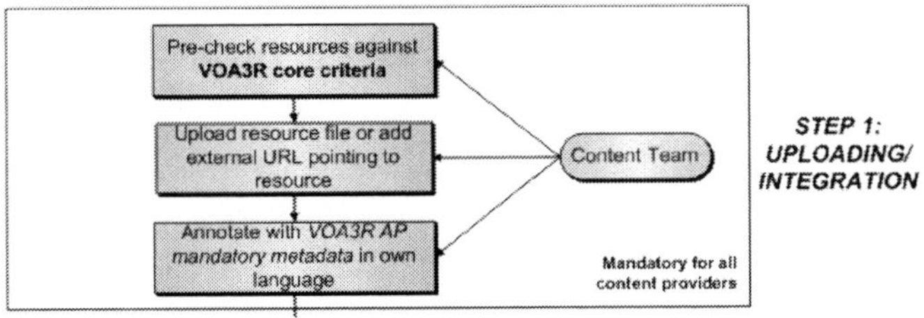


**Fig. 2.** Step 1 of the VOA3R content integration process

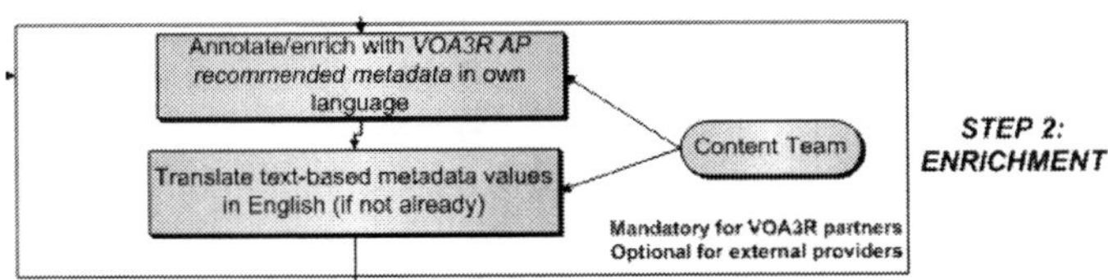


**Fig. 3.** Step 2 of the VOA3R content integration process

## 3. Validation

Before the publication of the resources, a validation of the metadata is necessary. Both the metadata records and the resource are checked against the VOA3R core criteria mentioned above. In case the validation is successful, the metadata records can be published on the VOA3R web portal. The check against the VOA3R core criteria in this step is not done by the content team member but by the content validator.

If the tool used by the partners that are willing to offer their content is not one of the tools already used by the VOA3R federation, it should have the ability to expose metadata through the OAI-PMH standard and map these metadata to the VOA3R application profile. The exposure of the metadata through the aforementioned standard should be done in this step.

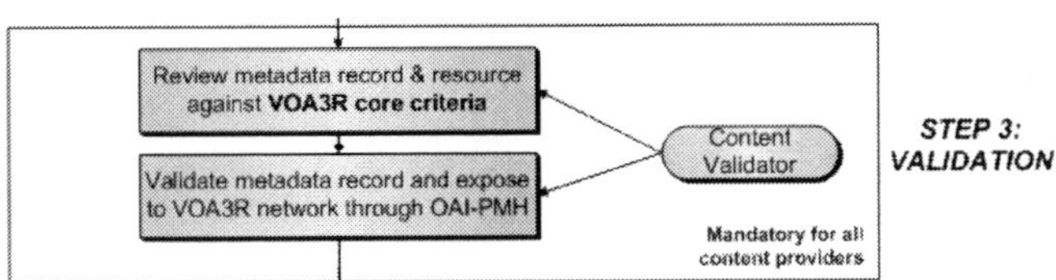


**Fig. 4.** Step 3 of the VOA3R content integration process

## 4. Quality Review/Assessment

After the resources are published, a number of them will be evaluated in terms of the quality of both their content and their metadata.. This evaluation will be performed with the use of predefined check grids in order to review the quality of the content and the metadata, as well as their relevance with the content.

According to the proposed methodology, the quality of the metadata is reviewed manually by a peer-reviewer (an expert in the field of the resource belonging to one of the participating organizations which is). The internal quality review schema will involve the project partners, who will cross-check each other's metadata records without having the permissions to change it. As regards the completeness of the metadata description, an automatic mechanism will check this aspect and will warn in case the mandatory fields of the metadata of any resource are not filled in.

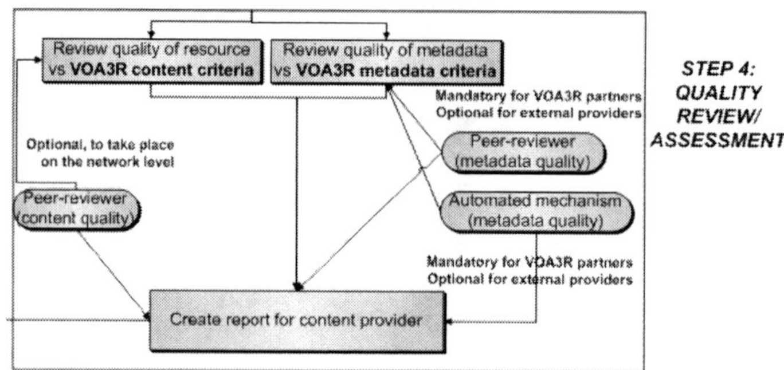


**Fig. 5.** Step 3 of the VOA3R content integration process

The aforementioned four steps are for all the content providers, using any system as a Content Management System to connect to VOA3R repository.

## 4 Scenario of Use: The Testing Phase

In this Chapter we will discuss only the Testing Phase, as the other three Phases could be subject for another scientific paper. The Testing Phase includes all steps mentioned in the previous Chapter. The testing Phase it is addressed to all the content providers using Confolio. Confolio (http://voa3r.confolio.org) is a repository tool, which is used for creating, storing, indexing and retrieving the metadata description for each resource which is stored in a networked repository.

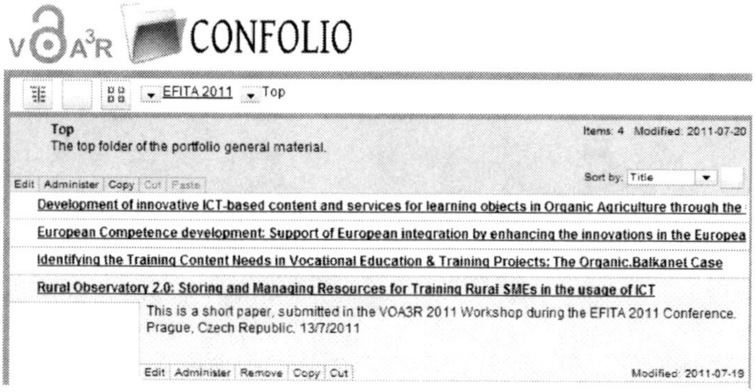

During the testing phase, the following steps will take place:

## 1 Uploading/Integration

Users will be asked to register in the VOA3R Confolio tool (http://voa3r.confolio.org). After they successfully log in, they will have to upload a number of digital resources, respecting the VOA3R Core Criteria Grid and provide the mandatory metadata description for each one of the resources.

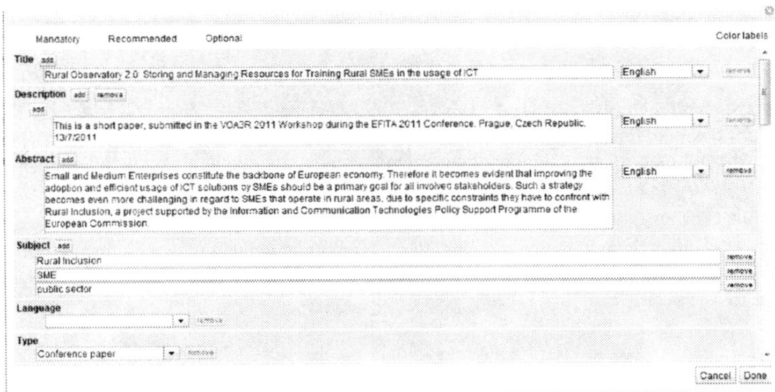


## 2 Enrichment

Partners will be asked to provide the recommended metadata description as well and translate in English all the metadata fields used.

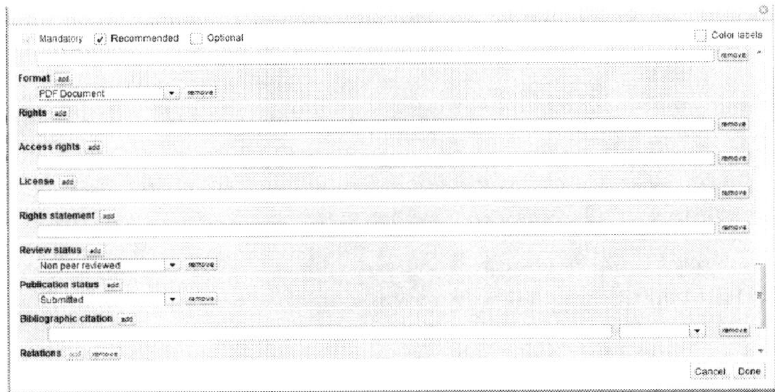


## 3 Validation

Uploaded resources and their metadata will be checked against the VOA3R Core Criteria and they will be published.

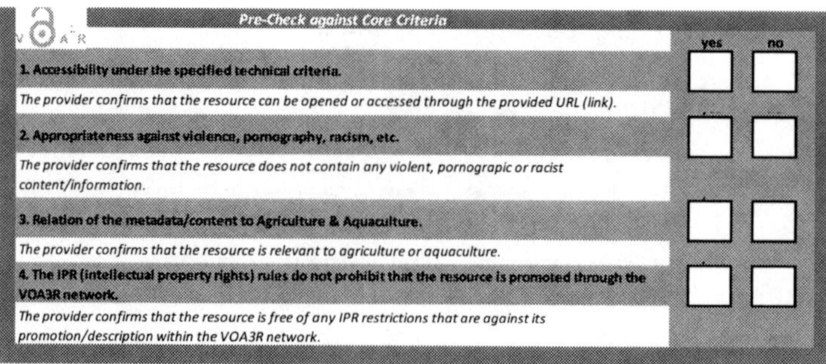

| Pre-Check against Core Criteria | yes | no |
|---|---|---|
| **1. Accessibility under the specified technical criteria.** | | |
| *The provider confirms that the resource can be opened or accessed through the provided URL (link).* | | |
| **2. Appropriateness against violence, pornography, racism, etc.** | | |
| *The provider confirms that the resource does not contain any violent, pornograpic or racist content/information.* | | |
| **3. Relation of the metadata/content to Agriculture & Aquaculture.** | | |
| *The provider confirms that the resource is relevant to agriculture or aquaculture.* | | |
| **4. The IPR (intellectual property rights) rules do not prohibit that the resource is promoted through the VOA3R network.** | | |
| *The provider confirms that the resource is free of any IPR restrictions that are against its promotion/description within the VOA3R network.* | | |

## 4 Quality Review/Assessment

Uploaded resources will be reviewed in terms of the content and the use of metadata by the WP5 leaders and a report will be sent back to the content teams for revising the metadata description.

| METADATA RECORD EVALUATION FORM | Evaluator | | | | |
|---|---|---|---|---|---|
| *Metadata record identifier:* | *Metadata record URI:* | | | | |
| | low → high | | | | |
| **1. In which degree is this metadata record completed?** | 1 | 2 | 3 | 4 | 5 |
| *Number of element values provided by the annotator in comparison to the total number of VOA3R recommended element values. All mandatory and recommended elements must be completed. Extra points are gained for optional elements provided. Points are subtracted if recommended elements are missing* | | | | | |
| **2. Please identify the overall accuracy of the metadata values provided** | 1 | 2 | 3 | 4 | 5 |
| *In an accurate metadata record, the data contained in the fields, correspond to the resource that is being described. Can you get the same information for the resource when looking at the resource itself and/or the metadata values? This question involves the task of checking the resource itself!* | | | | | |
| **3. Are the metadata values provided consistent with the metadata standard used?** | 1 | 2 | 3 | 4 | 5 |
| *Consistency measures the degree to which the metadata values provided are compliant to what is defined by the metadata application profile of VOA3R. Do they follow the definition of the element and the expected values?* | | | | | |
| **4. Do the metadata values describe the resource in an objective, unbiased way?** | 1 | 2 | 3 | 4 | 5 |
| *Degree in which the metadata values provided, describe the resource in an unbiased way, without undermining or promoting the resource in any way* | | | | | |
| **5. Are the metadata values provided, appropriate for the targeted use in the VOA3R network & platform?** | 1 | 2 | 3 | 4 | 5 |
| *Are the metadata values appropriate for helping users to find resources in the VOA3R network through the VOA3R Platform?* | | | | | |
| **6. Please define the degree of correctness of the language used** | 1 | 2 | 3 | 4 | 5 |
| *Is the language used in the metadata, syntactically and grammatically correct?* | | | | | |
| **7. Please provide an overall score for the metadata of this resource, based on your ratings in questions 1 to 6. The overall quality of the metadata record:** | 1 | 2 | 3 | 4 | 5 |
| **8. Do you consider the quality of the metadata record for this resource of an accessible level to be published in the VOA3R network?** | YES | NO | | | |
| **9. Comments** *Explanation of the review provided. Especially if the metadata record is rejected. Suggestions for improvement.* | | | | | |

The Testing Phase will take place in three periods:

- The first period is a trial annotation period when all the content providers will upload on Confolio a small number of sample resources in order to experiment and to get familiar with the proposed elements of the VOA3R metadata AP, as well as with the proposed value sets. At this current stage of the Testing Phase the annotations are only in English. The samples which will be uploaded are part of the promised content for the repository. The scope of this first period is to verify the design choices at an early stage and the feedback of the content providers is welcome and it will help in improving the process.
- In the second period, the common scheme and the metadata enrichment will be revised. The content providers that will adopt the common scheme will develop internally the appropriate infrastructure to make available the VOA3R profile through OAI-PMH protocol. The content providers that will not adopt the VOAR

profile will need to add extra annotations through Confolio. If in the first period there were tested a few samples, in this second period, there will be more samples. Moreover the annotations will not be only in English but in multiple languages.

- In the third period, final adjustments to the common metadata schema will be performed. Furthermore the final version in XML and RDF formats will be released. Content providers will focus their efforts on populating the content of their repositories with additional metadata. The content providers that will adopt the VOA3R metadata schema will have to populate the new elements using their own tools and procedures while the content providers that will not extend their metadata structure must provide with the additional metadata.

## 5 Conclusions

The need for working on standardized methodologies and procedures in the area of content integration from various repositories is constantly rising. What this paper has innovative is the process of integration of different repositories from diverse users, users with their own system or users with a system compatible with ours. Therefore this paper provides a methodology for content integration from various agricultural repositories to a single network, by suggesting a uniform method for the different phases of the process, as well as a common metadata AP to be used by all partners who integrate their content in the VOA3R collection. We would like to emphasize that the access to the metadata of the content will be open. Thus we are trying to head towards the trend of the society, open access. Regarding the content we cannot guarantee open access because of the copyrights issues.

**Acknowledgments.** The work presented in this paper was conducted under the framework of the VOA3R project funded by the European Commission through the ICT PSP Programme (ICT PSP), Theme 4 - Open access to scientific information.

## References

1. Organic.Edunet: Deliverable D4.1, Methodology for Content Population. A Multilingual Federation of Learning Repositories with Quality Content for the Awareness and Education of European Youth about Organic Agriculture and Agroecology (2009)
2. Patrikakis, C.Z., Koukouli, M., Papadopoulos, G.K., Sideridis, A.B.: Evaluating Behavioral Change in Multigroup Collaboration for Content Publishing Over the Web. Social Science Computer Review 27, 59 (2009) (originally published online June 5, 2008), doi:10.1177/0894439308319449
3. `http://ssc.sagepub.com/cgi/content/refs/27/1/59` (accessed June 26, 2011)
4. Palavitsinis, N., Manouselis, N., Sanchez, S.: Evaluation of a Metadata Application Profile for Learning Resources on Organic Agriculture. In: Sartori, F., Sicilia, M.Á., Manouselis, N. (eds.) MTSR 2009. CCIS, vol. 46, pp. 270–281. Springer, Heidelberg (2009)
5. Palavitsinis, N., Ebner, H., Manouselis, N., Sanchez, S., Naeve, A.: Evaluating Metadata Application Profiles Based on Usage Data. In: Proc. of theInternational Conference on Digital Libraries and the Semantic Web (ICSD 2009), Trento, Italy (2009)

# Publishing and Linking Semantically Annotated Agro-environmental Resources to LOD with AGROPub

Saša Nešić[1], Andrea Emilio Rizzoli[1], and Ioannis N. Athanasiadis[2]

[1] IDSIA, Manno, Switzerland
{sasa,andrea}@idsia.ch
[2] Democritus University of Thrace, Xanthi, Greece
ioannis@athanasiadis.info

**Abstract.** Publishing agro-environmental resources to a linked open data (LOD) cloud requires publishers to adopt a set of universally recognized linked data principles. These principles, along with semantic annotations based on shared domain ontologies can ensure the semantic integration of agro-environmental resources. In this paper we present a resource-publishing system, called AGROPub, that we developed to aid agro-environmental resource providers to annotate, publish and integrate their resources to LOD. The system comprises services and tools that enable resource providers to annotate their resources by relevant concepts from selected agro-environmental domain ontologies, to generate and publish RDF descriptions of the resources to LOD and to link the published resources to related resources from LOD. In addition to the services and tools dedicated to resource providers, AGROPub provides services and tools that enable consumers of the agro-environmental resources to search and annotate published resources by adding their own annotations as well as to evaluate them based on given criteria.

**Keywords:** agro-environmental resource publishing, annotation, linked-open data.

## 1 Introduction

The availability of high quality and up to date information on environmental conditions and agriculture has always been of major interest for many institutions, especially given the relevance of such information to our ability to secure food supply. Typical sources of information range from raw data provided by sensor networks and satellite imaging, to processed data stored in databases and catalogues, to information processing units such as mathematical models, software, web services, decision support systems, and to scholar papers, maps and reports, as those published in the JRC MARS bulletin[1]. In spite of being

[1] The European Joint Research Center periodically publishes the Bulletin of the Monitoring Agriculture ResourceS Unit (MARS) on agrometeorological analyses for Europe. For more see: `http://mars.jrc.it/mars/Bulletins-Publications`

E. García-Barriocanal et al. (Eds.): MTSR 2011, CCIS 240, pp. 478–488, 2011.

available on the Web, the great majority of the agro-environmental resources is typically accessed only by closed communities which are aware of them. Accessing agro-environmental resources effectively and efficiently still requires a deep understanding of both domain complexity and computer systems that provide them[14]. Moreover, related resources that are published and managed by different authorities are still greatly disconnected and unaware of each other. As a result, the existing agro-environmental resources on the Web look more as sets of disconnected information islands than an integrated information space.

To narrow the discovery gap that prevents a full and easy access to agro-environmental resources on the Web it is required to develop tools to organize the knowledge hidden in such resources. Recent research efforts [12,1,11] have shown that employing rich semantics in the form of domain ontologies is a promising solution. We rely on the TaToo approach[2] that proposes a comprehensive framework for ontology-based, semantic tagging (annotation) of environmental resources [11]. Among others, the TaToo framework provides a novel environmental resource model, called MERM (Minimal Environmental Resource Model) and a number of shared domain ontologies. Moreover, in order to successfully integrate agro-environmental resources on the Web into a globally unified agro-environmental information space, it is necessary to represent (i.e. identify and describe) the resources in accordance to some globally recognized principles. Linked open data (LOD) principles have been recognized as one of the most promising initiatives regarding this issue [4].

In the quest of this long-term vision, we present here AGROPub, a system that facilitates integration of agro-environmental resources into the semantically unified, global environmental information space [9] as a part of LOD. The system is designed as an extensible, service-oriented architecture that intends to provide a variety of services supporting both agro-environmental resource providers and consumers to contribute to this vision. The resource providers are responsible for the resource publishing to LOD as well as setting up some initial links between their resources and other related resources from LOD. However, the real resource integration into LOD happens over time as a result of the user interaction with the resources (e.g., searching and reusing, annotating, interlinking, and evaluating). In this paper our focus is on the AGROPub services and tools that enable agro-environmental resource providers to publish their resources to LOD, add initial semantic annotations, and setup initial links to existing, related resources from LOD.

The rest of the paper is organized as follows. In Section 2 we introduce the architecture of the AGROPub system and outline main features of the AGROPub services and tools. Then, in Section 3 and Section 4 we give detailed, step-wise descriptions of the AGROPub resource publishing and linking use-cases respectively. We conclude the paper with a brief overview of related work, and plans for future work.

[2] `http://www.tatoo-project.eu`

## 2 AGROPub: Agro- environmental Resource Publisher

Our main goal of developing the AGROPub system was to support agro-environmental resource providers in publishing and linking (integrating) their resources to the LOD cloud. Apart from that, we also wanted to provide services that would facilitate both resource providers and resource consumers in searching, navigating and evaluating linked agro-environmental resources. We designed the AGROPub system as a three-tier, client-server architecture (see Figure 1). It is composed of a data tier, a service tier, and a presentation tier.

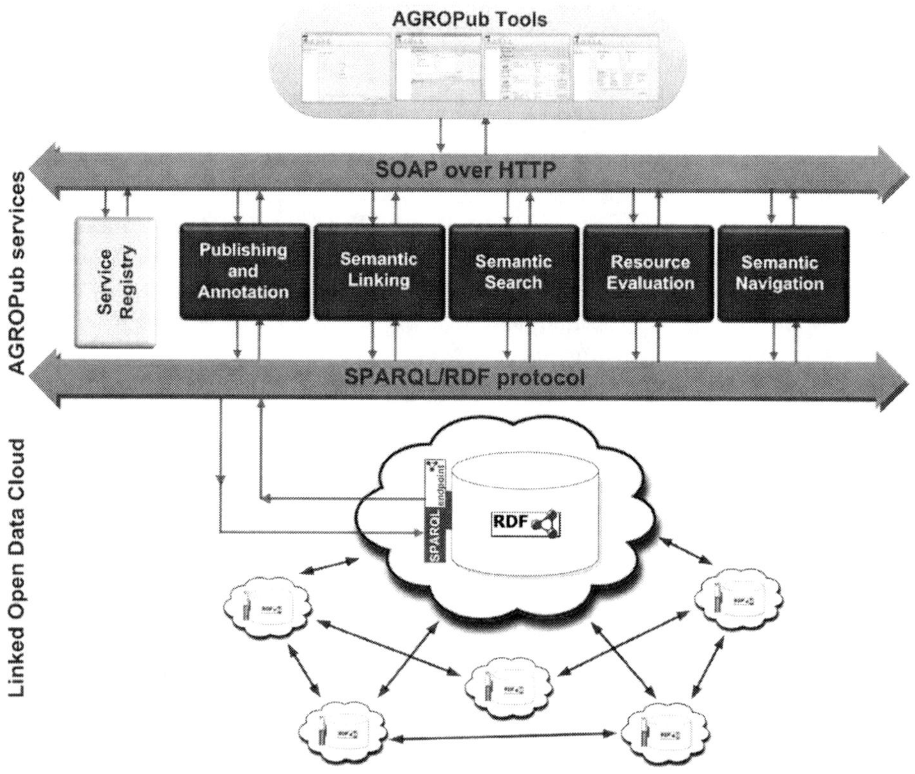


**Fig. 1.** Illustration of the AGROPub architecture

The **data tier** consists of RDF repositories storing descriptions (i.e. RDF triplets) of the published agro-environmental resources. Each of the RDF repositories from the data tier exposes an HTTP de-referenceable SPARQL endpoint, so that the published resources can be referred and linked to other resources from the LOD cloud.

The **service tier** offers the AGROPub system functionalities. The number of provided services can be extended by adding an arbitrary number of new services, which have to comply with the AGROPub design principles. Moreover, to be operational within the system each service has to be registered (through a service registry). The service registry provides registration and look up functionality for the AGROPub services as well as provides methods to enable and disable services. In a current design of the AGROPub system we provided a set of five services:

(a) *Publishing and Annotation*,
(b) *Semantic Linking*,
(c) *Semantic Search*,
(d) *Resource Evaluation*, and
(e) *Semantic Navigation*.

The services are designed as Windows Communication Foundation (WCF) Web services, all of them exposing a standardized Web-service interface.

The **presentation tier** is a top tier of the AGROPub architecture, which provides the user interface for the AGROPub services. The presentation tier is technology- and platform-independent according to the service-oriented nature of AGROPub. It may contain different kind of applications, including Web-based applications, desktop applications, and mobile phone applications. In the current version of the AGROPub prototype, as a part of the presentation tier we developed several tools and encapsulated them in a desktop application namely AGROPub Manager.

In the rest of this section, we report on the main features of the AGROPub services and the AGROPub Manager.

### 2.1 AGROPub Services

The **Publishing and Annotation** service provides with the functionality to generate and publish RDF descriptions of agro-environmental resources. The RDF resource descriptions are generated in accordance to the TaToo specifications of the Minimal Environmental Resource Model (MERM) [11]. The MERM model describes environmental resources by combining a set of basic resource information, such as a resource Web page, resource author and resource creation date, and a set of ontology-based annotation tags that represent domain-specific semantics of the resources. The annotation tags, actually, link concepts from a domain ontology to resources (in the form of URIs). As an example domain ontology, used to test our system, we used the JRC ontology [10]. This ontology provides concepts and properties for modeling different types of agro-environmental resources (e.g., agricultural models and agricultural software) as well as concepts and properties describing agricultural systems, processes and products. After generating the RDF descriptions the service stores them into the AGROPub RDF repository and finishes the publishing process. Besides the resource providers, the publishing and annotation service is supposed to be used by the resource consumers to add their own annotations to the linked agro-environmental resources.

The **Semantic Linking** service provides functionalities for associating related agro-environmental resources regardless of whether their RDF descriptions are stored into the same or different RDF repositories. The service is capable of generating two types of links: generic and property-based links [9]. The generic links are generated automatically whenever a new resource is published. When the resource publishing is finished, the publishing and annotation service triggers the semantic linking service which then crawls available RDF stores, identifies other resources related to the published one, and adds the appropriate generic links. The service determines the semantic relation between resources based on their ontology-based annotations. The property-based links are generated from information (i.e. resources to be linked and a linking property) provided manually by the AGROPub users.

The **Semantic Search** service implements a simple, concept-matching search [12] of linked agro-environmental resources. Similar to the resource annotation process, a user specifies a query by selecting concepts from the domain ontologies used by the AGROPub system. The system takes the user query and invokes the service, which in turn creates a corresponding SPARQL query, and executes it against available RDF repositories. The search result contains resources that are annotated by the query concepts. In the next service version, we plan to enhance the search algorithm by taking into account structure (i.e. ontology-graph traversal) [3] and formal semantics of the underlying domain ontologies (i.e. RDFS/OWL reasoning) [1].

The **Semantic Navigation** service provides functionalities that enable users to navigate across linked agro-environmental resources by following links among them. The navigation process requires the existence of a navigation user interface that is a part of the AGROPub presentation tier, through which the user can interact with the navigation service. The navigation starts by the user selecting a resource from the search results and then clicking on one of the resource's links. This user action invokes the semantic navigation service, which takes as input the resource's URI and the link's data, forms a navigational SPARQL query, and executes the query against the available RDF repositories.

The **Resource Evaluation** service enables AGROPub users to express their opinion on linked agro-environmental resources that are accessible trough the AGROPub system. A user is enabled to evaluate not only the actual resources but also the resource annotations added by others. Both resource and annotation evaluations are characterized by an evaluation criterion and a corresponding evaluation value. The service generates and stores the evaluations as RDF instances of the resource evaluation schema, which is defined as a part of the MERM model [11].

### 2.2 AGROPub Manager

The AGROPub Manager is a Windows-based, desktop application that comprises a set of tools enabling users to interact with the AGROPub services. Figure 2 shows a screenshot of the application main menu ribbon tab. The tools are accessible through five menu buttons namely: Resource Publishing, Resource Annotation, Resource Search and Resource Navigation. As their names suggest the first two tools enable users to publish their resources as well as to annotate already published resources. The next tool, Resource Search, enables users to specify their queries and search for desired resources. The tool also provides a detailed view of resource information and annotations. Among others, the detailed resource view contains user-interface elements that enable the resource evaluation and linking. The last tool, Resource Navigation, enables users to navigate across linked resources by clicking on links connecting them. Users can specify the initial resource by entering the resource's URI or by selecting one of the resources from the list of previously obtained search results.

**Fig. 2.** The main menu tab of the AGROPub Manager

In the following two sections we provide a closer look at the user actions composing the AGROPub resource publishing and linking use-cases. For each user action we refer to the employed tools, user-interface elements, and services of the AGROPub system.

## 3 Publishing Agro-environmental Resources

In this section, we present the resource publishing use-case of the AGROPub system. The resource publishing starts by the user selecting the Resource Publishing button from the main menu ribbon tab of the AGROPub Manager. The user is then faced to the new window with two panes (see Figure 3). The left pane contains a navigation bar with a list of supported agro-environmental resource types (e.g., agricultural models, software components and remote sensing measurements). After selecting a resource type, the right pane displays a resource-publishing control tab, containing forms for specifying a resource description.

In case the user wants to publish an agricultural model, s/he will follow the following steps:

(a) **Specifying resource info:** The first step in publishing the resource is to specify a set of basic information about the resource including the resource's URI, the URL of the Web page describing the resource, the resource's author, the resource's owner, and the resource's creation date. The only mandatory information to be provided in this step is the resource URI, which the only mandatory information in a resource description. The resource URI must

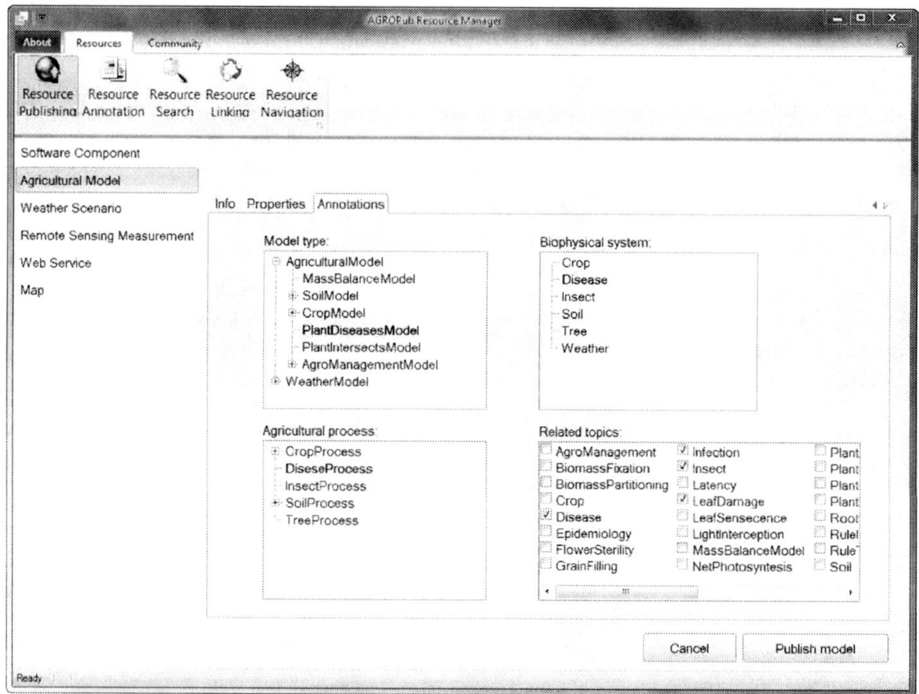


**Fig. 3.** Resource annotation and publishing with the AGROPub Manager

be globally unique, which we ensure by validating the entered resource URI against the URI schema that we proposed in [9]. Only in case of resources whose content can be obtained trough HTTP, the resource URIs and URLs are identical.

(b) **Specifying resource properties:** In this step the user specifies values for the properties that are specific to the selected resource type. In case of the agricultural-model, among others, the user specifies the inputs and outputs of the model, the model's algorithm and the software components that implement the model.

(c) **Specifying resource annotations:** This is the last step in which the user specifies the ontology-based annotations for the resource to be published. Basically, the user selects concepts from a domain ontology [2] that is loaded in the AGROPub manager. The right pane of Figure 3 shows concepts from the JRC ontology that we used in this example to annotate an agricultural model resource. The concepts from the ontology are shown into four boxes each of which is dedicated to a specific aspect of the resource annotation: agricultural model types, related biophysical systems, related agricultural process, and related agricultural topics.

Having passed trough these three steps, the user finishes with the specification of the resource description and can initiate resource publishing. By clicking on the "*Publish model*" button, the AGROPub Manager collects the user-specified model description and calls the publishing and annotation service. When the service finishes the publishing process, the user is notified by a confirmation message. If the user attempts to publish a resource that has already been published, the AGROPub Manager displays an appropriate warning message. Since each resource published to the LOD has to be unique, it is not possible to publish the same resource (i.e. the resource with the same resource URI) more than once.

## 4 Linking Agro-environmental Resources

After having published a resource to the LOD, it is also recommended that the resource publisher provides an initial set of the resource's outgoing links, connecting it to some related resources. Along with the initial resource annotations, specified during resource publishing, the initial set of outgoing links will enhance the resource discoverability and foster further integration of the resource in the LOD. Note that the links we consider in this section are the property-based links which are supposed to be added manually by the AGROPub users. Linking resources by the generic links [9], which are generated automatically, is not in the scope of this paper. To demonstrate the linking use case, consider the same agricultural model used in the resource publishing use-case (Section 3). The AGROPub user is offered with the following options:

1. **Browsing resource details:** The AGROPub Manager offers two ways to access the resource to be linked. One is to manually specify the resource URI and request its details and the other one is to select the resource from the previously obtained search results. The resource detailed view offers a resource-linking user interface, which is realized as a cascading popup with two cascades (see Figure 4). The first cascade contains a combo-box filled with a list of predefined linking properties. The second cascade contains a list of cached resources. Initially, this list is empty.
2. **Caching related resources:** After observing the resource's details the user needs to find resources that s/he might consider for linking afterwards. Therefore, the next step in the linking use-case is to explore available linked resources and to identify potential candidates for linking. To perform this step the user utilizes the search and navigation utilities of the AGROPub Manager. When the user identifies a resource that is related to the considered resource, the AGROPub Manager enables him/her to cache the resource into its internal memory. The user can keep on searching and navigating among available resources as long as s/he caches a satisfactory number of resources.
3. **Selecting a linking property:** Having cached related resources the user moves back to the resource's detailed view and opens the linking popup. The next step is to select the linking property over which the link will

be established. Currently, the user is constrained to select the property from the predefined list of properties such as `owl:sameAs`, `skos:related`, `skos:closeMatch` and `skos:exactMatch`. All these properties come with formal specifications from standard vocabularies. For the next version of the AGROPub System we plan to consider the possibility of using custom, user-defined properties for linking resources.

4. **Selecting a resource:** The last step in the linking process is to select a related resource from the list of cached resources. After the user selects the resource and pushes the "Create Link" button the AGROPub Manager calls the linking service, which generates and stores the link (i.e. the RDF triple) in the RDF repository. If the link between the two resources already exists in the repository, it will not be generated again.

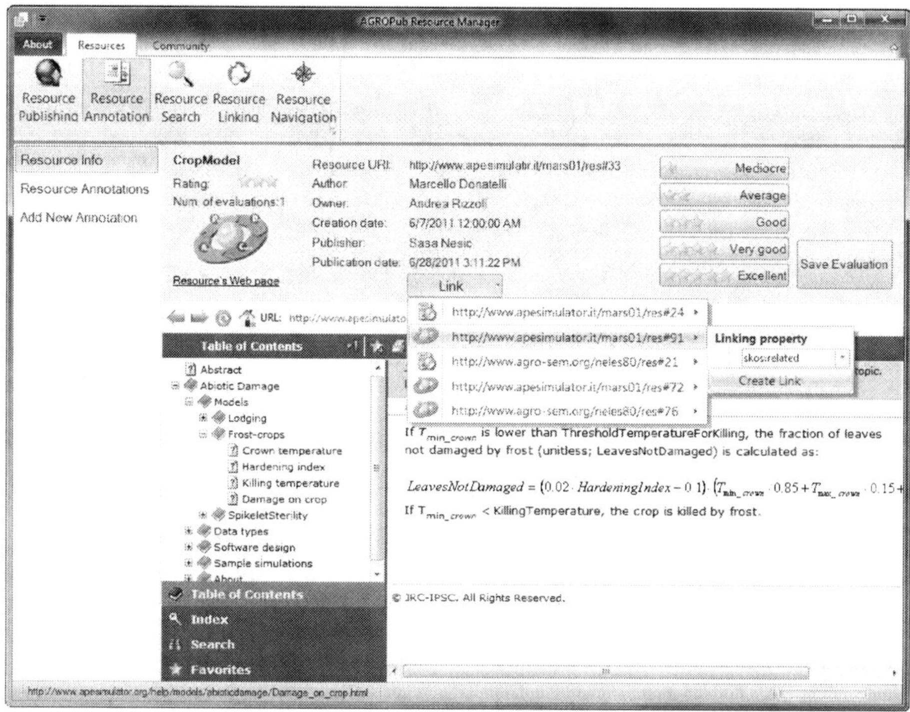


**Fig. 4.** Resource linking with the AGROPub Manager

## 5 Related Work and Discussion

While there are several generic-purpose tools for annotating web resources (as Annotea[3] and Delicious[4]), there has been a transition towards domain-specific

[3] `http://www.w3.org/2001/Annotea`

[4] `http://www.delicious.com`

tools that are able to incorporate functionality required by certain communities. For example, SemTag & Seeker have demonstrated how to perform automated semantic tagging of large corpora [5]. Connotea[5], and Mendeley[6] focus on the scientific publications lifecycle. Confolio[7] have been used for annotating and organizing educational resources [6].

Related to the agro-environmental domain there are several services that have been recently deployed, as those for open sensor architectures [13], reusable geospatial services [8], and modelling water resources [7]. However, there are not any services or tools available for annotating and evaluating such services. This need is addressed by AGROPub system presented in this paper, which aims to narrow the search and discovery gap of agro-environmental information.

In this paper, we presented the development of the AGROPub system, which is a result of our ongoing efforts to facilitate integration of agro-environmental resources into the semantically-unified environmental information space. The system enables agro-environmental resource providers to annotate their resources by ontology-based, semantic annotations and then publish and link resource descriptions to LOD. Besides the resource providers, the system provides services that enable resource consumers to search, navigate and evaluate linked agro-environmental resources. We designed the system to be extensible in terms of adding new services as well as independent from applied domain ontologies. Therefore, as a part of our feature work we plan to investigate on potential new services and to test the system with different domain ontologies. In order to evaluate the system and the underlying annotation, linking, search, evaluation and navigation approaches, we plan to conduct two evaluation studies. The first study will focus on a usability evaluation of the system's services and tools. The main focus of the second study will be on a discoverability of the linked agro-environmental resources within LOD.

**Acknowledgements.** The research leading to these results has received funding from the European Community's Seventh Framework Programme (FP7/2007-2013) under Grant Agreement Number 247893.

## References

1. Anyanwu, K., Maduko, A., Sheth, A.P.: Semrank: ranking complex relationship search results on the semantic web. In: Proceedings of the 14th International WWW Conference, pp. 117–127 (2005)
2. Athanasiadis, I.N., Rizzoli, A.E., Donatelli, M., Carlini, L.: Enriching environmental software model interfaces through ontology-based tools. International Journal of Applied Systemic Studies 4(1/2), 94–105 (2011)
3. Auer, S., Dietzold, S., Riechert, T.: Ontowiki - a tool for social, semantic collaboration. In: Cruz, I., Decker, S., Allemang, D., Preist, C., Schwabe, D., Mika, P., Uschold, M., Aroyo, L.M. (eds.) ISWC 2006. LNCS, vol. 4273, pp. 736–749. Springer, Heidelberg (2006)

[5] `http://www.connotea.org`
[6] `http://www.mendeley.com`
[7] `http://www.confolio.org`

4. Bizer, C., Heath, T., Berners-Lee, T.: Linked Data - The story so far. Int. Journal on Semantic Web and Information Systems 5(3), 1–22 (2009)
5. Dill, S., Eiron, N., Gibson, D., Gruhl, D., Guha, R., Jhingran, A., Kanungo, T., Rajagopalan, S., Tomkins, A., Tomlin, J.A., Zien, J.Y.: Semtag and seeker: bootstrapping the semantic web via automated semantic annotation. In: Proceedings of the 12th International Conference on World Wide Web, WWW 2003, pp. 178–186. ACM, New York (2003)
6. Ebner, H., Manouselis, N., Palmer, M., Enoksson, F., Palavitsinis, N., Kastrantas, K., Naeve, A.: Learning object annotation for agricultural learning repositories. In: Proc. 9th Intl. Conf. on Advanced Learning Technologies (ICALT 2009), pp. 438–442 (2009)
7. Goodall, J.L., Robinson, B.F., Castronova, A.M.: Modeling water resource systems using a service-oriented computing paradigm. Environmental Modelling and Software 26, 573–582 (2011)
8. Granell, C., Diaz, L., Gould, M.: Service-oriented applications for environmental models: Reusable geospatial services. Environmental Modelling and Software 25(2), 182–198 (2010)
9. Nešić, S., Rizzoli, A.E., Athanasiadis, I.N.: Towards a semantically uni ed environmental information space. In: Hřebíček, J., Schimak, G., Denzer, R. (eds.) Environmental Software Systems. IFIP AICT, vol. 359, pp. 407–418. Springer, Heidelberg (2011)
10. Nešić, S., Rizzoli, A.E., Athanasiadis, I.N., Donatelli, M.: Publishing agro-environmental data to a semantically unified environmental information space: a case study. In: Proceedings of the 25th International Conference on Informatics for Environmental Protection - EnviroInfo 2011 (in press, 2011)
11. Pariente, T., Fuentes, J.M., Sanguino, M.A., Yurtsever, S., Avellino, G., Rizzoli, A.E., Nešić, S.: A model for semantic annotation of environmental resources: The taToo semantic framework. In: Hřebíček, J., Schimak, G., Denzer, R. (eds.) Environmental Software Systems. IFIP AICT, vol. 359, pp. 419–427. Springer, Heidelberg (2011)
12. Rocha, C., Schwabe, D., de Aragão, M.P.: A hybrid approach for searching in the semantic web. In: Proceedings of the 13th International WWW Conference, pp. 374–383 (2004)
13. Uslander, T., Jacques, P., Simonis, I., Watson, K.: Designing environmental software applications based upon an open sensor service architecture. Environmental Modelling and Software 25(9), 977–987 (2010)
14. Villa, F., Athanasiadis, I.N., Rizzoli, A.E.: Modelling with knowledge: a review of emerging semantic approaches to environmental modelling. Environmental Modelling and Software 24(5), 577–587 (2009)

# Thesaurus Maintenance, Alignment and Publication as Linked Data: The AGROOVOC Use Case

Caterina Caracciolo[1], Ahsan Morshed[1], Armando Stellato[2], Gudrun Johannsen[1], Yves Jaques[1], and Johannes Keizer[1]

[1] Food and Agriculture Organization of the United Nations (FAO of the UN), v.le Terme di Caracalla 1, 00154 Roma, Italy
{caterina.caracciolo,ahsan.morshed, gudrun.johannsen,yves.jaques,johannes.keizer}@fao.org
[2] ART Group, Dept. of Computer Science, Systems and Production, University of Rome, Tor Vergata, Via del Politecnico 1, 00133 Rome, Italy
stellato@info.uniroma2.it

**Abstract.** The AGROVOC multilingual thesaurus maintained by the Food and Agriculture Organization of the United Nations (FAO) is now published as linked data. In order to reach this goal AGROVOC was expressed in Simple Knowledge Organization System (SKOS), and its concepts provided with dereferenceable URIs. AGROVOC is now aligned with ten other multilingual knowledge organization systems related to agriculture, using the SKOS properties *exact match* and *close match*. Alignments were automatically produced in Eclipse using a custom-designed tool and then validated by a domain expert. The resulting data is publicly available to both humans and machines using a SPARQL endpoint together with a modified version of Pubby, a lightweight front-end tool for publishing linked data. This paper describes the process that led to the current linked data AGROVOC and discusses current and future applications and directions.

**Keywords:** AGROVOC, Mapping, Agriculture, linked data.

## 1 Introduction

AGROVOC is a multilingual thesaurus covering all areas of interest to the Food and Agriculture Organization of the UN (FAO of the UN), including agriculture, fisheries, forestry, environment, etc. First developed in the 1980's, AGROVOC is now available in 19 languages, with an average of 40,000 terms in each language. AGROVOC is managed by FAO, and owned and maintained by an international community of individual experts and institutions active in the area of agriculture. It is used worldwide by researchers, librarians, and information managers for indexing, retrieving, and organizing data in agricultural information systems.

FAO moved to linked data expressed in SKOS due to the advantages inherent in using a widely implemented and standard model that is both human and machine-readable. In particular, its advantages for librarians promise to be of great value, as

E. García-Barriocanal et al. (Eds.): MTSR 2011, CCIS 240, pp. 489–499, 2011.

once thesauri are linked, the resources they index are linked as well. Also, linked data publishing offers the advantage of a single point of access using standard query languages such as SPARQL that are already widely deployed in computing applications.

This paper presents the result of this work and the process followed to achieve it. It presents in a single picture the current product, its past development, and its social and historical use context. As for any foundational information resource used and maintained by a geographically distributed community, and exploited over the years by hundreds of different applications, innovation is not only a matter of technical research and development; it also requires careful attention to service continuity and data evolution. Therefore this paper also describes the salient aspects of publishing AGROVOC as linked data side by side with previous AGROVOC versions expressed in relational models and consumed by legacy software applications.

The rest of this paper is organized as follows. Section 2 describes the evolution of the AGROVOC model and content following the advent of the Semantic Web. Section 3 presents VocBench, the editing and workflow management tool for AGROVOC. Section 4 concerns itself with the conversion of AGROVOC into an RDF/SKOS-XL resource. Section 5 is about publishing AGROVOC as linked data, and Section 6 presents the process followed to generate candidate links from AGROVOC to other thesauri. Section 7 summarizes and discusses the entire process of generating a linked data version of AGROVOC. Section 8 concludes.

## 2 Evolution of the AGROVOC Model and Content

The first attempt to bring AGROVOC to the Semantic Web dates to 2004 [1], and was based on Ontology Web Language (OWL). OWL was chosen because it allows for rich domain specification in its distinction between objects and classes of objects. However, as thesauri do not recognize a difference between object and class some forcing was made, which in turn made it problematic to use editing tools such as Protégé [2,3]. Fig. 1 provides a sketch of how AGROVOC content was organized in an OWL model. In that model, concepts were organized in a hierarchy defined through the classical `rdfs:subClassOf`. Thesaurus relations "broader term" and "narrower term" (BT/RT) were rendered by means of ad hoc OWL object properties, and their properties were attached to singleton instances of each class, which represented the concept itself. So, actually, each concept was represented though two resources: a class, organized in the hierarchy, and its associated singleton instance, filled with property values. This choice was made to remain inside the boundaries of OWL DL. Also, labels were managed by introducing a notion of lexicalization, which forced each concept to be explicitly linked to its name, or label. The consequences of this modeling style were that the original AGROVOC hierarchy of terms was visually lost to editors, while the modeling power of OWL was not exploited. In short, OWL was too strict to render a thesaurus resource, but at the same time it was too simplistic to model multilingual resources.

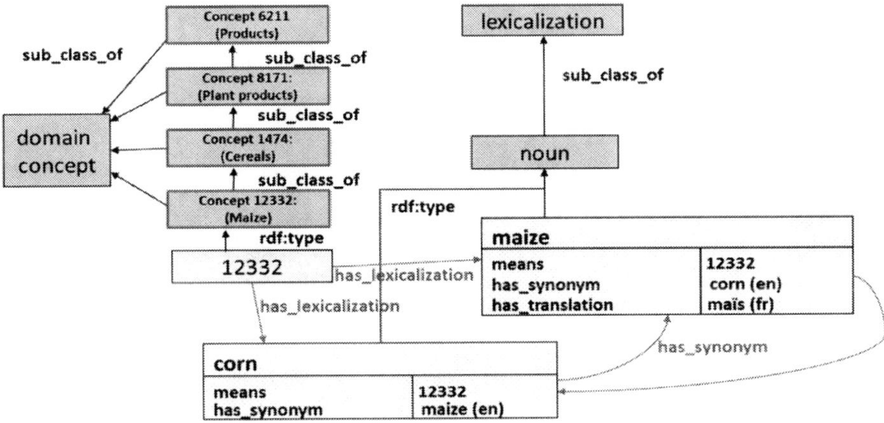


**Fig. 1.** AGROVOC Legacy Model based on OWL (2004)

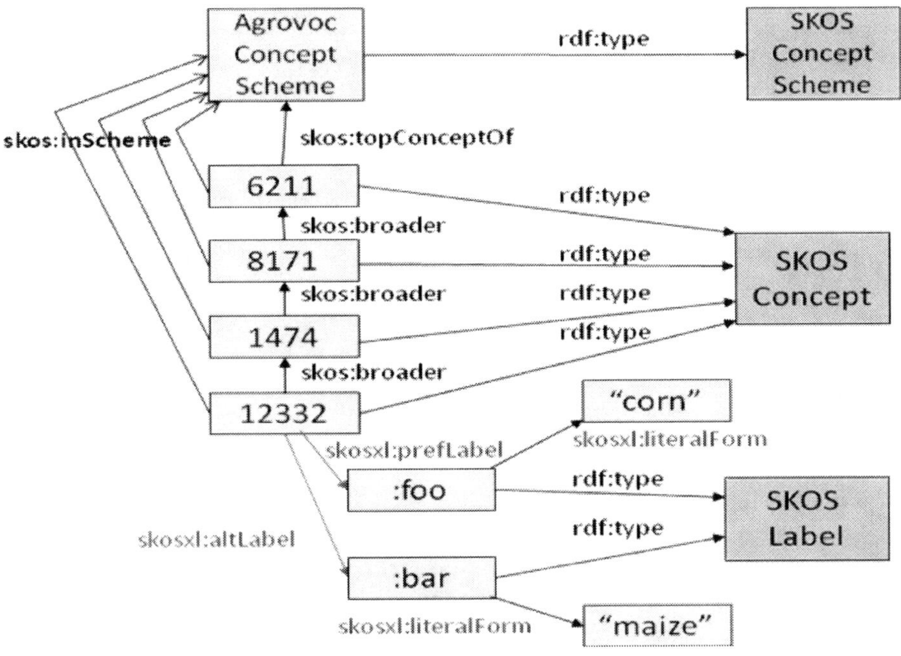


**Fig. 2.** AGROVOC Current Model (plain SKOS-XL)

In 2009, the W3C recommended the Simple Knowledge Organization System (SKOS) [4] for the rendering of resources such as thesauri over the web. As SKOS is a vocabulary for RDF specifically tailored to express thesauri, a looser semantics than that embodied by OWL is imposed on the resource. SKOS is the right choice when there is no need for formal semantics and reasoning (in particular, for classification of instances, possible in OWL thanks to the notion of object and class). Moreover,

SKOS includes two properties (`skos:broader, skos:narrower`) to express the general thesauri relations BT/NT. In this way it is possible to directly ground relationships over concepts, whereas OWL imposes that instances must be described through properties (a constraint of the OWL DL species), while being classified through classes.

In the same year, W3C also recommended a SKOS extension for managing labels, called SKOS-XL [5]. SKOS-XL offers a mechanism for treating labels (i.e., thesaurus terms) as first class objects. Labels are reified and given URIs (as opposed to being simple literals in RDF). The consequence of this approach is that with SKOS-XL, it is possible to keep track of various pieces of information about labels (e.g., date of creation and modification, editorial notes, etc.) that could not be expressed in SKOS.

In short, SKOS offers a standard vocabulary to express thesauri within RDF. With the SKOS-XL extension an appropriate linguistic characterization of thesaurus terms can also be provided. This is the reason why the previous attempt to express AGROVOC in OWL has been superseded by a SKOS-XL modeling.

In parallel with the definition of the most appropriate formal representation of AGROVOC for web consumption, AGROVOC also underwent a massive revision of its content. The number of top concepts was reduced to 25, and the hierarchies were reorganized accordingly. Also, a number of domain specific relations between concepts were added, defined globally for the entire AGROVOC. A future improvement in this direction is to simplify where possible, and standardize the domain specific relations introduced.

## 3 Support for AGROVOC Editorial Maintenance: VocBench[1]

From its inception, the AGROVOC thesaurus was stored in a relational format. In its relational model, AGROVOC was treated as a purely terminological resource, with no notion of concepts. Local identifiers were used to connect terms used in different languages to express the same meaning. Data maintenance was possible through a web application, developed in PHP and connected to the master database. Such a maintenance system was designed for use by one user at a time, and did not embody any notion of editorial workflow (including change validation), which was managed informally outside the tool.[2]

With AGROVOC's shift to the Semantic Web, the need emerged for an adequate way to manage its content. Due to the specificity of the OWL modeling adopted at the time, the use of traditional ontology editing tools (e.g., Protégé) was cumbersome. Moreover, given the multilingual, and therefore intrinsically collaborative nature of AGROVOC, there was a need for more sophisticated functionalities than those supported by the PHP application. In particular support was now required for distributed and collaborative editing as well as change validation within a formalized editorial workflow. Special attention to user roles and edit rights on languages was also required. This led to the development of the AGROVOC Concept Server Workbench, a web application meant to serve as the web-based platform for AGROVOC maintenance.

---

[1] `http://aims.fao.org/tools/vocbench-2`

[2] Note that that tool is still in use, as discussed in Sec. 8.

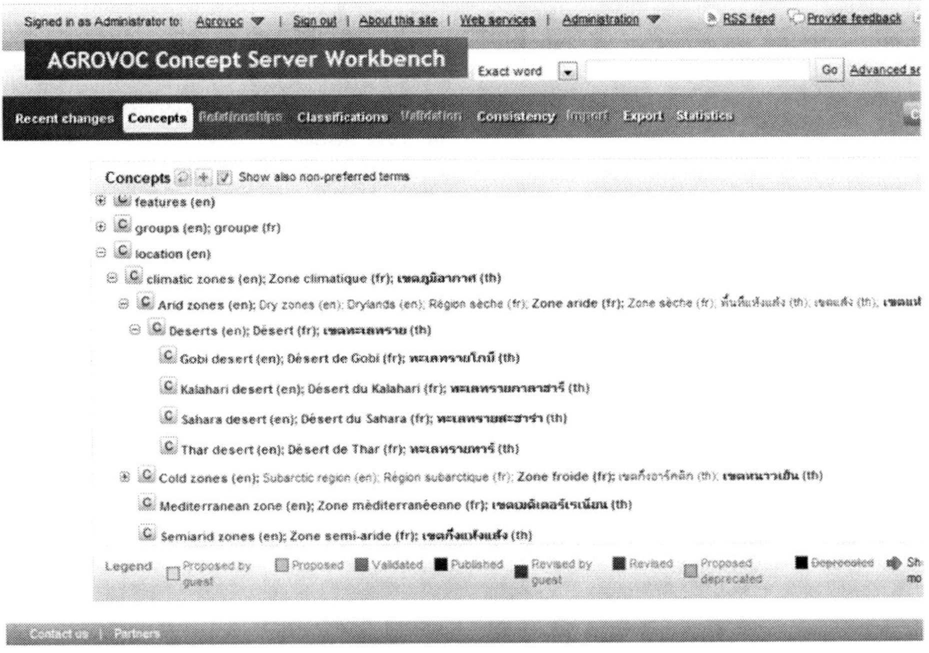


**Fig. 3.** User interface of VocBench v1.1, visualizing a fragment of AGROVOC

When AGROVOC was re-modeled in SKOS, the AGROVOC Concept Server Workbench followed, evolving into a general purpose (i.e., no longer exclusively AGROVOC based), SKOS-compliant platform for collaborative knowledge management. It was thus renamed VocBench.

Fig. **3** presents a screenshot of the VocBench user interface showing a fragment of AGROVOC.[3] VocBench improves on its predecessor in that it fully supports a formalized workflow, by user role and by language. Moreover, functionalities related to change tracking, translation of element names, and search across/within languages are fundamental to VocBench. VocBench still internally relies on the customized OWL model discussed in the previous section. However, it also enables data import and export to SKOS/SKOS-XL, which makes the application usable with other data sets.[4]

In order to improve its generality, VocBench's next major release (2.0) will feature a native interface for SKOS and SKOS-XL based on the OWL ART API[5] abstraction layer and SKOS-XL interfaces. This library for RDF provides a middle layer over different triple store technologies, so that applications exploiting its API may rely on a homogeneous and stable bus in which different, scenario-dependent technological choices can be taken. As an example, part of the current VocBench has already been

[3] At the time of writing VocBench is released as version 1.1, while version 1.2 is in phase of beta testing.

[4] Internally to FAO, VocBench is also used for the management of the Biotech Glossary. See `http://www.fao.org/biotech/biotech-glossary/en/`

[5] `http://art.uniroma2.it/owlart/`

switched to the OWLART API through its Protégé wrapper, code which will remain stable and thus seamlessly ported to the 2.0 version. At the same time, testing of VocBench on smaller portions of AGROVOC is conducted with in-memory models provided by Sesame, while performance and scalability tests are conducted on high performance triple stores (which will probably back the deployed VocBench). OWLART also features – as for the Jena API [6] and the Manchester OWL API [7] used in Protégé 4 – high level access methods specifically tailored for the various vocabularies of the RDF family. Currently supported vocabularies are RDF, RDFS, OWL (1st version), SKOS and SKOS-XL. These vocabulary APIs hide most of the triple management and provide abstract methods tightly connected with the specific RDF interpretation: for instance, in SKOS they manage much of the work which is necessary in order to avoid breaking the formal modeling constraints expressed in the specifications.

Support for generic OWL ontologies is also on the roadmap for future VocBench releases. As noted earlier, OWL is useful when a clear distinction between individual and classes is needed, as in the case of the FAO Journal Authority Data Collection (JAD), the next in line to be maintained through VocBench.

Given that VocBench still internally relies on the customized OWL model for AGROVOC, its native format is not suitable for linked data publication as-is. Periodical conversions are made into SKOS-XL format.

## 4 Conversion from VocBench Internal Model into SKOS-XL

As previously noted, SKOS-XL is used for publishing AGROVOC as linked data, while VocBench still relies internally on the legacy customized-OWL model for AGROVOC. Given that this internal data model will be in use until a fully SKOS-compliant release of VocBench is developed, a conversion process is needed in order to make AGROVOC easily available as linked data.

The conversion is performed by exploring AGROVOC concept by concept (by navigating the concept tree) and then properly converting all associated elements (the class realizing the concept in the tree, the associated singleton instance realizing the concept as an editable object, and its relationships). Another possible approach would be to perform a triple-by-triple based conversion, which was avoided because:

1. According to the Model translation directives: the same predicate may not always be translated the same way, but depends on its context (subject and object)
2. VocBench internally uses the Protégé API [2] backed by the Protégé DB, which does not allow for easy processing of triples. The Protégé DB (which allows for storage of Protégé resources over a relational database) uses an extension of the old Protégé Frame model as an inner model, which is based on a purely object-oriented paradigm. The difficulty in a triple-by-triple conversion lies in this model, which uses different “bags” for classes, instances and properties. Their role is not inferred by their role in RDF triples, but by their explicit membership to one of these bags. For this reason, Protégé does not allow an easy processing of triples, and mostly relies on a live-export of the model as a Jena read-only triple store. This export is known to be problematic so the conversion process natively uses Protégé’s API to access AGROVOC resources.

To summarize the process, the Protégé API (with DB backend) are used to read the legacy OWL version of the data and the OWLART API (by adopting the `SKOSXLModel` interface and the Sesame2 [8] implementation for the API) is used to convert the data in an NTRIPLES and RDFXML file, which is then used for linked data publication.

## 5 Technical Setup of Publishing AGROVOC as Linked Data

The linked data version of AGROVOC is now available online owing to collaboration between FAO and MIMOS Berhad[6]. Data is stored in an RDF triple store (Allegrograph[7]) hosted on a high-performance server in Kuala Lumpur. A SPARQL endpoint, combined with an http resolution of its entities, allows for publication as linked data. The HTML representation of linked data is made available through a version of Pubby[8] with customized velocity templates, providing more readable labels for properties in some cases, hiding redundant data, etc. As an example of the human readable visualization of an AGROVOC concept in linked data, see http://aims.fao.org/ aos/agrovoc/c_330892.

## 6 Linking AGROVOC to Other Resources

AGROVOC entered the linked data cloud with links to some ten resources (mostly thesauri, already available as RDF/SKOS resources, some of them also published as linked data) relevant to the domains covered by AGROVOC. Others are in progress[9]. This section describes the process adopted to identify those links: see Fig. **4** for a schematic view of the process. A detailed description of the process of providing AGROVOC with links to other thesauri, in the linked data style, can be found in [9].

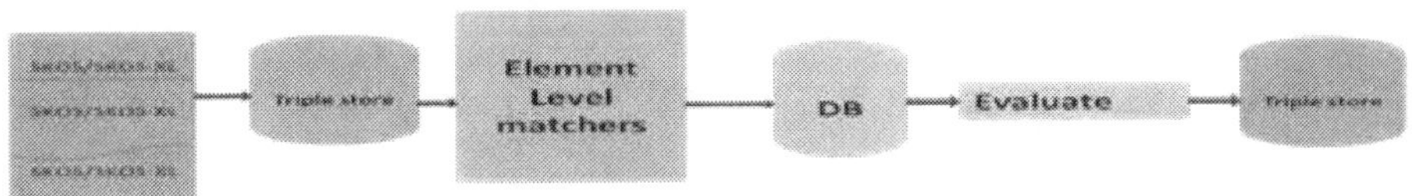


**Fig. 4.** Matching Process

All data repositories considered for alignment with AGROVOC are available as SKOS-RDF, and could be loaded in a local triple-store (in this case Sesame[10]).[11] All

[6] `http://www.mimos.my/`
[7] `http://www.franz.com/agraph/allegrograph/`
[8] `http://www4.wiwiss.fu-berlin.de/pubby/`
[9] For an updated list of resources linked to AGROVOC, see `http://aims.fao.org/standards/agrovoc/linked-open-data`
[10] `http://www.openrdf.org/`
[11] The entire thesauri were considered in all cases except in the case of RAMEAU, for which agriculture related concepts were considered (amounting to some 10% of its 150 thousand concepts).

possible pairs of concepts were considered, where the first concept in the pair comes from AGROVOC, and the second concept comes from one of the other thesauri. For each of the pairs of concepts so extracted, one preferred label per concept was selected (for the language being matched) and string similarity measures between labels was applied. Note that in this process only preferred labels in one language were considered as the matching methods used did not support more than one language label at a time. The single language in common was English in all cases except one, where French was the common language.

A selection of the most common string similarity measures was used [10], as implemented in the Alignment API[12] [11]. In order to combine these similarity values into a single number, an arithmetic average of all similarity values was computed, which seemed appropriate for a first attempt. Finally, an empirically identified threshold was applied to select candidate matches for further evaluation.

The candidate matches were presented to a domain expert for evaluation in the form of a spreadsheet. Once validated the mappings were loaded in the same triple store where the linked data version of AGROVOC is stored. This allows AGROVOC data to also display its outbound links in the style of linked data publishing[13].

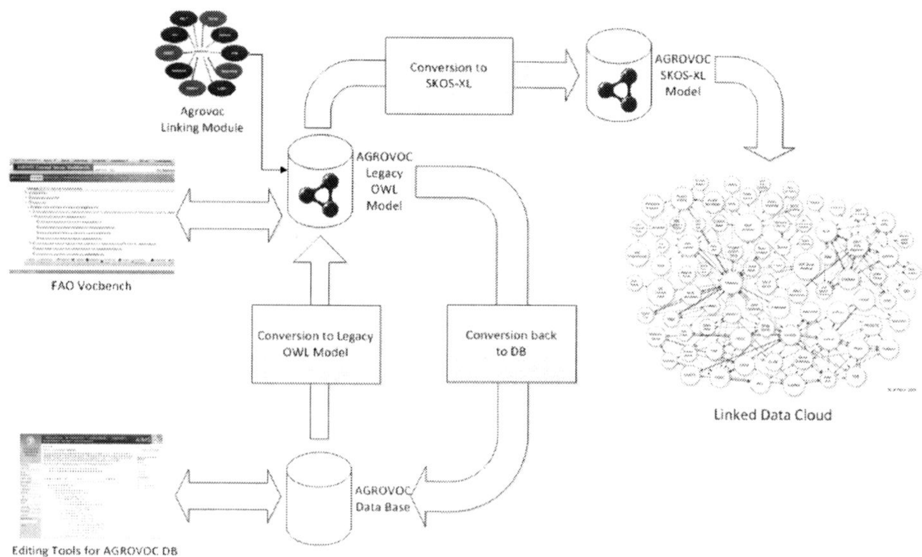


**Fig. 5.** Overview of the process for publishing AGROVOC as linked data

## 7 Overview of the Maintenance Process of AGROVOC LD

Fig. 5 provides a high-level view of the entire AGROVOC maintenance process, and its publication as linked data. On the left side of the picture, the tools in use for

[12] `http://alignapi.gforge.inria.fr/`

[13] `http://aims.fao.org/standards/agrovoc/linked-open-data`

AGROVOC data maintenance are represented. At the bottom left the legacy web tool based on the relational database is shown, while at the top left there is VocBench. Each allows interaction with a data repository: a relational database (the historical information management system for AGROVOC) and an RDF triple store based on the Protégé API, backed up by the Protégé DB backend in which the data is modeled according to the legacy OWL Model described in section 2.

Note that the relational database is still in use (it serves as a master repository of AGROVOC for many existing applications) and is periodically synchronized with the data repository corresponding to VocBench (see Fig. 5, arrow labeled "conversion back to DB"). However, the data for linked data publication comes from a conversion to SKOS-XL of the data stored in the Protégé DB, according to the legacy OWL model (see Sec. 4).

Given the current situation, publishing AGROVOC as linked data implies a series of steps, many dedicated to data conversion. This duplication of data repository, and consequent data conversions is obviously not ideal, and in principle it should be limited as much as possible. Since its first appearance in 1980s, AGROVOC has supported a worldwide community of users (people and institutions), who have developed a number of applications relying on the legacy relational model. These applications require support and so some of these conversion steps are unavoidable.

Another reason for keeping the relational format and its corresponding applications is that not all editors are able to immediately adopt VocBench. In some cases, this is due to scarce bandwidth, in which case a local copy of VocBench can be used, with batch inclusion in the master copy. In other cases, this is due to the fact that editors continue to use the old tool because they are already well acquainted with it and training efforts for a globally dispersed group of users are complex and resource-intensive.

## 8 Conclusion

AGROVOC's maintenance, alignment with other thesauri and publication as linked data is supported by an entire publishing chain, consisting of users engaged in a workflow supported by specialized tools. In particular, the re-modeling of AGROVOC using OWL and SKOS and it eventual publication as linked data implies a series of discrete steps requiring a mixture of domain experts, terminologists, ontologists and software developers. These roles must in turn be supported by a set of precise tools: editors and workflow managers such as VocBench, triple stores and SPARQL endpoints such as Allegrograph, RDF visualizers such as Pubby, and exotic APIs such as OWLART and Alignment API. In addition, careful attention must be paid to managing the support and migration of legacy applications tied to non-RDF models.

In the current maintenance process, both historical information management systems and new semantically-aware systems play a role. A sequence of conversion steps, some of which could in principle be streamlined, is not ideal. But support for previous versions and their user base is a business process requirement that cannot be ignored. Work is ongoing to provide training to AGROVOC editors, organizing workshops for data managers, and in improving the functionalities of the VocBench environment so that it can be used by all.

In this light, the immediate issues to address include the improvement of off-line VocBench editing (to address the needs of low-bandwidth users), continual VocBench usability improvements (which includes adapting its user interface to various language communities), and the completion of the revision and standardization of the AGROVOC model. This final point is expected to improve the efficiency of VocBench, and to streamline editors' work.

In consideration of the rising importance of linked data, development continues on VocBench so that it may natively support RDF/SKOS. This will have several beneficial effects: a single triple store can then be used to both edit and disseminate linked data, removing the need for tedious conversions. Secondly, the tool will be of use to any community organizing their data in SKOS. Another planned development is the integration within VocBench of the alignment functionalities that are currently hosted in Eclipse and used to extract and validate links to other resources. This will integrate the alignment workflow with the overall AGROVOC editing workflow.

The process followed to maintain, align and publish AGROVOC as linked data is repeatable. It is hoped that this overview can be useful to others with similar goals or problems.

**Acknowledgments.** The work described in this paper could have not been possible without the collaboration of a number of people. We wish to thank our colleagues Lim Ying Sean, Sachit Rajbhandari, Prashanta Shrestha, Lavanya Neelam, Jérôme Euzenat, Stefan Jensen, Antoine Isaac, Søren Roug, Thomas Baker, and Mary Redahan.

## References

1. Soergel, D., Lauser, B., Liang, A., Fisseha, F., Keizer, J., Katz, S.: Reengineering Thesauri for New Applications: The AGROVOC Example. Journal of Digital Information - JODI 4 (2004)
2. Gennari, J., Musen, M., Fergerson, R., Grosso, W., Crubézy, M., Eriksson, H., Noy, N., Tu, S.: The evolution of Protégé-2000: An environment for knowledge-based systems development. International Journal of Human-Computer Studies 58(1), 89–123 (2003)
3. Knublauch, H., Fergerson, R., Friedman Noy, N., Musen, M.: The Protégé OWL Plugin: An Open Development Environment for Semantic Web Applications. In: McIlraith, S.A., Plexousakis, D., van Harmelen, F. (eds.) ISWC 2004. LNCS, vol. 3298, pp. 229–243. Springer, Heidelberg (2004)
4. W3C: SKOS Simple Knowledge Organization System Reference. In: World Wide Web Consortium (W3C), http://www.w3.org/TR/skos-reference/ (accessed August 18, 2009)
5. W3C: SKOS Simple Knowledge Organization System eXtension for Labels (SKOS-XL). In: World Wide Web Consortium (W3C), http://www.w3.org/TR/skos-reference/skos-xl.html (accessed August 18, 2009)
6. McBride, B.: Jena: Implementing the RDF Model and Syntax Specification. In: Semantic Web Workshop, WWW 2001 (2001)

7. Bechhofer, S., Lord, P., Volz, R.: Cooking the Semantic Web with the OWL API. In: Fensel, D., Sycara, K., Mylopoulos, J. (eds.) ISWC 2003. LNCS, vol. 2870, pp. 659–675. Springer, Heidelberg (2003)
8. Broekstra, J., Kampman, A., van Harmelen, F.: Sesame: A Generic Architecture for Storing and Querying RDF and RDF Schema. In: Horrocks, I., Hendler, J. (eds.) ISWC 2002. LNCS, vol. 2342, pp. 54–68. Springer, Heidelberg (2002)
9. Morshed, A., Caracciolo, C., Gudrun, J., Keizer, J.: Thesaurus alignment for Linked Data publishing. In: Proc. of Dublin Core 2011 (forthcoming, 2011)
10. Cohen, W., Ravikumar, P., Fienberg, S.: A comparison of string distance metrics for name-matching tasks. In: IJCAI 2003 (2003)
11. Euzenat, J.: An API for Ontology Alignment. In: McIlraith, S., Plexousakis, D., van Harmelen, F. (eds.) ISWC 2004. LNCS, vol. 3298, pp. 698–712. Springer, Heidelberg (2004)
12. van Assem, M., Malaisé, V., Miles, A., Schreiber, G.: A Method to Convert Thesauri to SKOS. In: Sure, Y., Domingue, J. (eds.) ESWC 2006. LNCS, vol. 4011, pp. 95–109. Springer, Heidelberg (2006)
13. Neubert, J.: Bringing the "Thesaurus for Economics" on to the Web of linked data. In: Proc. WWW Workshop on linked data on the Web (LDOW 2009), Madrid (2009)
14. Zapilko, B., Sure, Y.: Converting the TheSoz to SKOS. GESIS Technical Report 2009/07. GESIS-Leibniz Institute for the Social Sciences (2009) ISSN: 1868-905, http://www.gesis.org/fileadmin/upload/forschung/publikationen/gesis_reihen/gesis_methodenberichte/2009/TechnicalReport_09_07.pdf

# Subject Classification with DITA Markup for Agricultural Learning Resources: A Case Example in Agroforestry

Thomas Zschocke

World Agroforestry Centre, P.O. Box 30677, Nairobi, 00100, Kenya
t.zschocke@cgiar.org

**Abstract.** Technical documentation and training materials are important elements in helping to accelerate the use and impact of agricultural research for development. The creation and delivery of these resources can be enhanced through content enrichment and the production and reuse of modular components. This process can be further improved by integrating rich semantic descriptions with resource metadata and domain-specific markup combined with the consistent use of controlled vocabulary. The Darwin Information Typing Architecture (DITA) supports both the integration of metadata as well as markup for technical as well as learning and training content. DITA also includes mechanisms for adding the semantics of taxonomy and ontology definitions for classifying content. This paper explores the potential use of enriching agroforestry learning resources with DITA markup through descriptive metadata and subject classification.

**Keywords:** DITA, markup languages, metadata, thesauruses, agroforestry.

## 1 Introduction

Modern information and communication technologies (ICTs) help to strengthen all parts of the agricultural research for development continuum by better connecting agricultural scientists and researchers as well as producers, service pro-viders, educators, and policy makers with new agricultural knowledge and information to help achieve development goals. ICTs have the potential to enhance the global public good nature of research content by increasing their availability, accessibility and applicability, and also help to broadening access to learning and improving the learning experience [1].

As a member of the Consultative Group for International Agricultural Research (CGIAR) since 1991, the World Agroforestry Centre (ICRAF) shares its research outputs as global public goods with its stakeholders in developing countries. ICRAF makes its scientific knowledge available in different types of scholarly, technical and educational publications as part of its strategy to enhance the impact of its research [2]. The Center encourages the uptake of its knowledge content especially within the context of capacity development. It

E. García-Barriocanal et al. (Eds.): MTSR 2011, CCIS 240, pp. 500–513, 2011.

makes more innovative use of ICTs in order to reach a larger number of people with more targeted information. Recently, ICRAF has joined the AgriDrupal community[1] to increase the reach of its content knowledge using an open source content management solution, Drupal, for agricultural information management and dissemination [3]. In the wider ICT for development context, the AgriDrupal initiative builds on the work of the United Nations Food and Agriculture Organization (FAO) in increasing the accessibility, availability and interoperability of exchanging agricultural research outputs in open network environments [4].

As a relatively young interdisciplinary field, agroforestry has yet to achieve a common, standardized terminology. Technical terms related to agroforestry retrieved from various knowledge organization systems (KOS) such as glossaries, dictionaries, thesauri or ontologies appear at first sight unambiguous. But, often these terms are not defined the same way across the various KOS. However, consistency in controlled vocabulary is needed to effectively search and retrieve, reuse and deliver content dynamically within an organization [5]. In order to improve this situation, the World Agroforestry Centre is applying more innovative approaches to achieve more unified semantics in agroforestry by mapping different terminology bases. The Center explores the use of the Darwin Information Typing Architecture (DITA), a XML-based architecture for authoring, organizing, producing, and delivering topic-oriented, information-typed content that can be reused and single-sourced in many different ways [6]. The current version of DITA 1.2 [7] contains specifications for content architectures for both the technical as well as the learning and training domain. DITA supports metadata descriptions, many of which are mapped to the Dublin Core metadata element set [8]. They can be included in the two basic building blocks of DITA, that is, DITA topic (<prolog> element) and DITA map (<topicmeta> element). In addition, DITA provides support for learning metadata based on a subset of the IEEE standard for learning object metadata (LOM) [9] for use in both learning topics and learning maps.

Metadata schemes such as the Dublin Core Metadata Element Set or the IEEE LOM as well as markup languages such as DITA were developed to enhance the discovery function of information services by supporting the creation and rich description of information containers consisting of different granular elements. While metadata is a formal scheme of structured information for the description of any type of resource [10], markup refers to information about the appearance, function or meaning of a document that is embedded as tags or labels in the text of documents intended to clearly separate the document structure from its presentation [11]. The form that this structural information takes is specified by the particular markup language being used. Metadata schemes can be encoded in any definable syntax with specific markup. For instance, DITA provides a set of tags and rules that not only specify how to encode the text of a document, but also the expression of corresponding metadata.

[1] Food and Agriculture Organization of the United Nations (FAO), AgriDrupal, `http://aims.fao.org/tools/agridrupal/`

The combination of metadata and markup helps to disclose both the content or topic of a resource on the granular or item level through descriptive metadata as well as on the content level through enriched metadata with subject-related elements and domain-oriented descriptors [12]. The further integration of component-level metadata descriptions with markup languages has the potential to greatly enhance the organization, discovery, access, use and reuse of content knowledge [13]. DITA has the ability to support the expression and description of both granular content elements as well as content level through corresponding markup for content structure, subject classification and metadata description

In order to fully utilize the possibilities of markup and associated metadata for resource discovery and content reuse, a controlled vocabulary is needed to achieve an effective formal description of information resources as well as the information components within them. This equally applies to the creation and sharing of interoperable learning objects where metadata and controlled vocabulary are needed for both resource description and discovery [14]. This notion is supported in the DITA specialization for learning and training content and the inclusion of metadata.

The paper presents initial results from ongoing work on using DITA to enhance the production of technical documentation and learning materials on agroforestry. Section 2 of this paper describes the steps taken in matching controlled vocabularies in the agroforestry domain. Section 3 looks at subject classification and DITA markup in the context of generating agroforestry-related content. The final section 4 provides an outlook of future work.

## 2 Identifying Controlled Vocabulary Terms in the Agroforestry Domain

The detail and precision of metadata depend to a large extent on the consistent use of controlled vocabularies. Descriptive metadata support the discovery and identification of information resources. They help to narrow the scope of the descriptions contained in metadata and refine their preciseness, which leads to an improved subject access of the resources [15]. AGROVOC[2] is an example of a multilingual thesaurus specifically designed for subject descriptions in the agricultural domain. It has been used as a KOS in, for instance, the HIVE initiative for dynamically integrating multiple controlled vocabularies [16],[3] the Organic.Edunet Web portal for learning resources on organic agriculture [17][4], and the agricultural learning repository of the CGIAR [18].[5]

---

[2] Food and Agriculture Organization of the United Nations (FAO), AGROVOC, `http://aims.fao.org/standards/agrovoc/`

[3] Helping Interdisciplinary Vocabulary Engineering (HIVE) vocabulary server, `http://hive.nescent.org/home.html`

[4] Organic.Edunet, `http://portal.organic-edunet.eu/`

[5] CGIAR Online Learning Resources (OLR) repository, `http://learning.cgiar.org`

### 2.1 Defining Agroforestry

Agroforestry, which evolved as a concept beginning in the 1970s, has now become a recognized land-use approach. Various conceptualizations and definitions of the concept have been proposed over the years [19], [20], [21]. While some argue that there is still no general consensus about the concept and classification of agroforestry [22], there is, however, agreement that agroforestry is an interdisciplinary approach to systems of land use that is practiced for various objectives by addressing the complex relationships between agriculture and forestry [23].

The World Agroforestry Centre [24] developed a first definition of agroforestry in its original charter from 1978 based on a conceptualization used in a report from 1977 [25] for a project, which was sponsored by the International Development Research Centre (IDRC) that had helped to prepare the establishment of ICRAF in the following year.[6] In the early 1980s, the definition was refined and seems to have been commonly accepted since then. It reads as follows [28]: "Agroforestry is a collective name for land-use systems and technologies where woody perennials (trees. shrubs, palms, bamboos, etc.) are deliberately used on the same land-management unit as agricultural crops and/or animals, either on the same form of spatial arrangement or temporal sequence. In agroforestry systems there are both ecological and economical interactions between the different components" (p. 37).

This definition, though slightly modified, was included by ICRAF in the mid-1990s in a specialized glossary for agroforestry [29] and also adapted in a German-English dictionary and glossary on agroforestry [30]. It is also used in abbreviated form in the Agricultural Thesaurus and Glossary of the Natural Agricultural Library of the United States Department of Agriculture (NALT) [31].[7] The SilvaTerm terminological database of the International Union of Forest Research Organizations (IUFRO)[8] lists short definitions of agroforestry adapted from the Society of American Foresters,[9] which are also modifications of ICRAF's definition from the 1980s. The definitions contained in both the multilingual thesauri AGROVOC and EuroVoc[10] are also similar modifications of the above. The definition contained in the GEMET Thesaurus,[11] however, limits the meaning of

---

[6] The Merriam-Webster's Collegiate Dictionary [26] probably refers to this IDRC project report from 1977 as the earliest recorded use of agroforestry in English in its entry of the same term as "land management involving the growing of trees in association with food crops or pastures". Surprisingly, the Oxford English Dictionary [27] has no entry for agroforestry.

[7] United States Department of Agriculture, Agricultural Thesaurus and Glossary (NALT), `http://agclass.nal.usda.gov/`

[8] International Union of Forest Research Organizations (IUFRO), SilvaTerm Database, `http://www.iufor.org/science/special/silvavoc/silvaterm/`

[9] Society of American Foresters, Dictionary of Forestry, `http://www.dictionaryofforestry.org/`

[10] EuroVoc, Multilingual Thesaurus of the European Union, `http://eurovoc.europa.eu/drupal/`

[11] GEneral Multilingual Environmental Thesaurus (GEMET), `http://www.eionet.europa.eu/gemet/`

agroforestry to "the interplanting of farm crops and trees", excluding the role of livestock and pasture, and categorizes it under the broader term of forestry instead of farming systems like in all the other cases above.

Today, the World Agroforestry Centre [2] has adopted the following definition: "Agroforestry is a dynamic, ecologically sound system of natural resource management. By integrating trees on farms and in the agricultural landscape, it helps diversify and sustain production for enhanced economic, environmental and social benefits" (p. 1). However, the defintion is not reflected in the KOS consulted above. According to the ISO 704 standard [32], the definition also appears not to be accurate in that it contains non-delimiting characteristics, which result in an extension of the definition and make it too broad.

## 2.2 Classifying Agroforestry Systems

Similar to the lack of unified semantics in defining agroforestry, there are different notions to classify agroforestry system and practices. According to Sinclair [33] and Nair [34], the general purpose of classifying these systems is to identify and group the different types of practicing agroforestry as a land use approach. This effort of the World Agroforestry Centre dates back to the 1980s, when the Center developed the Agroforestry Systems Inventory (AFSI) [35]. The Center has collected and made available this information in the corresponding AFSI database [36] and the related Multipurpose Tree and Shrub (MPTS) database [37].[12] According to the main literature, the following major types constitute the basic classification of agroforestry systems:

- Agrosilvicultural systems: Trees, including shrubs and vines, with crops
- Silvopastoral systems: Trees with pastures and livestock
- Agrosilvopastoral systems: Trees with crops and pasture and livestock
- Other systems: Multipurpose tree lots / woodlots with multipurpose management, trees with insects (entomoforestry), beekeeping (apiculture) with trees, trees with fisheries (acquaforestry, acquaculture), etc.

This basic classification integrates the three potential components in an agroforestry system, which include woody perennials, agricultural or horticultural crops, and animals [38]. Both, AGROVOC and NALT contain many of the terms related to agroforestry, its systems and practices mentioned above. However, the usage of the terms in these KOS is not always equivalent. These thesauri classify "agroforestry" as a type of farming systems. However, in the corresponding classification systems, the term is categorized differently: In the case of the AGRIS/CARIS Categorization Scheme of FAO [39] as "agro-forestry" [sic] is listed in the category "F08 Cropping Patterns and Systems" under the primary subject category "F Plant Science and Production" , and in the case of

[12] Today, ICRAF maintains the Agroforestree Database, which contains information about a wide range of tree species that can be used in agroforestry, `http://worldagroforestry.org/resources/databases/agroforestree/`

NALT [31] as "KA 125. Agroforestry" under the category "Forest and Range Resources". There are also additional differences in the treatment of agroforestry and related terms in these two KOS (see Table 1 of the Appendix on page 513). For instance, while both systems use agroforestry instead of "agro(-)/agri(-)silvicultural systems", AGRIS/CARIS labels agroforestry to be used for "farm forestry", whereas NALT refers to the latter as a related term. Also, NALT provides a number of narrower terms for "agroforestry", wheras AGROVOC has not included them at all or lists some of them as related terms of "agroforestry systems", for instance, agrosilvopastoral systems, which is replaced by "agroforestry" in NALT altogether. There are also differences in the definition/scope note of "agroforestry" in both KOS. Neither of these definitions, though, seem to comply with prescriptions in preparing intensional definitions of concepts according to the ISO 704 standard [32]. For instance, NALT uses "agroforestry systems" as the broader generic (superordinate) concept in the definition, when it actually uses agroforestry as the preferred term ("circular definition"). The definition of agroforestry contained in AGROVOC—"[s]imultaneous production, temporary or permanent, of forest trees with agricultural crops or animals in the same place"—uses the subordinate term "forest trees", which is actually limiting the scope of the concept, when it should rather include a broader range of woody perennials / plants than just trees, and fails to distinguish that crops "and" animals can be both present in this system, not just one or the other ("inaccurate definition").

## 3 Utilzing Agroforestry-Related Terms in DITA Markup

### 3.1 Resolving Equivalence among Terms

Noticing the differences in conceptualizing and classifying agroforestry as well as the disparate treatment of related terms in various KOS, it is a challenging task to provide efficient information services for end users to easily find the information relevant to their needs without additional effort. In order to address this terminology problem that limits the interoperability between different systems and their discovery function, terminology mapping can be used for improved distributed search and retrieval by matching terms and imposing links between them in different subject schemes [40]. While direct mapping in terms of "establishing equivalences between terms in different controlled vocabularies" ([41], p. 383) has been a very popular approach, it requires a lot of intellectual effort and resources because of the complexity of the task [40]. Examples of this method in the agricultural domain are the mappings of AGROVOC with NALT as part of the Ontology Alignment Evaluation Initiative (OAEI) [42],[13] and with the Chinese Agricultural Thesaurus (CAT) [43]. Both exercises conclude that automatic mapping is certainly helpful and effective. But, especially Lauser et al. [42] state that this approach works best mostly with simple lexical mappings. They argue

[13] OAEI 2007 Food Thesaurus Mapping Task, `http://oaei.ontologymatching.org/2007/food/`

that manual mapping is required for more complex categories and relationships where more background knowledge and intellectual effort is needed.

As an alternative approach, McCulloch and Mcgregor [44] suggest the use of the switching model. This model is based on a single terminology, which serves as an intermediary. Each of the terminologies under consideration for a retrieval system is mapped to a common terminology, which is sufficiently broad to include most, if not all, the categories required by the scheme for which it is used. This mechanism groups terms from various vocabularies around an anchor term or a notation, which facilitates hierarchical browsing and the display of thesaurus terms for user queries. The matching of the terms is based on the notion of equivalence, which is specified in ISO 5964 [45] and further discussed, for instance, by Doerr [46] and IFLA [47] as follows:

- Exact equivalance: One preferred term exists in all subject vocabularies
- Inexact (or near) equivalance: There is an equivalent term in each subject vocabulary for each preferred term that express the same general concept with which they overlap, but the meanings of the terms are not exactly identical
- Partial equivalence: For each preferred term in any of the subject vocabularies there exists an equivalent term with either a broader or narrower meaning
- Single-to-multiple equivalence: In order to express the meaning of the preferred term in one vocabulary, two or more preferred terms are needed in the other vocabulary
- Non-equivalent: For a preferred term in one subject vocabulary there is no term with an equivalent meaning in the other vocabulary.

In the context of this paper, we used a small subset from AGROVOC to serve as a more specialized vocabulary on agroforestry. We manually selected a core set of terms and matched them with the corresponding terminology in NALT. The preliminary results of the matching exercise are presented in Table 1 of the Appendix on page 513. This exercise illustrates the challenges to arrive at a shared set of controlled vocabularies: While there are equivalent terms in both KOS, in some cases there are only partial or inexact equivalences while other equivalent terms are missing altogether. By using the switching model based on an intermediary terminology certain gaps can be resolved, and users would have a greater choice in selecting search terms in an information system that may be more appropriate to their knowledge and understanding of a domain.

## 3.2 Handling Controlled Vocabulary in DITA Markup

We have explored in what ways controlled vocabulary and the relationship and association between subject terms can be managed in DITA markup. The main building blocks of DITA markup are topics and maps. DITA topics contain a single subject and form the basic unit of authoring and reuse. They can be of a specialized information type, such as task, concept, or reference; learning and training content constitutes another specialized information type in DITA. By default, DITA supports the integration of information topics (e.g., concept, task) as resources in learning and training content, thus allowing the reuse of content from the technical domain in an educational setting.

As a basic mechanism to manage terminology-related tasks, DITA includes a glossary specialization <glossentry> that supports terminological information and the definition of terms, acronyms, and abbreviations. Multiple glossary entries can be incorporated in a single collection through the glossary group <glossgroup> document type. DITA provides various elements that can be used to include more detailed information about the glossary term. For instance, the definition of a term <glossdef> can be described in more detail by including notes on the scope of usage <glossScopeNote> or supplementary information about the correct use of the term with <glossUsage>. At the same time, variants of the base term can be included through the <glossAlt> element with optional information about a synonym of the term <glossSynonym>, the association or cross-reference with another alternative term <glossAlternateFor>, the status of the variant <glossStatus> or notes on the usage of the variant term <glossUsage>. Using this mechansims, AGROVOC terms related to agroforestry can be replicated in DITA and matched with an alternative term in NALT. The following is example a single glossary entry for "taungya" from AGROVOC with related information from NALT:

```
<glossentry id="taungya.ags xml:lang="en">
 <glossterm>taungya</glossterm>
 <glossdef>A form of agroforestry system in which short term
 crops are grown in the early years of the plantation of a woody
 perennials species [...]
 </glossdef>
 <glossBody>
  <glossScopeNote>The defintion is based on NALT because
  AGROVOC does not include an explanation of the term.
  </glossScopeNote>
  <glossAlt id="taungyasystem.nalt">
  <glossSynonym>taungya system</glossSynonym>
  <glossUsage>NALT uses taungya system as the preferred
  term</glossUsage>
  </glossAlt>
  <glossAlt id="taungyapractice.nalt">
  <glossSynonym>taungya practice</glossSynonym>
  <glossUsage>NALT uses taungya system as the preferred
  term</glossUsage>
  </glossAlt>
 </glossBody>
</glossentry>
```

Individual glossary terms can be referenced from <keyword> or <term> elements inline with the text by creating associated keys for each term, e.g.

<topicref keys="agroforestry.ags" href="agroforestry.dita">, which can be used to link to the appropriate glossary entry, e.g.: <keyword keyref="agroforestry .ags"> agroforestry </keyword>. The <keywords> element can contain lists of a subject vocabulary, which can be included by referencing the glossary entry as illustrated above in the <metadata> and/or <prolog> elements. The elements inside of the <metadata> section of the <prolog> element contain information about the content and subject of a topic, e.g., audience, category, etc. The <prolog> elements outside of <metadata> provide lifecycle information for the content unit, e.g., author, copyright information, permissions, etc. Most of the prolog (metadata) elements are mapped to the Dublin Core Metadata Element Set [8]. The same principle applies to including metadata or rather controlled vocabulary that can be used to reference content as part of a learning object by changing the type attribute value, e.g. <glossentry>. Glossary terms can also be referenced in the <prolog> section of the <learningAssessment>, <learningContent>, <learningOverview>, <learningPlan>, <learningSummary> elements or the corresponding <topicmeta> element in (learning) maps.

While the use of controlled vocabulary can facilitate the organization of topics and other resources into structured information in DITA maps by using a glossary-based term listz, the classification of content into precise relationships among subjects and hierarchies is achieved through classification elements. In correspondence with the Simple Knowledge Organization System (SKOS) [48] specification, a subject classification in DITA is created by (1) a subject scheme (<subjectScheme>) to specify the relationships between subjects, which contains subject definitions (<subjectDef>) for the expression the meaning of a formal subject by defining both the category and a list of controlled values, and (2) the classification domain elements that identify the subject matter of content (<subjectref>) based on the subject definition in a subject scheme map, that is, the actual subject matter of the resource. While the <subjectdef> element explains what the subject covers within a <subjectScheme>, the latter organizes the subject documents into a hierarchical and associative relationship. The corresponding taxonomy map, which is equivalent to a SKOS scheme, expresses the relationships by using the same labels as in SKOS, that is, <hasNarrower>, <hasPart>, <hasKind>, <hasInheritance>, and <hasRelated>. Through this mechanism DITA allows to bring the semantics and the content closer together instead of managing them separately. Thus, DITA enables the creation of relationships between subjects, classification, and topics [49]. As both AGROVOC and NALT make their term entries available as RDF/XML, the can easily be processed and transformed to the corresponding elements in DITA.

The mechanism for subject classification complements the <lcLom> element for educational metadata. While the <lcLom> elements are only mapped to selected elements in the technical and educational categories of IEEE LOM, the actual subject classification is handled through the classification components as indicated above, which are actually part of the DITA base elements. This allows the creation of a singular classification system that can be applied to both technical as well as learning and training content topics rather than duplicating the same process within the <lcLom> element. In sum, DITA markup has the ability to maintain a taxonomy and classification as part of a content collection. Because of the topic structure, DITA markup is suited for semantic processing by providing a formal declaration of the topic's subject matter as indicated above.

## 4 Conclusion and Outlook

Our ongoing activity is related to similar efforts of another CGIAR Center, the International Crops Research Institute for the Semi-Arid Tropics (ICRISAT), which uses topic maps and controlled vocabulary based on AGROVOC to share and manage agricultural information [50]. In our current work we focus on testing further the integration of descriptive metadata with DITA markup that could make the production and reuse of agroforestry technical documentation and learning content more efficient. For instance, DITA topics have also been used to define the content objects within the Abstract Learning Object Content Model (ALOCoM) ontology for the reuse and repurposing of learning objects and their components, which are further enriched by metadata [51]. DITA can be utilized to facilitate the management of content within a social Web platform as shown in a case study about Autodesk, Inc., by the Gilbane Group [52]. The open source CMS Drupal can also be customized in such a way that it can process DITA markup, similar to the example on using Drupal for publishing documents based on Text Encoding Initiative (TEI) XML as presented by Bodine and Schlitz [53]. The future goal of our project is to develop a corporate Wiki (e.g., [54]) based on DITA XML as an innovative way to generate and share technical and learning resources on agroforestry through distributed collaboration with partners in the context of the AgriDrupal initiative.

## References

1. Ballantyne, P., Maru, A., Porcari, E.M.: Information and Communication Technologies—Opportunities to Mobilize Agricultural Science for Development. Crop. Sci. 50, S-63–S-69 (2010)
2. World Agroforestry Centre: Transforming Lives and Landscapes. Strategy 2008-2015. World Agroforestry Centre, Nairobi (2008), `http://www.worldagroforestry.org/publications/publicationsdetails?node=51106` (retrieved July 8, 2011)
3. Pesce, V., Subirats, I., Picarella, A., Keizer, J.: AgriDrupal: Repository Management Integrated into a Content Management System. Poster, Open Repositories (2011), `http://www.fao.org/docrep/article/am642e.pdf` (retrieved July 7, 2011)
4. Subirats, I., Onyancha, I., Salokhe, G., Kaloyanova, S., Anibaldi, S., Keizer, J.: Towards an Architecture for Open Archive Networks in Agricultural Sciences and Technology. Online Inform Rev. 32, 478–487 (2008)
5. Rockley, A.: Managing Enterprise Content: A Unified Content Strategy. New Riders, Indianapolis (2003)
6. Priestley, M., Hargis, G., Carpenter, S.: DITA: An XML-based Technical Documentation Authoring and Publishing Architecture. Tech Comm. 48, 352–367 (2001)

7. Eberlein, K.J., Anderson, R.D., Joseph, G. (eds.): Darwin Information Typing Architecture (DITA) Version 1.2. Organization for the Advancement of Structured Information Standards (OASIS), Burlington, MA (December 2010), http://docs.oasis-open.org/dita/v1.2/spec/DITA1.2-spec.html (accessed July 24, 2011)
8. ISO 15836:2009: Information and Documentation—The Dublin Core Metadata Element Set (2nd ed.). International Organization for Standardization (ISO), Geneva (February 15, 2009)
9. IEEE Std 1484.12.1$^{TM}$-2002:IEEE Standard for Learning Object Metadata. Institute of Electrical and Electronics Engineers (IEEE), New York, NY (June 2002)
10. NISO: Understanding Metadata. National Information Standards Organization (NISO) Press, Bethesda, MD (2004), http://www.niso.org/ (retrieved July 22, 2011)
11. Ressler, S.: Markup languages. In: Ralston, A., Reilly, E.D., Hemmendinger, D. (eds.) Encyclopedia of Computer Science, 4th edn., pp. 1080–1084. John Wiley, Chichester (2003)
12. Zeng, M.L.: Domain-Specific Markup Languages and Descriptive Metadata: Their Functions in Scientific Resource Discovery. Revista Eletrônica de Biblioteconomia e Ciência da Informação 15, 164–176 (2010), http://www.periodicos.ufsc.br/index.php/eb/article/view/16890 (retrieved on July 18, 2011)
13. Shreve, G.M., Zeng, M.L.: Integrating Resource Metadata and Domain Markup in an NSDL Collection. In: Proceedings of the International Conference on Dublin Core and Metadata Applications (DC 2003), pp. 223–229. Dublin Core Metadata Initiative (DCMI), Dublin (2003), http://dcpapers.dublincore.org/ojs/pubs/article/viewArticle/750 (retrieved July 18, 2011)
14. Qin, J., Hernádez, N.: Building Interoperable Vocabulary and Structures for Learning Objects. J. Am. Soc. Inform Sci. Tech. 57, 280–292 (2006)
15. Duval, E., Hodgins, W., Sutton, S., Weibel, S.L.: Metadata Principles and Practicalities. D-Lib. Mag. 8, http://dlib.org/dlib/april02/weibel/04weibel.html (retrieved July 22, 2011)
16. Greenberg, J., Losee, R., Agüera, J.R.P., Scherle, R., White, H., Willis, C.: HIVE: Helping Interdisciplinary Vocabulary Engineering. B. Am. Soc. Inform Sci. Tech. 37, 23–26 (2011)
17. Sánchez-Alonso, S., Sicilia, M.-Á.: Using an AGROVOC-based Ontology for the Description of Learning Resources on Organic Agriculture. In: Sicilia, M.-Á., Lytras, M.D. (eds.) Metadata and Semantics, pp. 481–492. Springer, Heidelberg (2009)
18. Zschocke, T., Beniest, J., Paisley, C., Najjar, J., Duval, E.: The LOM Application Profile for Agricultural Learning Resources of the CGIAR. Int. J. Metadata, Semantics and Ontologies 4, 13–23 (2009)
19. Lundgren, B.: What is Agroforestry? Agrofor. Sys. 1, 7–12 (1982)
20. Somarriba, E.: Revisiting the Past: An Essay on Agroforestry Definition. Agrofor. Sys. 19, 233–240 (1992)
21. Leakey, R.: Definition of Agroforestry Revisited. Agroforestry Today, 5–7 (1996)
22. Torquebiau, E.F.: A Renewed Perspective on Agroforstery Concepts and Classification. In: Comptes Rendus de l'Académie des Science. Series III, Science de la Vie, vol. 323, pp. 1009–1017. Académie des Science, Paris (2000)

23. Nair, P.K.R.: An Introduction to Agroforestry. Kluwer, Amsterdam (1993)
24. International Council for Research in Agroforestry: Charter of the International Council for Research in Agroforestry (ICRAF). Mimeograph, International Council for Research in Agroforestry (ICRAF), Nairobi, 10 p. (1978)
25. Bene, J.G., Beal, H.W., Coé, A.: Trees, Food, and People: Land Management in the Tropics. International Development Research Centre (IDRC), Ottawa, Canada (1977), `http://idl-bnc.idrc.ca/dspace/bitstream/10625/930/1/23537.pdf` (retrieved July 11, 2011)
26. Mish, F.C., et al. (eds.): The Merriam-Webster Collegiate Dictionary, 11th edn. Merriam-Webster, Springfield, MA (2008)
27. Oxford English Dictionary Online. Oxford University Press, New York (2011), `http://www.oed.com/` (accessed July 11, 2011)
28. Lundgren, B., Raintree, J.B.: Agroforestry. In: Nestel, B. (ed.) Agricultural Research for Development: Potentials and Challenges in Asia, pp. 37–49. International Service for National Agricultural Research (ISNAR), The Hague (1983)
29. Huxley, P., van Houten, H.: Glossary for Agroforestry. International Centre for Research in Agroforestry (ICRAF), Nairobi (1997)
30. Von Maydell, H.-J.: Agroforstwirtschaft / Agroforestry: Lexikon und Glossar / Dictionary and Glossary Deutsch / Englisch - English / German. Mitteilungen der Bundesforschungsanstalt für Forst- und Holzwirtschaft Nr. 173. Kommissionsverlag Max Wiedebusch, Hamburg (1993)
31. United States Department of Agriculture: Manual of Classification for Agricultural and Forestry Research, Education, and Extension. Revision VII. Classifications Used in the Current Research Information System. United States Department of Agriculture, Washington, D.C. (2005)
32. ISO 704:2009. Terminology Work—Principles and Methods. International Standardization Organization (ISO), Geneva (2009)
33. Sinclair, F.L.: A General Classification of Agroforestry Practice. Agrofor. Sys. 46, 161–180 (1999)
34. Nair, P.K.R.: Classification of Agroforestry Systems. Agrofor. Sys. 3, 97–128 (1985)
35. Nair, P.K.R.: Agroforestry Systems Inventory. Agrofor. Sys. 5, 301–317 (1987)
36. Nair, P.K.R.: The Agroforestry Systems Database at ICRAF. Agrofor. Sys. 6, 253–270 (1988)
37. Huxley, P.A., Westley, S.B. (eds.): Multipurpose Trees: Selection and Testing for Agroforestry. International Council for Research in Agroforestry (ICRAF), Nairobi (1989)
38. McAdam, J.H., Burgess, P.J., Graves, A.R., Rigueiro-Rodríguez, A., Mosquera-Losada, M.R.: Classifications and Functions of Agroforestry Systems in Europe. In: Rigueiro-Rodríguez, A., McAdam, J.H., Mosquera-Losada, M.R. (eds.) Agroforestry in Europe: Current Status and Future Prospects, pp. 21–41. Springer, Heidelberg (2009)
39. Prince-Perciballi, I. (ed.): AGRIS/CARIS: Categorization Scheme (Rev. 5.1). Food and Agriculture Organization of the United Nations (FAO), Rome (August 1998), `http://www.fao.org/docrep/003/U1808E/U1808E00.htm` (retrieved July 20, 2011)
40. McCulloch, E., Shiri, A., Nicholson, D.: Challenges and Issues in Terminology Mapping: A Digital Library Perspective. Electron Libr. 23, 671–677 (2005)
41. Zeng, M.L., Chan, L.M.: Trends and Issues in Establishing Interoperability Among Knowledge Organization Systems. J. Am. Soc. Inform. Sci. Tech. 55, 377–395 (2004)

42. Lauser, B., Johannsen, G., Caracciolo, C., van Hage, W.R., Keizer, J., Mayr, P.: Comparing Human and Automatic Thesaurus Mapping Approaches in the Agricultural Domain. In: Proceedings of the International Conference on Dublin Core and Metadata Applications (DC 2008), pp. 43–53. Dublin Core Metadata Initiative (DCMI), Dublin (2008), http://dcpapers.dublincore.org/ojs/pubs/article/view/918 (retrieved July 26, 2011)
43. Liang, A.C., Sini, M.: Mapping AGROVOC and the Chinese Agricultural Thesaurus: Definitions, Tools, Procedures. New Review of Hypermedia and Multimedia 12, 51–62 (2006)
44. McCulloach, E., Macgregor, G.: Analysis of equivalence mapping for Terminology Services. J. Inform. Sci. 34, 70–92 (2008)
45. ISO 5964-1985: Documentation — Guidelines for the Establishment and Development of Multilingual Thesauri (1st ed.). International Organization for Standardization (ISO), Geneva (1985)
46. Doerr, M.: Semantic Problems of Thesaurus Mapping. J. Digit Inform 1 (2001), http://journals.tdl.org/jodi/article/view/31 (retrieved July 20, 2011)
47. IFLA Working Group on Guidelines for Multilingual Thesauri: Guidelines for Multilingual Thesauri. IFLA Professional Reports, No. 115. International Federation of Library Associations and Institutions (IFLA), The Hague (2009), http://archive.ifla.org/VII/s29/pubs/Profrep115.pdf (retrieved July 21, 2011)
48. Miles, A., Bechhofer, S.: SKOS Simple Knowledge Organization System: Reference. World Wide Web Consortium (W3C), Cambridge, MA (August 2009), http://www.w3.org/TR/skos-reference/ (accessed July 24, 2011)
49. Hennum, E., Anderson, R., Bird, C.: Subject Classification with DITA and SKOS. International Business Machines Corp. (IBM), Armonk, NY (October 25, 2005), http://www.ibm.com/developerworks/xml/library/x-dita10/ (accessed July 24, 2011)
50. Patwar, S., Kaur, P., Sylvester, A.G., Balaji, V.: Towards a Novel Content Organisation in Agriculture Using Semantic Technologies: A Study with Topic Maps as a Tool. Int. J. Metadata, Semantics and Ontologies. 4, 65–71 (2009)
51. Verbert, K., Duval, E., Meire, M., Jovanović, J., Gašević, D.: Ontology-Based Learning Content Repurposing: The ALOCoM Framework. Int. J. on E-Learning 5, 67–74 (2006)
52. Bock, G., Waldt, D.: Managing Content for Continuous Learning at Autodesk. Case study, Outsell's Gilbane Group, Cambridge, MA (2011), http://cdn.mindtouch.com/pdfs/autodesk-case-study-gilbane.pdf (retrieved July 26, 2011)
53. Bodine, G., Schlitz, S.: Developing Drupal Publications to Support Standards-based XML: Customize Your Drupal Installation to Support Publication of TEI (or other) XML Documents. Tutorial, IBM, Armonk, NY (2011), http://www.ibm.com/developerworks/xml/tutorials/x-drupalxmltut/authors.html (retrieved July 7, 2011)
54. Meloche, J.A., Hasan, H., Willis, D., Pfaff, C.C., Qi, Y.: Cocreating Corporate Knowledge with a Wiki. Int. J. Knowl. Manag. 5, 33–50 (2009)

# Appendix

**Table 1.** Equivalence types of selected agroforestry-related terms in AGROVOC and NALT

| AGROVOC Code | Term | Equivalence Type | NALT Number | Term |
|---|---|---|---|---|
| 207 | agroforestry<br>BT farming systems | exact | 5256 | agroforestry<br>BT farming systems |
| 16097 | silvopastoral systems<br>BT agroforestry systems | exact | 5586 | silvopastoral systems<br>BT agroforestry |
| 33512 | multipurpose trees<br>RT agroforestry | exact | 130685 | multipurpose trees<br>RT agroforestry |
| 33456 | agrosilvicultural systems<br>USE agroforestry | exact | 71097 | agrosilvicultural systems<br>USE agroforestry |
| 330982 | agroforestry systems | inexact | 127652 | agroforestry systems<br>USE agroforestry |
| 28065 | farm forestry<br>USE agroforestry | inexact | 132682 | farm forestry<br>RT agroforestry |
| 24044 | line planting<br>BT planting | partial broader | 131733 | line planting<br>BT agroforestry |
| 33452 | alley cropping<br>UF hedgerow intercropping<br>BT intercropping | partial narrower | 5586 | alley cropping<br>UF hedgerow cropping<br>BT intercropping<br>BT agroforestry |
| 16613 | taungya<br>BT reforestation | single-to-multiple | 130727 | taungya system<br>UF taungya<br>BT agroforestry |
| 37590 | home gardens (agroforestry)<br>Status: Deleted descriptor | non-equivalent | 34347 | home gardens<br>BT gardens |
| 12217 | hedges<br>UF living fences<br>BT fencing | non-equivalent | 127350 | live fences<br>BT agroforestry |
| 16096 | agrosilvopastoral systems<br>BT agroforestry systems | non-equivalent | — | — |
| — | — | non-equivalent | 127350 | box planting<br>BT agroforestry |
| — | — | non-equivalent | 127350 | multistrata agroforestry systems<br>BT agroforestry |
| — | — | non-equivalent | 131728 | parkland agroforestry systems<br>UF miombo ecosystems<br>UF miombo forests<br>BT agroforestry |
| — | — | non-equivalent | 131643 | shade agroforestry systems<br>BT agroforestry |

# POWDER as Enabling Technology for the Semantic Interoperability of Agricultural Repositories

Pythagoras Karampiperis, Stasinos Konstantopoulos, and Vangelis Karkaletsis

National Center of Scientific Research "Demokritos", Athens, Greece
{pythk,constant,Vangelis}@iit.demokritos.gr

**Abstract.** Current approaches to the interoperability of heterogeneous resources typically maintain coordinated clones over which querying infrastructure operates. In the case of large-scale repositories, and especially when no single schema is clearly established so that the problem can be reduced to transforming legacy data, a more dynamic approach would be a benefit. In many real-world situation, however, useful queries need to combine information from different sources that are actively maintained in incompatible schemata and are too large to systematically clone. Examples include the various agricultural resource repositories and databases such as meteorological archives and GIS. In this position paper we explore the applicability of the W3C Protocol for Web Description Resources (POWDER) as infrastructure for the efficient and distributed retrieval of the meta-information needed to dynamically re-write queries in one schema to the (set of) semantically equivalent queries that need to be executed over the various heterogeneous schemata. This enables information providers to publish information in any schema as long as POWDER is used to annotate their repository with the coordination-related meta-information. Querying engines can then exploit existing and established resource discovery mechanisms implemented over the POWDER protocol to retrieve the meta-information pertinent to any single triple pattern in a query and use that to dynamically perform query rewriting.

## 1 Introduction

The high rate of evolution of Web 2.0 applications implies that on the one hand, increasingly complex and dynamic web-based learning infrastructures need to be managed more efficiently, and on the other hand, new type of learning services and mechanisms need to be developed and provided.

The digital resources that are developed to support teaching and learning activities have to be easily located and retrieved. For this purpose, database systems that facilitate their storage, location and retrieval have been developed and deployed online (Holden, 2003). Such systems, termed as repositories, are used to store any type of digital material. However, repositories for learning resources are considerably more complex, both in terms of what needs to be stored and how it may be delivered. The purpose of a repository with learning resources is not simply safe storage and delivery of the resources, but mainly the facilitation of their reuse and sharing (Duncan, 2002). Therefore, the repositories that are developed to provide access to digital learning

E. García-Barriocanal et al. (Eds.): MTSR 2011, CCIS 240, pp. 514–519, 2011.

resources are termed as learning repositories or LRs (Holden, 2003). There is a foreseen potential for agricultural stakeholders, from having access to online learning resources and repositories (Manouselis et al., 2010).

Current approaches to the interoperability of heterogeneous resources typically maintain coordinated clones over which querying infrastructure operates. In this position paper we explore the applicability of the W3C Protocol for Web Description Resources (POWDER) as infrastructure for the efficient and distributed retrieval of the meta-information needed to dynamically re-write queries in one schema to the (set of) semantically equivalent queries that need to be executed over the various heterogeneous schemata. This enables information providers to publish information in any schema as long as POWDER is used to annotate their repository with the coordination-related meta-information.

The paper is structured as follows: In section 2, we discuss the landscape of Organic Educational Resource Repositories, focusing on the challenges that should be met. In section 3, we present W3C POWDER as an enabling technology for these challenges. Finally, we discuss issues related to dynamic schema coordination and querying, and discuss conclusions that can be offered.

## 2 Organic Educational Resource Repositories

Currently, there exist several Learning Repositories (LRs) offering services for organic resources, such as the Organic.Edunet, and the FAO Capacity Building Portal (Manouselis et al., 2010).

Interoperability issues in this context are relevant for organizing learning resources into digital repositories, interconnecting learning repositories and exchanging metadata and queries, creating and exchanging open learning activities, as well as ensuring the quality of the published content.

There are two levels of interoperability required for the exchange of metadata records. *Structural interoperability* ensures that the structure of metadata instances produced by a repository conforms to a common base schema (Duval et al., 2006), while *semantic interoperability* concerns the meaning of the metadata records (Euzenat,2002). Metadata records that conform to a base schema are processed by services in different (but compatible) repositories (Heery,2002).

Although significant research effort has been devoted on *schema coordination* (either structural or semantic coordination), aiming to enable sharing of organic educational resources across different LRs, limited work has been done to enable linking of organic educational resources to heterogeneous information sources adhering to these (coordinated) schemata. As a result, complex user queries that use contextual information cannot be fully supported. For example, Meteorological archives and GIS information sources could be additionally used to support queries like: "Find resources that explain what is appropriate for my case, in my location, in this time of year".

In the case of large-scale repositories, and especially when no single schema is clearly established so that the problem can be reduced to transforming legacy data, a more dynamic approach would be a benefit.

## 3 The POWDER Specification

The Protocol for Web Description Resources (POWDER) is a general purpose content and quality labelling protocol based on Semantic Web technologies, developed by the POWDER W3C Working Group.[1]

POWDER is designed as a practical means of authoring and publishing semantic annotations of large resource collections; and, furthermore, that such annotations are attributed so that trust policies may be implemented. To take a simple example, POWDER expresses statements such as "everything on example.com is red and square; this statement was asserted by the entity identified as http://authority.example.org/company.rdf#me on 14th December 2007." This statement is exemplified in Figure 1.

At the core of POWDER is the Description Resource (DR), an association between:

- the scope of the DR, a set of resources. Scope is expressed as an *iriset*, a series of restrictions on the IRIs of resources that are in scope (lines 7-9 in Fig. 1).
- the formal semantic annotation that is assigned to all resources in scope (lines 13-14 in Fig. 1), assigning RDF descriptions chosen from formal semantic vocabularies; and
- a pre-formal description, such as free keywords, textual comments, or pictorial summaries meant to be displayed in, for example, presentations of search results (lines 11-12 in Fig. 1).

```
 1 <powder xmlns="http://www.w3.org/2007/05/powder#"
   xmlns:ex="http://example.org/vocab#">
 2   <attribution>
 3     <issuedby
   src="http://authority.example.org/company.rdf#me"/>
 4     <issued>2007-12-14T00:00:00</issued>
 5   </attribution>
 6   <dr>
 7     <iriset>
 8       <includehosts>example.com</includehosts>
 9     </iriset>
10     <descriptorset>
11       <displaytext>
   Everything on example.com is red and square
         </displaytext>
12       <displayicon
   src="http://authority.example.org/icons/red-square.png"/>
13       <ex:colour rdf:resource="http://rgb.org/vocab#red" />
14       <ex:shape>square</ex:shape>
15     </descriptorset>
16   </dr>
17 </powder>
```

**Fig. 1.** A simple POWDER document. *See text for detailed description*

---

[1] The POWDER Working Group was active from March 2007 until March 2009. More details on the WG and the specification are available at `http://www.w3.org/2007/powder/`

Furthermore, POWDER documents associate attribution blocks (lines 2-5 in Fig. 1) with one or more DR blocks. Attribution blocks provide information that can be used assess the trustworthiness of the DRs in a document, including the labelling authority that created it, creation time, validity period, and so on. Fig. 1 exemplifies the above showing a simple POWDER document corresponding to the December 14th assertion above.

### 3.1 Processing POWDER Documents

The processing model for POWDER is that POWDER processors implement a *describe (u,D)* function that returns an RDF description of resource *u* given a set *D* of POWDER documents.

POWDER documents can, to a large degree, be processed entirely as XML but are, in fact, transporting OWL/RDF graphs. More specifically, DRs can be expressed in OWL/RDF, under the extension that bridges abstract (but named) resources and the string representations of their IRIs. This extension reflects the basic premise of POWDER, absent from RDF semantics, that the structure of a resource's IRI is a property of the corresponding resource, upon which inference can be drawn. This allows POWDER documents to assert propositions about *"all resources on example.com"*.

The POWDER document as a whole is an owl: Ontology instance containing a number of such assertions, and the attribution block expresses owl: Annotation Property triples of the owl: Ontology instance itself.

### 3.2 Attribution and Authentication

Trust is a human quality and is rarely deterministic in an absolute sense; whether trust is conferred on any data is always a balance between the likelihood of it being true and the consequence of it being false. If a DR says that a particular cake recipe is really good but following it produces an inedible lump, little harm is done. If a DR states that some medical advice is peer reviewed and can be used as the basis for diagnosis and treatment, then the consequences of falsehood are clearly much more serious and a more robust authentication method is appropriate.

Although the POWDER specification is very detailed on the form and meaning of POWDER documents and the output of conforming processors, it is deliberately non-prescriptive about trust and authentication methods and method can be defined by a DR publisher that is appropriate for the particular context.

What POWDER offers is a detailed specification of how to endow DRs with attribution and authentication credentials, the raw information upon which a variety of trust methodologies can operate.

The *issuedby* attribution element is mandatory for all POWDER documents and provides the IRI that identifies the entity that created the document and is therefore responsible for the claims made in the DR(s) it contains. The strong recommendation is that such descriptions are provided as instances of the Agent class in either of the widely used FOAF or Dublin Core vocabularies.[2]

An RDF-aware system can use the IRI to retrieve the properties of the creator and use them to decide if the document is trustworthy. Besides whatever properties are defined by the vocabulary used to describe the creator, POWDER also defines the authenticate property that points to a resource that gives information on how to authenticate any POWDER documents that were created by this Agent.

There are other features of POWDER designed to facilitate trust in the data. These include certification (a DR that certifies the accuracy of another) and external supporting evidence (i.e. a link to any form of data published by a third party that agrees with the assertions in the DR).

### 3.3 Application and Deployment

Trust is a human quality and is rarely deterministic in an absolute sense; whether trust is conferred on any data is always a balance between the likelihood of it being true and the consequence of it being false. If a DR says that a particular cake recipe is really good but following it produces an inedible lump, little harm is done. If a DR states that some medical advice is peer reviewed and can be used as the basis for diagnosis and treatment, then the consequences of falsehood are clearly much more serious and a more robust authentication method is appropriate.

## 4 Dynamic Schema Coordination and Querying

The main advantage of a coordinated querying end-point is that all data is described by a single schema, so that one needs to be aware of a single schema or terminology to be able to access all data. Furthermore, there can be different querying end-points over the same data, each exposing a different *perspective* of the data; that is, present the data under a different schema, each with their own advantages for some communities or applications.

The results of state-of-the-art approaches for schema coordination (Euzenat and Shvaiko, 2007) are employed by repository federation (Correndo, 2010) to perform dynamic query rewriting to partially accomplish this. The coordinated resources, however, are *statically federated* with the intelligence behind the delegation of sub-queries among repositories being customized for each federation.

We envisage a *dynamically federated system* where the various repositories publish information about the kinds of data they hold and the schema they follow, so that a query-relegating system can automatically transform and forward sub-queries and collect and aggregate results.

In order to achieve this, we will exploit established *resource discovery* mechanisms implemented over the POWDER protocol. These will allow for a concise, universally shared description of meta-information about properties and types, including the schema where they appear and the repositories that hold pertinent information in this schema.

---

[2] foaf:Agent is defined at `http://xmlns.com/foaf/spec/#term Agent`. dcterms: Agent is defined at `http://dublincore.org/documents/dcmi-terms/#classes-Agent`

As at least one of the three elements of a triple pattern must be bound (or be a constant) in order for the pattern to represent a meaningful constraint, these known element or elements can be used to infer the resource and associated schema where more information about the known elements of the pattern can be retrieved from.

## 5 Conclusions

In this position paper we explore the applicability of the W3C Protocol for Web Description Resources (POWDER) as infrastructure for the efficient and distributed retrieval of the meta-information needed to dynamically re-write queries in one schema to the (set of) semantically equivalent queries that need to be executed over the various heterogeneous schemata. This enables information providers to publish information in any schema as long as POWDER is used to annotate their repository with the coordination-related meta-information. Querying engines can then exploit existing and established resource discovery mechanisms implemented over the POWDER protocol to retrieve the meta-information pertinent to any single triple pattern in a query and use that to dynamically perform query rewriting.

## References


Correndo, G., Salvadores, M., Millard, I., Glaser, H., Shadbolt, N.: SPARQL query rewriting for implementing data integration over linked data. In: Proc. EDBT 2010, Lausanne, March 22-26 (2010)

Duncan, C.: Digital repositories: the back-office of e-Learning or all e-Learning?. In: Proc. of ALT-C 2002, September 9-11 (2002)

Duval, E., Smith, N., Van Coillie, M.: Application profiles for learning. In: Proc. of the IEEE International Conference on Advanced Learning Technologies (ICALT), pp. 242–246 (2006)

Euzenat, J.: An infrastructure for formally ensuring interoperability in a heterogeneous semantic web. In: Cruz, et al. (eds.) The Emerging Semantic Web, Selected Papers from the 1st Semantic Web Working Symposium. IOS Press, Amsterdam (2002)

Euzenat, J., Shvaiko, P.: Ontology Matching. Springer, Heidelberg (2007)

Heery, R.: Application profiles: interoperable friend or foe? In: TEL Milestone Conference, The European Library, pp. 34–39 (2002)

Holden, C.: From Local Challenges to a Global Community: Learning Repositories and the Global Learning Repositories Summit. Version 1.0. Academic ADL Co-Lab, November 11 (2003)

Manouselis, N., Najjar, J., Kastrantas, K., Salokh, G., Strack, C.M., Duval, E.: Metadata interoperability in agricultural learning repositories: An analysis. Computers and Electronics in Agriculture 70, 302–320 (2010)

# A Model for Skills and Competences and Its Application in the Agricultural Sector

Simon Grant[1], Cleo Sgouropoulou[2], and Charalampos Thanopoulos[3]

[1] University of Bolton, Bolton, United Kingdom
[2] Hellenic Organisation for Standardisation, Athens, Greece
[3] Agro-Know Technologies, Athens, Greece
asimong@gmail.com, csgouro@cs.ntua.gr, cthanopoulos@agroknow.gr

**Abstract.** The evolving needs within the European learning and employment landscape set clear and mature requirements for the combination and integration of European policies and metadata standards towards a European Competence Model that will be able to support technology enhanced learning, training and employment, thus increasing mobility, and employability opportunities for European citizens. Starting with generic domain modelling, the eCOTOOL project provides formal structure for representing skills and competences distinguishing between and responding to the needs of two important user groups: those whose work relates to vocational education and training (VET) or employment; and those who build the tools. This paper presents and explains the "high-level competence model" that works on paper for general professionals, the "technical" competence model that provides the ontological foundation for effective interoperable tool-building, together with the methodology and rationale followed. In addition, it discusses the model's exploitation and usage in instruments like the Europass Certificate Supplement (ECS), a European paper format for presenting the skills and competences acquired through VET, providing concrete examples for the field of agriculture.

**Keywords:** Competence model, Skills, competence and Europass Certificate Supplement.

## 1 Introduction

One of the key aspects of professional learning in the workplace is establishing a closer link between, on the one hand, what is learned through the learning opportunities undertaken, and on the other hand, the skills and competences required on the job or for relevant professional advancement. To establish these links, there is a need for tools such as competence models and related instruments and standards, which can be used for enhancing the expressiveness of the key constituents of training and employment (i.e. job opportunities, individual skill profiles, training curricula), and facilitating the development of a number of valuable services, e.g. personal and professional development, action planning and target setting, competence-based learning and employment opportunity exploration. This promotes transparency,

E. García-Barriocanal et al. (Eds.): MTSR 2011, CCIS 240, pp. 520–534, 2011.

portability, and comparability of citizens' lifelong achievements, and provides the basis for the effective connection of their skills and competences with the employability needs of workplaces and employers.

This article discusses the major changes and challenges regarding the need for continuous development of individual and organisational skills and competences to respond to a globally increasing competition.

Within this context, European educational and training systems have undergone profound transformations aiming at the enhancement of the competence-based character of vocational education and training [1]. The article reports on the formal structure for representing of the skills and competences, providing by the European initiative eCOTOOL in order to facilitate the transparent representation and comparability of qualifications and competences through European policies and instruments, like the Europass Certificate Supplement (ECS). These instruments, however, can reach their full potential only by means of ICT systems and services that will support the management and exchange of competence information within and amongst organizations, government departments, and educational institutions [4]. On this basis, the article provides an overview of existing technical specifications or standards concerning competence information and data models, developed in different professional environments.

The evolving needs within the European learning and employment landscape set clear and mature requirements for the combination and integration of European policies and learning technology standards towards a European Competence Model that will be able to support Technology-Enhanced Learning, Training and Employment, thus increasing learning, mobility, and employability opportunities for European citizens.

The e-Competences Tools (eCOTOOL) European project outlined in the article (http://www.competencetools.eu/), targets these requirements by carrying out significant activities towards the electronic representation of competence descriptions, their integration to European instruments and the implementation of European-wide IT services for the management and exchange of competence information related to professional learning opportunities. eCOTOOL in close collaboration with the European Learner Mobility (ELM) initiative of the European Standardization Committee (CEN), takes forward the agenda towards a European Competence Model.

## 2 Changes in Requirements for Vocational Education and Training

The enhancement of mobility and employability undoubtedly remains a high priority action item within Europe [7]. The EU is now confronted with the challenge to accomplish the initiated shift to a competitive and knowledge-based economy in order to ensure growth and employment in the long term, taking into account job market demand for new skills and competences. The achievement of this ambitious goal has been linked to a fundamental transformation of learning, education, and training, from an input-based to an outcome-based paradigm throughout Europe. Nowadays, technological, economic, and organizational changes impose new demands for a Europe more responsive to the labour market's requirements, where education and training systems should be able to equip European citizens with knowledge, skills, and competences to meet the challenges of globally increasing competition; to generate new skills, in order to respond to the nature of the new jobs, as well as to improve the adaptability and employability of adults already in the labour force. The continuous development of knowledge, skills, and competences at the individual and organizational levels, the strategic management of human resources, encouraging a more dynamic and future-oriented interaction between labour supply and demand, are clearly requirements and challenges for the EU and member states, education and training providers, companies, workers and learners.

In order to meet the challenge of incorporating current and future skills in the labour market, the EU strategies and initiatives [6] focus on the enhancement of human capital, the promotion of employability by upgrading skills, and improvements in matching skills supply to labour market needs and demand. The New Skills for New Jobs initiative [9] stresses that education, training and employment policies of the member states must focus on increasing and adapting skills and providing better learning opportunities at all levels, to develop a workforce that it is highly skilled and responsive to the needs of the economy. Similarly, businesses have a keen interest in investing in human capital and improving their human resource management.

In addition, more transparent information on labour market trends and skills requirements, as well also the removal of obstacles to free movement of workers in the EU, including administrative barriers, would help achieve this goal, and improve occupational, sector and geographical mobility. Finally, the initiative stresses the need to improve the EU's capacity for skills assessment, anticipation and matching with existing vacancies [1].

## 3 The Need for the Competence Concepts

The current European setting with regard to the evolution of qualification standards reveals that there is a general shift towards the use of outcome-based standards independent from the type (occupational or educational) of qualification.

Learning outcomes are generally seen as facilitating the link between employment and education; they may be formulated in terms of competences, a concept shared by both systems.

According to a recent research from Cedefop [2] "*most European countries are planning or making a marked shift in this direction and learning outcomes feature as a component of lifelong learning strategies and mechanisms for implementation*", even though the regulation of learning inputs (duration, contents, learning arrangements, etc.) still plays an important role in most qualification systems. However, despite these common developments and some formal similarities in formulating occupational standards, a detailed comparison of outcome-oriented standards shows persisting differences which can be traced back to different understandings of "competence" and different goals ascribed to vocational education and training.

As reported in the research, there are two different contrasting understandings of competence in Europe: the functionalist concept of competence (used in the UK system of National Vocational Qualifications, which are based on National

Occupational Standards) and an understanding of competence, developed in Germany, referring to the implicitly assumed knowledge and skills in the competence, the role of the learner in the context of a profession, and in society as a whole. In a third group of countries, there is a tendency for reconciliation of the concept of competence as a divisible entity and a more holistic approach integrating the different dimensions of competence.

The cited Cedefop research presents the similarities and differences in the formulation of learning outcomes emanating from different understanding of competences. In these different approaches, learning outcomes are used for different purposes (qualifications, programmes, assessment) and. can be distinguished as: (i) based on a theoretical or research formulation; (ii) based on negotiation between stakeholders; and (iii) borrowed/adapted from elsewhere.

The same research points out the paramount need for a change in the traditional approaches to learning, towards a paradigm promoting the idea of putting the learner at the centre of the learning process and of concentrating on the intended learning outcomes of this process instead of what has been provided as input by teachers and instructors. "*Learning based uniquely on input will not respond adequately to future challenges for individuals, society or the economy. The trend is to rely, increasingly, on the identification of learning outcomes. This trend is recognised as critical in many different contexts across education and training systems.*" Learning outcome development might typically start in the initial education context, which includes traditional school settings, and work through to Higher Education (HE) and Vocational Education and Training (VET) opportunities. However, dynamic changes in the society and economy demand more and more lifelong learning and development of knowledge, skills and competences in the work context, encompassing both informal learning activities (i.e. intentional, but not accredited) as well as non-formal ones (i.e. as a side effect of other activities).

A most important, urgently needed harmonisation instrument for making competence concepts more tangible within the European continuing training and labour space, regards the representation of learning outcomes and related concepts as knowledge, skills, and competences, associated with learner information as well as with vocational education and training and employment opportunities. The development of a competence model including the expression and exchange of skills and competence information is an important and necessary step towards outcome-based vocational education and training practice that ensures higher transparency, mobility and employability of lifelong learners.

## 4 European Competence Related Policies and Instruments

Following the Lisbon meeting in March 2000, the Commission has taken initiatives to establish synergies between processes for the design of national frameworks of qualifications and an overarching European Qualifications Framework (EQF) for Lifelong Learning, taking into account the work done in the Bologna and Copenhagen context.

The Europass instrument, as well EQF and ECVET, has been developed at different periods of time. They aim at similar purposes of supporting mobility in

Europe, while providing information and transparency on qualifications to education and training stakeholders, to individuals, and to stakeholders on the labour market. These instruments are geared towards learning outcomes, and operate in the context of mobility at large (geographical and professional), not only in formal education and training settings, but also for the recognition of non-formal and informal learning achievements; thus, they are tools to support lifelong learning.

### 4.1 Europass Certificate Supplement

The Europass Certificate Supplement (issued by the authorities that award vocational educational and training certificates, to add detail and make them more easily understandable especially by employers or institutions outside the issuing country).

Though yet not widely deployed, the Europass Certificate Supplement (CS) is intended as the core instrument for professional training [14]. The CS concisely describes the skills and competences typically acquired by the holders of a certificate associated with a professional learning opportunity. It is not intended as a personal document, but as a reference for all holders of a relevant certificate. Section 3 of the ECS is headed "Profile of Skills and Competences" and contains a list of items using the action verb to describe the skills and competences [15].

## 5 Competence Related Technology Specifications and Standards

European frameworks and instruments need to be supported by interoperable technological systems and services in order to achieve their full potential.

Several international and national organisations have attempted to develop information technology specifications or standards concerning competence information and competence data models in different professional environments:

- IMS Global Learning Consortium Inc., with its Reusable Definition of Competency or Educational Objective (RDCEO) [12];
- IEEE-LTSC, with its Data Model for Reusable Competency Definition (RCD) [11];
- the HR-XML consortium, with its HR XML Competencies specifications ("Competencies schemas") [10];
- The German DIN, with its PAS 1093 Human Resource Development with Special Consideration of Learning, Education and Training - Competence Modelling In Human Resource Development [5].

International standardization bodies (appearing in the above list) have recently started to foster work around information models for learning outcomes. In particular, the ISO/IEC JTC1 SC36 - Information technology for learning, education and training has approved and started a new multi-part standardization project on competencies [13].

At a European level, the success of the first part of the multi-part standardization project "European Learner Mobility (ELM)" [4] demonstrates the needs and demands for European policies, proper standards and solutions for the practice to support the

EU facing the challenges and opportunities for future innovation, inclusion, growth and business.

Beyond the related upcoming European Standard EN 15981 on European Learner Achievement Information (EuroLMAI) [3], the ELM project aims at integrating learning outcomes and competences into existing European policies, namely EQF, Europass and ECVET.

Therefore, it is an opportune moment to design and develop a European competence model that takes European needs into account and enables the focus of learning outcomes and competences in life-long learning and professional training across sectors and systems boundaries, domains and institutions.

# 6 Implementing Competence Modelling in Europe

The eCOTOOL European project seeks to take forward this whole agenda by providing advanced and sustainable instruments and tools for competence modeling in Vocational Education and Training (VET) and in particular for the Europass Certificate Supplement (ECS).

A key objective of eCOTOOL has been to develop a generic, European skills and competence model that can be integrated with existing European policies (Europass, EQF, EQAVET, ECTS, and ECVET), as well as related "eco-tools" that are not yet available anywhere in Europe. The overall goal of the eCOTOOL project is to improve the development, exchange, and maintenance of VET certificates and their accessibility and transparency, by harmonizing Europass with other European instruments (EQF, ECVET), such that ICT tools representing competences, can then increase European mobility and transparency regarding work-related competence requirements.

The eCOTOOL modelling process aspires to introduce a well-defined methodology for creating definitions of competence concepts and structures, or frameworks, within particular occupational domains, taking into account the needs and requirements of the different European stakeholders as well as existing policies, practices and approaches influencing its application and usage within diverse contexts throughout whole Europe. The main steps of this methodology are briefly presented as follows.

## 6.1 Requirements Elicitation and Analysis

Competence models underlie the coherent and consistent construction of competence structures or frameworks for a broad range of stakeholder groups: Employers can make use of them in order to identify specific competences relevant to their organisations; perform job analysis and profiling; undertake gap analysis of competences within their work-force. Work-based learners, individually or within enterprises, can claim acquired competences and/or identify desired competences, skills, and knowledge that are addressed and supported by a VET offering. On that basis, learners are able to select the VET that best fits their personal needs. VET providers, as well as employers, can benefit from easier access to their opportunities by learners from other countries, through the more integrated competence market, allowing employers as well as individuals to compare more easily existing professional training and certificates, in particular from abroad. Professional or other bodies can use competence models to ensure that the occupational frameworks they create are consistent with other frameworks, and have the structure and coherence needed for use with ICT systems.

eCOTOOL has realised many initiatives and means for receiving and identifying the needs and demands of all European stakeholders concerning competences: First, the eCOTOOL partners have organized several surveys including interviews, discussions and online communications to directly get in contact with the stakeholders and their needs. Second, eCOTOOL has prepared and published reports based on desk-top research and questionnaires for European creators, publishers and users of the Europass CS with a special focus on the competence and skills descriptions. Finally, eCOTOOL has realized an internal workshop with the representatives from five Directorate Generals of the European Commission to discuss the requirements from the view point of the European Commission and their existing and upcoming policies.

All results of the eCOTOOL surveys, reports and internal workshops are reflected and integrated into the development of the eCOTOOL competence model.

### 6.2 eCOTOOL Competence Model

The aforementioned initiatives and communication with the stakeholder communities have clearly outlined an urgent demand for concrete, harmonised outcomes in the field of competence modelling that can be adapted and implemented into sectors ranging from vocational education and training to human resource and European policy development.

Responding to this demand the eCOTOOL competence model has been designed in terms of a two-tier approach comprising:

- the *eCOTOOL High-Level Competence Model,* developed to be used in introductory situations with people who may be involved in the structuring of competence definitions. This version includes an appropriate list of pilot terms and definitions related to core competence modelling concepts, as well as guidelines and tools for constructing competence structures.
- The *eCOTOOL Technical Competence Model*, providing the detailed underlying connections and the information model, to help with the design and interoperability of ICT-based competence services. This version comprises a fuller set of terms and definitions, conceptual models thoroughly representing the domain's concepts, processes, artifacts, relationships and constraints, as well as the proposed information models.

### 6.3 Use Cases of the eCOTOOL Competence Model

Competences define the competence concepts that are used in many ways beyond the ECS Section 3. ECS Section 3 ability item short definitions are one kind of competence definition.

Many principal uses of competence definitions can prove the importance of the Competence Models:

***Uses by employers.*** One can easily imagine employers using wording similar to ECS Section 3 ability item short definitions in a description of what a job involves. The wording may be rather too detailed for a job advertisement, but it could still be read by an applicant for a post to help them understand what the job involved, and whether they were in fact up to it. The use of competence definitions throughout large organisations is widespread, but there is less public awareness of it, partly because businesses tend to regard their own frameworks of skill and competence as commercially confidential, rather than making them public. If a business does manage its workforce competences, they will need definitions of those competences that are understandable by all concerned. Often these are discussed in HR circles under the term "competencies", plural of "competency". There is often a distinction drawn between "competency" and "competence", but this will not be followed up here.

***Individuals claiming abilities and competence.*** Individuals need be clear, in their CVs or portfolios, about the abilities or competence they are claiming. While these descriptions would typically not be skill headings in a CV, as they are again too detailed, similar wording might appear in a detailed description of a job that a person has done in the past.

***Assessing ability and competence.*** When recruiting, it is vital that employers can assess the abilities of potential employees. It is also very helpful if individuals can assess themselves, so that they can plan for the development of their abilities and competence. In each case, one can imagine an expert observing someone less experienced in a role, and assessing whether they are competent at performing these activities, and perhaps acting as part of a checklist to inform a junior employee about where they need to get more experience and develop their competence.

***Occupational framework.*** Competence concepts appear in occupational frameworks. In practice, when frameworks are being devised, often the individual definitions are written at the same time. It is also possible that a framework could include previously established competence definitions, including those from other frameworks.

***Learning, education and training courses.*** In each case, one can imagine courses whose syllabus covers these areas. However, they are not well adapted to form the kind of learning outcomes that are typical, at least in higher education.

***Comparing and contrasting uses.*** The fact that this wording could potentially be useful in several different contexts does confirm that we are dealing with an important topic. However, the more precise requirements for each use do differ. Therefore, the wording of the ECS ability item short definitions should not in itself be taken as definitive of the underlying concept, but simply as a form that is taken as suitable for the particular application of the ECS itself.

### 6.4 Application of eCOTOOL Competence Model into the ECS

The eCOTOOL Competence Model, and specifically the eCOTOOL High-Level Competence Model is designed to help people more effectively in writing ability for the Europass CS section 3. The section 3 of the Europass CS, "Profile of Skills and Competences" is produced to help communicate what has been covered in a professional training course.

## 7 The eCOTOOL High-Level Competence Model

The high-level version of the eCOTOOL Competence Model is developed to be used with and by people involved with competence definitions for any practical reason [16]. Because the needs of these stakeholders differ from the needs of technical systems developers, further technical details are presented only in the technical version, which follows this high-level model. This high-level model is high-level in that it omits much lower-level detail that is relevant to technical systems developers, but not directly relevant to stakeholders with a direct practical involvement (Fig. 1).

### 7.1 eCOTOOL High-Level Competence Model Structure

#### 7.1.1 Basic Components of the eCOTOOL Competence Model

According to the eCOTOOL competence model (high-level), an ability item short description has two parts: a) an action verb (or verbs) and b) the rest of the ability item's title or short description.

When composing an ability item, it may be a good idea to start by writing a first short description representing briefly the essence of what the ability is. If this description does not already start with an action verb, it is recommended in most cases to search for a suitable one. Some more advice can be found below to ensure good quality items with appropriate action verbs (Fig. 1).

The high-level model allows the classification of each ability item into one of the "KSC" categories: a) Knowledge; b) Skill and c) Competence;

#### 7.1.2 Unique Id Codes for Cross-Referencing Ability Items

Working with many ability items, it is probably recognisable that the same narrower ability may play a vital part in more than one broader ability. It is also convenient to separate the definition of individual ability items from their place in structures. For both of these reasons, it is very useful the creation a unique id code for each ability item.

In some instances, for existing competence structures, codes have already been devised. For instance, the LANTRA NOSs have their own short codes for each unit and element within the units. "Set out and establish crops" has code PH2, while the element "Set out crops in growing medium" is PH2.1 and "Establish crops in growing medium" is PH2.2. LANTRA does not give codes to any of the narrow individual ability components of an element.

| **Form A: eCOTOOL high-level model item definition table** | | |
|---|---|---|
| **ability item short description** | **action verb(s)** | **rest of short description** |
| | | |
| **KSC classification** | *Knowledge, or Skill, or Competence* | |
| **unique id code** | | |
| **author/authority** | | |
| **level assignments** | **level scheme** | **level** |
| | | |
| *(repeat as needed)* | | |
| **Categorisation** | **classification scheme** | **Term** |
| | | |
| *(repeat as needed)* | | |
| **full description** | | |

**Fig. 1.** eCOTOOL High-level Competence Model structure

### 7.1.3 Assignment of Levels to Ability Items

Often, abilities are described as being at specified levels. It is often the case that education, training, and professional development result in learners progressing from lower levels of ability to progressively higher levels. However, there is no uniformity, either in educational or occupational level schemes, about the number of levels, or terms used to describe them. In recent years, the European Qualifications Framework for Lifelong Learning (EQF) has been introduced in an attempt to aid the cross-mapping of different level schemes. It defines eight levels in each of the three areas of knowledge, skills and competence. The EQF is intended to serve more as a neutral reference point than a single common standard.

A person defining a competence structure or framework may use the EQF to state that they have judged a particular ability item as fitting best with the selected EQF level. In the eCOTOOL high-level model of levels, users can judge the level of an ability item in any level scheme or framework that is familiar to them.

### 7.1.4 Further Classifying Ability Items

Ability items can be categorised for any of several purposes, beyond the KSC categories. It may, for example, be useful to record where appropriate: Standard

Industry Codes; Standard Occupation Codes; library-oriented subject matter classifications; any education or training classifications; professional body classifications, etc. Levels are not dealt with as categories.

For any categorisation, you will need to specify the classification scheme, and the term from that scheme. An item may be classified under any number of schemes, or none.

#### 7.1.5 Constructing High-Level Model Ability Items

A table with the recommended elements is given here as Table 1: Form A. This form can be filled in for as many ability items as are relevant to the person or organisation documenting their competence concepts. Guidance on when to stop is given later.

A full description of the ability should be given in any case, even if this simply repeats the set of definitions of narrower abilities that make up the ability being defined.

### 7.2 Exploitation of the eCOTOOL High-Level Competence Model

The eCOTOOL high-level model is designed as easy-to-use, allowing the majority of current occupational frameworks to be given a common structure.

There are two principal alternative approaches to creating competence structures. One can either start at the top with a job requirement, progressively breaking that down into the constituent (subsidiary, narrower) parts; or identify abilities at all levels, and work out which ones are parts of which other ones. In order to allow both approaches, the eCOTOOL high-level competence model keeps separate track of the ability items on the one hand, and the structure information on the other. Every relevant item at every granularity (except perhaps the narrowest) needs to have a definition outlined as in the previous table. The unique id codes, along with short descriptions for readability, are then used in table to show what narrower abilities are parts of what broader ability.

## 8 eCOTOOL Technical Competence Model

The eCOTOOL Technical Competence Model is a complex model that provides the detailed underlying connections and the information model, to help with the design and interoperability of ICT-based competence services.

### 8.1 eCOTOOL High Level Competence Model Structure

#### 8.1.1 Concepts Surrounding Ability or Competence Concepts

Ability or competence concepts are represented as having levels; they may be part of broader abilities or have narrower sub-abilities; they may have criteria or conditions; and they may use terms from some kind of domain vocabulary or ontology. Together with the standard metadata, these relationships represent key components for a model, which are presented in the following schema (Fig. 2).

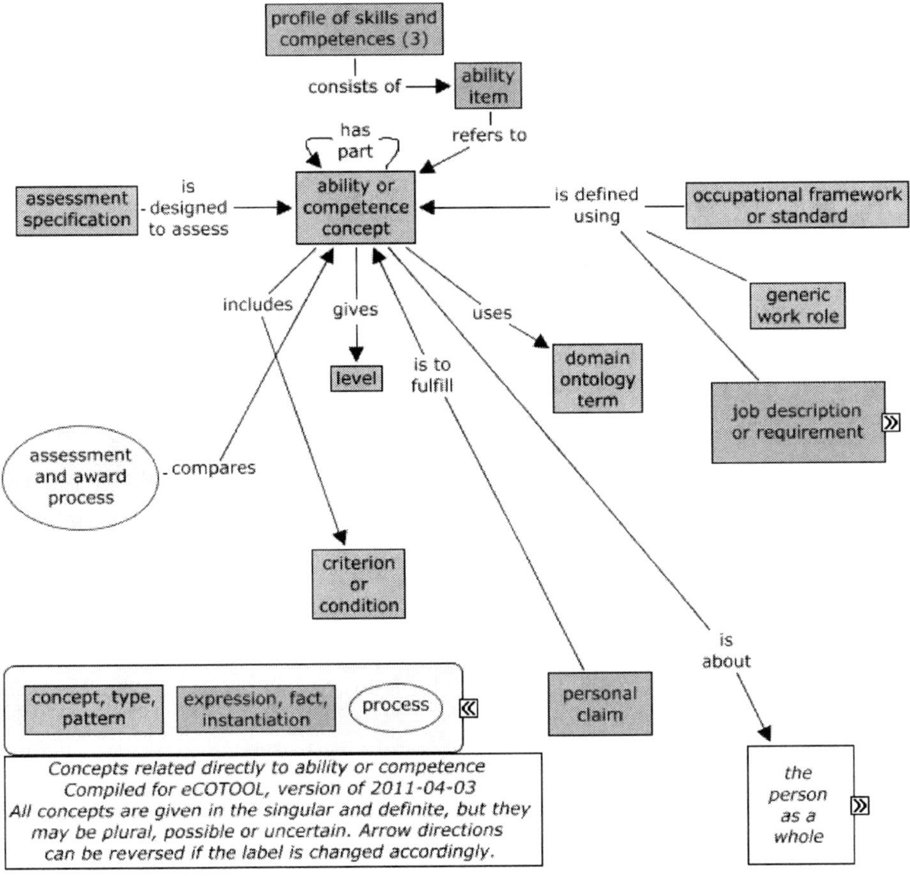


**Fig. 2.** Concepts related directly to ability or competence

Apart from the use in the ECS Section 3, the maps indicate that ability or competence concepts can be used: (a) in assessment, in the sense that an assessment is designed to assess a person's ability or competence in the given area, and the assessment process compares the evidence of a person's activity with the concepts; (b) in personal claims; (c) in job descriptions or requirements; and (d) in occupational frameworks or standards (which themselves may mention generic work roles). Not illustrated in the maps is their potential use (e) in courses of learning, education or training.

### 8.1.2 Concepts Surrounding Occupational Frameworks

An occupational framework or standard, in contrast, is defined largely in terms of the ability or competency concepts it includes, and the relationships between these concepts. Levels are defined within frameworks – in fact levels make no sense without explicit or implicit reference to a framework in which they are defined. It seems entirely plausible that occupational frameworks also may include criteria or conditions, perhaps providing context for the ability or competence concepts, and may use other terms from the domain, including generic work role terms (Fig. 3).

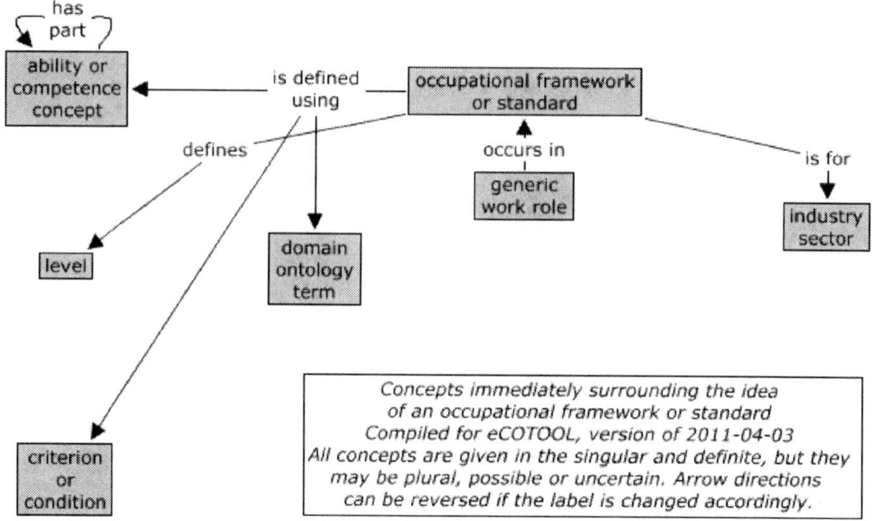


**Fig. 3.** Concepts surrounding the idea of an occupational framework or standard

### 8.2 Exploitation of the eCOTOOL Technical Competence Model

The competence concepts can be reused in different frameworks, and definitions that have already been written using a different specification (such as RDCEO) can be incorporated into a framework, without the necessity to restructure them.

A complete framework consists of one or more competence concept definitions, and any number of relationships between these concept definitions, and other concepts, whether or not included in the framework.

The eCOTOOL Technical Competence Model can be identified as an information model structure for the representation of information about competence concept definitions and frameworks. It might be useful at the same time to consider the general way in which a typical occupational standard might be structured in terms of the concepts contained.

## 9 Conclusion

Making lifelong learning and learner mobility a reality continues to remain a challenge and a key strategic priority for the European Education and Training Programme 2020 [9].

The European Commission affirms the growing success of technology-enhanced learning: "*A decade of experience in Europe has proven its value as an innovative tool for education and training*" and that it "*reflects the growing complexity of e-learning and its role as a basic tool for education and training as well as it concludes that e-learning should be seen as an important part of learning in general*" [8]. That is in line with the decision of the European Commission that technology-enhanced learning has become one of the four transversal lines of the Lifelong Learning

Programme and a general priority in the four vertical programmes [7]. In summary, competence orientation and competence models can be seen as one of the main drivers and facilitators of the new knowledge and information society and technology-enhanced learning as a key instrument with growing impact and usage.

The importance of competence models is linked to the increase of competition, innovation and rising consumer demand, forcing enterprises and industries to search for more skilled and competent employees. On the other hand learners as employees and individuals look for vocational educational and training programs offering the relevant knowledge, skills and competences for the new challenges at their workplace [6]. Competence modelling and technology-enhanced learning can go hand in hand for facilitating and achieving the required change towards competence orientation.

Progress has been made as a result of the European Qualifications Framework which triggered development work on national qualification frameworks in many countries, including a new focus on learning outcomes. By increasing the transparency of qualifications, the EQF will facilitate the mobility of learners and workers between countries and their access to lifelong learning.

However, much remains to be done to establish flexible learning pathways, e.g. through better transitions between VET, higher education, and professional learning, opening universities to non-traditional learners, and the validation of non-formal learning. Learning must be attractive and accessible for all citizens, independent of age, educational attainment, employment or social status. Better coordination between different education and training sectors, institutional commitment and partnership with all stakeholders is required.

Learner mobility between countries is a vital enhancement to building people's employability and adaptability. Evaluations of the EU programmes show that mobility breaks down barriers between people and groups, makes the benefit of European citizenship more tangible and helps people become more adaptable and open to mobility when they enter the labour market. Cross-border learner mobility should become the norm, rather than the exception that it is today. The focus on learning outcomes and in particular the development of a generic competence model facilitated by the eCOTOOL project will strengthen both the integration and implementation of the European policies as well as the mobility and life-long learning in whole Europe.

**Acknowledgements.** This work has been sponsored by the eCOTOOL (e-competences tools) project (Project number: 504614 - LLP - 1 - 2009 - DE - Leonardo - LMP), that is funded by the European Commission in the frame of the Lifelong Learning Program (http://www.competencetools.eu/). This publication reflects the views only of the authors, and the Commission cannot be held responsible for any use which may be made of the information contained therein. The authors would like to thank all the eCOTOOL consortium partners for their contribution in the design and implementation of this initiative.

## References

1. Begg, I., Erhel, C., Mortensen, J.: Medium-term Employment Challenges, Centre for European Policy Studies Special Report (2010) ISBN: 978-92-9079-977-1

2. Cedefop: The shift to learning outcomes - Conceptual, political and practical developments in Europe (2008), http://www.cedefop.europa.eu/EN/Files/4079_en.pdf
3. CEN Technical Committee 353: prEN 15981 – European Learner Mobility Achievement Information Model (EuroLMAI) (2010)
4. CEN Workshop on Learning Technologies: Guidelines on a European Learner Mobility Model, CWA 16133:2010 (2010)
5. Deutsches Institut für Normung: Publicly Available Specification (PAS) 1093 Human Resource Development with special consideration of learning, Education and Training - Competence Modeling in Human Resource Development (2009), http://www.qedinfo.de/downloads
6. European Commission: e-Skills in the 21st Century: Fostering Competitiveness, Growth and Jobs (2007), http://eurlex.europa.eu/LexUriServ/LexUriServ.do?uri=COM:2007:0496:FIN:EN:PDF
7. European Commission: An updated strategic framework for European cooperation in education and training (2008), http://eurlex.europa.eu/LexUriServ/LexUriServ.do?uri=COM:2008:0865:FIN:EN:PDF
8. European Commission: The use of ICT to support innovation and lifelong learning for all - A report on progress (2008), http://ec.europa.eu/education/lifelonglearningprogramme/doc/sec2629.pdf
9. European Commission: New Skills for New Jobs – Anticipating and matching labour market and skills needs (2008), http://eurlex.europa.eu/LexUriServ/LexUriServ.do?uri=SEC:2008:3058:FIN:EN:PDF
10. HR-XML Consortium: Competencies (Measurable Characteristics) (2006), http://ns.hr-xml.org/2_4/HR-XML-2_4/CPO/Competencies.html
11. IEEE Learning Technology Standards Committee: IEEE 1484.20.1 - Data Model for Reusable Competency Definitions (RCD) (2007)
12. IMS Global Consortium: Reusable Definition of Competency or Educational Objective (RDCEO) (2002), http://www.imsglobal.org/competencies/
13. ISO/IEC JT1 SC36: ISO/IEC NP TR 20006 Information Technology for Learning, Education and Training- Information Model for Competency Information Technology for Learning, Education and Training- Information Model for Competency (2010)
14. Otero, M.S., Fernández, J., Pijuan, A., Torrecillas, C., Romijn, C., Uwland, A.: First Evaluation of Europass - A Final Report to the Directorate General Education and Culture of the European Commission (2008), http://ec.europa.eu/dgs/education_culture/evalreports/training/2007/europass/report_en.pdf
15. The Europass Certificate Supplement (2007), http://europass.cedefop.europa.eu/europass/home/vernav/InformationOn/EuropassCertificateSupplement.csp
16. eCOTOOL High-Level Competence Model Explanation (2011), http://www.competencetools.eu/wp-content/uploads/2011/07/eCOTOOL_High_Level_Competence_Model_EN.pdf

# Author Index